Springer Geology

The book series Springer Geology comprises a broad portfolio of scientific books, aiming at researchers, students, and everyone interested in geology. The series includes peer-reviewed monographs, edited volumes, textbooks, and conference proceedings. It covers the entire research area of geology including, but not limited to, economic geology, mineral resources, historical geology, quantitative geology, structural geology, geomorphology, paleontology, and sedimentology.

More information about this series at http://www.springer.com/series/10172

Soumyajit Mukherjee
Editor

Tectonics and Structural Geology: Indian Context

Springer

Editor
Soumyajit Mukherjee
Department of Earth Sciences
Indian Institute of Technology Bombay
Powai, Mumbai, Maharashtra, India

ISSN 2197-9545 ISSN 2197-9553 (electronic)
Springer Geology
ISBN 978-3-030-07591-0 ISBN 978-3-319-99341-6 (eBook)
https://doi.org/10.1007/978-3-319-99341-6

This Springer imprint is published by the registered company Springer Nature Switzerland AG
The registered company address is: Gewerbestrasse 11, 6330 Cham, Switzerland

...geology contents in geological text books for compulsory education is not regularly updated, so new paradigms are included belatedly (as happened, e.g., with plate tectonics), and this is one of the reasons why younger students lag behind in Geosciences.

—Brusi et al. (2016)

Brusi D, Calonge A, Souza E (2016) Textbooks: A tool to support geosciences learning. In: Vasconcelos C (ed) Geoscience education: Indoor and outdoor. Springer, pp 173–206. ISBN: 978-3-319-43318-0.

Acknowledgements

Annett Buettener and **Helen Ranchner** (Springer) are thanked for handling this book proposal positively. The **Springer proofreading team** is acknowledged for support. I thank the **authors** and the **reviewers** for their participation. Thesis students, interns and visitors in the Geodynamics lab during 2017–2018: **Narayan Bose**, **Dripta Dutta** and **Tarunkanti Das** (IIT Bombay), **Prof. Seema Singh** and **Ajay Kumar** (Panjab University), **Swagato Dasgupta** (Haliburton), **Troyee Dasgupta** (Reliance Industries Limited), **Chandan Majumdar** (Schlumberger), **Tuhin Biswas** (ONGC), **Rajkumar Ghosh** (Geological Survey of India), **Chanel Vidal** (Iowa State University), **Saber Idriss** (University of SFax), **Puja Banerjee** (Institut De Physique Du Globe De Paris), **Ishiqua Agarwal** (IIT Kharagpur), **Naimisha Vanik** and **Haroon Saikh** (MS University Baroda), **Shiba Nikalje** (St. Xavier's College, Mumbai), **Amey Dashputre** and **Renuka Kale** (Fergusson College), **Rucha Kanchan** and **Samidha Shinde** (Pune University), **Lokesh Tayade** (IISER Pune), **Rohit Shaw, Madhurima Bose, Anuva Chowdhury** and **Jayesh Mukherjee** (Presidency University, Kolkata) helped in various ways. A **research sabbatical** provided by IIT Bombay to me for the year 2017 helped much to edit this book.

Mumbai, India
June 2018

Soumyajit Mukherjee
soumyajitm@gmail.com
smukherjee@iitb.ac.in

Contents

Introduction to Tectonics and Structural Geology: Indian Context

Soumyajit Mukherjee

1 Summary of Different Chapters

Tectonics and structural geology of Indian terrain is of great interest to the Government and a number of private exploration agencies that are working presently. This edited volume aims to meet this requirement. In addition, B.Sc. and M.Sc. geoscience students undergoing geohistory and/or tectonic courses would benefit using this book.

This edited volume brings 16 research papers (Chaps. 2–17) from both academia and industry.

Mukherjee et al. (2019) in Chap. 2 present an exhaustive review on the geology and the geochronology and of the Chotanagpur Granite Gneissic Complex (CGGC). They classify the CGGC into three domains, and also comment on the India-Antarctica reconstruction.

Padmalal et al. (2019) in Chap. 3 perform morphologic dating of the seismogenic Allah Bund Fault scarp as 208, 200, and 193 yrs B.P. These dates establish reliably that those scarps were produced by the 1819 earthquake.

Patil Pillai and Kale (2019) in Chap. 4 detail the sedimentation and the tectonic histories of the Kaladgi Purana (Proterozoic) basin. The basin in the first stage underwent sagging. A nested continental sag basin formed afterward.

Babar et al. (2019) in Chap. 5 describe with several field photographs the deformation features near the basement granites around Degloor (Maharashtra). They work out the stress regime and the stress axes orientations. One can compare these findings with the Deccan tectonics as well by going through Misra et al. (2014, 2015), Misra and Mukherjee (2015, 2017), etc.

S. Mukherjee (✉)
Department of Earth Sciences, Indian Institute of Technology Bombay, Powai, Mumbai 400076, Maharashtra, India
e-mail: smukherjee@iitb.ac.in; soumyajitm@gmail.com

S. Mukherjee (ed.), *Tectonics and Structural Geology: Indian Context*,
Springer Geology, https://doi.org/10.1007/978-3-319-99341-6_1

In their very detailed review on the Bengal basin, Hossain et al. (2019) in Chap. 6 present the basic division of this basin, fault distribution, and how these divisions evolved temporally with or without volcanism.

Goswami and Upadhyay (2019) in Chap. 7 study the structural geology and geochemistry of the Kadiri schist belt (Cuddapah) and decipher an ocean-continent subduction tectonics and a volcanic arc setting of the terrain.

Detailed field investigation of the structural geology of the Nallamalai Fold Belt (Cuddapah) by Tripathy et al. (2019) in Chap. 8 reveals a Pan-African thin-skinned tectonics, which link with the tectonics of the East Gondwana fragments.

Multi disciplinary geoscientific studies by Mazumder et al. (2019) in Chap. 9 reveal that a number of E trending steeply dipping shear zones pass through the northern part of the Cauvery Basin that was later reactivated.

Dasgupta (2019) in Chap. 10 reviews the Cauvery basin's tectonics. Half grabens in its all the three sub basins signify a rift origin of the basin. This article analyzes the transfer zone geometries from the Cauvery basin that are crucial in developing hydrocarbon trap conditions.

Misra et al. (2019) in Chap. 11 study the field structural geology of the Ramgarh impact structure (SE Rajasthan), and especially its fracture patterns. They conclude that impacting happened at the palaeo-channel of the river Parvati.

Dinkar et al. (2019) in Chap. 12 describe in detail field structural geology from the Lalitpur district (Uttar Pradesh). The notable information are E/ENE trending axial traces and Proterozoic to Neoproterozoic reactivation plausible in the southern part of the study area.

Singh and Awasthi (2019) in Chap. 13 discuss the tectonics of the Kangra region (Himachal Pradesh), which is presumably devoid of any weak layer below itself. Overpressure condition at depth possibly due to fluid activity had helped to propagate this crustal wedge towards the foreland side.

Kumar et al. (2019a) in Chap. 14 describe from the field along with attractive photographs the damage zone associated with the Munsiari Thrust, a strand of the Main Central Thrust, from the Mandakini river section, Higher Himalaya. The authors document more landslides from the damage zone and perform engineering geological studies from such zones.

Mahato et al. (2019) in Chap. 15 perform detailed field studies from the Mussoorie syncline and the nearby regions from the Uttarakhand Lesser Himalaya. Top-to-N/NE back shear and Himalayan arc-parallel shears (such as top-to-NW) are the new meso scale findings in this work.

Banerjee et al. (2019) in Chap. 16 too document orogen-parallel shear from the Darjeeling Group of rocks from the Sikkim Lesser Himalaya. A more detail work from the same research group has been submitted in a journal where such deformation is reported from the Siwalik Himalaya (Dutta et al. submitted).

Kumar et al. (2019b) in Chap. 17 discuss the database of lead (Pb) content in the Indian Gondwana coal ($^{207}Pb/^{206}Pb$ = 0.7150–0.8845; $^{208}Pb/^{206}Pb$ = 1.9484–2.2231; Pb concentration = 3.2–566 mg kg^{-1}). This study will have a far-reaching implication in India-Antarctica plate reconstruction.

Readers without any instructors, especially students (in some unfortunate cases), are requested to go through few recent books on structural geological and tectonic principles and Indian case studies (e.g., Sharma 2010; Mukherjee 2013a, b, 2014, 2015a, b; Mukherjee et al. 2017; Mukherjee and Mulchrone 2015; Mukherjee et al. 2015, 2017; Valdiya 2016; Bose and Mukherjee 2017; Dasgupta and Mukherjee 2017; Chetty 2018; Misra and Mukherjee 2018; Roy and Purohit 2018; Acharyya, in press) before going through this book.

Refer this book as follows:

- Mukherjee S (2019) Tectonics and Structural Geology: Indian Context. Springer International Publishing AG, Cham. ISBN 978-3-319-99340-9. pp. 1–455.

Refer individual chapters of this book as follows:

- Banerjee S, Bose N, Mukherjee S (2019) Field structural geological studies around Kurseong, Darjeeling-Sikkim Himalaya, India. In: Mukherjee S (ed) Tectonics and Structural Geology: Indian context. Springer International Publishing AG, Cham. ISBN 978-3-319-99340-9. pp. 425–440.

References

Acharyya SK (in press) Tectonic setting and Gondwana Basin architecture in the Indian Shield. In: Mukherjee S (eds) Developments in structural geology and tectonics. Elsevier. ISBN: 978-0-12-815218-8

Babar MD, Kaplay RD, Mukherjee S, Mahato S, Gurav C (2019) NE-SW strike-slip fault in the granitoid from the margin of the South East Dharwar Craton, Degloor, Nanded district, Maharashtra, India. In: Mukherjee S (ed) Tectonics and structural geology: Indian context. Springer International Publishing AG, Cham, pp 115–134. ISBN 978-3-319-99340-9

Banerjee S, Bose N, Mukherjee S (2019) Field structural geological studies around Kurseong, Darjeeling-Sikkim Himalaya, India. In: Mukherjee S (ed) Tectonics and structural geology: Indian context. Springer International Publishing AG, Cham, pp 425–440. ISBN 978-3-319-99340-9

Bose N, Mukherjee S (2017) Map interpretation for structural geologists. In: Mukherjee S (ed) Developments in structural geology and tectonics. Elsevier, Amsterdam. ISBN: 978-0-12-809681-9. ISSN: 2542-9000

Chetty TRK (2018) Proterozic orogens of India: a critical Window to Gondwana. Elsevier, Amsterdam. ISBN: 9780128044414

Dasgupta S (2019) Implication of transfer zones in rift fault propagation: example from Cauvery basin, Indian east coast. In: Mukherjee S (ed) Tectonics and structural geology: Indian contexts. Springer International Publishing AG, Cham, pp 313–326. ISBN 978-3-319-99340-9

Dasgupta S, Mukherjee S (2017) Brittle shear tectonics in a narrow continental rift: asymmetric non-volcanic Barmer basin (Rajasthan, India). The Journal of Geology 125, 561–591

Dinkar GK, Bhattacharya AR, Verma AK, Sharma P (2019) Geology, structural architecture and tectonic framework of the rocks of Southern Lalitpur District, Uttar Pradesh, India: an epitome of the Indian Peninsular Shield. In: Mukherjee S (ed) Tectonics and structural geology: Indian context. Springer International Publishing AG, Cham, pp 353–379. ISBN 978-3-319-99340-9

Dutta D, Biswas T, Mukherjee S (Submitted) Orogen-parallel compression in NW Himalaya: evidence from structural and paleostress studies of brittle deformation from the pebbles of Upper Siwalik conglomerates, Uttarakhand, India. Journal of Earth System Science

Goswami S, Upadhyay PK (2019) Tectonic history of the granitoids and Kadiri schist belt in the SW of Cuddapah basin, Andhra Pradesh, India. Tectonic history of the granitoids and Kadiri schist belt in the SW of Cuddapah basin, Andhra Pradesh, India. In: Mukherjee S (ed) Tectonics and structural geology: Indian context. Springer International Publishing AG, Cham, pp 253–278. ISBN 978-3-319-99340-9

Hossain MS, Khan MSH, Chowdhury KR, Abdullah R (2019) Synthesis of the tectonic and structural elements of the Bengal Basin and its surroundings. In: Mukherjee S (ed) Tectonics and structural geology: Indian context. Springer International Publishing AG, Cham, pp 135–218. ISBN 978-3-319-99340-9

Kumar M, Joshi RC, Dutt Pant P (2019a) Impact of structural damage zones on slope stability: a case study from Mandakini Valley, Uttarakhand state (India). In: Mukherjee S (ed) Tectonics and structural geology: Indian context. Springer International Publishing AG, Cham, pp 397–410. ISBN 978-3-319-99340-9

Kumar R, Ghosh JG, Patel SS, Das A, Sengupta S, Krishna KVSS, Guha D (2019b) Pb—isotopic characterization of major Indian Gondwana Coalfields: implications for environmental fingerprinting and Gondwana reconstruction. In: Mukherjee S (ed) Tectonics and structural geology: Indian context. Springer International Publishing AG, Cham, pp 441–455. ISBN 978-3-319-99340-9

Mahato S, Mukherjee S, Bose N (2019) Documentation of brittle structures (back shear and arc-parallel shear) from Sategal and Dhanaulti regions of the Garhwal Lesser Himalaya (Uttarakhand, India). In: Mukherjee S (ed) Tectonics and structural geology: Indian context. Springer International Publishing AG, Cham, pp 411–423. ISBN 978-3-319-99340-9

Mazumder S, Tep B, Pangtey KKS, Mitra DS (2019) Basement tectonics and shear zones in Cauvery Basin (India): implications in hydrocarbon exploration. In: Mukherjee S (ed) Tectonics and structural geology: Indian context. Springer International Publishing AG, Cham, pp 279–311. ISBN 978-3-319-99340-9

Misra AA, Bhattacharya G, Mukherjee S, Bose N (2014) Near N-S paleo-extension in the western Deccan region in India: Does it link strike-slip tectonics with India-Seychelles rifting? International Journal of Earth Sciences 103:1645–1680

Misra AA, Mukherjee S (2015) Tectonic inheritance in continental rifts and passive margins. Springer Briefs in Earth Sciences. ISBN 978-3-319-20576-2

Misra AA, Mukherjee S (2017) Dyke-brittle shear relationships in the Western Deccan Strike Slip Zone around Mumbai (Maharashtra, India). In: Mukherjee S, Misra AA, Calvès G, Nemčok M (eds) Tectonics of the Deccan large igneous province. Geological Society, London, pp 269–295 (Special Publications 445)

Misra AA, Mukherjee S (2018) Atlas of structural geological interpretation from seismic images. Wiley Blackwell. ISBN: 978-1-119-15832-5

Misra AA, Sinha N, Mukherjee S (2015) Repeat ridge jumps and microcontinent separation: insights from NE Arabian Sea. Marine and Petroleum Geology 59, 406–428

Misra S, Srivastava PK, Arif MD (2019) Remote sensing, structural and rock magnetic analyses of the Ramgarh structure of SE Rajasthan, Central India—further clues to its impact origin and time of genesis. In: Mukherjee S (ed) Tectonics and structural geology: Indian context. Springer International Publishing AG, Cham, pp 327–352. ISBN 978-3-319-99340-9

Mukherjee S (2013a) Channel flow extrusion model to constrain dynamic viscosity and Prandtl number of the Higher Himalayan Shear Zone. International Journal of Earth Sciences 102, 1811–1835

Mukherjee S (2013b) Deformation microstructures in rocks. Springer Geochemistry/Mineralogy, Berlin, pp 1–111. ISBN 978-3-642-25608-0

Mukherjee S (2014) Atlas of shear zone structures in meso-scale. Springer Geology, Cham, pp 1–124. ISBN 978-3-319-0088-6

Mukherjee S (2015a) Petroleum geosciences: Indian contexts. Springer Geology. ISBN 978-3-319-03119-4
Mukherjee S (2015b) Atlas of structural geology. Elsevier, Amsterdam. ISBN 978-0-12-420152-1
Mukherjee S, Mulchrone KF (2015) Ductile shear zones: from micro-to macro-scales. Wiley Blackwell. ISBN: 978-1-118-84496-0
Mukherjee S, Carosi R, van der Beek PA, Mukherjee BK, Robinson DM (2015) Tectonics of the Himalaya: an introduction. In: Mukherjee S, Carosi R, van der Beek P, Mukherjee BK, Robinson D (eds) Geological Society, London, pp 1–3. ISBN 978-1-86239-703-3. ISSN 0305-8719 (Special Publications 412)
Mukherjee S, Misra AA, Calvès G, Nemčok M (2017) Tectonics of the Deccan large igneous province: an introduction. In: Mukherjee S, Misra AA, Calvès G, Nemčok M (eds) Tectonics of the Deccan large igneous province. Geological Society, London, pp. 1–9 (Special Publications 445)
Mukherjee S, Dey A, Sanyal S, Sengupta P (2019) Proterozoic crustal evolution of the Chotanagpur Granite Gneissic complex, Jharkhand-Bihar-West Bengal, India: current status and future prospect. In: Mukherjee S (ed) Tectonics and structural geology: Indian context. Springer International Publishing AG, Cham, pp 7–54. ISBN 978-3-319-99340-9
Padmalal A, Khonde N, Maurya DM, Shaikh M, Kumar A, Vanik N, Chamyal LS (2019) Geomorphic characteristics and morphologic dating of the Allah Bund Fault scarp, Great Rann of Kachchh, Western India. In: Mukherjee S (ed) Tectonics and structural geology: Indian context. Springer International Publishing AG, Cham, pp 55–74. ISBN 978-3-319-99340-9
Pillai SP, Kale VS (2019) Interplay between tectonics & eustasy in a Proterozoic epicratonic, polyhistoric basin: Kaladgi basin: North Dharwar craton. In: Mukherjee S (ed) Tectonics and structural geology: Indian context. Springer International Publishing AG, Cham, pp 75–114. ISBN 978-3-319-99340-9
Roy AB, Purohit R (2018) Indian shield: Precambrian evolution and Phanerozoic reconstitution. Elsevier, Amsterdam. ISBN: 9780128098394
Sharma RS (2010) Cratons and fold belts of India. Springer, Berlin. ISBN: 978-3-642-01459-8
Singh T, Awasthi AK (2019) Deformation in the Kangra Reentrant, Himachal Pradesh of NW-Sub Himalaya of India: a paradox. In: Mukherjee S (ed) Tectonics and structural geology: Indian context. Springer International Publishing AG, Cham, pp 381–396. ISBN 978-3-319-99340-9
Tripathy V, Satyapal, Mitra SK, Sai BBS (2019) Fold-thrust belt architecture and structural evolution of the Northern part of the Nallamalai Fold Belt, Cuddapah basin, Andhra Pradesh, India. In: Mukherjee S (ed) Tectonics and structural geology: Indian context. Springer International Publishing AG, Cham, pp 219–252. ISBN 978-3-319-99340-9
Valdiya KS (2016) The making of India, 2nd edn. Springer, Berlin. ISBN: 978-3-319-25029-8

Proterozoic Crustal Evolution of the Chotanagpur Granite Gneissic Complex, Jharkhand-Bihar-West Bengal, India: Current Status and Future Prospect

Subham Mukherjee, Anindita Dey, Sanjoy Sanyal and Pulak Sengupta

1 Introduction

We presently believe that the continental crust grew episodically and that smaller continents were united and disintegrated several times in the past ~4000 Ma (Hoffmann 1989; Rogers 1996). This process of plate jostling and its subsequent destruction, commonly termed as 'supercontinental cycle', eventually controls the harmonic interactions among lithosphere, hydrosphere and atmosphere over geological time (Worsley et al. 1985, 1986; Piper 2013). The antiquity and the petrological diversity of the rocks in the Indian shield offer unique opportunity to configure the supercontinents that existed in the geological past (Acharyya 2003; Meert et al. 2010). One of the outstanding and hotly debated problems is the position of the Indian shield in the proposed Precambrian supercontinents in general and timing of suturing of India and east Antarctica in particular (Torsvik et al. 2001; Dasgupta and Sengupta 2003; Pisarevsky et al. 2003; Bhowmik et al. 2012). With regard to the latter, two competing views exist. One view is India and Antarctica were united at least from ~1000 Ma (Hoffmann 1989; Dalziel 1991; Li et al. 2008). The disclaimers of this view propose that the two continents got juxtaposed not earlier than ~900 Ma (Bhowmik et al. 2012) if not after ~750 Ma (Merdith et al. 2017; Torsvik et al. 2001). In the reconstructed Rodinia, a Proterozoic supercontinent, the CGGC and the Eastern Ghats Mobile Belt juxtaposed against the east Antarctic Precambrian basement (Dasgupta and Sengupta 2003; Chatterjee et al. 2010; Mukherjee et al. 2017a). Therefore, rocks of these two areas of the Indian shield are likely to provide the clinching evidence about the timing of Indo-Antarctic suturing.

Till the end of twentieth century, limited petrological information and rudimentary geochronological data were available from the CGGC that is positioned

S. Mukherjee · A. Dey · S. Sanyal · P. Sengupta (✉)
Department of Geological Sciences, Jadavpur University, Kolkata 700032, India
e-mail: pulaksg@gmail.com

S. Mukherjee (ed.), *Tectonics and Structural Geology: Indian Context*,
Springer Geology, https://doi.org/10.1007/978-3-319-99341-6_2

against the Precambrian Vestfold block of east Antarctica in the reconstructed Rodinia (Fig. 1). In the past one decade several studies in the light of modern petrology and robust geochronology have been published. In this work we have reviewed the published information on the CGGC with the following aims:

(1) To establish an event stratigraphy showing the magmatic and tectonic pulses that shaped the rocks of the CGGC during the Precambrian Era
(2) Comparisons of the thermo-tectonic pulses that are recorded in the CGGC with the adjoining crustal domains in the Indian shield
(3) Influence of the Precambrian supercontinental cycles on the CGGC
(4) Timing of amalgamation of the Indo-Antarctic landmasses
(5) To delineate the gaps in knowledge and scope of future study.

2 Extent and the Boundary Relations

The CGGC is an east-west trending mobile belt that belongs to the east Indian Shield and is exposed across the states of Jharkhand, Bihar, West Bengal and Chhattisgarh covering an area of over 100,000 km^2 (reviewed in Mahadevan 2002; Acharyya 2003). The northern margin of the CGGC is covered by quaternary sediments of Gangetic alluvium (Fig. 2a). Sediments of the Bengal Basin mark the eastern boundary of the terrain and Mesozoic volcanics of Rajmahal Trap covers the northeastern fringe of the terrain. The western margin of CGGC is dominantly covered by Gondwana deposits of Permian to mid-Cretaceous age (Mahadevan 2002). However, in the northwestern part Vindhyan sediments and the Mahakoshal group of rocks are in contact with the Proterozoic rocks of the CGGC. Towards the south the contact between CGGC and Proterozoic rocks of the North Singhbhum Fold Belt (NSFB) is marked by the east-west trending crustal scale shear zone called the South Purulia Shear Zone (SPSZ) or the Tamar-Porapahar-Khatra Shear zone.

3 Classification of the CGGC on the Basis of Extant Petrological and Geochronological Data

Scarcity of detail petrological, lithological and geochronological data is the major hindrance for a reasonable classification of the CGGC. Scattered distribution of exposures, owing to tropical weathering and urbanization, further complicate the problem. In earlier studies (Mahadevan 2002; Sanyal and Sengupta 2012), the CGGC were divided into five N-S blocks (Fig. 2b). They considered the Chotanagpur plateau and the Gondwana deposits (Permian-mid Cretaceous) as the basis of classification. The authors of this communication are of the view that Precambrian CGGC

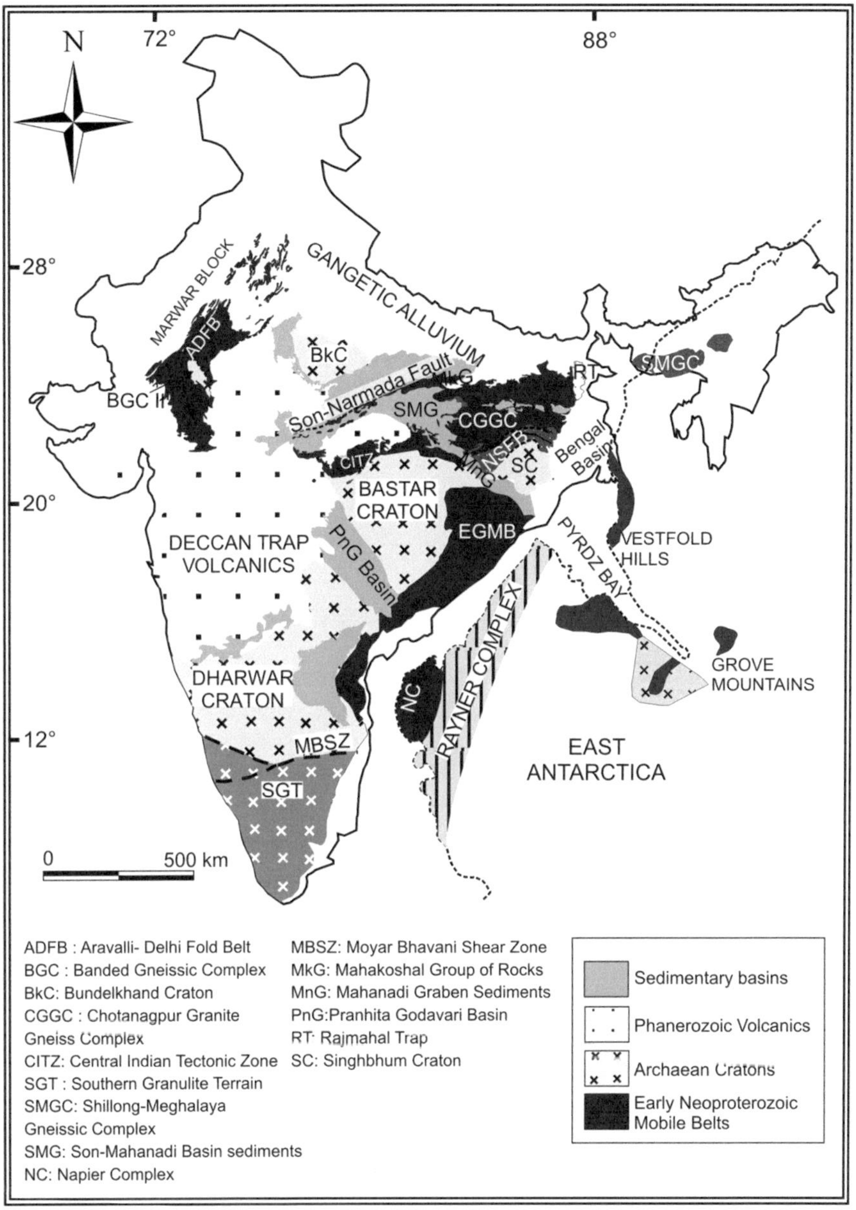

Fig. 1 Proposed configuration of India and Antarctica during the assembly of Rodinia; modified after Dasgupta and Sengupta (2003). The schematic map of India and it does not represent the erstwhile configuration as northeastern and northwestern margin (Tethyan sequence) formed later

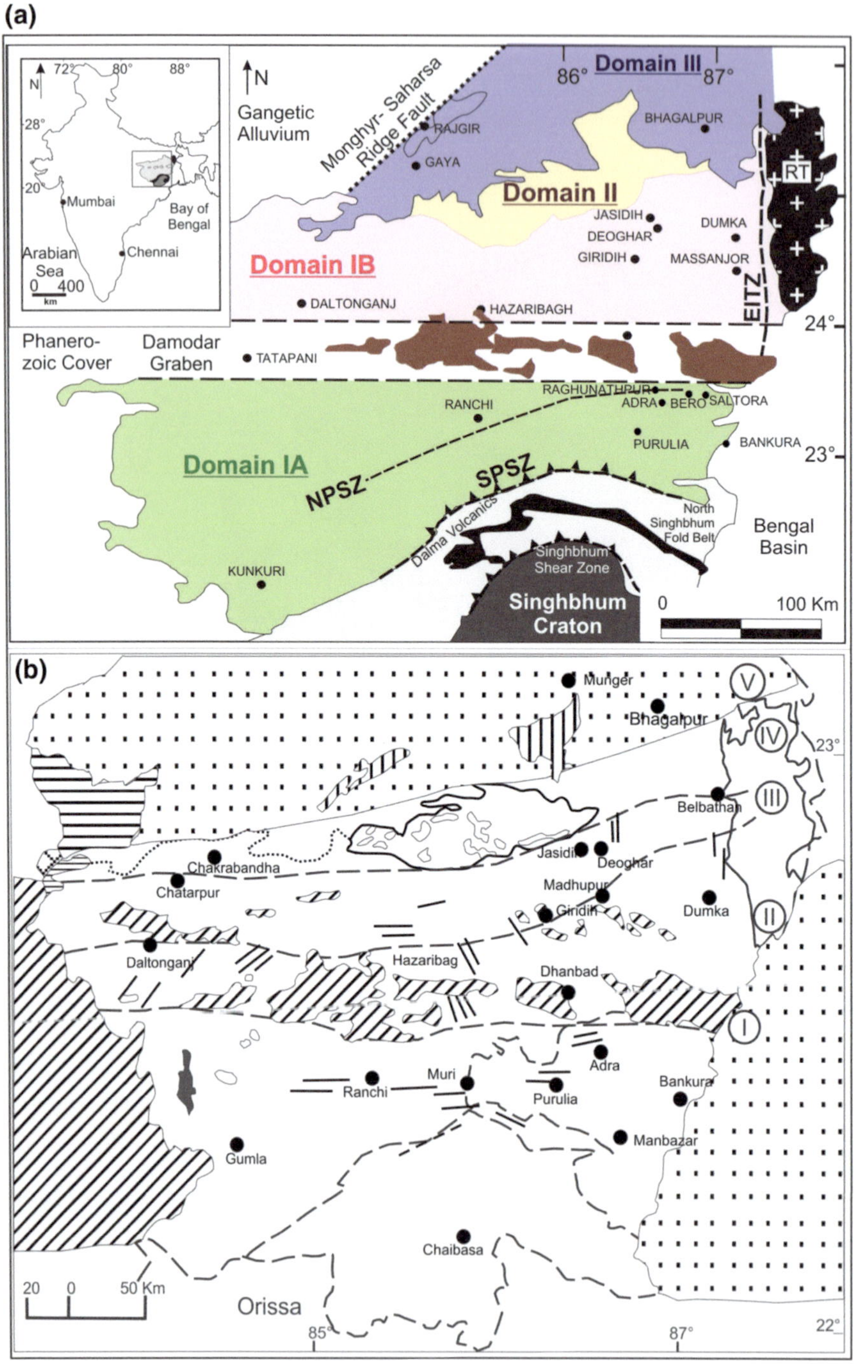
(a)
Domain III
Domain II
Domain IB
Domain IA
Monghyr- Saharsa Ridge Fault
Gangetic Alluvium
RAJGIR
GAYA
BHAGALPUR
JASIDIH
DEOGHAR
DUMKA
GIRIDIH
MASSANJOR
DALTONGANJ
HAZARIBAGH
EITZ
RT
Phanero-zoic Cover
Damodar Graben
TATAPANI
RAGHUNATHPUR
ADRA
BERO
SALTORA
RANCHI
PURULIA
BANKURA
NPSZ
SPSZ
Dalma Volcanics
North Singhbhum Fold Belt
Singhbhum Shear Zone
Singhbhum Craton
Bengal Basin
KUNKURI
0 100 Km
86° 87° 24° 23°
Mumbai
Chennai
Bay of Bengal
Arabian Sea
0 400
(b)
Munger
Bhagalpur
V
IV
III
II
I
Belbathan
Jasidih
Deoghar
Chakrabandha
Chatarpur
Madhupur
Giridih
Dumka
Daltonganj
Hazaribag
Dhanbad
Adra
Ranchi
Muri
Purulia
Bankura
Manbazar
Gumla
Chaibasa
Orissa
20 0 50 Km
85° 87° 23° 22°

◀**Fig. 2** **a** Geological map of the Chotanagpur Granite Gneiss Complex (CGGC) and the showing major domains (modified after Acharyya 2003). EITZ: Eastern Indian Tectonic Zone, NPSZ: North Purulia Shear Zone, SPSZ: South Purulia Shear Zone, RT: Rajmahal Trap. **b** Geological map of the Chotanagpur Granite Gneiss Complex (CGGC) showing five subdivisions proposed by Mahadevan (2002) and Sanyal and Sengupta (2012), modified after Sanyal and Sengupta (2012)

should not be divided using geomorphic feature of younger sedimentary rocks of Gondwana.

In the past few years, a large amount of petrological and geochronological information has been published on the rocks of the CGGC. Integrating all these information the CGGC has been divided into three roughly east-west domains each with characteristics lithology, metamorphic history and geochronological information. Each of the domains has broadly E-W trend and their disposition from south to north, are Domain I, Domain II and Domain III (Fig. 2a). The CGGC is dissected by three major lineaments (Fig. 2a). The South Purulia Shear Zone (SPSZ) and Monghyr-Saharsa Ridge Fault roughly coincide with the southern and northern boundaries of the CGGC respectively. The lineament, that bound the exposure of Gondwana deposits of Damudar valley run through the Domain I of the CGGC (Mandal 2016) and hereafter termed as Gondwana Boundary Faults (GBF; Fig. 2a). Using the GBF as a marker, Domain I is further divided into two geographic sub-domains viz. Domain IA (south) and IB (north). No major lithological/geochronological break has been noted across the GBF. The subdivision of Domain I only help synthesize the published geological and geochronological data. No tectonic lineament separate Domain II from the adjoining Domain III and Domain I. Preponderance of mica-bearing pegmatite intrusive (the Bihar Mica Belt; BMB) make Domain II a distinct lithounit (Fig. 2a).

It may be mentioned here that intra- and inter-domain correlation of the metamorphic and structural events are fraught with following problems:

(a) The areas from where the geological information is available are not continuous and scattered over large areas of the CGGC. Practically no information is available from a vast expanse of the CGGC.
(b) Lack of precise geochronological data renders correlation of geological events reported from different parts of the CGGC difficult.

Nevertheless, the salient lithological, structural, petrological and geochronological features as reported in the published work from the three domains are presented.

3.1 *Domain IA*

This domain covers rocks exposed in the southernmost part of the CGGC and is bounded by GBF and the SPSZ (Fig. 2a). The geological information albeit sparse, clusters around the Bankura-Saltora-Bero area in the east, the Raghunathpur-Adra-Ranchi areas in the central part and Raikera-Kunkuri region in the western part of the domain (Fig. 2a). Petrology and geochronology of the rocks exposed around Bankura-Saltora-Bero—have been extensively studied by several workers (Manna and Sen 1974; Roy 1977; Bhattacharyya and Mukherjee 1987; Sen and Bhattacharya 1993; Mukherjee et al. 2005; Chatterjee et al. 2008; Maji et al. 2008), which reveals that variably deformed migmatitic felsic orthogneisses, holding dismembered rafts of mafic granulite and calc-silicate gneisses is the dominant rocktype intruded by massif type anorthositic rocks (Bengal anorthosite). Metamorphic grade varies from amphibolite to granulite grade conditions. Documentation of granulite and amphibolite facies rocks mostly coming from the eastern and the western part of the domain respectively (Karmaker et al. 2011; Goswami and Bhattacharya 2010, 2013; Maji et al. 2008; Chatterjee et al. 2008; Sanyal and Sengupta 2012). However the published information does not support any systemic variation in the estimated metamorphic conditions along any geographic direction. The rocks in this region are folded to east-west closing folds and an E-W trending axial planar fabric dipping steeply towards north (Maji et al. 2008).

In the easternmost parts, near Bero-Saltora-Santuri, high grade metapelitic rocks and migmatitic quartzofeldspathic rocks are exposed. Three deformation ($D_{1\text{-}3}$) phases accompanied by four metamorphic events ($M_{1\text{-}4}$) have been inferred from the area by Maji et al. (2008). Earliest tectonothermal event (M_1), considered to be of granulite grade (minimum P–T estimates of 5–6 kbar and 750–850 °C), produced the migmatitic banding of the orthogneisses (S_1) (Sen and Bhattacharya 1993; Maji et al. 2008) and occurred at ca. 1.70 Ga, inferred from chemical dating of monazite (Chatterjee et al. 2010). Both the subsequent deformations ($D_{2\text{-}3}$) occurred under amphibolite facies ($M_{2\text{-}3}$) between 1.3 and 1.1 Ga and replaces the older granulite mineralogy to variable extant (Maji et al. 2008). Towards the eastern fringe of the domain, near Saltora (West Bengal) (Fig. 2a) intrusion of massif anorthosite, called the Bengal anorthosite, occurred at ca. 1.55 Ga between the D_1 and D_2 (Bhattacharyya and Mukherjee 1987; Chatterjee et al. 2008; Maji et al. 2008). The most pervasive metamorphic event is inferred to have occurred between 1.0 and 0.95 Ga, that has been designated as M_4 by Maji et al. (2008). However the P–T conditions of the event is debated. Maji et al.(2008) inferred that the metamoprhism culminated at 650 ± 50 °C at 4–5 kbar whereas Chatterjee et al. (2008) recovered a high grade conditions (850–900 °C and 8.5–11 kbar) from gabbro anorthositic rocks. Near Kankarkiari (West Bengal), migmatitic felsic orthogneisses has been intruded by nepheline-bearing syenite and subsequently got deformed and metamorphosed (Das et al. 2016; Goswami and Bhattacharyya 2010). No geochronological data is available to constrain this tectonothermal event experienced by the syenitic rocks. However, field observations suggest that they have intruded after the Grenvillian metamorphic event and subsequently got deformed and metamorphosed under amphibolite grade yielding a P–T condition of 700–750 °C and ~10 kbar, associated with development of foliation and folding (Das et al. 2016). Younger Neoproterozoic ages (ca. 900–820 Ma) recovered from the overgrowths on older monazite grains have been documented by several workers (Maji et al. 2008; Chatterjee et al. 2010), which further attest for the Late Tonian–Early Cryogenian tectonothermal event. One of the source of heating could be deformation itself (Mukherjee and Mulchrone 2013; Mulchrone and Mukherjee 2015, 2016; Mukherjee 2017).

The area around Raghunathpur (West Bengal), west of Bero, reveals an ensemble of different generations of felsic orthogneisses (porphyritic and deformed granitoids), metapelitic and calcareous enclaves (Dunn 1929; Baidya et al. 1987, 1989; Ray Barman and Bishui 1994; Goswami and Bhattacharyya 2010, 2013; Karmakar et al. 2011). Three major deformational events (D_1-D_3) have been identified from the area associated with two major metamorphic phases (M_1-M_2). Non-porphyritic granite intrusion occurred at 1178 ± 61 Ma (Rb–Sr whole rock isochron, Ray Barman and Bishui 1994) and is synchronous with development of S_1 during D_1 (Goswami and Bhattacharyya 2010). A porphyritic charnockite that was emplaced prior to D_2 shows a Rb–Sr whole rock age of 1071 ± 64 Ma (Ray Barman and Bishui 1994; Goswami and Bhattacharya 2010). These porphyritic granites have been geochemically classified as Shoshonitic to high K-calc alkalic intrusives that were formed via mixing of mantle-derived mafic magma and crustal melts followed by fractional crystallization in continent-continent collisional settings (Goswami and Bhattacharyya 2013). Towards the southeast of Raghunathpur, near Adra (West Bengal), migmatitic quartzofeldspathic gneiss contains enclave of Mg–Al granulite and mafic granulites. Chemical ages of monazite from these Mg–Al pelite and migmatitic gneiss indicate that the most pervasive and prominent tectonothermal event of the area occurred between ca. 990–940 Ma that culminated at ~870 °C and 11 kbar pressure followed by a steeply decompressive path (Karmakar et al. 2011). Goswami and Bhattacharyya (2010) determined a similar temperature (~800 °C) but lower pressure (6.5–7.5 kbar) and argued that M_1 spanned over during both D_1 and D_2. Youngest monazite age population of ca. 850–775 Ma recovered by Karmakar et al. (2011) are consistent with the Neoproterozoic dates (870 ± 40 Ma: K–Ar biotite of porphyritic granite and 810 ± 40 Ma: K–Ar muscovite of leucogranite) reported by Baidya et al. (1987) from the western part of the area near Jaipur, West Bengal. Although petrological manifestations are lacking, the youngest age clusters presumably constrains the third tectonothermal event (D_3).

Litho-package exposed in the south-central of the terrain, south of Ranchi, resembles rocktypes of the eastern part, containing garnetiferous migmatitic felsic gneiss, pelitic schist and minor calc-silicate bodies that have been intruded by porphyritic granite (Sarkar and Jha 1985; Rekha et al. 2011). Zircon and monazite geochronological studies of both the older migmatitic gneiss and metapelite reveals Neoproterozoic ages (944 ± 9 and 921 ± 18 Ma), inferred to be metamorphic, whereas younger granites yield an emplacement age of 928 ± 23 Ma with older inherited components of 1072 ± 17 and 1239 ± 66 Ma (Rekha et al. 2011).

At the south-western part of the terrain, near Raikera–Kunkuri region (Chhattisgarh), different generation of granite bodies are associated with pelitic schist (with chlorite, biotite and hornblende schist), quartzite and dolerite dykes/sills (Singh and Krishna 2009). Two-mica bearing grey granites, derived from juvenile crustal sources (SrI ≈ 0.7047 and low high field strength elements), intruded the crust at 1005 ± 51 Ma (Singh and Krishna 2009). On the other hand, younger Rb–Sr isochron age of 815 ± 47 Ma and high SrI (0.7539) determined from the pink granite are inferred to be the late metasomatic event associated with the Y-mineralization in the area (Singh and Krishna 2009).

3.2 Domain IB

The E-W trending Domain IB is sandwiched between GBF and Domain II (Fig. 2a). From east to west, the geological information from this domain in concentrated around the cities named Masanjor, Dumka, Deoghar, Jasidih, and Daltonganj. Amongst them, most of the granulite grade enclave rocks are exposed around Masanjor, Dumka and Deoghar in Jharkhand (Mahadevan 2002). The general lithology of the domain is felsic gneiss of varied mineralogy and composition which suffered granulite grade metamorphism and anatexis. Variably metamorphosed gneisses and schists of supracrustals and basic rocks occur as enclaves within the felsic gneiss. The general strike of the domain is E-W to NW-SE except its northeastern part where the strike becomes N-S. At least three stages of folding has been identified by different workers (Ghosh and Sengupta 1999; Sanyal and Sengupta 2012; Mukherjee et al. 2017a; Dey et al. *under review a*; Dey et al. *under review b*) throughout the domain.

The eastern extremity of this domain (in and around Masanjor and Dumka) exposes gneisses of variable composition. Migmatitic felsic gneiss constitutes the dominant rock type of this area. Compositionally it varies from migmatitic charnockite (orthopyroxene + K-feldspar + plagioclase + quartz + garnet + ilmenite) to amphibole-biotite gneiss (amphibole ± biotite + K-feldspar + plagioclase + quartz + garnet + ilmenite). Based on the on-going study of the authors it is presumed that the latter is the retrogressed counterpart of the former. Using unpublished U–Pb zircon dates of Ray Barman, Acharyya (2003) constrained the protolith ages of the migmatitic charnockite to be 1624 ± 5 Ma. The host migmatitic felsic gneiss contains rafts/enclaves of khondalite (quartz + K-feldspar + plagioclase + sillimanite + garnet + ilmenite), Mg–Al granulite (orthopyroxene + sapphirine + spinel + sillimanite + quartz + perthite + plagioclase + garnet + ilmenite + biotite + cordierite), calc-silicates (clinopyroxene + garnet + quartz + sphene + K-feldspar + plagioclase) and mafic granulite (plagioclase + clinopyroxene + garnet + hornblende + ilmenite + titanite). A large body of porphyritic charnockite (quartz + K-feldspar + plagioclase + orthopyroxene + clinopyroxene + garnet + hornblende + ilmenite + biotite + magnetite) intruded the host felsic gneiss and contains enclaves of the rocks it intruded. From mutual field relations between the litho-units, three stages of deformation have been established for the country rock (Mukherjee et al. 2017a; Dey et al. *under review a*; Sanyal and Sengupta 2012). The enclave rocks developed a gneissic banding that predates the foliation of the migmatitic felsic gneiss, formed during D_1. The foliation in the host gneiss represents the dominant foliation of the area and strikes E-W. They refolded during subsequent D_2 and D_3 deformation (Sanyal and Sengupta 2012). Protolith of porphyritic charnockite emplaced in between D_1 and D_2 at 1515 ± 5 Ma (Acharyya 2003). A swarm of mafic dykes (plagioclase + amphibole + clinopyroxene + chlorite + epidote + calcite + quartz + ilmenite) cut across the foliation of host gneiss but is folded by open D_3 folds. The Mg–Al granulitic enclave rocks develop a mineral assemblage of aluminous (7 wt% Al_2O_3) orthopyroxene + magnetite-hercynite + sillimanite +quartz + garnet + melt which is a good indicator of ultrahigh-temperature metamorphism (>900 °C) at a pressure in excess of ~8 kbar (Sanyal and Sengupta 2012). From conventional geothermobarometry of the mafic enclaves, Sanyal and Sengupta (2012) has obtained distinctly lower temperatures 825–850 °C at 8–9 kbar indicating subsequent cooling followed by the UHT metamorphism. Geothermobarometry of the porphyritic charnockite constrain the conditions of M_3 metamorphism synchronous to D_3, at 700 ± 50 °C and 6.5 ± 1 kbar (Sanyal and Sengupta 2012).

Maximum petrological and geochronological information is available from further north-west, in between Dumka and Deoghar town (Jharkhand); especially from northern part of Dumka. The country rock of felsic orthogneiss hosts km to cm scale rafts of pelitic rocks (garnet-sillimanite-biotite-K-feldspar-plagioclase-quartz ± spinel), mafic rocks (plagioclase + clinopyroxene + orthopyroxene + garnet + hornblende + ilmenite + rutile), calc silicates (clinopyroxene + plagioclase + titanite ± garnet ± amphibole ± scapolite ± calcite), granulites and augen gneisses (K-feldspar + plagioclase + quartz + biotite + hornblende + apatite). Mineralogically the host gneiss varies from charnockitic (orthopyroxene + clinopyroxene + garnet + plagioclase + K-feldspar + quartz + hornblende + biotite + ilmenite) to biotite-hornblende gneiss to hornblende-biotite gneiss (garnet + plagioclase + K-feldspar + quartz +hornblende + biotite + ilmenite) (Mukherjee et al. 2017a). Geochemical and isotopic studies of the host felsic gneiss confirm that they have a ferroan (A type) character (Mukherjee et al. 2017a, 2018). The ortho-gneisses show a prominent N-S trending migmatitic banding and this regional fabric locally swerves around the enclaves. The pelitic enclaves contain voluminous (>30%) leucosomal segregations (S_1; Fig. 3a). The S_1 which are discordant to and are dragged to parallelism with the pervasive foliation (S_2) of the host felsic gneiss (Fig. 3a). Numerical modelling with appropriate bulk for the metapelites constrain an early high grade event M_1 occurred at 7 ± 1 kbar and 1000 ± 50 °C i.e. at MP (medium pressure)-UHT condition which generated voluminous S_1 leucosomal foliation (Dey et al. *under review*, a). This event was followed by another metamorphism whereby the felsic orthogneisses and the meta-sedimentary gneisses developed a prominent migmatitic foliation that are currently ~N-S trending. Numerical modelling combined with conventional geo-thermobarometry of the host gneiss as well as the pelitic enclaves constrain the peak of this metamorphism at 770 ± 50 °C and 9 ± 1 kbar (Dey et al. *under review*, a; Chatterjee et al. 2008; Mukherjee et al. 2017a). However, petrological study of a suite of mafic enclave reveals much higher pressure (12 ± 1 kbar and 800 ± 50 °C) for the same (Dey et al. *under review*, b). In all these studies

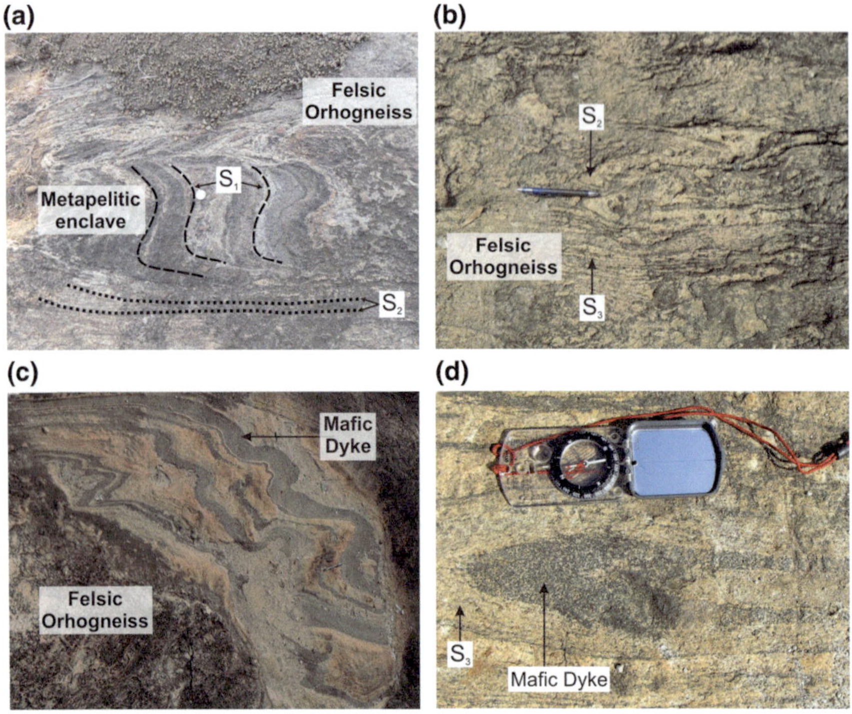

Fig. 3 **a** Internal foliation (S_1) of metapelitic enclaves and external foliation (S_2) within host felsic gneisses; **b** folding of S_2, designated by coarse leucosomes, and development of axial planar S_3 within felsic orthogneiss; **c** folding of the mafic dyke during D_2 and **d** development of S_3 along the axial planes of folded mafic dyke

this HP-MT metamorphism was followed by a steep decompressive path indicating that the peak condition was attained through a continent-continent collisional event. Towards the western part of this domain, in a stretch from Dumka to Jasidihi (Jharkhand) (Fig. 2a), a swarm of mafic dykes (now amphibolites with orthopyroxene + clinopyroxene + hornblende + biotite + plagioclase + titanite + quartz) occur within the felsic orthogneiss (Fig. 3c). The mafic dykes cut the S_2 fabric of the host felsic orthogneiss. Along with S_2, the mafic dykes are folded by two sets of co-axial folds with N-S closure (D_3; Ghosh and Sengupta 1999; Ray et al. 2011a, b; Mukherjee et al. 2017a; Dey et al. *under review* a; Dey et al. *under review* b). A prominent N-S trending planar fabric (S_3) that is defined by hornblende developed along the axial plane of the earliest fold (Ghosh and Sengupta 1999; Sanyal and Sengupta 2012; Mukherjee et al. 2017a). Locally S_3 is folded by an open fold with nearly vertical axial plane (Ghosh and Sengupta 1999; Sanyal and Sengupta 2012; Mukherjee et al. 2017a). Deformation and metamorphism of the mafic dykes and their host felsic gneiss (M_3-D_3) mark the latest major tectonothermal event of Domain I. The host felsic gneiss and the metapelitic granulite enclave recorded a

P–T of 7.3 ± 0.1 kbar, 615 ± 15 °C and 4.3 ± 0.7 kbar, 600 ± 60 °C, respectively for this metamorphism (Dey et al. *under review*, a). Bhattacharjee et al. (2012) reported a gabbro-anorthosite body (Hizla anorthosite) near Dumka area, Jharkhand. This anorthosite body intruded the charnockitic country rock and is deformed and metamorphosed along with the latter rock. These rocks that are likely to be the product of D_2-M_2 tectonothermal event recorded an unusually wide range of pressure and temperature of (511–915 °C and 5.0–7.5 kbar) for the metamorphism of the anorthosite (Bhattacharjee et al. 2012). The published petrological information is not robust enough for draw any conclusion about this wide range of metamorphic P–T values. Multiple generations of pegmatitic veins criss-cross all the lithounits. No structural data on the orientation of these veins are available. Eastern margin of the Domain I is marked by a highly tectonized N- to NNE-zone shear deformation. Shear driven non-cylindrical folds (D_2-D_3 fold interference) has been described by Chatterjee et al. (2010). This tectonic zone is termed as the Eastern Indian Tectonic Zone (EITZ; Chatterjee et al. 2010). A linear gravity high follows the trend of EITZ over a length of approximately 400 km (Singh et al. 2004). From a quartzo-feldspathic gneiss from northern part of Dumka, Chatterjee et al. (2010) has quantified the peak conditions of the shearing and anatexis along EITZ at ~11 kbar and <800 °C followed by isothermal decompression.

U–Pb zircon and monazite age of the metapelitic enclaves attests the age of M_1 metamorphism at 1630 ± 50 Ma (Dey et al. *under review*, a). Similar Paleoproterozoic metamorphic ages has also been obtained by Rekha et al. (2011) and by Chatterjee et al. (2010) from metapelites and anatectic quartzofeldspathic gneiss near Dumka. From detrital zircon study of the metapelitic enclaves, Dey et al. (2017) has constrained the age of deposition of the sedimentary protoliths of these enclaves within 1700–1680 Ma and has ascribed that the sediments were derived from Archean and Paleoproterozoic sources (Rekha et al. 2011; Dey et al. 2017). The protolith of the felsic orthogneiss was derived via high temperature partial melting of Paleoproterozoic crustal source (~1800–1600 Ma) with limited mantle input (Lu–Hf model age: Mukherjee et al. 2018) and was presumably emplaced in a continental rift setting at ~1450 Ma (U–Pb zircon dating; Mukherjee et al. 2017a, 2018). Mesoproterozoic dates have also been reported from U–Pb dating of zircon from metapelites and anatectic quartzofeldspathic gneiss in this domain (Rekha et al. 2011; Dey et al. *under review*, a). Rb–Sr whole-rock date the emplacement of a syenitic rock (1457 ± 63 Ma) and a charnockite (1331 ± 125 Ma) in Dumka-Jamua, Jharkhand area of Jharkhand (Ray Barman et al. 1994). U–Pb zircon and Th–U–Pb monazite age constrained the age of the granulite grade metamorphism at ~1000–930 Ma (Dey et al. *under review*, a; Chatterjee et al. 2008, 2010; Rekha et al. 2011; Mukherjee et al. 2017a). Chatterjee et al. (2010) has retrieved a younger age population of ~870–650 Ma from the monazite dates, which they linked with the tectonothermal activity and sinistral shearing along the EITZ. Zircon and monazite from metapelite and felsic gneiss yield ages for D_3 and M_3, ranging within 900–800 Ma (Chatterjee et al. 2010; Dey et al. *under review,* a; Mukherjee et al. 2017a) with very few ages as low as ~650 Ma (Chatterjee et al. 2010). These age of the D_3 deformation of the area have been correlated with the age duration of shearing along the EITZ.

Not much geological and geochronological information are available from further west of Jasidih, in and around Dhanbadand and Hazaribag (Jharkhand). It is noteworthy that the strike of the schists and gneisses of these areas varies between E-W to NW-SE. Parts of the Dhanbad district (Jharkhand) exposes banded gneisses of amphibolite facies (Roy Chowdhury 1979) with enclaves of hornblende schists and gneisses, olivine-orthopyroxene bearing metanorites and quartzites. The rocks folded thrice with development of E-W to NW-SE trending foliation axial planar to the second generation folding (Sarangi and Mohanty 1998).

Further west, near Hazaribagh (Jharkhand), amphibolites, pelitic and calcareous rocks dominate with a persistent E-W trending schistosity. The rocks have also been folded thrice with first two folds being tight and coaxial, and the last one being open warp (Roy Chowdhury 1979; Mahadevan 2002). The whole rock Rb–Sr isochron date of biotite-K-feldspar from a migmatite record one of the youngest ages (481 ± 18 Ma; Pandey et al. 1986b).

The area around Daltonganj (Jharkhand) exposes granite gneisses and migmatites interbanded with Graphite-bearing pelitic schists, limestones, quartzites and intruded by mafic-ultramafic, anorthositic-komatiitic rocks (Bhattacharya et al. 2010). These ultramafic rocks host magnetite (Ghose 1983; Sinha and Bhattacharya 1995), fluorite (Soni et al. 1991) and base metal mineralization (Ghose 1992; Bhattacharya et al. 2010). Thermometry of granite gneiss confirms that the gneissic foliation formed under high grade condition (850–800 °C; Chatterjee and Ghosh 2011). The granite gneisses are intruded by non-foliated granitic to granodioritic rocks (Rode, 1948; Ghose 1983; Mazumdar 1988). Anatexis of granulite to form leucogranite, and partial melting of metapelite forming these non-foliated granitoids probably occurred at 650–800 °C and 2–6 kbar (Srivastava and Ghose 1992). Whole-rock Rb–Sr dating of the granite gneiss suggests an event of granitic magmatism at 1741 ± 65 Ma (Ray Barman and Bishui 1994) followed by high grade metamorphism and anatexis. Chatterjee and Ghosh (2011) has dated this high grade metamorphism from xenotime at 975 ± 67 Ma. The granite gneiss and metapelitic sequence deformed thrice (D_1-D_3) (Lahiri and Das 1984; Patel 2007). Interference of D_1 and D_2 produced NW-SE trending domal structure with development of prominent axial planar schistosity. Post-D_3 pegmatite veins and mafic dykes are common in the area. In the Tatapani area, the calc-silicate rocks develop the rare assemblage vesuvianite + grossular garnet + diopside + wollastonite + quartz. Patel (2007) estimated the conditions of vesuvianite formation at <4 kbar and 590–650 °C in presence of a highly aqueous fluid.

3.3 Domain II

Sandwiched between Domain III (in the north) and GBF (in the south), this domain is the ENE-WSW trending the Bihar Mica belt (BMB) exposes a distinct lithological ensemble in the CGGC (Fig. 2a). The geological formation in this belt include meta-sedimentary sequence dominantly constituting muscovite-biotite schist interbedded with micaceous quartzites, conglomerates, calc-silicate rocks and hornblende schists. The carbonate metasediments are occasionally found associated with economic grade base-metal mineralization (Ghose 1992). These metasedimentary rocks of BMB rest on the high grade rocks of CGGC with an erosional conglomerate base (Ghose and Mukherjee 2000). The Pb/Pb age of galena from base-metal deposits brackets the sedimentation age in the BMB within the time span of 1700–1650 Ma (Singh et al. 2001). Rb–Sr whole rock dating of a migmatitic granite gneissic basement from south of BMB yields an age of 1717 ± 102 Ma (Mallik et al. 1991). The meta-sedimentary package recorded three phases of folding (D_1, D_2 and D_3), where first two folds are most dominant and produce E-W trending axial planar foliation while the third fold is weak and less pervasive. The whole litho-unit is extensively intruded by large bodies of granitoid rocks, dolerite dykes, gabbro anorthosite and mica pegmatites (sometimes REE and rare metal rich). Granitoids were emplaced syn- to post-D_2 folding with ages of most of the granite intrusives clustering within 1300–1100 Ma and age of oldest granite intrusion around 1600 Ma (Whole-rock and Rb–Sr mineral isochron ages; Pandey et al. 1986a; Misra and Dey 2002). K–Ar dating of granitoids yields lower ages (1080–850 Ma) (Sarkar 1980). The BMB plutons characteristically show very high SrI ratio which suggests their derivation from sialic crustal sources (Misra and Dey, 2002). Different generations of mica-bearing granite–pegmatites emplaced pre-D_2 to post-D_3 folding (Mahadevan 2002). The conditions of emplacement as well as formation of the BMB granite plutons is estimated ~5 kbar and slightly >1000 °C in a relatively anhydrous condition within a possible post-orogenic setting (Misra and Dey 2002). The age of emplacement of the oldest REE-rare metal-bearing mica pegmatites of the BMB has been restricted within 960 ± 50 Ma (Pb/Pb age; Vinogradov et al. 1964) and 910 ± 20 Ma (U–Pb and Pb/Pb ages of the Columbite–Tantalite minerals; Krishna et al. 2003). Fission-track ages of garnet (830 Ma), muscovite (760 Ma), biotite (590 Ma) and apatite (590 Ma) date the final exhumation of the mica pegmatites of this region (Lal et al. 1976). It suggests that the rocks of this area had not received any significant thermal pulse, which by implication means cratonization, since 590 Ma (Lal et al. 1976). In summary, this domain bears history of Paleoproterozoic and Mesoproterozoic granitic activities. Evidence of Paleoproterozoic sedimentation is noteworthy. The abundant intrusion of pegmatites took place in Neoproterozoic which cooled gradually.

3.4 Domain III

The northern fringe of CGGC, north of the Bihar mica Belt, exposes an ensemble of migmatitic quartzofeldspathic gneisses and supracrustals. Sedimentary rocks of the Munger Group and the Rajgir Group showing basement–cover relationship with the

CGGC gneisses demarcated by a conglomerate horizon (Ghose and Mukherjee 2000). Supracrustals are dominated by quartzite and phyllite, metamorphosed under low-grade in association with mafic-ultramafic and felsic intrusives.

The basement of these sedimentary units is marked by the porphyritic granite (K-feldspar + plagioclase + biotite + quartz) near Gaya (Bihar) and biotite granite in the Bhagalpur area (Bihar) (Chatterjee and Ghose 2011). The intrusion age of the granitoids are constrained at ca. 1697 ± 17 Ma from U–Th–Pb monazite (Chatterjee and Ghose 2011). U–Pb xenotime age of 1583 ± 50 Ma is inferred to be the cooling age of the porphyritic granitoids (Chatterjee and Ghose 2011). Detailed geochemical study of the porphyritic granites reveal an A-type character and suggest that they have been derived from predominant crustal source in an extensional (Yadav et al. 2014). In and around Bathani village, near Gaya (Bihar) a volcanic and volcano sedimentary sequence comprises of banded iron formation (BIF), garnet-mica schist, chert bands and mafic-intermediate volcanics (Saikia et al. 2014). Detailed geochemical studies suggest that the intrusion of bimodal volcanics, tholeiitic affinity of the basalt and calc-alkaline trend of the intermediate-felsic magmatism corroborate well with an island-arc setting (Saikia et al. 2014, 2017). U–Pb zircon geochronology constrains the intrusion of the felsic magmatism between ca. 1750–1660 Ma. Subsequently the lithounit got metamorphosed at ca. 930 Ma and at 768 Ma. Earlier metamorphism occurred under granulite grade whereas the younger age associates with a shearing linked with the Eastern Indian tectonic Zone (Chatterjee et al. 2010).

4 Discussions

4.1 Tectonothermal Events at CGGC

This section collates the geological and geochronological information (Table 1) and reconstructs a plausible event stratigraphy that shaped the CGGC.

4.1.1 Sedimentation History of the Supracrustal Enclaves

Timing of sedimentation and provenance of the precursor sediments of metasedimentary rocks provide critical information about the tectonic history of a terrane (Sengupta et al. 2015; Dey et al. 2017). In CGGC, the supracrustal enclave rocks with quartzite, metapelite sand metacarbonate are likely to correspond to of sandstone, shale and limestone found in a stable continental shelf (Malone et al. 2008; Basu et al. 2008, 2014; Sengupta et al. 2015; Dey et al. 2017). Strong structural and metamorphic overprints virtually obliterate the sedimentary features of these supracrustal rocks. This renders reconstruction of sedimentary process in high grade rocks difficult. However, using the dates obtained from detrital zircon or other

Table 1 Available geochronological data from different domains of CGGC; ages are in Ma

Age	Method	Type of age	References
Domain IA			
595	U–Pb monazite	Alteration	Chatterjee et al. (2010)
784	U–Pb monazite	Metamorphism	Chatterjee et al. (2010)
810 ± 40	K–Ar muscovite	Magmatism	Baidya et al. (1987)
815 ± 47	Rb–Sr whole rock	Metasomatism	Singh and Krishna (2009)
825–818	U–Pb monazite	Metamorphism	Maji et al. (2008)
850–775	U–Pb monazite	Deformation event	Karmakar et al. (2011)
859	U–Pb monazite	Metamorphism	Chatterjee et al. (2010)
860–830	U–Pb monazite	Metamorphism	Chatterjee et al. (2010)
870 ± 40	K–Ar biotite	Magmatism	Baidya et al. (1987)
921	U–Pb monazite	Metamorphism	Rekha et al. (2011)
921	U–Pb monazite	Metamorphism	Rekha et al. (2011)
923	U–Pb zircon	Metamorphism	Rekha et al. (2011)
928	U–Pb zircon	Metamorphism	Rekha et al. (2011)
937	U–Pb monazite	Metamorphism	Chatterjee et al. (2010)
944	U–Pb monazite	Metamorphism	Rekha et al. (2011)
946	U–Pb monazite	Metamorphism	Chatterjee et al. (2010)
947 ± 27	U–Pb zircon	Metamorphism	Chatterjee et al. (2008)
965	U–Pb monazite	Metamorphism	Chatterjee et al. (2010)
990–940	U–Pb monazite	Metamorphism	Karmakar et al. (2011)
1005 ± 51	Rb–Sr whole rock	Magmatism	Singh and Krishna (2009)
1021–967	U–Pb monazite	Metamorphic/magmatic	Maji et al. (2008)
1025	U–Pb zircon	Metamorphism	Rekha et al. (2011)
1059 ± 104	Rb–Sr whole rock	Magmatism	Krishna et al. (1996)
1065 ± 74	Rb–Sr whole rock	Magmatism	Krishna et al. (1996)
1071 ± 64	Rb–Sr whole rock	Retrogression	Ray Barman et al. (1994)
1072	U–Pb zircon	Metamorphism	Rekha et al. (2011)
1138 ± 193	Rb–Sr whole rock	Magmatism	Krishna et al. (1996)
1176	U–Pb monazite	Metamorphism	Maji et al. (2008)
1178 ± 61	Rb–Sr whole rock	Retrogression	Ray Barman et al. (1994)
1200–1100	U–Pb monazite	Older thermal events	Karmakar et al. (2011)
1239	U–Pb zircon	Metamorphism	Rekha et al. (2011)
1331 ± 42	Rb–Sr whole rock		Krishna et al. (1996)
1422–1305	U–Pb monazite	Older thermal events	Chatterjee et al. (2010)
1550 ± 12	U–Pb zircon	Magmatism	Chatterjee et al. (2008)
1717–1446	U–Pb monazite	Older thermal event??	Chatterjee et al. (2010)
1800	U–Pb monazite	Inherited	Karmakar et al. (2011)
1870–1691	U–Pb monazite	Older thermal event??	Chatterjee et al. (2010)

(continued)

Table 1 (continued)

Age	Method	Type of age	References
Domain IB			
481 ± 18	Rb–Sr Bt, Kfs	?	Pandey et al. (1986a)
649	U–Pb monazite	Metamorphism	Chatterjee et al. (2010)
850–750	U–Pb monazite	Metamorphism	Sanyal et al. (2007)
872–838	U–Pb monazite	Metamorphism	Chatterjee et al. (2010)
876	U–Pb monazite	Metamorphism	Chatterjee et al. (2010)
902	U–Pb zircon	Metamorphism	Mukherjee et al. (2017a)
907 ± 49	U–Pb zircon	Metamorphism	Mukherjee et al. (2018)
910	U–Pb monazite	Metamorphism	Rekha et al. (2011)
937	U–Pb monazite	Metamorphism	Chatterjee et al. (2010)
943	U–Pb zircon	Metamorphism	Mukherjee et al. (2017a)
945	U–Pb monazite	Metamorphism	Rekha et al. (2011)
948 ± 22	U–Pb zircon	Metamorphism	Mukherjee et al. (2018)
950 ± 20	U–Pb monazite	Retrogression	Chatterjee et al. (2008)
954	U–Pb monazite	Metamorphism	Chatterjee et al. (2010)
965–930	U–Pb monazite	Metamorphism	Chatterjee et al. (2010)
975 ± 67	U–Pb monazite	Metamorphism	Chatterjee and Ghose (2011)
995 ± 24	U–Pb monazite	Metamorphism	Chatterjee et al. (2008)
1009	U–Pb zircon	Metamorphism	Rekha et al. (2011)
1100–930	U–Pb monazite	Metamorphism	Sanyal et al. (2007)
1119	Rb–Sr whole rock	?	Sarkar et al. (1986)
1183	U–Pb monazite	Older thermal events	Chatterjee et al. (2010)
1190 ± 26	U–Pb monazite	Metamorphism	Chatterjee et al. (2008)
1270	U–Pb zircon	?	Rekha et al. (2011)
1272	U–Pb monazite	Older thermal events	Chatterjee et al. (2010)
1278	U–Pb monazite	Older thermal events	Chatterjee et al. (2010)
1331 ± 125	Rb–Sr whole rock	Cooling age	Ray Barman et al. (1994)
1333	U–Pb zircon	?	Rekha et al. (2011)
1377	U–Pb zircon	?	Rekha et al. (2011)
1435	U–Pb zircon	?	Rekha et al. (2011)
1446 ± 7	U–Pb zircon	Magmatism	Mukherjee et al. (2018)
1447	U–Pb zircon	Magmatism	Mukherjee et al. (2017a)
1457 ± 63	Rb–Sr whole rock	Older thermal events	Ray Barman et al. (1994)
1462	U–Pb zircon	?	Rekha et al. (2011)
1470 ± 2	U–Pb zircon	Magmatism	Mukherjee et al. (2018)
1480	U–Pb monazite	?	Rekha et al. (2011)
1515 ± 5	U–Pb zircon	Magmatism	Acharyya (2003)

(continued)

Table 1 (continued)

Age	Method	Type of age	References
1522 ± 71	Rb–Sr whole rock	Magmatism	Mallik et al. (1991)
1580 ± 33	Rb–Sr whole rock	Magmatism	Mallik et al. (1991)
1599 ± 33	Rb–Sr whole rock	Magmatism	Mallik et al. (1991)
1624 ± 5	U–Pb zircon	Magmatism	Acharyya (2003)
1649	U–Pb zircon	Metamorphism	Rekha et al. (2011)
1660–1270	U–Pb monazite	Metamorphism	Sanyal et al. (2007)
1720	U–Pb monazite	Older thermal event	Chatterjee et al. (2010)
1741 ± 65	Rb–Sr whole rock	?	Ray Barman and Bishui (1994)
1824–1659	U–Pb monazite	Older thermal event	Chatterjee et al. (2010)
1870–1720	U–Pb monazite	UHT metamorpism	Sanyal et al. (2007)
2600–1900	U–Pb zircon	Detrital grains	Rekha et al. (2011)
Domain II			
590	Apt Fission Track	Cooling age	Lal et al. (1976)
595	Bt Fission Track	Cooling age	Lal et al. (1976)
760	mus Fission Track	Cooling age	Lal et al. (1976)
830	grt Fission Track	Cooling age	Lal et al. (1976)
855 ± 25	Mica Rb–Sr	Magmatism	Pandey et al. (1986b)
910 ± 19	U–Pb, Pb–Pb	Magmatism	Krishna et al. (2003)
960 ± 50	Pb–Pb	Magmatism	Vinogradov et al. (1964)
1086–850	K–Ar	?	Sarkar (1980)
1020 ± 46	Rb–Sr whole rock	Magmatism	Mallik (1993)
1100–700	Fission track mica	Cooling age	Lal et al. (1976)
1242 ± 34	Rb–Sr whole rock	?	Pandey et al. (1986a)
1238 ± 33	Rb–Sr whole rock	Magmatism	Mallik (1993)
1285 ± 108	Rb–Sr whole rock	Magmatism	Mallik (1993)
1300–1100	Rb–Sr whole rock	?	Pandey et al. (1986a, b)
1590 ± 30	Rb–Sr whole rock	Magmatism	Pandey et al. (1986b)
Domain III			
557 ± 99	U–Pb xenotime	?	Chatterjee and Ghose (2011)
768 ± 11	U–Pb uraninite	Metamorphism	Chatterjee and Ghose (2011)
929	U–Pb xenotime	Metamorphism	Chatterjee and Ghose (2011)
1044 ± 35	U–Pb zircon	Magmatism	Wanjari et al. (2012)
1337 ± 26	Rb–Sr whole rock	Magmatism	Wanjari et al. (2012)
1583 ± 50	U–Pb xenotime	Cooling age	Chatterjee and Ghose (2011)
1697 ± 17	U–Pb Monazite	Magmatism	Chatterjee and Ghose (2011)
1737–1664	U–Pb zircon	Magmatism	Saikia et al. (2017)

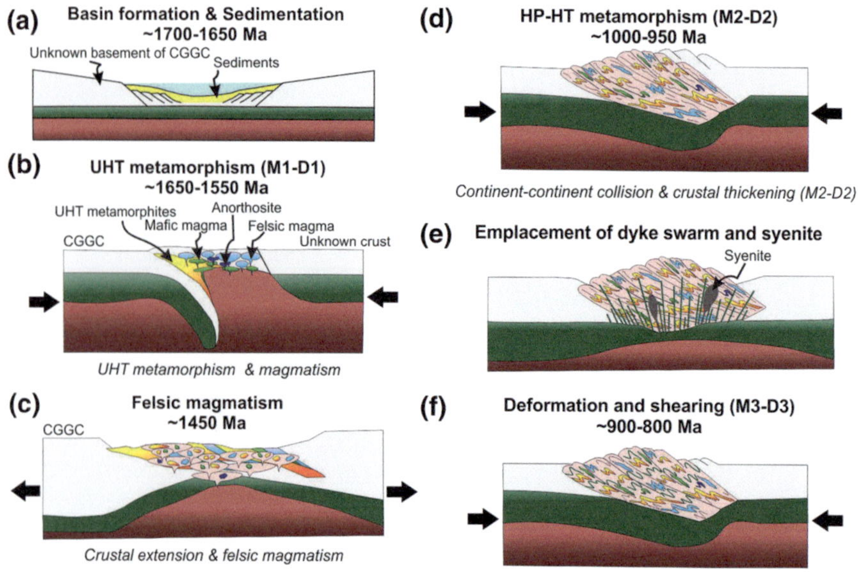

Fig. 4 **a** Basin formation and sedimentation over an unknown basement of CGGC during ~1700–1650 Ma. **b** Moderate pressure ultra-high temperature metamorphism and deformation (M_1-D_1) induced by continent-continent collision, lithospheric delamination and mantle upwelling during ~1650–1550 Ma and intrusion of syn- to post-collisional granitoids and anorthosites. **c** Intrusion of ferroan granites in an extensional setting at ~1450 Ma. **d** High pressure-granulite grade metamorphism and deformation (M_2-D_2) in continent-continent collisional setting at 1000–950 Ma. **e** Intrusion of syenite and mafic dykes within the granulitised crust of CGGC. **f** Deformation and shear related folding accompanied by metamorphism (M_3-D_3) in CGGC

authigenic datable minerals, few studies from northern part of CGGC (domain IB & II) have attempted to constrain the timing of sedimentation of the protolith of meta-sedimentary rocks. From the Pb–Pb dating of galena Singh et al. (2001) has inferred a sedimentation age of 1700–1650 Ma in parts of BMB. Dey et al. (2017) has analysed and studied $^{207}Pb/^{206}Pb$ apparent dates of the detrital zircons from UHT pelitic granulite from Dumka and Deoghar (Jharkhand) (Fig. 2a). The ages of the youngest analysed detrital zircon cores and the oldest metamorphic overgrowth constrain the age of sedimentation of the precursor of metapelites within a narrow age bracket of ~1700–1680 Ma. From the detrital zircon dates, the authors suggested that protoliths of these sediments were sourced from the ~2700–1700 Ma old domains in the adjoining cratonic areas and also from the rocks of similar age now occurring in the Lesser Himalayan region. Both the studies are consistent with the interpretation that during ~1700–1650 Ma the (unknown) basement of CGGC experienced a phase of basin formation and sedimentation (Fig. 4a). A part of these sediments got subsequently metamorphosed by later tectonothermal events.

4.1.2 Magmatic History

The oldest tectonothermal event reported so far from northern and central CGGC occurred in the late Paleoproterozoic (ca. 1750–1640 Ma) (Mallik et al. 1991; Ray Barman and Bishui 1994; Chatterjee and Ghose 2011; Saikia et al. 2017). U–Pb zircon ages distinctly indicates the intrusion of arc-related bimodal volcanics in the northern part (Saikia et al. 2017). However, contemporaneous (1697 Ma; Chatterjee and Ghose 2011) porphyritic granites from the northern domain are inferred to be of anorogenic affinity (Yadav et al. 2014). In the eastern fringe of the terrain, intrusion of massif anorthosite body, occurred at 1550 Ma and inferred to be the manifestation of post-orogenic magmatism (Fig. 4b) (Chatterjee et al. 2008). Subsequently in the eastern part of the domain Mesoproterozoic (ca. 1450 Ma) ferroan granites have intruded the Paleoproterozoic crust in an extensional setting (Fig. 4c) (Mukherjee et al. 2017a, 2018). A body of syenite near Dumka (Jharkhand) yields a Rb–Sr whole rock age of 1457 $\pm$ 63 Ma (Ray Barman and Bishui 1994). Early Neoproterozoic felsic magmatism so far has been reported only from southern part of the CGGC (Singh and Krishna 2009). However, post-Grenvillian (<950 Ma) magmatic pulses, including granite, syenite, mafic dykes and pegmatite intrusions have been documented from different parts the terrain (Fig. 4e). Biotite bearing granites have considered to intrude the early crust $\sim$870–810 Ma (Baidya et al. 1987). Although no geochronological dates are available, a body of syenite intruded the granulitised crust in the southwestern part of the terrain, associated with NPSZ (Das et al. 2016, 2017c). A set of mafic dykes also intruded the granulitised crust in the northeastern part of the CGGC (Sanyal and Sengupta 2012; Mukherjee et al. 2017a). Numerous mica-bearing pegmatites with intrusion age of 960–855 Ma have been reported from the northern part of the terrain (Pandey et al. 1986b; Krishna et al. 2003; Vinogradov et al. 1964).

4.1.3 Metamorphism and Deformation: Summary of the Events

The oldest metamorphic event recorded in the CGGC is characterised by an ultra-high temperature metamorphism (M_1) at $\sim$1640 Ma (Dey et al. 2017). Similar ages (ca. 1850–1600 Ma; U–Pb zircon and Th–U–Pb monazite dates) have been recorded by several workers and inferred to be the manifestation of high grade thermal event (Chatterjee et al. 2010; Rekha et al. 2011). Peak metamorphic conditions recovered from the metapelitic rocks yield a P–T condition of 1000–950 °C at 7–5 kbar pressure (Sanyal and Sengupta 2012; Dey et al. *under review*, a). Moderate pressure ultra-high temperature metamorphism and a clockwise P–T–*t* path, inferred from the mineral phase equilibria of the metapelite, presumably reflect continent-continent collision followed by lithospheric delamination and mantle upwelling (Fig. 4b) (Dey et al. *under review*, a). Mesoproterozoic (ca. 1450–1300 Ma) ages, recovered by several workers, are inferred to be the manifestation of thermal imprint (Maji et al. 2008; Chatterjee et al. 2010; Karmakar et al. 2011; Rekha et al. 2011). The most pervasive tectonothermal event recorded from different parts of the CGGC occurred at

1000–950 Ma (M_2). From the northeastern margin of the terrain, several workers (Mukherjee et al. 2017a, Dey et al. *under review*, a, Dey et al. *under review,* b) have inferred that the metamorphism culminated at 770–800 °C and 9–12 kbar pressure, followed by a steep decompressive path. Similar P–T path has been recorded from the southern part of the terrain where the peak culminated at 870 °C and 11 kbar followed by decompression (Karmakar et al. 2011). Such high pressure event associated with clockwise path is inferred to be the manifestation of continent-continent collision (Fig. 4d). Post-Grenvillian metamorphic event (M_3) has been reported from north-eastern margin of the terrain where U–Th–Pb monazite constrains the age of the event between 900 and 780 Ma (Chatterjee et al. 2010; Mukherjee et al. 2017a, Dey et al. *under review*, a). Chatterjee et al. (2010) estimated a high-pressure peak for M_3 culminating at 820 °C and 11 kbar which correlates with the development of Eastern Indian tectonic Zone (Fig. 4f).

Structural analyses from different parts of the terrane reveal that the CGGC has experienced multiple deformational events (Mahadevan 2002; Maji et al. 2008; Goswami and Bhattacharyya 2010; Karmakar et al. 2011; Sanyal and Sengupta 2012; Mukherjee et al. 2017a). However, structural data from scattered places and scarcity of detailed geochronological information are some of the major hindrance for correlating the different deformational and metamorphic events reported by different workers over the entire CGGC. Unfortunately, sizable amount of structural and geochronological information are available only from the rocks of Domain I. In the following section an attempt has been made to develop a structural history of the CGGC using the structural, metamorphic and geochronological information from Domain I. Three major deformational phases seem to have affected the rocks of Domain I. The earliest deformational event (D_1) is preserved in the metasedimentary enclaves within the felsic orthogneiss. This event is marked by a migmatitic banding (S_1) which is inferred to be associated with M_1 metamorphic event that is dated to be ~1650 Ma (Fig. 4a) (Sanyal and Sengupta 2012; Mukherjee et al. 2017a; Dey et al. *under review* a; Dey et al. *under review*, b). The next major deformational event (D_2) is manifested by the development of S_2 foliations within the felsic orthogneisses and transposition of S_1 within the metasedimentary enclaves. The D_2 is associated with ~1000–950 Ma old granulite facies metamorphism (M_2) (Fig. 4c), especially from the eastern margin of the terrane (Maji et al. 2008; Karmakar et al. 2011; Mukherjee et al. 2017a; Dey et al. *under review* a; Dey et al. *under review*, b). Folding of the late intrusives (mafic dykes in the northeastern part and nepheline syenite in the southern part), along with the S_2 of the host felsic orthogneisses, depicts the third major deformational event (D_3). There is no detail information on the age of emplacement of the mafic dykes and subsequent D_3 deformation. The EITZ and with its characteristic asymmetrical folds, N-NNE trending planar fabric and high pressure metamorphism appear to be coeval with the D_3-M_3 tectonothermal event (Fig. 4f). Similar shear-related folds also reported by Mukherjee et al. (2015b) and Mukherjee (submitted manuscript).

4.2 Evaluation of Shear Zones

Several prominent crustal-scale shear zones cross through the CGGC. The 'Tamar-Porapahar-Khatra' shear zone is the southernmost lineament affecting the Proterozoic gneissic rocks of CGGC and separating it from the low grade metamorphites of North Singhbhum Fold Belt (NSFB). This ENE-WSW trending 150 km long shear zone passes through the south of the Purulia town (West Bengal) and also known as 'South Purulia Shear Zone' (SPSZ). ~150 km long, the 'North Purulia Shear Zone (NPSZ) is another prominent E-W to NE-SW trending lineament that affects the CGGC rocks occurring north of Purulia.

Approximately 150 km long arcuate shear zone which is referred to as the South Purulia Shear Zone (SPSZ: Mazumdar 1988), or Northern Shear Zone (NSZ: Kumar et al. 1978), or Tamar-Porapahar-Khatra Fault Zone (TPKF: Mahadevan 2002), is a crustal-scale ENE-WSW trending brittle-ductile shear zone that marks the southern margin of the CGGC and separates the terrain from the low to medium grade meta-sedimentary and meta-igneous rocks of the North Singhbhum Fold Belt (Fig. 2a; Mahadevan 1992; Acharyya et al. 2006; Mahato et al. 2008; Sanyal and Sengupta 2012). This shear zone has been extensively deformed during several tectonothermal events (Bhattacharyya et al. 1992; Sengupta et al. 2005) and represents an ensemble of a nepheline-bearing syenite, tuffaceous rocks, tourmalinite, carbonatite, apatite bearing rocks and granitoids altered hydrothermally (Banerji 1985; Basu 1993; Acharyya et al. 2006; Chattopadhyay et al. 2015a). Abundance of carbonatite and alkaline connotes crustal rifting (Acharyya et al. 2006; Chakrabarty and Sen 2010). Intrusion of the alkaline rocks at ca. 922 Ma, evident from the U–Pb zircon geochronology (Reddy et al. 2009), constrains the timing of the extensional event, which was followed by another tectonothermal event that deformed the carbonatite-alkaline rocks forming DARCs (Deformed Alkaline Carbonatite Rocks) (Chattopadhyay et al. 2015a). Economically viable apatite deposit, tuffaceous rocks and Fe-ore are reported from the area (Banerji 1985).

The NPSZ is an E-W to ENE-WSW trending crustal-scale shear zone with a prominent steep northerly dip traversing the metamorphites of CGGC. This lineament passes through Jhalda-Jaypur in the west through Raghunathpur in the central part and can be traced up to north of Murlu near Saltora (West Bengal) in the east (Baidya et al. 1989; Mahadevan 1992; Dasgupta et al. 2000; Som et al. 2007; Maji et al. 2008; Goswami and Bhattacharyya 2010). The shear zone is best developed in the granitoids where an early mylonitic fabric has developed in deep structural level in amphibolite facies condition. The crystal plastic deformation of quartz and feldspar indicates the temperature ~650–700 °C during this early ductile deformation (Vernon 2004). The NPSZ has the signatures of protracted period of ductile deformation where the early shear foliation folded. Later stage of brittle deformation has affected the early mylonitic fabric of the NPSZ. Discrete bodies of nepheline syenite is found to be emplaced along this crustal lineament near Kankarkiari–Kusumda area near Saltora (West Bengal) cross-cutting the gneissic fabric of felsic gneiss and khondalite and get metamorphosed and deformed when NPSZ formed (Goswami and Bhattacharyya

2010; Das et al. 2017b). Das et al. (2017a) has also reported carbonatite in the central part of NPSZ where apatite-Fe-oxide/hydroxide-silica bearing veins coexist with alkali metasomatised granitoids. In the western part of NPSZ late stage pegmatite veins within granite and calc-silicate rocks have been explored for Cs, Li and other rare elements (Som et al. 2007).

The Eastern Indian Tectonic Zone (EITZ), roughly N- to NNW trending crustal-scale shear zone, is observed towards the eastern margin of the CGGC (Fig. 2a) (Chatterjee et al. 2010). The shear fabric, manifested by asymmetric folds, developed over the older lithounits and is traceable along CGGC, North Singhbhum Fold Belt and Singhbhum Craton. Detailed petrological studies by Chatterjee et al. (2010) have shown that the metamorphism associated with the development of the shear zone culminated at ~11–12 kbar and 800 °C. Monazite geochronology constrains the timing of the of the tectonothermal event at ca. 870–780 Ma, which agrees with the field observations that EITZ developed after the major tectonic event at CGGC i.e. granulite grade metamorphism dated ~950 Ma.

4.3 *Extension of the CGGC*

4.3.1 Northern Boundary of CGGC

Quaternary sediments of Gangetic alluvium cover the northern margin of the CGGC. Further north, the Late Paleoproterozoic rocks of the Lower Lesser Himalaya occurs interleaved with the Phanerozoic rocks (Mukherjee et al. 2013, 2015a; Mukherjee 2013, 2015). These units expose metamorphosed and unmetamorphosed sedimentary sequence and granitic plutons of age ~1900–1700 Ma (Fig. 5). Geochemical signatures of the granitoids point towards their emplacement in a continental arc setting (Kohn et al. 2010) existing along the northern boundary of India at that time. Detrital zircon populations of the sedimentary units are characterized by Paleoproterozoic–Late Archean ages (2.6–1.8 Ga), with no younger population (Richards et al. 2005). According to Richards et al.(2005) "Detrital zircon ages are younger than their respective Hf-isotope derived crustal formation ages by 0.7–2.1 Ga, indicating that the source regions of the detrital zircons consisted of older terranes with considerable amounts of reworking and renewed magmatism". Many workers have suggested that these Proterozoic rocks of the Lesser Himalaya constitute the remnants of the Greater Indian Landmass, subducted below the Himalayan orogenic belt (DeCelles 2000). Petrological and limited geochronological studies suggest that the Paleoproterozoic felsic orthogneisses exposed within the quaternary alluvium towards the north of the CGGC resembles the basement orthogneisses of the CGGC. Felsic magmatism in the northern part of the CGGC, near Bathani, shows enrichment of Th, U, and Pb over Nb, Ta and Ti which corroborates with the geochemical characters of the Paleoproterozoic granitoids of the lower Lesser Himalaya. Elevated magmatic temperature, derived from the zircon saturation thermometry and predominance of crustal source also poses the possibility that the Paleoproterozoic basement orthogneisses presumably continued beneath the Gangetic alluvium to the lower Lesser Himalaya.

4.3.2 Western Boundary of CGGC

The Mohakoshal Belt (MB) and the Sausar Mobile Belt (SMB) occur to the NW and SW of CGGC respectively (Fig. 2a) and constitute the northern and the southern structural blocks of Central Indian Tectonic Zone (CITZ) that sutures two Achaean blocks: Bundelkhand at the north and Bastar at the south (Acharyya 2003; Jain et al. 1991; Mishra et al. 2000; Radhakrishna and Naqvi 1986; Yedekar et al. 1990).

The Mahakoshal belt is composed of intercalated riftogenic volcano-sedimentary successions (Roy and Prasad 2003 and the references therein), metamorphosed and intruded with post-orogenic granitoids aging 1800–1700 Ma (Rb/Sr whole rock age and U/Pb zircon age; Bora et al. 2013; Pandey et al. 1998; Sarkar 1998; Srivastava et al. 2000). The sedimentary sequence was metamorphosed in low pressure–medium temperature conditions in a possible back arc setting (Roy and Prasad 2003; Wani and Mondal 2016). On contrary, a recent study proposed that the sediments were metamorphosed in moderate pressure–low temperature conditions in a collisional setting followed by rapid exhumation (Deshmukh et al. 2017).

The Sausar Mobile Belt occurs at the southern margin of CITZ and is a collage of different litho-tectonic components. The southern Bhandara-Balaghat granulite (BBG) domain exposes enclaves of supracrustal and meta-igneous granulites occurring within felsic gneisses. They have been metamorphosed under lower crustal, ultra-high temperature (UHT) granulite facies conditions along a counter-clockwise P–T path (Bhowmik 2006; Bhowmik et al. 2005) in a possible back arc setting associated with emplacement of arc magma (Bhowmik et al. 2011). Although early workers predicted an Archean–Early Paleoproterozoic age (Bhowmik 2006; Bhowmik et al. 2005; Roy et al. 2006), precise U–Pb zircon and texturally controlled monazite dating constrains the timing of this metamorphism and magmatism at ~1600 Ma (Bhandari et al. 2011; Bhowmik et al. 2011, 2014). This event was followed by several events of granulite-amphibolite grade overprinting (BasuSarbadhikari and Bhowmik 2008; Bhowmik et al. 2014) with a terminal phase of metamorphism at ~1400 Ma (Bhandari et al. 2011; Bhowmik et al. 2011). The central Sausar Group of rocks deformed multiply and metamorphosed in greenschist to amphibolite facies during ~1060–950 Ma in a collisional orogeny (Bhowmik et al. 1999, 2011, 2012; Bhowmik and Roy 2003; Chattopadhyay et al. 2015b; Pal and Bhowmik 1998) accompanied by felsic magmatism (Chattopadhyay et al. 2015b; Pal and Bhowmik 1998). The northern Ramakona-Katangi granulite (RKG) domain contains mafic, felsic and pelitic rocks as rafts, which are metamorphosed in high pressure upper amphibolite to granulite facies along a clockwise P–T path indicating a continent-continent collision (Bhowmik and Roy 2003; Bhowmik and Spiering 2004) at 1040–950 Ma (Bhowmik et al. 2012). This was followed by emplacement of post-orogenic

Fig. 5 Timing of magmatic and metamorphic events in different Proterozoic mobile belts of India, Lower Lesser Himalaya and east Antarctica. Age scale changes at 1000 Ma. References: **CITZ (Central Indian Tectonic Zone)**: Bhandari et al. (2011), Bora et al. (2013), Bhowmik et al. (2005), Roy et al. (2006), Panigrahi et al. (2004), Deshmukh et al. (2017), Pandey et al. (1986b), Bhowmik et al. (2011); **Aravalli-Delhi Fold Belt**: Kaur et al. (2009), (2011), (2017a, b); Buick et al. (2006), Chatterjee et al. (2017), Deb et al. (2002), Biju-Sekhar et al. (2003), Pandit et al. (2003), Bhowmik et al. (2010), Wiedenbeck et al. (1996), Dharma Rao et al. (2011, 2013), Mukhopadhyay et al. (2000), Gupta et al. (1998), Sivaraman and Raval (1995); **North Singhbhum Fold Belt**: Bhattacharya et al. (2015), Mahato et al. (2008), Rekha et al. (2011), Ramakrishnan and Vaidyanadhan (2008), Roy et al. (2002); **CGGC**: Chatterjee et al. (2008), Chatterjee and Ghose (2011), Karmakar et al. (2011), Rekha et al. (2011), Chatterjee et al. (2010), Mallik et al. (1991), Ray Barman and Bishui (1994), Dey et al. (2017), Mukherjee et al. (2017a, 2018), Saikia et al. (2017), Sanyal et al. (2007); **Lower Lesser Himalaya**: Kohn et al. (2010), Sakai et al. (2013), Larson et al. (2016), Chambers et al. (2008), Treloar and Rex (1990), Richards et al. (2005), Miller et al. (2000), Célérier et al. (2009), DiPietro and Isachsen (2001), Yin et al. (2010), Liao et al. (2008), Upreti et al. (2003), Decelles et al. (2000), Long et al. (2008), Daniel et al. (2003); **EGMB (Eastern Ghats Mobile Belt)**: Dobmeier and Simmat (2002), Simmat and Raith (2008), Bose et al. (2011), Das et al. (2011), Upadhyay et al. (2006), Upadhyay and Raith (2006), Sarkar and Schenk (2016), Das et al. (2017a, b, c), Rickers et al. (2001), Korhonen et al. (2013), Krause et al. (2001), Dharma Rao et al. (2012), Mezger and Cosca (1999), Dobmeier and Raith (2003), Henderson et al. (2014); **SMGC (Shillong-Meghalaya Gneissic Complex)**: Kumar et al. (2017), Chatterjee et al. (2007); East Antarctica: Kelly et al. (2002), Corvino et al. (2008), Liu et al. (2013, 2017, 2009), Morrissey et al. (2015), Tsunogae et al. (2014, 2016), Owada et al. (2003), Zhang et al. (2012b), Asami et al. (2002), Elburg et al. (2015), Goodge et al. (2008) ▶

granites at ~940 Ma derived from a Paleo-Proterozoic crustal source (Bhowmik et al. 2012). Two suites of intrusive granite from Betul metamorphic belt, adjacent to Sausar Mobile Belt, yield Rb–Sr whole rock-mineral ages of ~1550 Ma and ~850 Ma (Roy and Prasad 2003).

Based on all these study (summarised in Fig. 4), the workers have concluded that the CITZ indicates amalgamation of northern and southern Indian blocks and formation of a proto-greater-Indian landmass since Paleo-Mesoproterozoic. The Mahakoshal belt is a product of northward subduction or subduction to collisional event prevailing in Early Paleoproterozoic along southern margin of Bundelkhand craton (Deshmukh et al. 2017; Roy and Prasad 2003; Wani and Mondal 2016). The southern Sausar Mobile Belt bears signature of arc setting resulting from probable subduction of north Indian blocks under the Bastar craton in the Late Paleoproterozoic (Bhandari et al. 2011; Bhowmik 2006; Bhowmik et al. 2005). It is noteworthy that evidence of Neoproterozoic (~1050–950 Ma) collision, which is widespread in the Sausar belt is yet to be recorded from the Mahakoshal belt. Based on meagre data base Mesoproterozoic to Early Neoproterozoic tectonothermal events that is recorded in the CGGC (D_2-M_2) has been correlated with the tectonothermal events recorded in the Sausar mobile belt and the gneissic complex of Shillong Meghalaya. (Acharyya 2003; Bhandari et al. 2011; Bhowmik et al. 2005, 2012). More detail studies are required to support (or reject) this proposition.

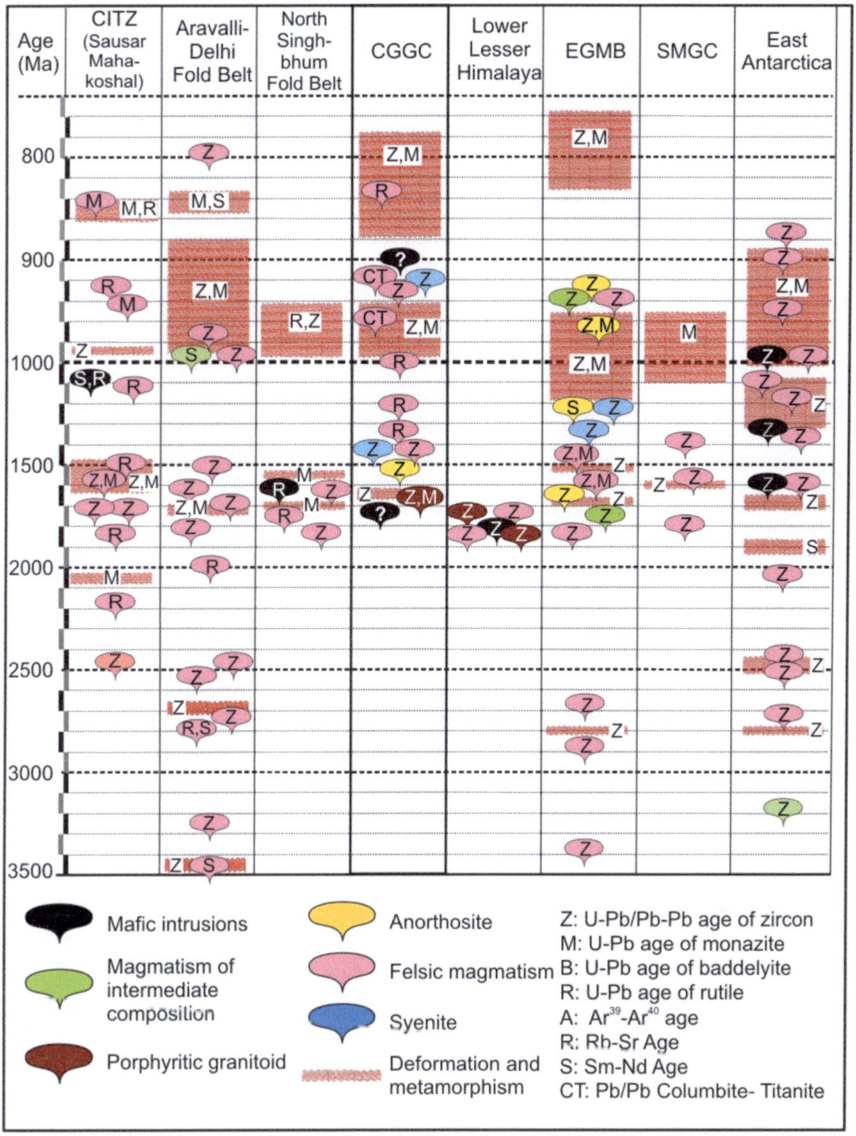

Age (Ma)
CITZ (Sausar Maha-koshal)
Aravalli-Delhi Fold Belt
North Singh-bhum Fold Belt
CGGC
Lower Lesser Himalaya
EGMB
SMGC
East Antarctica
800
900
1000
1500
2000
2500
3000
3500
Mafic intrusions
Magmatism of intermediate composition
Porphyritic granitoid
Anorthosite
Felsic magmatism
Syenite
Deformation and metamorphism
Z: U-Pb/Pb-Pb age of zircon
M: U-Pb age of monazite
B: U-Pb age of baddelyite
R: U-Pb age of rutile
A: Ar39-Ar40 age
R: Rb-Sr Age
S: Sm-Nd Age
CT: Pb/Pb Columbite- Titanite

4.3.3 Eastern Margin

In the eastern margin, the CGGC is bordered by the Cenozoic sediments of Bengal basin that ranges more than 200,000 km^2 comprising the largest alluvium delta in the world covering parts of West Bengal and Tripura and Bangladesh (Hossain et al. 2019). Sedimentation occurred in the continental passive margin (pre-Oligocene) to a remnant ocean basin (beginning of Miocene) (Alam et al. 2003). Basement rocks below the Phanerozoic sedimentary cover compositionally ranges from tonalite, diorite to granite (Ameen et al. 1998, 2007; Kabir et al. 2001; Hossain et al. 2007). U–Pb zircon geochronology of basement rocks recovered from subcrop investigations near Maddhapara, in the northeastern Bangladesh, reveals an intrusion ages of 1722 ± 6 Ma from a tonalite suite (Ameen et al. 2007) and 1730 ± 11 Ma from a diorite suite (Hossain et al. 2007). Limited geochemical data suggest 'pre-plate collision' affinity for the tonalite and 'syn-collisional' affinity for granitic basements (Ameen et al. 2007) whereas contemporaneous diorite are inferred to have a calc-alkaline origin associated with a subduction zone settings (Hossain and Tsunogae 2014).

Towards the northeastern part of the Bengal basin, the Shillong-Meghalaya Gneissic Complex (SMGC) comprises an ensemble of basement gneisses, metamorphosed to amphibolite-granulite grade, overlain by the Shillong Group of sediments (Ahmed 1983; Nandy 2001). Different generations metamorphosed granite and garnodiorites and porphyritic granitoids (Fig. 4), having a wide range of isotopic age (ca. 1700–550 Ma), are reported from the terrane (Mazumdar 1976; Ghosh et al. 1994; Bhattacharya and Ray Barman 2000; Ghosh et al. 2005; Kumar et al. 2017). Lal et al. (1978) calculated P–T conditions of 750 °C and 5 kbar from the metapelitic rocks near Sonapahar, eastern SMGC, whereas Chatterjee et al. (2007) determined a higher P–T of 850 °C and 7.5 kbar associated with a counter-clockwise P–T–*t* path. Chemical monazite dating constrains the age of the metamorphism at ~1596 ± 15 Ma whereas in the western part near Garo-Goalpara, the dominant metamorphism is dated to be 500 ± 14 Ma (Chatterjee et al. 2007). Mitra (1998) determined Pb–Pb ages of 1550–1530 Ma from the detrital zircons from the supracrustals. In a recent publication Kumar et al. (2017) documented that granitoids from Rongjeng and Guwahati (Assam) intruded the Neo-Archean crust at 1778 ± 37 and 1630 ± 16 Ma respectively. Detailed geochemical studies reveal dominant crustal source and intrusion of the granitoids during syn-collisional settings.

Lithological and geochronological similarities indicate that the SMGC has been a part of the Indian Shield (Evans 1964; Crawford 1974) or extension of CGGC (Desikachar 1974; Chatterjee et al. 2007). Considering the distinct geochronological similarity, it is also suggested that basement of the Bengal basin is the extension of the Paleoproterozoic crust of CGGC and CITZ (Hossain et al. 2007; Hossain and Tsunogae 2014). All these inferences further confirms the idea of Paleoproterozoic crust o CGGC being extended up to SMGC as the Garo-Rajmahal gap represents a shallow basement ridge (Desikachar 1974).

4.4 Southern Margin: North Singhbhum Fold Belt & Singhbhum Craton

The contact between the Paleoproterozoic CGGC and Archean Singhbhum craton is demarcated by the North Singhbhum Fold belt (NSFB), a 200 km long and 50 km wide curvilinear orogenic belt. Extant geological information suggest that the NSFB represents Proterozoic rift-related sedimentation along major suture zones (Sarkar 1982; Sarkar and Saha 1983; Gupta and Basu 2000; Sarkar 2000; Bhattacharya and Mahapatra 2008; Bhattacharya et al. 2015). The Dalma Ophiolite Belt (DOB), an ensemble of dominantly mafic- ultramafic rocks with minor felsic components representing bimodal mafic-felsic volcanics, forms the most conspicuous part of the NSFB. Geochemical evidences suggest that the magmatism occurred at 1619 $\pm$ 38 Ma (Rb–Sr whole rocks isochron age: Roy et al. 2002) under an extensional regime, presumably in a back-arc like settings (Chakrabarti 1985; Bose et al. 1989). Metasedimentary rocks of the NSFB have been deformed-metamorphosed in at least three tectonothermal events (Mahato et al. 2008) representing a greenschist facies that grades into amphibolite facies at the contact of the CGGC (Bose 1954; Ray and Gangopadhyay 1971; Mahato et al. 2008). Available geochronological study from the southern part of the NSFB (S-NSFB) reveals that the metamorphism occurred between 1.72 and 1.55 Ga in the southern part of the Dalma Volcanics (Mahato et al. 2008; Chatterjee et al. 2010; Rekha et al. 2011). In the northern part (N-NSFB) geochronological data are scarce and reveals an Early Neoproterozoic tectonothermal event ($\sim$0.95 Ga; Rekha et al. 2011).

Discordant structural analyses, mismatching metamorphic grade and distinctly different geochronological records suggest that Singhbhum Craton and CGGC evolved separately in distant geological time and got amalgamated afterwards. The contact between the Singhbhum Craton and the CGGC has been severely debated and scarcity of geochronological and petrological data further blurs the nature and the timing of the amalgamation and in this perspective the NSFB plays an important role in the formation of the East Indian Shield (EIS). Several tectonic models have been proposed to construct the evolutionary history of the EIS (Mahadevan 2002), which mostly revolves around two major ideas: (1) subduction of the Singhbhum Craton beneath the CGGC followed by collision between two continents (Sarkar and Saha 1962, 1977; Sarkar 1982; Mahato et al. 2008; Rekha et al. 2011). and (2) plume-driven basin formation accompanied by mafic magmatism and crustal shortening (Mukhopadhyay 1990; Gupta and Basu 2000). Former model (Sarkar and Saha 1962, 1977; Sarkar 1982; Mahato et al. 2008; Rekha et al. 2011) considers the Singhbhum Craton and the CGGC as separate blocks that collided together and the NSFB represented a milieu of arc magmatism, sedimentation and obducted oceanic crust. Latter model indicates the presence of a unified crustal section suggesting a prolonged connection between the Singhbhum Craton and the CGGC where NSFB represented basin formed via incipient rifting. In a recent publication Bhattacharya et al. (2015) argued that different sedimentary basins and magmatic bodies in the NSFB did not evolve synchronously suggesting that the juxtaposition of the CGGC and the Singhbhum Craton is not straight forwards and warrants further geochronological and petrological investigations.

4.5 CGGC and the Proterozoic Supercontinental Cycles

4.5.1 Columbia Supercontinent

Late Paleoproterozoic history of the Earth has been largely influenced by the presence of Columbia Supercontinent that was inferred to be amalgamated between 2.1 and 1.8 Ga (Rogers and Santosh 2009). However, several recent publications argued for more prolonged amalgamation and placed the timing of final packing at ca. 1450 Ma (Sarkar and Schenk 2016; Meert and Santosh 2017). The configuration of the supercontinent is also debated owing to the scarcity of adequate petrological, geochronological and paleomagnetic data. Abundance of Late Paleoproterozoic magmatism and metamorphism (Figs. 5 and 6) and available paleomagnetic data indicate that Indian plate played an important role during the formation of Columbia supercontinent (Rogers and Santosh 2002; Zhao et al. 2002; Zhang et al. 2012a). Apart from its position in the supercontinent assembly, the configuration of Greater Indian Landmass raised considerable debate as the timing of the CITZ, through which different Archean-Early Paleoproterozoic cratons of India (Yedekar et al. 1990; Acharyya 2003; Roy and Prasad 2003) amalgamated remained unclear. Initial notion of a Late Paleoproterozoic-Mesoproterozoic ($\sim$1.8–1.5 Ga) suturing (Acharyya 2003; Roy and Prasad 2003; Bhandari et al. 2011) has been challenged by the proposition of Neoproterozoic (1.0–0.95 Ga) assembly of the CITZ (Chattopadhyay and Khasdeo 2011; Bhowmik et al. 2012). However, high-grade metamorphic events at ca. 1800–1600 Ma reported from the northern and southern part of CITZ (Bhandari et al. 2011; Deshmukh et al. 2017) and arc magmatism at ca. 1750 Ma (Bora et al. 2013) points towards an accretion orogen.

In the Ongle domain (Andhra Pradesh) of Eastern Ghats Mobile Belt (EGMB) arc magmatism occurred at 1750–1710 Ma, followed by an UHT metamorphism at

Fig. 6 Regional diagram summarizing Paleoproterozoic magmatic and metamorphic events in ▶ India and Antarctica. The schematic configuration is drawn after Dasgupta and Sengupta (2003). The pressure–temperature trajectories of different domains are presented with their geochronological data using the following references: a. Bhowmik et al. (2005, 2014); b. Bhandari et al. (2011); c. Chatterjee et al. (2007); d. Mahato et al. (2008); e. Dey et al. *under review* a; f. Sarkar et al. (2014); g. Sengupta et al. (1999); h. Roy et al. (2005); i. Kelly et al. (2002); j. Halpin et al. (2013); k. Halpin et al. (2013). Data for the magmatic events are collected from: Saikia et al. (2017), Chatterjee and Ghose (2011), Treloar and Rex (1990), DiPietro and Isachsen (2001), Zeitler et al. (1989), Miller et al. (2000), Singh et al. (1994), Richards et al. (2005), Chambers et al. (2008), Célérier et al. (2009), Decelles et al. (2000), Kohn et al. (2010), Daniel et al. (2003), Long et al. (2008), Yin et al. (2010), Mandal et al. (2016), Ameen et al. (2007), Hossain et al. (2007), Bora et al. (2013), Bhowmik et al. (2011), Sarkar et al. (2015), Henderson et al. (2014), Kovach et al. (2001), Kumar et al. (2017), Kaur et al. (2009, 2017a), Bhattacharya et al. (2015)

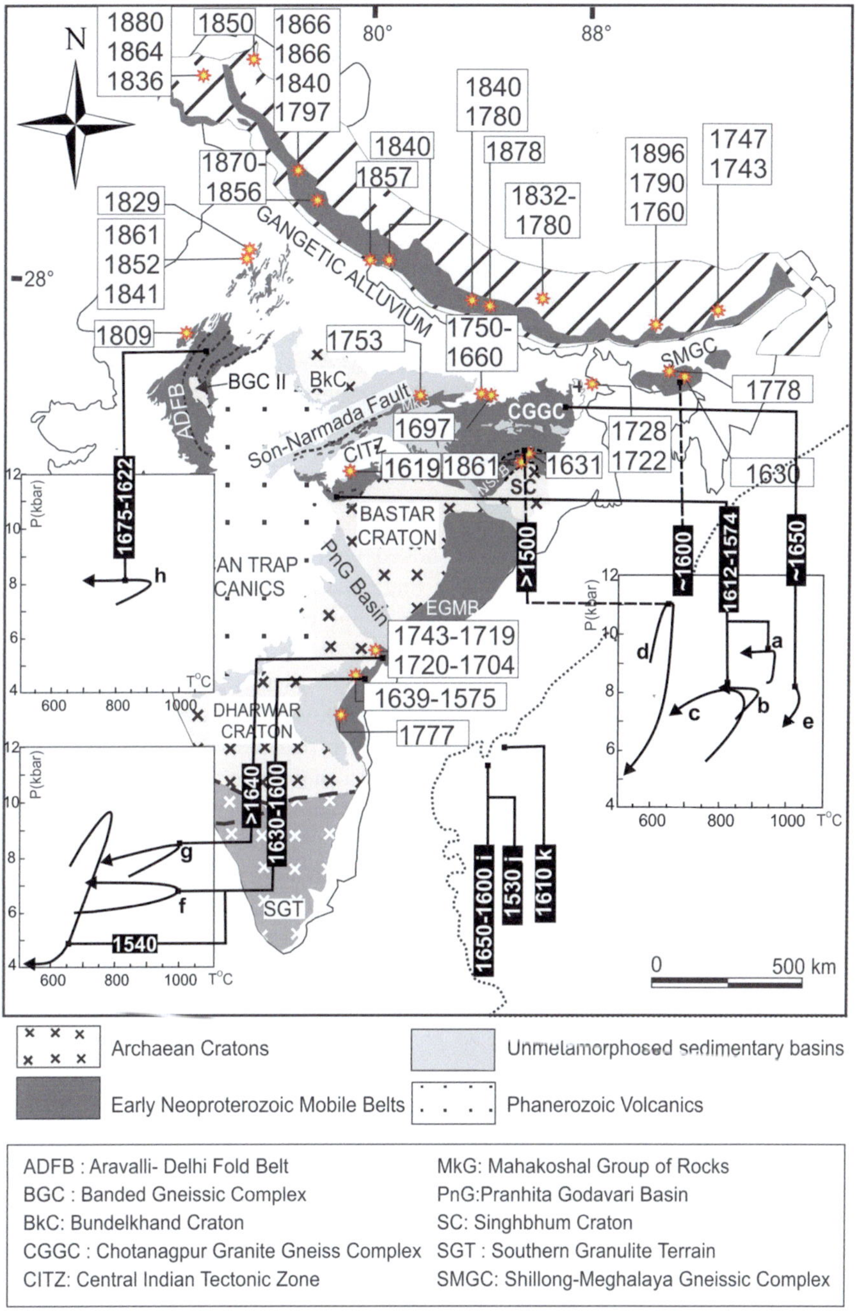
N
80°
88°
28°
1880
1864
1836
1850
1866
1866
1840
1797
1840
1780
1840
1857
1878
1870-
1856
1832-
1780
1896
1790
1760
1747
1743
1829
1861
1852
1841
1809
GANGETIC ALLUVIUM
1753
1750-
1660
SMGC
1778
BGC II
BkC
ADFB
Son-Narmada Fault
MkG
CGGC
1697
CITZ
1619
1861
1631
SC
1728
1722
1630
1675-1622
BASTAR CRATON
>1500
~1600
1612-1574
~1650
h
DECCAN TRAP VOLCANICS
PnG Basin
EGMB
P(kbar)
T°C
a
b
c
d
e
1743-1719
1720-1704
1639-1575
1777
DHARWAR CRATON
>1640
1630-1600
g
f
1540
SGT
1650-1600 i
1530 j
1610 k
0
500 km
Archaean Cratons
Unmetamorphosed sedimentary basins
Early Neoproterozoic Mobile Belts
Phanerozoic Volcanics
ADFB : Aravalli- Delhi Fold Belt
BGC : Banded Gneissic Complex
BkC: Bundelkhand Craton
CGGC : Chotanagpur Granite Gneiss Complex
CITZ: Central Indian Tectonic Zone
MkG: Mahakoshal Group of Rocks
PnG:Pranhita Godavari Basin
SC: Singhbhum Craton
SGT : Southern Granulite Terrain
SMGC: Shillong-Meghalaya Gneissic Complex

~ 1610 Ma (Sarkar and Schenk 2014; Sarkar et al. 2015). Calc-alkaline magmatism of arc-affinity with intrusion age of 1791–1771 Ma is also reported from Vinjamuru domain of the EGMB (Vadlamani et al. 2013). Paleoproterozoic (~ 1860–1810 Ma) arc magmatism is also reported from Aravalli-Delhi Mobile Belt (Kaur et al. 2009, 2017a).

Although geochronological data are scarce from the vast terrain of the CGGC, Paleoproterozoic dates, corresponding to tectonothermal and magmatic events, are reported from different parts of the terrain (Fig. 6). Dey et al. (2017) and Dey et al. (*under review,* a) reported Late Paleoproterozoic (~ 1650 Ma) UHT metamorphism from the northeastern part of the terrane and the inferred P–T path corroborate with convergent margin settings. Saikia et al. (2017) reported Paleoproterozoic volcano-sedimentary sequence from the northern part of the terrain and the associated felsic magmatism (ca. 1750–1660 Ma) resembles an arc settings. Evidences of such tectonothermal events from parts of CGGC indicate its involvement in the formation of the Greater Indian landmass during the assembly of Columbia supercontinent.

The Mesoproterozoic extensional events throughout the world, manifested by rift-related magmatism, deemed to be linked with the breakdown of the supercontinent (Rogers and Santosh 2002). Contemporaneous rift-related magmatism, manifested by several ferroan granite and alkaline batholiths (Vinukonda granite: 1589 Ma: Dobmeier et al. 2006; Elchuru alkaline rocks: 1442 ± 30 Ma; Upadhyay et al. 2006; Errakonda and Uppalapadu alkaline rocks: 1352 ± 2 Ma; Vijaya Kumar et al. 2007) has been widely reported from western margin of the Eastern Ghats belt which reflect the extensional events. Despite their restricted occurrence, extension-related Mesoproterozoic (ca. 1450 Ma) anorogenic granites have intruded the Paleoproterozoic crust in the northeastern part of CGGC (Mukherjee et al. 2017a, 2018). The multiple tectonothermal events (~ 1.75–1.3 Ga) that are recorded in the EGMB and CGGC are also recorded from the fragments of the erstwhile Columbia Supercontinents (Australia: Hand et al. 2007; Cutts et al. 2013; East Antarctica: Kelly et al. 2002; Halpin et al. 2007b; Namibia: Kroner et al. 2010; Scotland : Friend and Kinny 2001). This observation points to the fact that the Indian shield was a part of the Columbia supercontinent.

4.5.2 Rodinia Supercontinent

Configuration and position of the Greater Indian landmass during the formation of Rodinia supercontinent remained highly debatable (e.g. Bhowmik et al. 2012). Widely inferred notion places the Greater Indian landmass adjacent to Australia and East Antarctica along the marginal part of the Rodinia supercontinent (Dalziel 1991; Hoffman 1991; Torsvik et al. 1996; Weil et al. 1998; Dasgupta and Sengupta 2003; Li et al. 2008). However, the view has been opposed by several workers (Fitzsimons 2000; Powell et al. 2001; Torsvik et al. 2001; Powell and Pisarevsky 2002; Merdith et al. 2017) based on available Paleomagnetic data. Numerous studies have shown that Early Neoproterozoic (ca. 1000–900 Ma) tectonothermal events experienced by the EGMB and the Rayner Complex of East Antarctica are very much similar, which lends the idea of the Greater Indian landmass being connected to Antarctica (Dasgupta and Sengupta 2003). Neoproterozoic tectonothermal events presented in this review is consistent with evolution of the CGGC in a continent-continent collision zone (Chatterjee and Ghose 2011; Karmakar et al. 2011; Sanyal and Sengupta 2012). The NSFB that is located to the south of the CGGC does not record strong impress of ~1000–950 Ma tectonothermal event (reviewed in Chatterjee et al. 2010). The pervasive planar fabric in most part of the CGGC is E-W (Mahadevan 2002). This feature and the

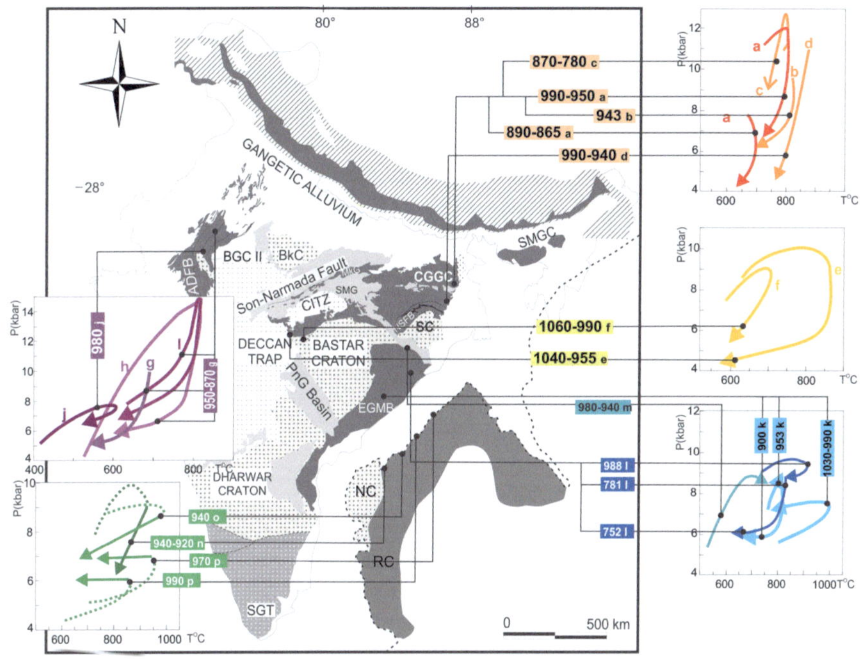

Fig. 7 Regional diagram summarizing pressure–temperature trajectories and the geochronological data of different Neoproterozoic metamorphic events in India and Antarctica. The schematic configuration is drawn after Dasgupta and Sengupta (2003). The other abbreviations are NC: Napier complex, RC: Rayner Complex, ADFB: Aravalli Delhi Fld belt, BnC: Bundelkhand Craton, CGGC: Chotanagpur Granite Gneissic Complex, SMGC: Shillong Meghalaya Gneissic Complex, SC: Singhbhum Craton, NSFB: North Singhbhum Fold Belt, EGMB: Eastern Ghats Mobile Belt, BC: Bastar Craton, SGT: Southern Granulite Terrain, CITZ: Central Indian Tectonic Zone. The references are as follows: a. Dey et al. (*under review*; b). Mukherjee et al. (2017a); c. Chatterjee et al. (2010); d. Karmakar et al. (2011); e. Bhowmik and Spiering (2004); Bhowmik et al. (2012); f. Pal and Bhowmik (1998); Bhowmik et al. (2012); g. Bhowmik et al. (2010); h. Bhowmik et al. (2009); i. Saha et al. (2008); j. Bose et al. (2017); k. Das et al. (2011), Das et al. (2017c); l. Bose et al. (2016); m. Chattopadhyay et al. (2015a); n. Halpin et al. (2005, 2007b, 2013); Kelly et al. (2002); Kelly and Harley (2004); o. Halpin et al. (2007b); p. Halpin et al. (2007a)

structural attributes supporting that the NSFB is overridden by the CGGC (reviewed in Mahadevan 2002). This deduction by implication points to the fact that the overthrusted unit possibly came from the north. The CGGC is surrounded by Phanerozoic sedimentary cover on its three sides (Fig. 2a). For this reason the nature of northern continent could not be studied.

Grenvillian tectonothermal events similar to CGGC are also recorded from the EGMB and the entire Rayner Complex of east Antarctica (reviewed in Mukherjee et al. 2017a; Li et al. 2008). Neoproterozoic clockwise P–T paths recovered from different parts of the CITZ further corroborates the reworking of the Paleoproterozoic mobile belts of India and amalgamation of the Greater Indian landmass (Fig. 7). However, geochronological studies have shown that although most of the cratons were amalgamated by $\sim$1000 Ma, India was not attached to the supercontinent assembly until $\sim$ 950 to 900 Ma (Li et al. 2008).

Timing of the break-up of Rodinia supercontinent has been controversial and mostly placed at ca. 825 Ma, as manifested by widespread plume-related magmatism throughout the world (reviewed in Li et al. 2008). However, late Tonian-Cryogenian (ca. 900–800 Ma) subduction-related magmatic activity has been reported from different peripheral parts of Rodinia (reviewed in Cawood et al. 2016). Post Grenvillian (<950 Ma) tectonothermal activity, reported from different parts of the CGGC, EGMB and East Antarctica suggest that the final assembly and breakup of the Rodinia was prolonged in these parts. Contemporaneous (ca. 870–780 Ma) high grade metamorphism event and shearing associated with the development of EITZ further poses the possibility that fragmentation of amalgamated East Antarctica and Greater Indian landmass did not initiate until ca. 780 Ma (Chatterjee et al. 2010).

5 The CGGC: Future Prospects

(A) Intra- and inter domain correlation of the magmatic, metamorphic and tectonic history is of paramount importance to understand the evolution of the CGGC. As mentioned above, the present data base are meagre (and imprecise too) to do this correlation. Strain patterns can change with space and metamorphic intensity may vary from place to place. Consequently, geometry of structure and estimated P–T values may vary from place to place. This may lead to a wrong interpretation about the tectonic evolution of the CGGC. It is not clear if the contrasting metamorphic and structural information that are reported from different parts of the CGGC reflects imprecision of the data base or they are indicative of special/temporal variation of strain pattern and metamorphic intensity. This problem can be approached by the following ways:

(i) combined petrological and structural studied along continuous segments of the CGGC
(ii) Precise dating of metamorphic and structural events. This can also help correlate the geological events over a large area.

(B) Studies in the light of modern petrology and geochronology are published mostly from the Domain I of the CGGC. The other two domains viz. Domain II and III are least studied. ~1700 to 1650 Ma old supracrustal rocks (in the jacket of low grade metamorphism) of the Domain II and III cannot be traced south of the BMB. Though similar sedimentation age (1700–1680 Ma) has been reported for some metapelitic enclaves from Domain I, it is not clear if the sedimentary protoliths of the metapelites (Domain I) are temporally and spatially linked to the supracrustal rocks of the Domain II and III of the CGGC. Basement of the low grade supracrustal rocks of Domain II and III are also not well studied. It is also not clear if the basement of the low-grade supracrustal rocks continued further south. The ~1000–850 Ma metamorphic events that are preponderant in Domain I are absent in Domain II-III. These contrasts in geology and geochronology raise the following questions:

(i) Do Domain II and III constitute a different crustal block that was sutured to Domain I after 850 Ma?
(ii) Is the oldest basement of all the three domains are same and that the Grenvillian front runs south of the BMB?

(C) Several workers (Chatterjee and Ghose 2011; Saikia et al. 2017) argued that the litho ensemble that are now exposed in Domains II and III form eastward extension of the Paleoproterozoic Mahakoshal Group of Central Indian Tectonic Zone (CITZ) where such rocks are abundantly found. If this be the case then only the Domain I should represent the CGGC. Detail petrological and geochronological information including delineation of the oldest components from the basement rocks of Domains II and III are likely to validate (or discard) the proposition.
(D) Studies have shown that nepheline syenite and carbonatite signifies crustal extension that are related to breakdown of supercontinental cycle (Upadhyay and Raith 2006). These rocks are present in NPSZ and SPSZ. It has been mentioned before in Sects. 3.1 and 4.2 that the alkaline rocks and carbonatite rocks are deformed and metamorphosed (DARC, Burke et al. 2003). Several studies have shown that DARCs represent opening and closing stage of the Wilson cycle of plate tectonics. Detail petrogenesis and geochronology of the DARCs may provide valuable information on the evolution of the CGGC.
(E) The extent of the ~1650 Ma old UHT metamorphism as has been documented from the NW part of the Domain I should be evaluated through rigorous studies in other parts of Domain I.
(F) A few reports (cf. Misra and Dey 2002) show intrusion of voluminous hot granitic magma (~1000 °C) within low grade supracrustal rocks in the BMB. Though heat rendering capacity of felsic magmas are certainly lower than those

of basaltic compositions, yet some contact effects are expected in view of the large volume of the felsic magma in the BMB.

(G) Felsic orthogneisses constitute the major component of the Domain I. Existing petrological and geochronological information is available from a few scattered areas. It is not clear if the magmatic protoliths of the entire felsic orthogneisses were emplaced within a short time during crustal extension (Mukherjee et al. 2018). Or else, different pulses of felsic magmatism were emplaced in diverse tectonic setting over a protracted period of time.

(H) The relation between the CGGC (Domain I) and the NSFB require to be studied in great detail. In view of the last high grade event at 1000–950 Ma in Domain I and its absence in NSFB, suturing of the two unit during Paleo-Mesoproterozoic time seems implausible. Timing and mechanism of juxtaposition between CGGC and NSFB are to be studied.

Acknowledgements S.M. and A.D. acknowledge the financial support in the form of research fellowships from the University Grant Commission, New Delhi and Council of Scientific and Industrial Research, New Delhi respectively. P.S. and S.S. acknowledge the grants received from the programs awarded to the Department of Geological Sciences, Jadavpur University: University Potential for Excellence (UPE-Phase II), Promotion of University Research and Scientific Excellence and Fund for Improvement of Science and Technology (FIST-Phase II) from Department of Science and Technology and Center of Advance Studies (CAS-phase VI).We express our sincere thanks to Soumyajit Mukherjee for providing very constructive detailed review. Mukherjee (2019) summarizes this work.

References

Acharyya A, Roy S, Chaudhuri BK et al (2006) Proterozoic rock suites along South Purulia Shear Zone, Eastern India: evidence for rift related setting. Journal of Geological Society of India 68, 1069–1086

Acharyya SK (2003) The nature of mesoproterozoic Central indian tectonic zone with exhumed and reworked older granulites. Gondwana Research 6, 197–214

Ahmed M (1983) Depositional environment of the basal conglomerate of the Barapani Formation, Shillong Group, Khasi Hills, Meghalaya. Quarternary Journal Geological, Mining, and Metallurgical Society of India 55, 62–68

Alam M, Alam MM, Curray JR et al (2003) An overview of the sedimentary geology of the Bengal Basin in relation to the regional tectonic framework and basin-fill history. Sedimentary Geology 155, 179–208

Ameen SMM, Khan MSH, Akon E, Kazi AI (1998) Petrography and major oxide chemistry of some Precambrian crystalline rocks from Maddhapara, Dinajpur. Bangladesh Geoscience Journal 4, 1–19

Ameen SMM, Wilde SA, Kabir Z et al (2007) Paleoproterozoic granitoids in the basement of Bangladesh : a piece of the Indian shield or an exotic fragment of the Gondwana jigsaw ? Gondwana Research 12, 380–387

Asami M, Suzuki K, Grew ES (2002) Chemical Th-U-total Pb dating by electron microprobe analysis of monazite, xenotime and zircon from the Archean Napier complex, East Antarctica: evidence for ultra-high-temperature metamorphism at 2400 Ma. Precambrian Research 114, 249–275

Baidya TK, Chakravorty PS, Drubetskoy E, Khiltova VJ (1987) New geochronologic data on some granitic phases of the Chotanagpur granite gneiss complex in the northwestern Purulia dist., West Bengal. Indian Journal of Earth Science 14, 136–141

Baidya TK, Maity N, Biswas P (1989) Tectonic phases and crustal evolution in a part of the Eastern Chotanagpur Gneissic Complex. Journal of Geological Society of India 34, 318–324

Banerji AK (1985) On the nature of a part of boundary between Chotanagpur plateau and Singhbhum orogenic belt and its role in mineralization. Bulletin of the Geology, Mining and Metallurgical Society of India 53, 171–180

Basu SK (1993) Alkaline-carbonatite complex in Precambrian of South Purulia Shear Zone, Eastern India: its characteristics and mineral potentialities. Indian Minerals 47, 179–194

Basu A, Patranabis-Deb S, Schieber J, Dhang PC (2008) Stratigraphic position of the ~1000 Ma Sukhda Tuff (Chhattisgarh Supergroup, India) and the 500 Ma question. Precambrian Research 167, 383–388

Bhandari A, Pant NC, Bhowmik SK, Goswami-Banerjee S (2011) 1.6 Ga ultrahigh-temperature granulite metamorphism in the Central Indian Tectonic Zone_insights from metamorphic reaction history, geothermobarometry and monazite chemical ages. Geological Journal 46, 198–216

Bhattacharjee N, Ray J, Ganguly S, Saha A (2012) Mineralogical study of gabbro-anorthosite from Dumka, Chhotanagpur Gneissic complex, Eastern Indian Shield. Journal of Geological Society of India 80, 481–492

Bhattacharya BP, Ray Barman T (2000) Precambrians of Meghalaya: a concept. Proc Dr MS Krishnan Birth Centen Sem Geol Surv, Ind Spec, Publ 55, 95–100

Bhattacharya DK, Mukherjee D, Barla VC (2010) Komatiite within Chhotanagpur Gneissic Complex at Semra, Palamau district, Jharkhand: petrological and geochemical fingerprints. Journal of Geological Society of India 76, 589–606

Bhattacharya HN, Mahapatra S (2008) Evolution of the Proterozoic rift margin sediments–North Singhbhum Mobile Belt, Jharkhand-Orissa, India. Precambrian Research 162, 302–316

Bhattacharya HN, Nelson DR, Thern ER, Altermann W (2015) Petrogenesis and geochronology of the Arkasani Granophyre and felsic Dalma volcanic rocks: implications for the evolution of the Proterozoic North Singhbhum Mobile Belt, east India. Geological Magazine 152, 492–503

Bhattacharyya HN, Chatterjee A, Chowdhury S (1992) Tourmalinite from Cu–U belt of Singhbhum, Bihar, India. Journal of Geological Society of India 39, 191–195

Bhattacharyya PK, Mukherjee S (1987) Granulites in and around the Bengal anorthosite, eastern India: genesis of coronal garnet and evolution of the granulite–anorthosite complex. Geological Magazine 124, 21–32

Bhowmik SK (2006) Ultra high temperature-metamorphism and its significance in the Central Indian Tectonic Zone. Lihos 92, 484–505

Bhowmik SK, Bernhardt HJ, Dasgupta S (2010) Grenvillian age high-pressure upper amphibolite-granulite metamorphism in the Aravalli-Delhi Mobile Belt, Northwestern India: new evidence from monazite chemical age and its implication. Precambrian Research 178, 168–184

Bhowmik SK, Pal T, Roy A, Pant NC (1999) Evidence for pre-Grenvillian high-pressure granulite metamorphism from the northern margin of the Sausar mobile belt in Central India. Journal of Geological Society of India 53, 385–399

Bhowmik SK, Roy A (2003) Garnetiferous metabasites from the Sausar Mobile Belt: petrology, P-T path and implications for the tectonothermal evolution of the Central Indian Tectonic Zone. Journal of Petrology 44, 387–420

Bhowmik SK, Sarbadhikari AB, Spiering B, Raith MM (2005) Mesoproterozoic reworking of Palaeoproterozoic ultrahigh-temperature granulites in the Central Indian Tectonic Zone and its implications. Journal of Petrology 46, 1085–1119

Bhowmik SK, Spiering B (2004) Constraining the prograde and retrograde P-T paths of granulites using decomposition of initially zoned garnets : an example from the Central Indian Tectonic Zone. Contributions to Mineralogy and Petrology 147, 581–603

Bhowmik SK, Saha L, Dasgupta S, Fukuoka M (2009) Metamorphic phase relations in orthopyroxene-bearing granitoids: implication for high-pressure metamorphism and prograde melting in the continental crust. Journal of Metamorphic Geology 27:295–315. https://doi.org/10.1111/j.1525-1314.2009.00818.x

Bhowmik SK, Wilde SA, Bhandari A (2011) Zircon U-Pb/Lu-Hf and monazite chemical dating of the Tirodi biotite gneiss: implication for latest Palaeoproterozoic to Early Mesoproterozoic orogenesis in the Central Indian Tectonic Zone. Geological Journal 46, 574–596

Bhowmik SK, Wilde SA, Bhandari A, et al (2012) Growth of the Greater Indian Landmass and its assembly in Rodinia : geochronological evidence from the Central Indian Tectonic Zone. Gondwana Research 22, 54–72

Bhowmik SK, Wilde SA, Bhandari A, Sarbadhikari AB (2014) Zoned monazite and zircon as monitors for the thermal history of granulite terranes: an example from the Central Indian Tectonic Zone. Journal of Petrology 55, 585–62

Biju-Sekhar S, Yokoyama K, Pandit MK et al (2003) Late Paleoproterozoic magmatism in Delhi Fold Belt, NW India and its implication: evidence from EPMA chemical ages of zircons. Journal of Asian Earth Sciences 22, 189–207

Bora S, Kumar S, Yi K et al (2013) Geochemistry and U–Pb SHRIMP zircon chronology of granitoids and microgranular enclaves from Jhirgadandi Pluton of Mahakoshal Belt, Central India Tectonic Zone, India. Journal of Asian Earth Sciences 70, 99–114

Bose M, Chakrabarti MK, Saunders AD (1989) Petrochemistry of the lavas from Proterozoic Dalma volcanic belt, Singhbhum, eastern India. Geologische Rundschau 78, 633–648

Bose PK, Sarkar S, Chakrabarty S, Banerjee S (2001) Overview of the Meso-to Neoproterozoic evolution of the Vindhyan basin, central India. Sedimentary Geology 141, 395–419

Bose RN (1954) The metamorphic rocks around Barabhum and Bunduan, south Manbhum. The Quarterly Journal of the Geological, Mining, and Metallurgical Society of India 29, 19–3

Bose S, Das K, Torimoto J et al (2016) Evolution of the Chilka Lake granulite complex, northern Eastern Ghats Belt, India: first evidence of ~780 Ma decompression of the deep crust and its implication on the India–Antarctica correlation. Lithos 263, 161–189

Bose S, Dunkley DJ, Dasgupta S et al (2011) India-Antarctica-Australia-Laurentia connection in the Paleoproterozoic-Mesoproterozoic revisited: evidence from new zircon U-Pb and monazite chemical age data from the Eastern Ghats Belt, India. Bulletin of Geological Society of America 123, 2031–2049

Bose S, Seth P, Dasgupta N (2017) Meso-neoproterozoic mid-crustal metamorphic record from the Ajmer–Shrinagar section, Rajasthan, India and its implication to the assembly of the Greater Indian Landmass during the Grenvillian-age orogenesis. In: Pant NC, Dasgupta S (eds) Crustal evolution of India and Antarctica: the supercontinent connection. Geological Society, London, Special Publications, 457, 291–318

Buick IS, Allen C, Pandit M et al (2006) The proterozoic magmatic and metamorphic history of the Banded Gneiss Complex, central Rajasthan, India: LA-ICP-MS U-Pb zircon constraints. Precambrian Research 151, 119–142

Burke K, Ashwal LD, Webb SJ (2003) New way to map old sutures using deformed alkaline rocks and carbonatites. Geology 31, 391–394

Cawood PA, Strachan RA, Pisarevsky SA et al (2016) Linking collisional and accretionary orogens during Rodinia assembly and breakup : Implications for models of supercontinent cycles. Earth and Planetary Science Letters 449, 118–126

Célérier J, Harrison TM, Webb AAG, Yin A (2009) The Kumaun and Garwhal Lesser Himalaya, India: part 1. Structure and stratigraphy. Bulletin of Geological Society of America 121, 1262–1280

Chakrabarti MK (1985) On the high-magnesian lavas of the Dalma metavolcanic sequence of Singhbhum District, Bihar. In: Indian National Science Academy, 51, pp 598–609

Chakrabarty A, Sen AK (2010) Enigmatic association of the carbonatite and alkali-pyroxenite along the Northern Shear Zone, Purulia, West Bengal: a saga of primary magmatic carbonatite. Journal of Geological Socoiety of India 76, 403–413

Chambers JA, Argles TW, Horstwood MSA et al (2008) Tectonic implications of Palaeoproterozoic anatexis and Late Miocene metamorphism in the Lesser Himalayan Sequence, Sutlej Valley, NW India. Journal of Geological Society 165, 725–737
Chatterjee N, Banerjee M, Bhattacharya A, Maji AK (2010) Monazite chronology, metamorphism–anatexis and tectonic relevance of the mid-Neoproterozoic Eastern Indian Tectonic Zone. Precambrian Research 179, 99–120
Chatterjee N, Crowley JL, Ghose NC (2008) Geochronology of the 1.55 Ga Bengal anorthosite and Grenvillian metamorphism in the Chotanagpur gneissic complex, eastern India. Precambrian Research 161, 303–316
Chatterjee N, Ghose NC (2011) Extensive early Neoproterozoic high-grade metamorphism in North Chotanagpur Gneissic Complex of the Central Indian Tectonic Zone. Gondwana Research 20, 362–379
Chatterjee N, Mazumdar AC, Bhattacharya A, Saikia RR (2007) Mesoproterozoic granulites of the Shillong-Meghalaya Plateau: evidence of westward continuation of the Prydz Bay Pan-African suture into Northeastern India. Precambrian Research 152, 1–26
Chatterjee SM, Roy Choudhury M, Das S, Roy A (2017) Significance and dynamics of the neoproterozoic (810 Ma) Phulad Shear Zone, Rajasthan, NW India. Tectonics 36, 1432–1454
Chattopadhyay A, Das K, Hayasaka Y, Sarkar A (2015b) Syn-and post-tectonic granite plutonism in the Sausar Fold Belt, central India: Age constraints and tectonic implications. Journal of Asian Earth Sciences 107, 110–112
Chattopadhyay A, Khasdeo L (2011) Structural evolution of Gavilgarh-Tan Shear Zone, Central India: a possible case of partitioned transpression during Mesoproterozoic oblique collision within Central Indian Tectonic Zone. Precambrian Research 186, 70–88
Chattopadhyay N, Ray S, Sanyal S, Sengupta P (2015a) Mineralogical, textural and chemical reconstitution of granitic rock in ductile shear zone: a study from a part of the South Purulia Shear Zone, West Bengal, India. Ductile Shear Zones from micro-to macro-scales Wiley, Chichester, pp 141–163
Corvino AF, Boger SD, Henjes-kunst F et al (2008) Superimposed tectonic events at 2450 Ma, 2100 Ma, 900 Ma and 500 Ma in the North Mawson Escarpment, Antarctic Prince Charles Mountains. Precambrian Research 167, 281–302
Crawford AR (1974) Indo-Antarctica, Gondwana Land and pattern of the distortion of a granulite belt. Tectonophysics 22, 141–157
Cutts KA, Kelsey DE, Hand M (2013) Evidence for late Paleoproterozoic (ca 1690–1665 Ma) high- to ultrahigh-temperature metamorphism in southern Australia: implications for Proterozoic supercontinent models. Gondwana Research 23, 617–640
Dalziel IWD (1991) Pacific margins of Laurentia and East Antarctica-Australia as a conjugate rift pair: evidence and implications for an Eocambrian supercontinent. Geology 19, 598–601
Daniel CG, Hollister LS, Parrish RR, Grujic D (2003) Exhumation of the main central thrust from lower crustal depths, eastern Buthan Himalaya. Journal of Metamorphic Geology 21, 317–334
Das E, Karmakar S, Dey A et al (2017c) Reaction textures, pressure—temperature paths and chemical dates of monazite from a new suite of sapphirine—spinel granulites from parts of the Eastern Ghats Province, India : insights into the final amalgamation of India and East Antarctica during the formation of Rodinia. Crustal Evolution of India and Antarctica: The Supercontinent Connection, London, Special Publications
Das K, Bose S, Karmakar S (2011) Multiple tectonometamorphic imprints in the lower crust : first evidence of ca. 950 Ma (zircon U-Pb SHRIMP) compressional reworking of UHT aluminous granulites from the Eastern Ghats Belt, India. Geological Journal 239, 217–239
Das S, Dasgupta N, Sanyal S et al (2017a) Dolomitic carbonatite from the Chotanagpur Granite Gneiss Complex : a new DARC (deformed alkaline rocks and carbonatite) in the Precambrian shield of India. Current Science 113, 4–6
Das S, Roy N, Sen P, Sanyal S, Karmakar S, Sengupta P, Sengupta SK (2017b) Evidence of extensive alkali metasomatism along North Purulia Shear Zone (NPSZ) near Lanka-Nawadih area in the southern parts of Chhotanagpur gneissic complex of East Indian shield. In:

Humboldt Kolleg “Earth and material sciences for sustainable societal developments” (13–15 Jan), Raichak, India. pp 50–51
Das S, Sanyal S, Karmakar S, Sengupta P (2016) Petrology of a suite of deformed alkaline rocks in Kankarkiari area of Purulia district, West Bengal: evidence of intra-continental suturing in parts of Eastern Indian Shield during Neoproterozoic time. In: International conference on electron microscopy & XXXVII annual meeting of EMSI (2–4 June), Varanasi, India, Abstract Volume, pp 85–86
Dasgupta S, Narula PL, Acharyya SK, Banerjee J (2000) Seismotectonic atlas of India and its environs. In: Narula PL et al. (ed), Geological survey of India. Geological Survey of India
Dasgupta S, Sengupta P (2003) Indo-Antarctic correlation: a perspective from the Eastern Ghats Granulite Belt, India. Geological Society London Special Publications 206, 131–143
Deb M, Thorpe R, Krstic D (2002) Hindoli group of rocks in the Eastern Fringe of the Aravalli-Delhi Orogenic Belt-Archean secondary Greenstone Belt or Proterozoic Supracrus tals? Gondwana Research 5, 879–883
Decelles PG, Gehrels GE, Quade J et al (2000) Tectonic implications of of U-Pb U-Pb Zircon ages of of the Himalayan orogenic belt in Nepal. Science 288, 497–499
Deshmukh T, Prabhakar N, Bhattacharya A, Madhavan K (2017) Late Paleoproterozoic clockwise P–T history in the Mahakoshal Belt, Central Indian Tectonic zone: implications for Columbia supercontinent assembly. Precambrian Research 298, 56–78
Desikachar SV (1974) A review of the tectonic and geological history of eastern India in terms of ‘plate tectonics’ theory. Journal of Geological Society of India 15, 137–149
Dey A, Karmakar S, Mauricio Ibanez-Mejia SM et al (under review, a) Petrology and geochronology of metapelitic enclaves within felsic orthogneiss from the NE Chotanagpur Granite Gneissic Complex, eastern India: evidence for Stenian-Tonian reworking of a late Paleoproterozoic crust. Precambrian Research
Dey A, Karmakar S, Mukherjee S et al (under review, b) High Pressure metamorphism of mafic granulites from the Chotanagpur Granite Gneissic Complex, India: evidence for collisional tectonics during assembly of Rodinia. Journal of Geodynnamics
Dey A, Mukherjee S, Sanyal S et al (2017) Deciphering sedimentary provenance and timing of sedimentation from a suite of metapelites from the Chotanagpur Granite Gneissic complex, India: implications for Proterozoic tectonics in the East-Central part of the Indian shield. In: Mazumder R (ed) Sediment Provenance. Influences on compositional change from source to sink, Elsevier Ltd, pp 453–486
Dharma Rao CV, Santosh M, Dong Y (2012) U-Pb zircon chronology of the Pangidi-Kondapalle layered intrusion, Eastern Ghats belt, India: constraints on Mesoproterozoic arc magmatism in a convergent margin setting. Journal of Asian Earth Sciences 49, 362–375
Dharma Rao CV, Santosh M, Kim SW, Li S (2013) Arc magmatism in the Delhi Fold Belt: SHRIMP U-Pb zircon ages of granitoids and implications for Neoproterozoic convergent margin tectonics in NW India. Journal of Asian Earth Sciences 78, 83–99
Dharma Rao CV, Santosh M, Purohit R et al (2011) LA-ICP-MS U-Pb zircon age constraints on the Paleoproterozoic and Neoarchean history of the Sandmata Complex in Rajasthan within the NW Indian Plate. Journal of Asian Earth Sciences 42, 286–305
DiPietro JA, Isachsen CE (2001) U-Pb zircon ages from the Indian plate in northwest Pakistan and their significance to Himalayan and pre-Himalayan geologic history. Tectonics 20, 510
Dobmeier C, Lütke S, Hammerschmidt K, Mezger K (2006) Emplacement and deformation of the Vinukonda meta-granite (Eastern Ghats, India)—Implications for the geological evolution of peninsular India and for Rodinia reconstructions. Precambrian Research 146, 165–178
Dobmeier C, Simmat R (2002) Post-Grenvillean transpression in the Chilka Lake area, Eastern Ghats Belt—implications for the geological evolution of peninsular India. Precambrian Research 113, 243–268
Dobmeier CJ, Raith MM (2003) Crustal architecture and evolution of the Eastern Ghats Belt and adjacent regions of India. Geological Society London Special Publications 206, 145–168
Dunn JA (1929) The geology of north Singhbhum including parts of Ranchi and Singhbhum districts. Memoirs of the Geological Survey of India 54, 1–280

Elburg M, Jacobs J, Andersen T et al (2015) Early Neoproterozoic metagabbro-tonalite-trondhjemite of Sør Rondane (East Antarctica): implications for supercontinent assembly. Precambrian Research 259, 189–206

Evans P (1964) The tectonic framework of Assam. Journal of Geological Society of India 5, 80–96

Fitzsimons ICW (2000) Grenville-age basement provinces in East Antarctica: evidence for three separate collisional orogens. Geology 28, 879–882

Friend C, Kinny P (2001) A reappraisal of the Lewisian Gneiss Complex: geochronological evidence for its tectonic assembly from disparate terranes in the Proterozoic. Contributions to Mineralogy and Petrology 142, 198–218

Ghose NC (1983) Geology, tectonics and evolution of the Chhotanagpur granite-gneiss complex, Eastern India. Recent Researches in Geology 10, 211–247

Ghose NC (1992) Chhotanagpur gneiss-granulite complex, Eastern India: present status and future prospect. Indian Journal of Geology 64, 100–121

Ghose NC, Mukherjee D (2000) Chotanagpur gneiss–granulite complex, Eastern India—a kaleidoscope of global events. In: Trivedi AN, Sarkar BC, Ghose NC (eds) Geology and mineral resources of Bihar and Jharkhand, Platinum Jubilee Commemoration Volume, Indian School of Mines, Dhanbad, monograph 2, Institute of Geoexploration and Environment, Patna, pp 33–58

Ghosh S, Chakraborty S, Paul DK et al (1994) New Rb–Sr isotopic ages and geochemistry of granitoids from Meghalaya and their significance in middle to late Proterozoic crustal evolution. Indian Mineralogist 48, 33–44

Ghosh S, Fallick AE, Paul DK, Potts PJ (2005) Geochemistry and origin of Neoproterozoic granitoids of Meghalaya, Northeast India: Implications for linkage with amalgamation of Gondwana supercontinent. Gondwana Research 8, 421–432

Ghosh SK, Sengupta S (1999) Boudinage and composite boudinage in superposed deformations and syntectonic migmatization. Journal of Structural Geology 21, 97–110

Goodge JW, Vervoort JD, Fanning CM et al (2008) A positive test of East Antarctica–Laurentia Juxtaposition within the Rodinia Supercontinent. Science 321, 235–240

Goswami B, Bhattacharyya C (2010) Tectonothermal Evolution of Chhotanagpur Granite Gneiss complex from Northeastern part of Puruliya District, West Bengal, Eastern India. Indian Journal of Geology 80, 41–54

Goswami B, Bhattacharyya C (2013) Petrogenesis of shoshonitic granitoids, Eastern India: implications for the late Grenvillian post-collisional magmatism. Geoscience Frontiers 5, 821–843

Gupta A, Basu A (2000) North Singhbhum Proterozoic Mobile Belt Eastern India—a review. Geological Survey of India Special Publications 55, 195–226

Gupta P, Guha DB, Chattopadhyay B (1998) Basement-cover relationship in the Khetri copper belt and the emplacement mechanism of the granite massifs, Rajasthan, India. Journal of Geological Society of India 52, 417–432

Halpin JA, Clarke GL, White RW, Kelsey DE (2007a) Contrasting P–T–t paths for Neoproterozoic metamorphism in MacRobertson and Kemp Lands, east Antarctica. Journal of Metamorphic Geology 25, 683–701

Halpin JA, Daczko NR, Clarke GL, Murray KR (2013) Basin analysis in polymetamorphic terranes: an example from east Antarctica. Precambrian Research 231, 78–97

Halpin JA, Gerakiteys CL, Clarke GL et al (2005) In-situ U – Pb geochronology and Hf isotope analyses of the Rayner Complex, east Antarctica. Contributions to Mineralogy and Petrology 148, 689–706

Halpin JA, White RW, Clarke GL, Kelsey DE (2007b) The Proterozoic P–T–t evolution of the Kemp Land coast, East Antarctica; Constraints from Si-saturated and Si-undersaturated metapelites. Journal of Petrolology 48, 1321–1349

Hand M, Reid A, Jagodzinski L (2007) Tectonic framework and evolution of the Gawler craton, southern Australia. Economic Geology 102, 1377–1395

Henderson B, Collins AS, Payne J et al (2014) Geologically constraining India in Columbia: the age, isotopic provenance and geochemistry of the protoliths of the Ongole Domain, Southern Eastern Ghats, India. Gondwana Research 26, 888–906

Hoffman PF (1991) Did the breakout of Laurentia turn Gondwanaland inside-out ? Hoffman PF, Published by : American Association for the Advancement of Science
Hoffmann P (1989) Speculations on Laurentia's first gigayear (2.0–1.0 Ga). Geology 17, 135–138
Hossain I, Tsunogae T (2014) Crystallization conditions and petrogenesis of the paleoproterozoic basement rocks in Bangladesh : an evaluation of biotite and coexisting amphibole mineral chemistry. Journal of Earth Sciemce 25, 87–97
Hossain I, Tsunogae T, Rajesh HM (2007) Palaeoproterozoic U – Pb SHRIMP zircon age from basement rocks in Bangladesh : a possible remnant of the Columbia supercontinent. Competes Rendus Geoscience 339, 979–986
Hossain Md S, Khan SH, Chowdhury KR, Abdullah R (2019) Synthesis of the tectonic and structural elements of the Bengal Basin (and its surroundings). In: Mukherjee S. (Ed) Tectonics and structural geology: Indian context. Springer International Publishing AG, Cham, pp 135–218. ISBN 978-3-319-99340-9
Jain SC, Yedekar DB, Nair KKK (1991) Central Indian shear zone: a major Pre-Cambrian crustal boundary. Journal of Geological Society of India 37, 521–531
Kabir MZ, Khalil RC, Akon E et al (2001) Petrogenetic study of Precambrian basement rocks from Maddhapara, Dinajpur, Bangladesh. Bangladesh Geoscience Journal 7, 1–18
Karmakar S, Bose S, Sarbadhikari AB, Das K (2011) Evolution of granulite enclaves and associated gneisses from Purulia, Chhotanagpur Granite Gneiss Complex, India: evidence for 990–940 Ma tectonothermal event(s) at the eastern India cratonic fringe zone. Journal of Asian Earth Sciences 41, 69–88
Kaur P, Chaudhri N, Raczek I et al (2009) Record of 1.82 Ga Andean-type continental arc magmatism in NE Rajasthan, India: insights from zircon and Sm-Nd ages, combined with Nd-Sr isotope geochemistry. Gondwana Research 16, 56–71
Kaur P, Chaudhri N, Raczek I, et al (2011) Zircon ages of late Palaeoproterozoic (ca. 1.72-1.70 Ga) extension-related granitoids in NE Rajasthan, India: regional and tectonic significance. Gondwana Research 19, 1040–1053
Kaur P, Zeh A, Chaudhri N (2017a) Palaeoproterozoic continental arc magmatism, and Neoproterozoic metamorphism in the Aravalli-Delhi orogenic belt, NW India: new constraints from in situ zircon U-Pb-Hf isotope systematics, monazite dating and whole-rock geochemistry. Journal of Asian Earth Sciences 136, 68–88
Kaur P, Zeh A, Chaudhri N, Eliyas N (2017b) Two distinct sources of 1.73–1.70 Ga A-type granites from the northern Aravalli orogen, NW India: constraints from in situ zircon U-Pb ages and Lu-Hf isotopes. Gondwana Research 49, 164–181
Kelly NM, Clarke GL, Fanning CM (2002) A two-stage evolution of the Neoproterozoic Rayner structural episode: new U-Pb sensitive high resolution ion microprobe constraints from the Oygarden group, Kemp Land, East Antarctica. Precambrian Res 116, 307 330
Kelly NM, Harley SL (2004) Orthopyroxene–corundum in Mg–Al-rich granulites from the Oygarden Islands, east Antarctica. Journal of Petrolology 45, 1481–1512
Kohn MJ, Paul SK, Corrie SL (2010) The lower lesser Himalayan sequence : a Paleoproterozoic arc on the northern margin of the Indian plate. GSA Bulletin 122, 323–335
Korhonen FJ, Clark C, Brown M et al (2013) How long-livedis ultrahigh temperature (UHT) metamorphism? Constraints from zircon andmonazite geochronology in the Eastern Ghats orogenic belt, India. Precambrian Research 234, 322–350
Kovach VP, Berezhnayal NG, Salnikoval EB, Kotovl AB (2001) The Western Charnockite zone of the Eastern Ghats Belt, India—an Laurentia-Siberia connection revisited again : an overview of U-Pb Zircon. 6–7
Krause O, Dobmeier C, Raith MM, Mezger K (2001) Age of emplacement of massif-type anorthosites in the Eastern Ghats Belt, India: constraints from U-Pb zircon dating and structural studies. Precambrian Research 109, 25–38
Krishna V, Prasad RN, Pandey UK, et al (1996) Rb-Sr geochronology of Chotanagpur gneiss-granulite complex around Kailashnathgufa area, Raigarh district, MP. In: Proceedings of seventh national symposium on mass spectrometry

Krishna V, Sastry D, Pandey BK, Sinha RP (2003) U-Pb and Pb-Pb ages on columbite-tantalite minerals from pegmatites of Bihar Mica Belt, Jharkhand, India. In: ISMAS silver jubilee symposium on mass spectrometry. V. 2: contributed papers

Kroner A, Rojas-Agramonte Y, Hegner E, et al (2010) SHRIMP zircon dating and Nd isotopic systematics of Palaeoproterozoic migmatitic orthogneisses in the Epupa metamorphic complex of northwestern Namibia. Precambrian Research 183, 50–69

Kumar MN, Das N, Dasgupta S (1978) Geology and mineralization along the Northern Shear Zone, Purulia District, West Bengal—an up-to-date appraisal. Records of the Geological Survey of India 133, 25–31

Kumar S, Rino V, Hayasaka Y et al (2017) Contribution of Columbia and Gondwana Supercontinent assembly- and growth-related magmatism in the evolution of the Meghalaya Plateau and the Mikir Hills, Northeast India: constraints from U-Pb SHRIMP zircon geochronology and geochemistry. Lithos 277, 356–375

Lahiri G, Das S (1984) Petrology of the area east of Daltonganj, Palamau district, Bihar. Journal of Geological Society of India 25, 490–504

Lal N, Saini HS, Nagpaul KK, K. Sharma K (1976) Tectonic and cooling history of the Bihar Mica Belt, India, as revealed by fission-track analysis. Tectonophysics 34, 163–180

Lal RK, Ackermand D, Seifert F, Haldar SK (1978) Chemographic relationships in sapphirine-bearing rocks from Sonapahar, Assam, India. Contributions to Mineral Petrol 67, 169–187

Larson KP, Kellett DA, Cottle JM et al (2016) Anatexis, cooling, and kinematics during orogenesis: miocene development of the himalayan metamorphic core, east-central nepal. Geosphere 12, 1575–1593

Li ZX, Bogdanova SV, Collins AS et al (2008) Assembly, configuration, and break-up history of Rodinia: a synthesis. Precambrian Research 160, 179–210

Liao QA, Li DW, Lu L et al (2008) Paleoproterozoic granitic gneisses of the Dinggye and LhagoiKangri areas from the higher and northern Himalaya, Tibet: geochronology and implications. Science China, Section D Earth Science 51, 240–248

Liu X, Zhao Y, Chen H, Song B (2017) New zircon U – Pb and Hf – Nd isotopic constraints on the timing of magmatism, sedimentation and metamorphism in the northern Prince Charles Mountains, East Antarctica. Precambrian Research 299, 15–33

Liu X, Zhao Y, Hu J (2013) The c. 1000–900 Ma and c. 550 – 500 Ma tectonothermal events in the Prince Charles Mountains—Prydz Bay region, East Antarctica, and their relations to supercontinent evolution. In: Harley SL, Fitzsimons ICW, Zhao Y (eds) Antarctica and supercontinent evolution. Geological Society, London, Special Publications, pp 95–112

Liu X, Zhao Y, Song B et al (2009) SHRIMP U-Pb zircon geochronology of high-grade rocks and charnockites from the eastern Amery Ice Shelf and southwestern Prydz Bay, East Antarctica: constraints on late Mesoproterozoic to Cambrian tectonothermal events related to supercontinent assembly. Gondwana Research 16, 342–361

Long SP, McQuarrie N, Tobgay T et al (2008) Tectonostratigraphy of the lesser Himalaya of Bhutan: deducing the paleostratigraphy of the northern Indian margin. In: AGU fall meeting abstracts

Mahadevan TM (1992) Geological evolution of the Chhotanagpur gneissic complex in parts of Purulia district. Indian Journal of Geology 64, 1–22

Mahadevan TM (2002) Geology of Bihar & Jharkhand. Geological Society of India, Bangalore

Mahato S, Goon S, Bhattacharya A et al (2008) Thermo-tectonic evolution of the North Singhbhum Mobile Belt (eastern India): a view from the western part of the belt. Precambrian Research 162, 102–127

Maji AK, Goon S, Bhattacharya A et al (2008) Proterozoic polyphase metamorphism in the Chhotanagpur Gneissic Complex (India), and implication for trans-continental Gondwanaland correlation. Precambrian Research 162, 385–402

Mallik AK (1993) Dating of the granite plutons in Bihar mica belt. Bihar Records of the Geological Survey of India 126:27–29

Mallik AK, Gupta SN, Barman Ray T (1991) Dating of early Precambrian granite-greenstone complex of the eastern Indian Precambrian shield with special reference to the Chotanagpur granite gneiss complex. Records of Geological Survey of India 125, 20–21

Malone SJ et al (2008). Paleomagnetism and Detrital Zircon geochronology of the upper Vindhyan sequence, Son Valley and Rajasthan, India: a ca. 1000 Ma Closure age for the Purana basins? Precambrian Research 164, 137–159

Mandal S, Robinson DM, Kohn MJ et al (2016) Zircon U-Pb ages and Hf isotopes of the Askot klippe, Kumaun, northwest India: implications for Paleoproterozoic tectonics, basin evolution and associated metallogeny of the northern Indian cratonic margin. Tectonics 35, 965–982

Mandal P (2016) Shear-wave splitting in Eastern Indian Shield: detection of a Pan-African suture separating Archean and Meso-Proterozoic terrains. Precambrian Research. Elsevier B.V. 275, 278–285

Manna SS, Sen SK (1974) Origin of garnet in basic granulites around Saltora, West Bengal, India. Contributions to Mineralogy and Petrology 44, 195–218

Mazumdar SK (1976) A summary of the Precambrian Geology of Khasi Hills, Meghalaya. Geologoical Survey of India Miscillineous Publications 23, 311–334

Mazumdar SK (1988) Crustal evolution of the Chotanagpur gneissic complex and the Mica Belt of Bihar. In: Mukhopadhyay D (ed) Precambrian of the Eastern Indian shield. Geological Society of India Memoir 8, 49–84

Meert JG, Pandit MK, Pradhan VR et al (2010) Precambrian crustal evolution of Peninsular India: a 3.0 billion year odyssey. Journal of Asian Earth Sciences 39, 483–515

Meert JG, Santosh M (2017) The Columbia supercontinent revisited. Gondwana Research 50, 67–83

Merdith AS, Collins AS, Williams SE et al (2017) A full-plate global reconstruction of the Neoproterozoic. Gondwana Research 50, 84–134

Mezger K, Cosca MA (1999) The thermal history of the Eastern Ghats Belt (India) as revealed by U-Pb and 40Ar/39Ar dating of metamorphic and magmatic minerals: implications for the SWEAT correlation. Precambrian Research 94, 251–271

Miller C, Klotzli U, Frank W et al (2000) Proterozoic crustal evolution in the NW Himalaya (India) as recorded by circa 1.80 Ga mafic and 1.84 Ga granitic magmatism. Precambrian Research 103, 191–206

Misra S, Dey S (2002) Bihar Mica Belt plutons-an example of post-orogenic granite from eastern Indian shield. Journal of Geological Society of India 59, 363–37

Mishra DC, Singh B, Tiwari VM, et al (2000) Two cases of continental collisions and related tectonics during the Proterozoic period in India - insights from gravity modelling constrained by seismic and magnetotelluric studies. Precambrian Research 99:149–169. https://doi.org/10.1016/S0301-9268(99)00037-6

Mitra SK (1998) Structure, sulphide mineralization and age of the Shillong group of rocks, Meghalaya. In: SK Krishnan centenary commemorative national seminar, p 118–119

Morrissey LJ, Hand M, Kelsey DE (2015) Multi-stage metamorphism in the Rayner—Eastern Ghats Terrane : P–T–t constraints from the northern Prince Charles Mountains, east Antarctica. Precambrian Research 267, 137–163

Mukherjee D, Ghose NC, Chatterjee N (2005) Crystallization history of a massif anorthosite in the eastern Indian shield margin based on borehole lithology. Journal of Asian Earth Sciences 25, 77–94

Mukherjee S (In press) Kinematics of pure shear ductile deformation within rigid walls: New analyses. In: Billi A, Fagereng A. (eds) Problems and solutions in structural geology and tectonics. Series editor: Mukherjee S, Developments in structural geology and tectonics book series. Elsevier. ISBN: 9780128140482. ISSN: 2542–9000

Mukherjee S (2013) Channel flow extrusion model to constrain dynamic viscosity and Prandtl number of the Higher Himalayan Shear Zone. International Journal of Earth Sciences 102, 1811–1835

Mukherjee S (2015) A review on out-of-sequence deformation in the Himalaya. In: Mukherjee S, Carosi R, van der Beek P, Mukherjee B, Robinson D (eds) Tectonics of the Himalaya. Geological Society, London, Special Publications 412, 67–109
Mukherjee S (2017) Shear heating by translational brittle reverse faulting along a single, sharp and straight fault plane
Mukherjee S (2019) Introduction to “Tectonics and Structural Geology: Indian Context”. In: Mukherjee S (ed) Tectonics and structural geology: Indian context. Springer International Publishing AG, Cham, pp 1–5. ISBN: 978-3-319-99340-9
Mukherjee S, Mulchrone KF (2013) Viscous dissipation pattern in incompressible Newtonian simple shear zones: an analytical model. International Journal of Earth Sciences 102:1165–1170
Mukherjee S, Carosi R, van der Beek P, Mukherjee B, Robinson D (2015a) Tectonics of the Himalaya: an introduction. In: Mukherjee S, Carosi R, van der Beek P, Mukherjee B, Robinson D (eds) Geological Society, London, Special Publications 412, 1–3
Mukherjee S, Dey A, Ibanez-Mejia M et al (2018) Geochemistry, U-Pb geochronology and Lu-Hf isotope systematics of a suite of ferroan (A-type) granitoids from the CGGC: Evidence for Mesoproterozoic crustal extension in the east Indian shield. Precambrian Research 305, 40–63
Mukherjee S, Dey A, Sanyal S et al (2017a) Petrology and U–Pb geochronology of zircon in a suite of charnockitic gneisses from parts of the Chotanagpur Granite Gneiss Complex (CGGC): evidence for the reworking of a Mesoproterozoic basement during the formation of the Rodinia supercontinent. In: Pant NC, Dasgupta S (Eds) Crustal evolution of India and Antarctica: the supercontinent connection. Geological Society of London, Special Publications 451, 197–231
Mukherjee S, Mukherjee B, Thiede R (2013) Geosciences of the Himalaya-Karakoram-Tibet Orogen. Thematic volume’s editorial. International Journal of Earth Sciences 102, 1757–1758
Mukherjee S, Punekar J, Mahadani T, Mukherjee R (2015b) A review on intrafolial folds and their morphologies from the detachments of the western Indian Higher Himalaya. In: Mukherjee S, Mulchrone KF (eds) Ductile shear zones: from micro- to macro-scales. Wiley Blackwell, pp 182–205
Mukhopadhyay D (1990) Precambrian plate tectonics in the Eastern Indian Shield. In: Sychanthavong SPH (ed) Crustal evolution and metallogeny. Oxford and IBH Publishing Co, New Delhi, pp 75–100
Mukhopadhyay D, Bhattacharyya T, Chattopadhyay N, et al (2000) Anasagar gneiss: a folded granitoid pluton in the Phanerozoic South Delhi Fold Belt, central Rajasthan. International Journal of Earth System Sciences 109, 22–38
Mulchrone K, Mukherjee S (2015) Shear senses and viscous dissipation of layered ductile simple shear zones. Pure and Applied Geophysics 172:2635–2642
Mulchrone K, Mukherjee S (2016) Kinematics and shear heat pattern of ductile simple shear zones with “slip boundary condition”. International Journal of Earth System Sciences 105:1015–1020
Nandy DR (2001) Geodynamics of the Northeastern India and the Adjoining Region. ACB Publications 209
Owada M, Osanai Y, Toyoshima T et al (2003) Early Proterozoic tectonothermal events in the Napier Complex, East Antarctica: implications for the formation of East Gondwana. Gondwana Research 6, 231–240
Pal T, Bhowmik SK (1998) Metamorphic history of Sausar Group of rocks. Geological Survey of India (Unpublished report)
Pandey BK, Gupta JN, Lall Y (1986a) Whole rock and Rb-Sr isochron ages for the granites from Bihar mica belt of Hazaribagh, Bihar, India. Indian Journal of Earth Science 12, 157–162
Pandey BK, Krishna V, Chabria T (1998) An overview of Chotanagpur Gneiss-Granulite Complex and adjoining sedimentary sequences, eastern and central India. In: International seminar on Precambrian crust in Eastern and Central India 1998. Abstract Volume UNESCO-IUGS-IGCP-368. pp 131–135
Pandey BK, Upadhayay DL, Sinha KK (1986b) Geochronology of Jajawal-Binda-Nagnaha granitoids in relation to uranium mineralisation. Indian Journal of Earth Science 13, 163–168

Pandit MK, Carter LM, Ashwal LD et al (2003) Age, petrogenesis and significance of 1 Ga granitoids and related rocks from the Sendra area, Aravalli Craton, NW India. Journal of Asian Earth Sciences 22, 363–381

Panigrahi MK, Bream BR, Misra KC, Naik RK (2004) Age of granitic activity associated with copper-molybdenum mineralization at Malanjkhand, Central India. Mineralium Deposita 39, 670–677

Patel SC (2007) Vesuvianite-wollastonite-grossular-bearing calc-silicate rock near Tatapani, Surguja district, Chhattisgarh. Journal of Earth System Science 116, 143–147

Piper JDA (2013) Continental velocity through Precambrian times: the link to magmatism, crustal accretion and episodes of global cooling. Geoscience Frontiers 4, 7–36

Pisarevsky SA, Wingate MTD, Powell CM et al (2003) Models of Rodinia assembly and fragmentation. Geological Society, London, Special Publications 206, 35–55

Powell CM, Jones DL, Pisarevsky S, Wingate MTD (2001) Palaeomagnetic constraints on the position of the Kalahari craton in Rodinia. Precambrian Research 110, 33–46

Powell CM, Pisarevsky SA (2002) Late neoproterozoic assembly of East Gondwana. Geology 30, 3–6

Radhakrishna BP, Naqvi SM (1986) Precambrian continental crust of India and its evolution. Journal of Geology 94, 145–166

Ramakrishnan M, Vaidyanadhan R (2008) Geology of India. Geological Society of India Bangalore

Ray Barman T, Bishui PK (1994) Dating of Chotanagpur gneissic complex of eastern Indian Precambrian shield. Records of Geological Survey of India 127, 25–27

Ray Barman T, Bishui PK, Mukhopadhyay K, Ray JN (1994) Rb-Sr geochronology of the high-grade rocks from Purulia, West Bengal and Jamua-Dumka sector, Bihar. Indian Minerals 48, 45–60

Ray J, Saha A, Ganguly S et al (2011a) Geochemistry and petrogenesis of Neoproterozoic Mylliem granitoids, Meghalaya Plateau, northeastern India. Journal of Earth System Science 120, 459–473

Ray S, Gangopadhyay PK (1971) Metamorphic belt of Singhbhum, Manbhum and Chottanagpur. Journal of Geological Society of India 12, 286–294

Ray S, Sanyal S, Sengupta P (2011b) Mineralogical control on rheological inversion of a suite of deformed Mafic Dykes from parts of the Chottanagpur Granite Gneiss complex of Eastern India, 263–276

Reddy S, Clarke C, Mazumder R (2009) Temporal constraints on the evolution of the Singhbhum Crustal Province from U-Pb SHRIMP data. In: Mazumder R, Saha D (eds) Paleoproterozoic Supercontinents and global evolution international association for Gondwana research conference series, 9 abstract volume. pp 17–18

Rekha S, Upadhyay D, Bhattacharya A et al (2011) Lithostructural and chronological constraints for tectonic restoration of Proterozoic accretion in the Eastern Indian Precambrian shield. Precambrian Research 187, 313–333

Richards A, Argles T, Harris N et al (2005) Himalayan architecture constrained by isotopic tracers from clastic sediments. Earth and Planetary Science Letters 236, 773–796

Rickers K, Mezger K, Raith MM (2001) Evolution of the continental crust in the Proterozoic Eastern Ghats Belt, India and new constraints for Rodinia reconstruction: implications from Sm-Nd, Rb-Sr and Pb-Pb isotopes. Precambrian Research 112, 183–210

Rode KP (1948) On charnockitic rocks of Palamau, Bihar, India. Schweiz, Min Petr Mitt 28, 288–307

Rogers JJW (1996) A history of continents in the past 3 billion years. Journal of Geology 104, 91–107

Rogers JJW, Santosh M (2002) Configuration of Columbia, a Mesoproterozoic supercontinent. Gondwana Research 5, 5–22

Rogers JJW, Santosh M (2009) Tectonics and surface effects of the supercontinent Columbia. Gondwana Research 15, 373–380

Roy A, Kagami H, Yoshida M et al (2006) Rb – Sr and Sm – Nd dating of different metamorphic events from the Sausar Mobile Belt, central India : implications for Proterozoic crustal evolution. Journal of Asian Earth Sciences 26, 61–76

Roy A, Prasad MH (2003) Tectonothermal events in Central Indian Tectonic Zone (CITZ) and its implications in Rodinian crustal assembly. Journal of Asian Earth Sciences 22, 115–129

Roy A, Sarkarl A, Jeyakumar S et al (2002) Mid-Proterozoic plume-related thermal event in Eastern Indian Craton : evidence from trace elements, REE geochemistry and Sr - Nd isotope systematics of basic-ultrabasic intrusives from Dalma volcanic belt. Gondwana Research 5, 133–146

Roy AB, Kroner A, Bhattachaya PK, Rathore S (2005) Metamorphic evolution and zircon geochronology of early Proterozoic granulites in the Aravalli Mountains of northwestern India. Geological Magazine 142, 287–302

Roy AK (1977) Structure and metamorphic evolution of the Bengal anorthosite and associated rocks. Journal of Geological Society of India 18, 203–223

Roy Chowdhury M (1979) Annual general report of the Geological survey of India, for the year 1972–73. Records of Geological Survey of India 107, 147–151

Saha L, Bhowmik SK, Fukuoka M, Dasgupta S (2008) Contrasting episodes of regional granulite-facies metamorphism in enclaves and host gneisses from the Aravalli–Delhi mobile belt, NW India. 49

Saikia A, Gogoi B, Kaulina T et al (2017) Geochemical and U – Pb zircon age characterization of granites of the Bathani volcano sedimentary sequence, Chotanagpur granite gneiss complex, Eastern India : vestiges of the Nuna supercontinent in the Central Indian Tectonic zone. In: Pant NC, Dsgupta S (eds) Crustal evolution of India and Antarctica: the supercontinent connection. Geological Society, London, Special Publications, 457, 233–252

Saikia A, Gogoi B, Ahmad M, Ahmad T (2014) Geochemical constraints on the evolution of mafic and felsic rocks in the Bathani volcanic and volcanosedimentary sequence of Chotanagpur granite gneiss complex. Journal of Earth System Science 123, 959–987

Sakai H, Iwano H, Danhara T et al (2013) Rift-related origin of the paleoproterozoic kuncha formation, and cooling history of the kuncha nappe and taplejung granites, eastern nepal lesser himalaya: a multichronological approach. Island Arc 22, 338–360

Sanyal S, Sengupta P (2012) Metamorphic evolution of the Chotanagpur granite gneiss complex of the East Indian Shield : current status metamorphic evolution of the Chotanagpur granite gneiss complex of the East Indian Shield : current status. Geological Society of London, Special Publications 365, 117–145

Sanyal S, Sengupta P, Goswami R (2007) Evidence of mesoproterozoic ultra-high temperature metamorphism from parts of CGGC, Jharkhand, India. In: International conference on Precambrian sedimentation and tectonics and Second GPSS meeting. Indian Institute of Technology, Bombay, pp 62–63

Sarangi S, Mohanty S (1998) Structural studies in the Chhotanagpur gneissic complex near Gomoh, Dhanbad district, Bihar. Indian Journal of Geology 70, 73–80

Sarbadhikari AB, Bhowmik SK (2008) Constraining the metamorphic evolution of a cryptic hot Mesoproterozoic orogen in the Central Indian Tectonic Zone, using P – T pseudosection modelling of mafic intrusions and host reworked granulites. 162, 128–149

Sarkar A (1998) Geochronology and geochemistry of mesoproterozoic intrusive plutonites from the eastern segments of the Mahakoshal greenstone belts, Central India. In: Seminar on Precambrian crust in Eastern and Central India, Bhubaneswar, IGCP, Bhubaneswar, pp 82–85

Sarkar AN (1982) Precambrian tectonic evolution of eastern India: a model of converging microplates. Tectonophysics 86, 363–397

Sarkar AN, Jha BN (1985) Straucture, metamorphism and and granite evolution of the Chottanagpur granite gneissic complex. Geological Survey of India Records 113, 1–12

Sarkar SC (2000) Crustal evolution and metallogeny in the eastern Indian craton. Geological Survey of India Special Publications 55, 195–226

Sarkar SN, Saha AK (1962) A revision of the Precambrian stratigraphy and tectonics of Singhbhum and adjacent regions. Quarternary Journal Geological Mining and Metallurgical Society of India 34, 97–136

Sarkar SN, Saha AK (1977) The present status of the Precambrian stratigraphy, tectonics and geochronology of Singhbhum-Keonjhar-Mayurbhanj region, eastern India. Indian Journal of Earth Sciences, S. R. Volume 37–65

Sarkar SN, Saha AK (1983) Structure and tectonics of the Singhbhum-Orissa iron ore craton, eastern India. In: Roy S (ed) Structure and tectonics of Precambrian rocks in India. Hindustan Publishing Corporation, New Delhi, Recent Researches in Geology 10, 1–25

Sarkar T, Schenk V (2014) Two-stage granulite formation in a Proterozoic magmatic arc (Ongole domain of the Eastern Ghats Belt, India): part 1. Petrology and pressure–temperature evolution. Precambrian Research 255, 485–509

Sarkar T, Schenk V (2016) Early mesoproterozoic (1.6–1.5 Ga) granulite facies events in the Ongole domain: geodynamic significance and global correlation. Journal of Metamorphic Geology 34, 765–784

Sarkar SN, Ghosh D, St Lambert RJ (1986) Rubidium-strontium and lead isotopic studies on the soda-granites from Mosaboni, Singhbhum Copper Belt, E. India. Indian Society of Earth Sciences 13:101–116

Sarkar T, Schenk V, Berndt J (2015) Formation and evolution of a Proterozoic magmatic arc: geochemical and geochronological constraints from meta-igneous rocks of the Ongole domain, Eastern Ghats Belt, India. Contributions to Mineralogy and Petrology 169, 1–27

Sarkar, SN(1980) Precambrian stratigraphy and geochronology of Peninsular India: a review. Indian Journal of Earth Sciences 7, 12–26

Sen SK, Bhattacharya A (1993) Post-peakpressure–temperature–fluid history of the granulites around Saltora, West Bengal. In: Proceedings of the National academy of sciences of India, 63(A), 282–306

Sengupta N, Mukhopadhyay D, Sengupta P, Radegund H (2005) Tourmaline-bearing rocks in the Singhbhum shear zone, eastern India: Evidence of boron infiltration during regional metamorphism. American Mineralogist 90, 1241

Sengupta P, Sen J, Dasgupta S et al (1999) Ultra-high temperature metamorphism of metapelitic granulites from Kondapalle, Eastern Ghats Belt: implications for the Indo-Antarctic correlation. Journal of Petrology 40, 1065–1087

Sengupta P, Raith MM, Kooijman E, Talukdar M, Chowdhury P, Sanyal S, Mezger K, Mukhopadhyay D (2015) Provenance, timing of sedimentation and metamorphism of metasedimentary rock suites from the Southern granulite terrane, India. In: Mazumder R, Eriksson PG (eds) Precambrian basins of India: stratigraphic and tectonic context. Geological Society, London, Memoirs 43, 297–308

Simmat R, Raith MM (2008) U-Th-Pb monazite geochronometry of the Eastern Ghats Belt, India: timing and spatial disposition of poly-metamorphism. Precambrian Research 162, 16–39

Singh AP, Kumar N, Singh B (2004) Magmatic underplating beneath the Rajmahal Traps: gravity signature and derived 3-D configuration. Journal of Earth System Science 113, 759–769

Singh RN, Thorpe R, Kristic D (2001) Galena Pb-isotope data of base metal occurrences in the Hesatu-Belbathan belt, eastern Precambrian shield. Journal of Geological Society of India 57, 535–538

Singh S, Claesson S, Jain AK, et al (1994) Geochemistry of the Proterozoic peraluminous granitoids from the Higher Himalayan Crystalline (HHC) and Jutogh Nappe, NW-Himalaya, Himachal Pradesh, India. Journal of Nepal Geological Society 10, 125

Singh Y, Krishna V (2009) Rb–Sr geochronology and petrogenesis of granitoids from the Chotanagpur granite gneiss complex of Raikera–Kunkuri region, Central India. Journal of Geological Society of India 74, 200–208

Sinha AK, Bhattacharya DK (1995) Geochemistry of the magnetite deposits around Sua, Palamau district, Bihar. Journal of Geological Society of India 46, 313–316

Sivaraman T V, Raval U (1995) U-Pb isotopic study of zircons from a few granitoids of Delhi-Aravalli belt. Journal of Geological Society of India 46, 461–475

Som SK, Bandyopadhyay KC, Prasad RK (2007) Cesium enrichment and resource evaluation in aplite and pegmatite at the Southern slope of cesium enrichment and resource evaluation in aplite and pegmatite. Journal of Geological Society of India 70, 273–281

Soni S, Mukherjee AB, Sengupta DK (1991) A new fluorite deposit in the Palamau District, Bihar and the associated iron-fluorine-tungsten skarns and hornfelses. Journal of Geological Society of India 38, 504–510

Srivastava AK, Pandey KH, Kumar G (2000) Geochemical characteristics of the Jhirgadandi granitoid, Sonbhadra district, Uttar Pradesh. In: Proceedings of national seminar on tectonomagamatism, geochemistry and metamorphism of Precambrian terrains, pp 189–199

Srivastava SC, Ghose NC (1992) Petrology of the highgrade gneisses and granites around Chianki, south of daltonganj, district Palamau, Bihar. Indian Journal of Geology 64, 122–142

Torsvik TH, Carter LM, Ashwal LD, Bhushan SK (2001) Rodinia refined or obscured : palaeomagnetism of the Malani igneous suite (NW India). Precambrian Research 108, 319–333

Torsvik TH, Smethurst MA, Meert JG et al (1996) Continental break-up and collision in the Neoproterozoic and Palaeozoic—a tale of Baltica and Laurentia. Earth-Science Reviews 40, 229–258

Treloar PJ, Rex DC (1990) Cooling and uplift histories of the crystalline thrust stack of the Indian Plate internal zones west of Nanga Parbat, Pakistan Himalaya. Tectonophysics 180, 323–349

Tsunogae T, Dunkley DJ, Horie K et al (2014) Petrology and SHRIMP zircon geochronology of granulites from Vesleknausen, Lützow-Holm complex, East Antarctica: Neoarchean magmatism and Neoproterozoic high-grade metamorphism. Geoscience Frontiers 5, 167–182

Tsunogae T, Yang QY, Santosh M (2016) Neoarchean–Early Paleoproterozoic and early Neoproterozoic arc magmatism in the Lützow–Holm complex, East Antarctica: insights from petrology, geochemistry, zircon U–Pb geochronology and Lu–Hf isotopes. Lithos 263, 239–256

Upadhyay D, Raith MM (2006) Intrusion age, geochemistry and metamorphic conditions of a quartz-monzosyenite intrusion at the craton-Eastern Ghats Belt contact near Jojuru, India. Gondwana Research 10, 267–276

Upadhyay D, Raith MM, Mezger K, Hammerschmidt K (2006) Mesoproterozoic rift-related alkaline magmatism at Elchuru, Prakasam Alkaline Province, SE India. Lithos 89, 447–477

Upreti BN, Rai SM, Sakai H et al. (2003) Early Proterozoic granite of the Taplejung Window, far eastern lesser Nepal Himalaya. Journal of Nepal Geological Society 28, 9–18

Vadlamani R, Kröner A, Vasudevan D et al (2013) Zircon evaporation ages and geochemistry of metamorphosed volcanic rocks from the Vinjamuru domain, Krishna Province: evidence for 1.78 Ga convergent tectonics along the southeastern margin of the Eastern Dharwar Craton. Geological Journal 48, 293–309

Vernon RH (2004) A practical guide to rock microstructure. 594 p. Cambridge University Press. ISBN 0 521 89133 7

Vijaya Kumar K, Frost CD, Frost BR, Chamberlain KR (2007) The Chimakurti, Errakonda, and Uppalapadu plutons, Eastern Ghats Belt, India: an unusual association of tholeiitic and alkaline magmatism. Lithos 97, 30–57

Vinogradov A, Tugarinov AL, Zhykov C et al (1964) Geochronology of Indian Precambrian. In: Report of the 22nd international congress, New Delhi. pp 553–567

Wani H, Mondal MEA (2016) Geochemical evidence for the Paleoproterozoic arc–Back arc basin association and its importance in understanding the evolution of the Central Indian Tectonic Zone. Tectonophysics 690, 318–335

Wanjari NR, Chaturvedi R, Mahanta DN (2012) Specialised thematic mapping in Munger–Rajgir Group of rocks to examine structural and stratigraphic set up in and around Gaya–Rajgir areas in parts of Gaya, Nawada and Jahanabad districts of Bihar

Weil AB, Van der Voo R, Mac Niocaill C, Meert JG (1998) The Proterozoic supercontinent Rodinia: paleomagnetically derived reconstructions for 1100 to 800 Ma. Earth and Planetary Science Letters 154, 13–24

Wiedenbeck M, Goswami JN, Roy AB (1996) An ion microprobe study of single zircons from the Amet granite, Rajasthan. Journal of Geological Society of India 48, 127–137

Worsley TR, Moody JB, Nance RD (1985) Proterozoic to recent tectonic tuning of biogeochemical cycles. Carbon cycle atmospheric CO natural variations archean to present, 561–572
Worsley TR, Nance RD, Moody JB (1986) Tectonic cycles and the history of the Earth's biogeochemical and paleoceanographic record. Paleoceanography 1, 233–263
Yadav BS, Wanjari N, Ahmad T, Chaturvedi R (2014) Geochemistry and petrogenesis of proterozoic granitic rocks from northern margin of the chotanagpur gneissic complex (CGC). Journal of Earth System Science 125, 1041–1060
Yedekar DB, Jain SC, Nair KKK, Dutta KK (1990) The Central Indian collision suture. Precambrian of Central India. Geological Survey of India Special Publications 28, 1–37
Yin A, Dubey CS, Kelty TK et al (2010) Geologic correlation of the Himalayan orogen and Indian craton: part 2. Structural geology, geochronology, and tectonic evolution of the eastern Himalaya. Bulletin of Geological Society of America 122, 360–395
Zeitler PK, Sutter JF, Williams IS et al (1989) Geochronology and temperature history of the Nanga Parbat–Haramosh massif, Pakistan. Geological Society of America Special Paper 232, 1–22
Zhang S, Li ZX, Evans DAD et al (2012a) Pre-Rodinia supercontinent Nuna shaping up: a global synthesis with new paleomagnetic results from North China. Earth and Planetary Sciences Letters 353–354, 145–155
Zhang SH, Zhao Y, Liu XC et al (2012b) U-Pb geochronology and geochemistry of the bedrocks and moraine sediments from the Windmill islands: implications for Proterozoic evolution of East Antarctica. Precambrian Research 206–207, 52–71
Zhao G, Cawood PA, Wilde AS, Sun M (2002) Review of global 2.1-1.8 Ga orogens: implications for a pre-Rhodinia supercontinent. Earth-Science Reviews 59, 125–162

Geomorphic Characteristics and Morphologic Dating of the Allah Bund Fault Scarp, Great Rann of Kachchh, Western India

Akash Padmalal, Nitesh Khonde, D. M. Maurya, Mohammedharoon Shaikh, Abhishek Kumar, Naimisha Vanik and L. S. Chamyal

1 Introduction

The 16 June 1819 Allah Bund earthquake is the largest known earthquake in western India that produced spectacular geomorphological changes in its epicentral area in the Great Rann of Kachchh (Lyell 1855). The earthquake though not instrumentally recorded, is described in several historical accounts (MacMurdo 1824; Baker 1846; Wynne 1872; Oldham 1926). The earthquake produced ~90 km long E-W trending and south facing low fault scarp to its north, blocking the distributary of the Indus river that flowed through the Kori creek into the Arabian sea and caused subsidence of the area around Sindri, which was submerged by a local tsunami. The scarp is located in the logistically challenging and largely inaccessible saline terrain of Great Rann, which shows peculiar geomorphic characters (Roy and Merh 1982; Merh 2005). This is the most likely reason for the fact that very few field-based studies have been carried out on the Allah Bund Fault scarp (Rajendran and Rajendran 2001; Thakkar et al. 2012). The Great Rann is a tectonically formed basin bounded by the Kachchh Mainland Fault (KMF) in the south and the Nagar Parkar Fault (NPF) in the north (Biswas 1987). Towards the west the basin opens up to the Arabian Sea and the Indus delta region. The basin formed a major sink for sediments during Pleistocene-Holocene that are dominantly fine-grained (clayey silt to silty clay) and presumably sourced from the Himalaya through the now extinct Vedic Saraswati river and the Indus delta and deposited in shallow marine marginal gulf environment (Valdiya 2002; Maurya et al. 2013; Khonde et al. 2017a, b). In this article, we describe the geomorphological characteristics along the Allah Bund Fault scarp in detail and demonstrate the appli-

A. Padmalal · N. Khonde · D. M. Maurya (✉) · M. Shaikh
A. Kumar · N. Vanik · L. S. Chamyal
Department of Geology, The M. S. University of Baroda,
Vadodara 390002, Gujarat, India
e-mail: dmmaurya@yahoo.com

S. Mukherjee (ed.), *Tectonics and Structural Geology: Indian Context*,
Springer Geology, https://doi.org/10.1007/978-3-319-99341-6_3

cation of morphological dating technique to determine the age of the scarp. We begin with a detailed description of the enigmatic landscape characteristics of the Great Rann to understand the geomorphic setting and evolution of the Allah Bund Fault scarp.

2 Great Rann of Kachchh

The major geomorphic components of the Great Rann are the almost flat and hyper-saline surface, hereafter described as the Rann surface, several islands (locally called 'bet') of different shapes and sizes and the $\sim$ E-W Allah Bund scarp (Fig. 2). Part of the extensive Rann surface is salt-encrusted while the remainder is free of salt crust though the sediments are inherently saline. The most overbearing part of the Rann is the flatness of the Rann surface that lies 2–6 m amsl. The Rann surface consists of several large to small islands that remain above the submergence level. There are several smaller islands rising up to 1–10 m above the Rann surface, especially in the northern part of the Great Rann and consist of sediments resembling the Rann surface raised to a higher level (Khonde et al. 2013). The top cover of these islands is usually made up of 1–2 m thick aeolian sediment blown from the wind-swept Rann surface.

The major factor responsible for the unique present day environmental conditions of the Rann surface is its periodic submergence by sea water and annual monsoon precipitation. The flat Rann surface and the negligible westward gradient allows extensive inundation by sea water from the Arabian sea in the west and by river waters from the northeast and south during the monsoon season (Glennie and Evans 1976; Roy and Merh 1982). The submergence pattern of the Rann surface is however not uniform as it shows very small variations in elevation which has resulted in variable geomorphic characteristics. The Great Rann is therefore geomorphologically divisible into four geomorphic units (Fig. 2 and 3) the Banni plain, Supra tidal salt flat, Inland saline flat and the Bet zone. Owing to imperceptible gradient, the boundaries between the various geomorphic units are gradational.

2.1 *The Banni Plain*

The Banni plain is a vast flat terrain, highly vegetated with thorny shrubs and grasses, and extending from the mainland Kachchh in the south and the Pachham island in the north (Fig. 1a). It is a distinct geomorphic surface of the Great Rann that occurs at the highest elevation and is consequently completely free of present day marine influence (Figs. 2, 3 and 4a). The entire terrain of the Banni plain is regarded as a vast raised mudflat (Kar 1995) that coincides with the subsurface

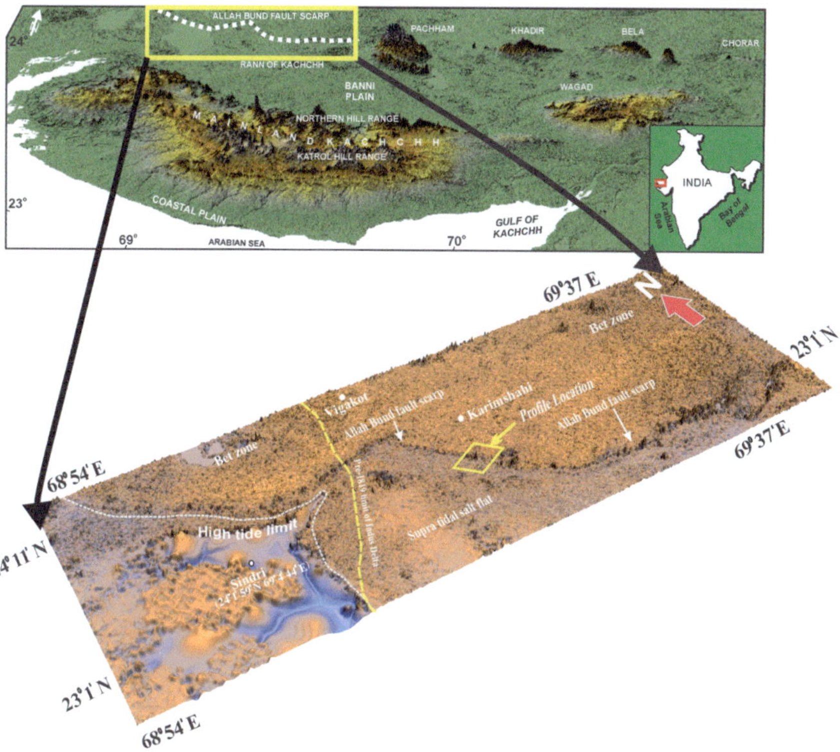

Fig. 1 **a** Digital elevation model (DEM) of Kachchh region prepared from SRTM data (http://srtm.csi.cgiar.org). DEM prepared by Patidar (2010). Inset-Location map. **b** Digital Elevation Model (DEM) of Allah Bund Fault Scarp prepared from SRTM data (http://srtm.csi.cgiar.org) showing the various geomorphological characteristics. Profile location indicated is for profiles given in Fig. 8

Median High. Based on variations in elevation, Kar (1995) has divided the Banni plain into three sub-units-high level mudflat, an undifferentiated and sloping low level mudflat and a residual depression.

2.2 *Supra Tidal Salt Flat*

This is a vast but linear and narrow E-W trending low lying zone with several centimeters thick salt crust between the Banni plain in the south and the Bet zone to the north (Fig. 2 and 4b–d). This zone occurs at the lowest elevation (~ 2 m) in the west and gradually rises to ~ 4 m towards east. Being at the lowest elevation, this zone forms the main pathway through which the saline waters of the Arabian sea in the west enter and spread out (Roy and Merh 1982) submerging about two-thirds of the Rann surface to varying degrees depending upon the volume/magnitude of the

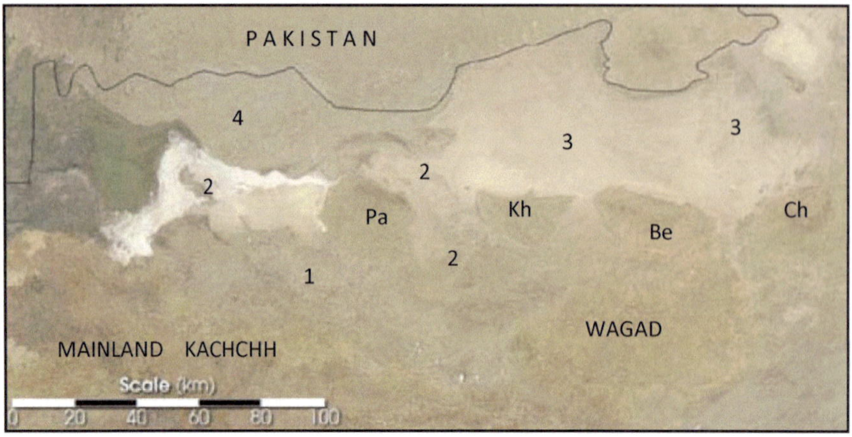

Fig. 2 Satellite image of Great Rann of Kachchh basin showing the variations in surface morphology. The image shows the completely dry Rann surface as it appeared in the extreme arid season in May, 2003. *Source* www.earthobservatory.nasa.gov. The geomorphological divisions of the Great Rann (1–4) are also indicated. 1-Banni plain, 2-Supra tidal salt flat, 3-Inland saline flats and 4-Bet zone. (Pa-Pachham island, Kh-Khadir island, Be-Bela island, Ch-Chorar island)

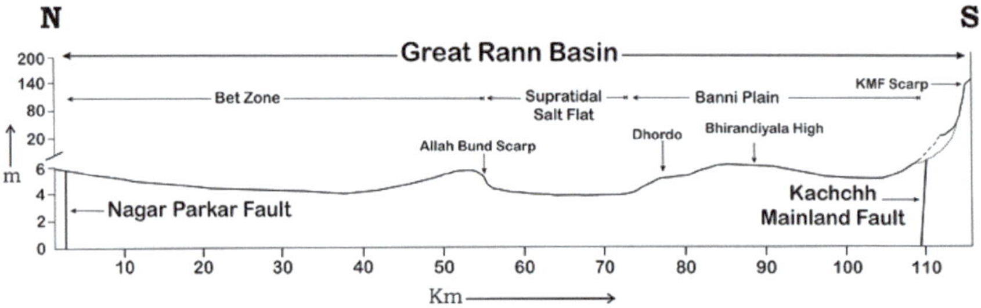

Fig. 3 N-S topographic profile across the Great Rann basin showing the geomorphic divisions. Vertical scale is highly exaggerated. The elevation data is based on the SOI topographical maps (survey years-1960–66)

Fig. 4 **a** Photograph showing the typical nature of the surface of the Banni plain. Sand storms as seen in the picture are a common sight during the peak summers. **b** Photograph showing the typical extensive flat surface of the supra tidal salt flat. The salt crust is the result of regular marine inundation of the surface. **c** Close view of the large polygonal cracks in the salt crust. **d** Photograph showing the thickness (~10 cm) of the salt crust. **e** View of the Inland saline flat to the north of Bela island. The scarp in the background marks the geomorphic expression of the Island Belt Fault (IBF). **f** View of the typical salt crust free surface of the Bet zone. **g** Northward view of the developing gullies in the northern most part of the Bet zone. The northward upslope nature of the surface is also clearly visible. **h** Surface of the Bet zone covered with numerous bivalve shells which thrive during periods of submergence ▶

(a)
(b)
(c)
(d)
(e)
(f)
(g)
(h)

ingression resulting in a thick salt crust (Fig. 4d). Evaporation to dryness results in several centimeteres thick, residual salt crusts, which characterize the surface of the supra tidal salt flats (Fig. 4b–d).

2.3 Inland Saline Flat

This zone comprises the easternmost part of the Great Rann (Fig. 2 and 4e) that is not influenced by marine submergence, but is inundated by monsoon precipitation and by the rivers from the east and north. The elevation rises towards the margins of the zone, giving it a shallow bowl like morphology and comprise inherently saline Rann sediments (Roy and Merh 1982).

2.4 The Bet Zone

It comprises the flat Rann surface in the northwestern part of the Great Rann shows several bets occurring few metres above the Rann surface (Fig. 4f–h). The Bet zone is delimited by the Kuar and Bedia bets in the east and the supra tidal salt flat in the south (Fig. 2). In addition to the bets with well defined margins, several elevated parts within the Rann surface also exist which gradually rise above the Rann surface. Many of these are comparable to the bets in terms of their size. Their surfaces comprise aeolian sediments and generally support small vegetation like scrubs and grasses of various size. Morphologically, the shapes of all bets and almost all elevated tracts described above are elongated in N-S direction, which, on a map gives the misleading impression of wide N-S trending channels separated by bets. However, these are not 'channels' in any sense as they are basically flat Rann surfaces several tens of kilometers wide comprising the inter-bet regions (Khonde et al. 2013). The surface of the Bet zone shows a very gradual northward slope away from the Allah Bund scarp which testifies to the uplift along the ABF (Fig. 3).

The western part of the Bet zone i.e. the area around Vigukot and further west shows submergence characteristics which is slightly different from the rest of the Bet zone and the Great Rann (Khonde et al. 2013). The 1819 Allah Bund earthquake caused drastic geomorphological changes along with subsequent flooding events of the now defunct distributary of Indus River (Burnes 1835; Oldham 1926; Bilham 1998). The now disconnected distributary of Indus—the Nara river/Pooran flowed into this region and joined Arabian sea through the Kori creek in the south before the earthquake. This part therefore formed the eastern margin of the Indus Delta till the 1819 earthquake changed the morphology of the area. Presently also, the region to the west and SW of Vigukot is prone to flooding frequently by river floods from the north and relatively less frequently by sea water influx from the south. In addition numerous small shallow ($\sim$0.5 m deep) channels of uncertain affinity also exist. Overall, the role of rivers is evidently more pronounced in the western most part than the rest of the Bet zone. Contrary to the rest of the Great Rann, the Bet zone shows wide variation in elevation due to the presence of bets and is characterized by several seasonal short distance channels, pools i.e. local depressions and elevated surfaces (Khonde et al. 2013).

The E-W trending intrabasinal structural features within the Great Rann like the Island Belt Fault (IBF), the Allah Bund Fault (ABF) and the subsurface Banni faults correlate remarkably with the geomorphic divisions described above. A critical evaluation of the elevation differences of the various geomorphic divisions of the Great Rann shows a strong correlation with the above mentioned regional structural elements worked out by Biswas (1987, 1993). All geomorphological units of the Great Rann show prominent structural control (Maurya et al. 2013). The Banni plain astrides the subsurface Median High, while the supra tidal salt flat and the inner saline flats occupy a structural depression to the north of the subsurface Bhirandiyala high and the Island Belt Fault (IBF). The supra tidal flats extends into a depression to the south of the island belt between the Pachchh and Khadir islands, the Wagad highland to the east and the mainland in the south. The Bet zone occurs on the upthrown northern block of the roughly E-W trending Allah Bund Fault.

Based on the distinct geomorphic divisions attributed to variable submergence pattern and their correlation with structural elements, it is inferred that the emergence of the Rann surface may have occurred gradually due to differential tectonic activity along the various subsurface intrabsinal faults in the recent past, which formed the morphologic units viz. the Banni plain, the Bet zone and the supra tidal salt flat (Maurya et al. 2013, 2016). The Banni plain astrides the subsurface Median high and is separated from the remainder of the Rann basin by the Banni Fault to the north (Biswas 1974). Similarly, the Bet zone is delimited by the Allah Bund Fault to its south (Roy and Merh 1982). The close association of these units with faults suggest differential tectonic activity along subsurface faults within the Great Rann basin may have played a major role in the emergence of various morphologic units at different times (Maurya et al. 2013). Based on elevation and present day submergence characteristics, the Banni plain appears to be the first to emerge followed by the Bet zone, the inner saline flat and the supra tidal salt flat, which still gets submerged by marine waters regularly.

3 The 1819 Allah Bund Earthquake

The 16 June, 1819 Allah Bund earthquake is considered as a major intraplate earthquake event (Johnston 1989) which produced remarkable surface deformation in the Great Rann of Kachchh. The shocks were felt almost all over the Indian subcontinent (MacMurdo 1824). The event is specifically well known for the extensive destruction and ground deformation. The event that killed 1543 inhabitant was estimated to be of intensity of XI (Oldham 1926). Intense damage was reported from all over Kachchh including Bhuj and Anjar towns. Far away cities such as Ahmadabad also suffered damage to buildings and monuments. Based on the intensity report, Gutenberg and Richer (1954) assigned a magnitude of 8.4 to the event. Later on Chandra (1977) and Johnston and Kanter (1990) estimated the magnitude to be 7.8 and epicenter on 23.6° N and 69.6° E, while Quittmeyer and Jacob (1979) estimated the epicenter at 24° N and 69° E. According to Bilham (1998), the 1819 event was a near the surface (0–10 km depth) reverse-slip rupture on a 90-km-long, 50° to 70° N-dipping fault plane which matches the measured elevation changes from the event. To et al. (2004) estimated a 50-km-long rupture dipping 45 to the north with 3–8 m slip.

Prior to the 1819 earthquake, the region was a flourishing trade route extending northward into Sind through Kori creek and the distributary of the Indus (Grindlay 1808). The Sindri fort was a major halting point and a revenue collection centre along this route. Extensive surface effects resulting due to the 1819 earthquake have been documented in several historical records (MacMurdo 1824; Baker 1846; Wynne 1872; Oldham 1926; Frere 1870. A detailed review of historical accounts for this earthquake is available in Bilham (1998). The earthquake produced an astonishing variety of surface effects that includes vertical movements of the ground, flooding of regions near sea level, widespread liquefaction, and a local tsunami and the complete damming of a distributary of the Indus river. Two significant changes occurred in the region around Sindri fort at the western part of Great Rann. The foundation of fort and surrounding Rann surface subsided by more than 1 m and the region 7 km north of fort got elevated to 3–6 m, shutting out northward navigation into the Sind province due to blocking of the distributary of Indus river. The northern region (the present Bet zone) was uplifted forming a 10–15 ft high south facing scarp which caused the complete damming of the river. The natural dam so formed was named by local people as the Allah Bund (dam of God). The scarp presently marks the sharp boundary between the Bet zone in the north and the active tidal flats of Kori creek and the Bet zone Notably all the above surface changes documented in the available historical records are from the western part of the Great Rann of Kachchh. However, it is also mentioned that several temporarily active rivers with subsequent sand venting caused transient flooding of major part of Rann surface (Baker 1846; Oldham 1926).

4 Nature and Morphology of Allah Bund Fault Scarp

The Allah Bund Fault scarp is an elongated steeply dipping south facing scarp bordered on the south by the salt encrusted surface of Rann. The Bet zone comprises the flat Rann surface in the northwestern part of the Great Rann shows several bets occurring few metres above the Rann surface. The Bet zone is delimited by the Kuar and Bedia bets in the east and the supra tidal salt flat in the south. To the north lie the sand ridges of Sind (Pakistan). Towards the west, the Bet zone imperceptibly merges with the Indus delta. Historical documents suggest that the south facing fault scarp was produced during the 1819 earthquake, causing uplift of the northern part of the Great Rann (Burnes 1835; Oldham 1926),

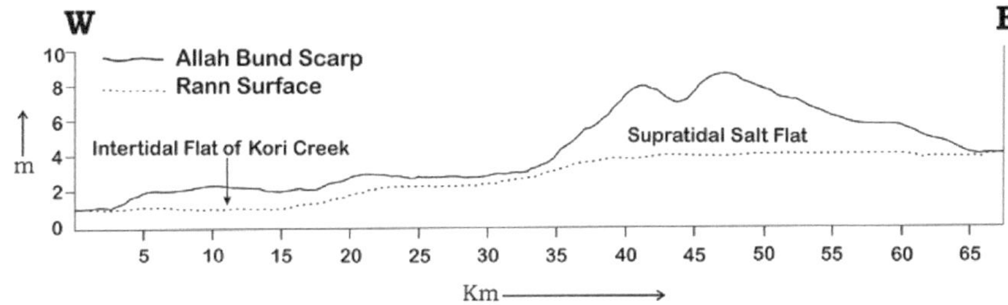

Fig. 5 Topographic cross sections drawn across the E-W trending Allah Bund scarp. The top profile is from the western extremity while the bottom one is from the eastern extremity of the scarp. Vertical scale is highly exaggerated. The elevation data is based on the SOI topographical maps (survey years-1960–66)

which corresponds with the present day Bet zone. The scarp trends E-W and laterally extends for ~90 kms (Oldham 1926). However, presently the scarp is visibly identifiable for about ~60 kms length in the western and central part of the extension as per Oldham (1926). In the westernmost part, the scarp rises gently northward rise of the ground above the intertidal flats of the Kori creek (Figs. 5 and 6). Further east, the scarp gains elevation and therefore is more distinctly identifiable (Fig. 5). The scarp height continues to increase eastward with the highest elevation recorded ~4–5 m from the supra tidal salt flat in the central part (Figs. 5 and 6). Beyond this the scarp again gradually reduces in height and finally disappears in the flatness of the Rann surface to the southeast of Shakti bet. As per Oldham (1926), the scarp continued eastward where it is presumably represented by the southern cliffy margins of the Gainda bet, Mori bet and the Kuar bet. Sediments of the bund consist of thinly laminated alternate layers of dark brown silt and clays encrusted with salt and ferruginous tubules; these differ from the land-derived fluvial bet sediments that impinge against the bund in the north (Rajendran and Rajendran 2001).

At places, the height of the scarp is accentuated by the deposition of 1–2 m thick wind-blown saline silty sediments. This especially observed in the central part where the scarp attains highest elevation. A major significant characteristic that defines the scarp as an erosional scarp is its deeply gullied nature (Thakkar et al. 2012). The gullies are 1–3 m deep and usually form a dendritic pattern over the crest of the scarp. Another significant characteristic feature of the scarp is the subvertical northward face of the scarp surface, which finally merges with the flat Rann surface. The northward slope developed over the scarp is attributed to the back tilting of the Rann surface due to upliftment along the scarp. The backtilting being subtle is not recognisable in the field at many places as the elevation drop is marginal (1–2 m) that occurs over the distance of few kilometers (Fig. 6). The back tilted surface over the scarp suggests that the scarp is tectonically formed and is in sharp contrast to the flat surface of the Great Rann.

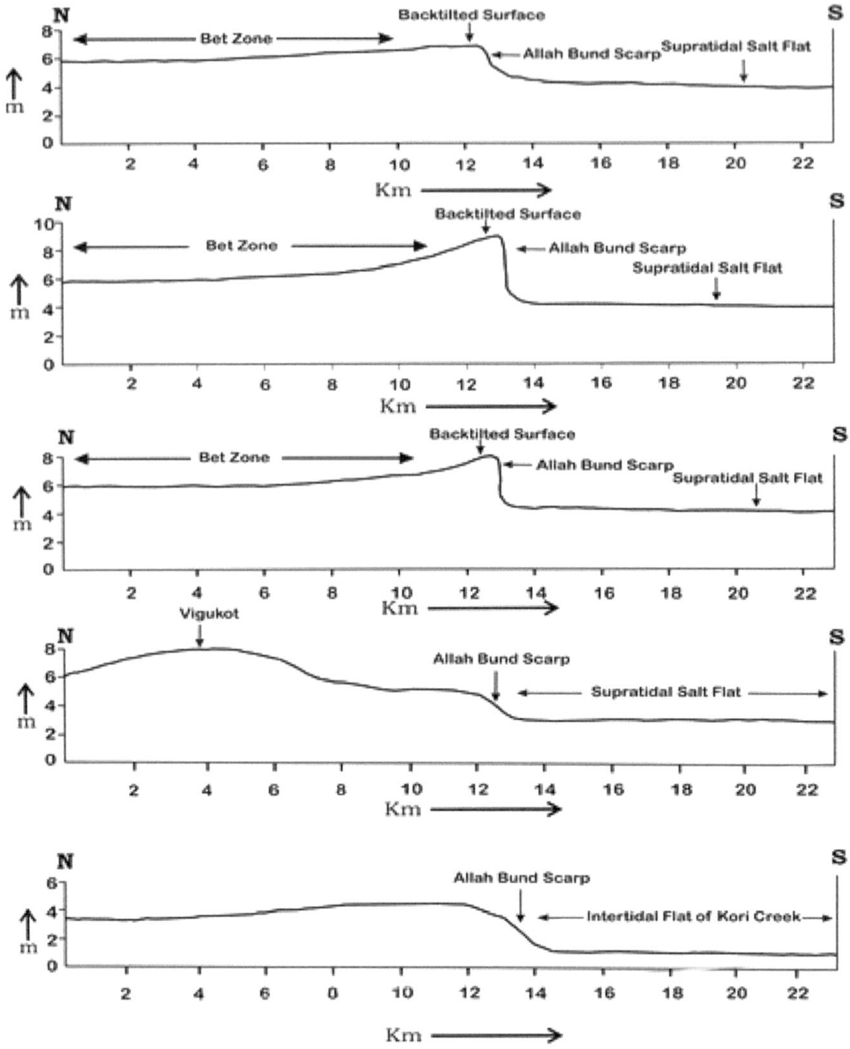

Fig. 6 N-S topographic profile across the Great Rann basin showing the geomorphic divisions. Vertical scale is highly exaggerated. The elevation data is based on the SOI topographical maps (survey years-1960–66)

5 Morphological Dating of the Allah Bund Fault Scarp

Numerical dating methods have opened a way for estimation of age of geomorphological features such as fault scarps and river terraces (Wallace 1977). This method has been successfully used in determining the age of the fault scarps in Southern Arava, Israel (Enzel et al. 1994), West Yellowstone, Montana (Nash 1984), Lost River fault in Custer County, Idaho (Hank and Schwartz 1987), and terrace risers in Rhine Graben (Niviere and Marquis 2000). In tectonically active areas, this method characterizes the degradation of a geomorphic marker based on rate of vertical movement and erosion (Carretier et al. 2002). The evolution of a cohesionless slope through time can be evaluated quantitatively by assuming that sediment transport is a diffusion process (Colman and Watson 1983). Generally, the word diffusion describes how chemicals in solution move from areas of high concentration to areas of low concentration, and how heat moves from areas of high temperature to areas of low temperature. The sediment diffusion indicates the gravity transport of sediment from an area of high elevation (the top of the slope) toward an area of low elevation (the base of the slope). Without fresh uplift or down cutting, diffusion tends to make slopes smoother and less steep temporally. Scarp evolution is controlled by climate and several other geological factors. Measure of age dependent parameters such as maximum scarp slope or the curvature of crest will point towards the age (Wallace 1977). Under conditions in which the transport efficiency is uniform and the flux is dictated linearly by the local slope, the combination of the mass flux and mass conservation equations results in the diffusion equation (Colman and Watson 1983).

$$\kappa t = \frac{d^2}{4\pi} \frac{1}{(\tan\theta - \tan\alpha)^2}, \quad \text{(Colman and Watson 1983)} \qquad (1)$$

This solution and model (Eq. 1) is applied as discussed in Colman and Watson (1983) is applied to determine the morphological age of the Allah Bund Fault scarp. The parameters applied (Fig. 7) in the equation are given in Table 1.

Over a time, the morphology of a scarp is altered by gravitational and slope processes (Enzel et al. 1994). For the most part of this time, it follows geomorphic process of slope modification i.e. mass wasting and wash processes (Summerfield 1991). Fault scarps that cut unconsolidated sediment or soil are very promising for

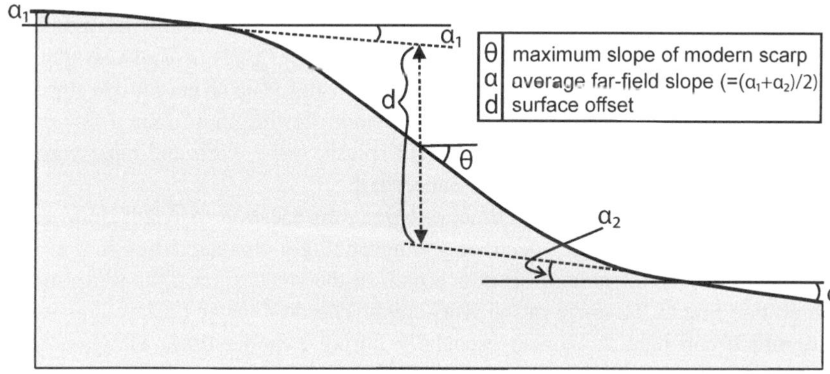

Fig. 7 Schematic model of a fault scarp showing the various input parameters to be measured for the solving the diffusion equation (after Colman and Watson 1983)

Table 1 Details of parameters utilized in solving diffusion equation (after Colman and Watson 1983)

Parameter	Explanation	Unit
κ	Diffusivity	m^2/year
t	Time	Years
d	Vertical displacement on a scarp	Meters
Π	Pi = 3.14159	None
θ	Maximum scarp slope angle	Degree
α	Average far field slope	Degree

diffusion modeling as they form instantly and then systematically degrade. In practice, several criteria must be met for a solution to be possible (Colman and Watson 1983):

(A) The scarp must be transport-limited.
(B) After the formation of scarp, it must degrade under constant conditions.
(C) Overall landscape level and constant scarp base level should be maintained.
(D) The scarp must have formed in a single rupture event.

Therefore, it is apparent that selecting an ideal value for the diffusivity constant or rate constant (κ) is the key aspect of morphologic dating of fault scarp. Several studies of morphologic dating in arid regions have implied lower values of κ and vice versa (Begin 1992). Altogether κ value should be a constant, which satisfies these slope degradation parameters.

5.1 *Morphological Dating Parameters and Site Selection*

For applying morphologic dating technique to the Allah Bund Fault scarp, it is essential to appreciate its peculiar climato-geomorphic setting. The Kachchh basin is located in the hyper-arid zone of western India that includes the Thar desert to its north. The region consequently receives very little rainfall during the monsoon season. This is testified by the rocky topography of other parts of Kachchh that are literally free of any vegetation except the thorny bushes. The rivers too are strongly ephemeral with water flow during monsoon seldom lasting more than a few days. We therefore consider negligible role of rain splash, slope wash and mass wasting in the degradation of the Allah Bund Fault scarp.

Apart from the extremely dry climatic regime, the location of Allah Bund Fault scarp in the Great Rann with peculiar geomorphologic characteristics is also significant as far scarp degradation is concerned. In the western part, the scarp marks the high tide line to the north of the Kori creek. The tidal water flushes against the scarp, which can lead to erosion especially during extreme tides. However, this process is limited to the western part only that is also the zone of the maximum recorded surface effects during 1819 earthquake. As a result, the scarp suffered maximum degradation in this part as evidenced by its very low height. The highly gullied nature of the scarp in this part also testifies to the relatively higher magnitude of scarp degradation.

The scarp in the central and eastern parts is located above the high tide line. Here, the scarp witnesses periodic submergence of the vast supratidal salt flat in front of it. The submergence however is generally < a meter and rarely reaches the base of the scarp. We therefore rule out erosion by the submerging waters of the scarp in this part. The Bet zone also gets submerged by thin sheet of rain water that flows along the general southward slope. These waters are responsible for the formation of gullies on top of the scarp. However, as mentioned above, the scarp is highly gullied in the western part, whereas the gullies are significantly less in the central and eastern part. This is due to the presence of rivers from the north dissecting through the scarp to drain into the intertidal flats of the Kori creek. However, there are no such drainages in the central and eastern parts resulting in significantly less number of gullies, which are also shallower and wide spaced.

Also, the height of the Allah Bund Fault scarp is the maximum in the central part. In the eastern part, the scarp height gradually reduces before finally disappearing in the flat Rann surface. Evidently therefore that the scarp has undergone least degradation in the central part.

Taking into account the climatic and geomorphologic factors as discussed above, ~2 km length in the central part of the Allah Bund Fault scarp is most suitable for morphologic dating. Moreover, the scarp in this part can be included in the transport limited category of slope classification i.e. transportation process occurs much slower than the weathering. For the same reasons, we assume negligible sediment transportation rate and diffusion constant for calculating the age of the Allah Bund Fault scarp. Previous studies show that faults that cut unconsolidated sediment are the most promising ones for diffusion modeling because it forms instantaneously and will systematically degrade (Colman and Watson 1983). This criterion is met in the case of Allah Bund scarp, which makes it relatively easier to quantify. Further, the scarp is not affected by any anthropogenic influences as the Great Rann is an inhospitable terrain due to its difficult environmental conditions.

Three sites were selected from the central part of the Allah Bund Fault scarp that shows maximum elevation, is sub-vertical and has clear-cut morphological expression. The general shape of the scarp is determinable by a single profile without the need of large sample sizes. The selected scarp face is away from the vicinity of notches, gullies and local channels. The base of the selected scarp locations are devoid of colluvium or small fan deposits, which can create irregularities in the scarp profile. The thickness of post-Allah Bund aeolian deposit is removed from the profile to get the original scarp height. The age of the scarp was calculated for these three sites as outlined by Colman and Watson (1983) and Nash (1987).

6 Dating Results

In this study the application of diffusion equation is used to estimate the age of Allah Bund scarp. Topographic profiles were prepared using the Survey of India topographical maps (Nos. 40 H/4, 40 H/8) and the SRTM Digital Elevation Model data (Fig. 8). The study site is located near the north of Rann, an area with extreme arid condition. The climate of the region is consistent since last ~2000 years after the marked withdrawal of high sea (Merh and Patel 1988). The salt encrusted sediment deposits are consistent with such conditions. The Late Holocene sediments, which make the scarp, consist of fine alluvial (silty clay and clayey silt). Maximum slope angles measured along the free face at all three places were above 80°. This is consistent with the almost vertical nature of the Allah Bund Fault scarp. Steeper slope angles further suggest a much younger age for the scarp (Wallace 1977; Bucknam and Anderson 1979). Average far field (α) is calculated

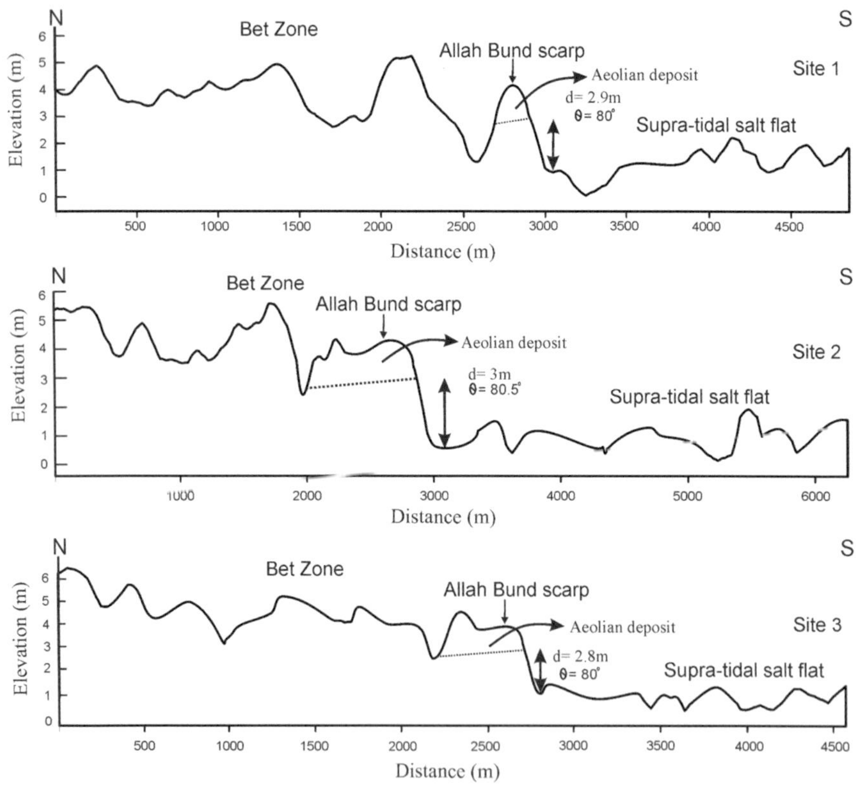

Fig. 8 Topographic profiles drawn at three sites across the Allah Bund Fault in the central part. Location of the profiles is marked in Fig. 1b. Vertical scale highly exaggerated. Profiles were drawn from SRTM data (http://srtm.csi.cgiar.org). The thickness of aeolian deposit capping the scarp is as measured in the field

by averaging the base and crest far field angles. Since there are no previous studies available on the rates of sediment deposition, transportation and erosion, we used a estimated regional diffusion constant generally applied to regions of high aridity (e.g. Northern Negev, Israel, Begin 1992). The minimum diffusion constant value of $1 \times 10^{-4} m^2 year^{-1}$ matches with the geomorphological scenario of the scarp as described above. The morphologic ages were calculated using Eq. 1.

Site 1

The topographic profile drawn over the Allah Bund Fault scarp along 24° 5′ 29″N, 69° 19′ 42″E to 24° 3′ 37.6″N, 69° 17′ 20.5″E is shown in Fig. 8. The 2.4 km long N-S transect shows a near vertical (80°) scarp with measured height of 2.9 m above the ground level. The thickness of the aeolian sediment cover is subtracted to get the actual height of the scarp. As described previously the average far field value is negligible ($\alpha = 0°$). The scarp marks the abrupt change in the topography from the supra-tidal flat to the elevated surface of the Bet zone. The morphologic age of the scarp is determined using Eq. 1 as below.

$$1 \times 10^{-4} \times t = \frac{8.41}{4 \times 3.14159} \times \frac{1}{(\tan 80)^2}$$

$$t = 208 \text{ years.}$$

Site 2

Figure 8 shows the topographic profile across the Allah Bund scarp at this site (24° 5′ 40.7″N, 69° 19′ 31.7″E to 24° 3′ 9.94″N, 69° 17′ 3.2″E). The near vertical scarp (80.5°) here gave a measured height of 3 m above the supra-tidal salt flat. The thickness of the aeolian sediment cover was substracted to get the correct scarp height. Taking the average far field value as negligible ($\alpha = 0°$), the morphologic age is determined using Eq. 1 as below.

$$1 \times 10^{-4} \times t = \frac{9}{4 \times 3.14159} \times \frac{1}{(\tan 80.5)^2}$$

$$t = 200 \text{ years.}$$

Site 3

The topographic profile was drawn along a N-S transect along 24° 5′ 38.5″N, 69° 19′ 8.60″E to 24° 3′ 22.95″N, 69° 17′ 22.69″E. The profile shows a 2.5 km long transect with well demarcated sharp contrast in the topography marked by the Allah Bund Fault scarp (Fig. 8). The scarp marks the abrupt change in elevation of Bet zone in the upthrown block and the flat monotonous supra-tidal salt flat in the downthrown block to the south. The total height of the scarp is 4 m, however, the correct scarp height is determined as 2.8 m after subtracting the thickness of the aeolian sediment cover. The near vertical scarp (80°) and the negligible average far field value is used to determine the morphological age using Eq. 1 as below.

$$1 \times 10^{-4} \times t = \frac{7.84}{4 \times 3.14159} \times \frac{1}{(\tan 80)^2}$$

$$t = 193 \text{ years.}$$

Here values of t obtained represent the morphological age i.e. the time since the fault scarp was formed. The morphological ages of the Allah Bund Fault scarp obtained at the above three sites are 208, 200 and 193 years before present. These ages indicate that the scarp was formed in 1809, 1817 and 1824 A.D. respectively. The results unequivocally suggest that the scarp was produced during the 1819 earthquake event.

7 Concluding Discussion

The landscape of the Great Rann of Kachchh shows unique geomorphological characteristics (Roy and Merh 1982). The high intensity earthquakes during historic and prehistoric times together with marine processes from the west have governed the geomorphic evolution of the Great Rann (Roy and Merh 1982; Glennie and Evans 1976). Looking to the several devastating earthquake that have occurred along the various faults in Kachchh during last 200 yrs viz. in 1819, 1845, 1956 and 2001 (Wynne 1872). It is apparent that the region may have been affected previously by several other high-magnitude earthquakes though there is no paleoseismic record nor historical accounts (MacMurdo 1824, Baker 1846, Oldham 1926). The available historical accounts suggest the 1819 earthquake formed a 10–15 ft high scarp that blocked the drainages from the north and simultaneously the area around Sindri subsided and was submerged by a local tsunami (Macmurdo 1824; Baker 1846). However; all contemporary accounts that describe the spectacular landscape changes are from the western part only, which formed the trade route from Lakhpat to Sind via Sindri and a part of the eastern fringe of Indus delta. According to Oldham (1926), the Allah Bund scarp extended for ~90 km, however presently the scarp is identifiable for ~60 km length only. Our investigations show the Allah Bund scarp is highly eroded in the western parts due to the tide from the Kori creek in the south and the river water from the north during monsoon. The eastern part of the scarp is also low and disappears completely in the Rann surface. The scarp shows the maximum height in the central part where it is locally covered by aeolian deposits. Modern investigations have indicated that the height of the fault scarp is a reflection of cumulative uplift during 1819 and previous earthquakes (Rajendran and Rajendran 1999). However, there is no such indication from the available historical accounts. In the present study morphologic dating was carried out to determine the age of the Allah Bund scarp.

The Allah Bund scarp though having low height and laterally extensive, is found suitable for morphologic dating in its central part. The scarp here fulfills the necessary geomorphic conditions for carrying out morphologic dating (Colman and

Watson 1983; Nash 1987). The basic parameters where measured in the field. Topographic profiles across the scarp at the selected sites were prepared from topographical map and SRTM data. The present geomorphological environment along scarp summarized earlier in the article suggests that the central part is a zone of minimum erosion and subsequently least amount of scarp degradation. The hyper arid climate and the non influence of anthropogenic activity means the diffusion constant for arid region can be applied to the Allah Bund scarp for determining its morphologic ages. We therefore selected the value of diffusion constant as 1×10^{-4} m^2 $year^{-1}$, which has been applied successfully in the morphologic dating of fault scarps located in the arid regions (Begin 1992). The morphological ages of the Allah Bund scarp can be calculated as per Colman and Watson (1983).

The morphological ages of the Allah Bund fault scarp at the three selected sites yielded an age of 208, 200 and 193 years B. P. Effectively these ages means that the scarp was produced during the 1819 Allah Bund earthquake, since no other major earthquake occurred in the Great Rann during the period indicated by the ages i.e. between 1809 and 1825.

Our studies substantiate the description in the historical accounts regarding the Allah Bund fault scarp. We therefore rule out the possibility of the scarp being a product of multiple earthquake events. Our study also suggests that the 1819 Allah Bund earthquake was the largest earthquake that occurred in the Great Rann after its emergence from the shallow marginal sea. This is evident from the large-scale landscape changes, that includes the formation of a ~90 km long E-W trending south facing scarp and also that no other comparable scarp occurs in the vast extent of the Great Rann. Other earthquake that may have occurred during the time possibly did not rupture the surface. However, the possibility of earthquakes of comparable magnitude in prehistoric times in Holocene cannot be ruled out. It is likely that some of them may have produced surface rupture, however these may be buried under thick sediments, as the Great Rann was a shallow marginal gulf during the Late Pleistocene to Holocene (Maurya et al. 2013; Khonde et al. 2017a, b).

It is pertinent to mention here that several large earthquakes have occurred in the Kachchh rift basin, including the 1819 Allah bund earthquake and 2001 Bhuj earthquake. The heightened level of seismicity is anomalous in view of the low level seismic activity witnessed in the surrounding regions viz. Saurashtra in south, Gujarat alluvial plain in the east and Barmer-Jaisalmer basins in the northeast (Maurya et al. 2017). It has been suggested that the stress changes related to large earthquakes in the Kachchh region affect the subsequent pattern of earthquakes. To et al. (2004) demonstrated that the 1819 Allah Bund earthquake induced stress changes which led to several moderate magnitude earthquake in the Great Rann. Post-seismic relaxation following the 1819 earthquake resulted in the loading of 2001 Bhuj earthquake to a significant amount (To et al. 2004). The 2001 event was followed by long aftershock activity and increased seismic activity along the South Wagad Fault (SWF), Gedi Fault (GF) and the Island Belt Fault (IBF) (Mandal 2009). Earthquake swarm activity in different part of Saurashtra since 2001 has been attributed to stress perturbation caused by 2001 Bhuj earthquake (Rastogi et al. 2013). Low magnitude shocks are repeated from the Gujarat alluvial plain in the last 200 years (Maurya et al. 2000). Reactivation of pre-existing lineaments and faults have resulted in moderate to low seismic activity in Barmer basin also in recent times (Dasgupta and Mukherjee in press). However, none of the known historical earthquakes are known to have produced surface rupture apart from the 1819 Allah Bund earthquake. Moreover, the co-seismic and post-seismic stress changes due to pre-instrumental events still remain a topic of debate (Rajendran and Rajendran 2001). Detailed palaeoseismic studies are necessary to unravel the seismic history of the Great Rann for which the Allah Bund fault scarp can serve as an analog.

Acknowledgments We are grateful to Soumyajit Mukherjee for inviting us to contribute this article and for providing two rounds of review comments. The present study was funded by the Ministry of Earth Sciences (MoES), Government of India in the form of a research project (Project No. MoES/P.O. (Seismo)/1(170)/2013) to DMM and LSC. Mukherjee (2019) summarizes this work.

References

Baker WE (1846) Remarks on the Allah Bund and on the drainage of the eastern part of the Sind basin. Transactions Bombay Geographic Society 7, 186–188

Begin ZB (1992) Application of quantitative morphologic dating to paleo-seismicity of northwestern Negev, Israel. Israel Journal of Earth Science 41, 95–103

Bilham R (1998) Slip parameters for the Rann of Kachchh, India, 16 June 1819, earthquake, quantified from contemporary accounts. Geological Society Special Publication 146, 295–319

Biswas SK (1974) Landscape of Kutch—a morphotectonic analysis. Indian Journal of Earth Science 1, 177–190

Biswas SK (1993) Geology of Kutch. K.D. Malaviya Institute of Petroleum Exploration, Dehradun, p 450

Biswas SK (1987) Regional tectonic framework, structure and evolution of the Western Marginal Basins of India. Tectonophysics 135, 307–327

Bucknam RL, Anderson RE (1979) Estimation of fault- scarp ages from a scarp-height-slope-angle relationship. Geology 7, 11–14

Burnes A (1835) Memoir of the eastern branch of the River Indus, giving an account of the alterations produced on it by an earthquake, also a theory of the formation of the Rann and some conjunctures on the route of Alexander the Great; drawn up in the years 1827–1828. Transactions of the Royal Asiatic Society of Great Britain and Ireland 3, 550–588

Carretier S, Ritz JF, Jackson J, Bayasgalan A (2002) Morphologic dating of cumulative reverse fault scarp: examples from the GurvanBogd fault system, Mongolia. Geophysical Journal International 148, 256–277

Chandra U (1977) Earthquakes of Peninsular India: a seismotectonic study. Bulletin of Seismological Society of America 67, 1387–1413

Colman SM, Watson K (1983) Ages estimated from a diffusion-equation model for scarp degradation. Science 221, 263–265

Dasgupta S, Mukherjee S (In press) Remote sensing in lineament identification: Examples from western India. In: Billi A, Fagereng A (Eds) Problems and solutions in structural geology and tectonics. Developments in structural geology and tectonics book series. Series Editor: Mukherjee S. Elsevier. ISSN: 2542–9000

Enzel Y, Amit R, Bruce J, Harrison J, Porat N (1994) Morphologic dating of fault scarps and terrace risers in the southern Arava, Israel: comparison to other age-dating techniques and implications for paleoseismicity. Israel Journal of Earth Science 43, 91–103
Frere HBE (1870) Notes on the Runn of Cutch and neighboring region, J. R. Geograph. Soc. London 40, 181–207
Glennie KW, Evans G (1976) A reconnaissance of the Great Rann of Kutch, India. Sedimentology 23, 625–647
Grindlay (1808) Diary reproduced in Burnes A., Travels into Bokhara, Volume 3 Appendix. 1834
Gutenberg B, Richer CF (1954) Seismicity of Earth. Princeton University Press, New Jersey, USA, p 310p
Hanks TC, Schwartz DP (1987) Morphologic dating of the pre-1983 fault scarp on the Lost River fault at Doublepring Pass Road, Custer County, Idaho, Bulletin of Seismological Society of America 77, 837–846
Johnston AC (1989) The seismicity of stable continental interiors. In: Gregersen S, Bhasham PW (eds), Earthquakes at North-Atlantic Passive Margins: Neotectonics and Postglacial Rebound, Kluver Academic Publishers, pp 299–327
Johnston AC, Kanter LR (1990) Earthquakes in stable continental crust. Scientific American 262, 68–75
Kar A (1995) Geomorphology of the western India. Memoir of Geological Society of India 32, 168–190
Khonde N, Maurya DM, Singh AD, Das A, Chamyal LS (2013) Sediment characteristics and foraminiferal distribution in the Bet Zone of the Great Rann of Kachchh, Western India. Geological Society of India, Special Publications 1, 1–15
Khonde N, Maurya, DM, Chamyal LS (2017a) Late Pleistocene-Holocene clay mineral record from the Great Rann of Kachchh, Western India: implication for palaeoenvironments and sediment sources. Quaternary International 443, 86–98
Khonde N, Singh SK, Maurya DM, Rai VK, Chamyal LS, Giosan L (2017b)Tracing the Vedic Saraswati River in the Great Rann of Kachchh. Scientific Reports 7, 5476
Lyell C (1855) A manual of elementary geology or, the ancient changes of the Earth and its inhabitants as illustrated by geological monuments. John Murray, London, p 655
MacMurdo J (1824) Papers relating to the earthquake which occurred in India in 1819. Philosophical Magazine 63, 105–177
Mandal P (2009) Estimation of static stress changes after the 2001 Bhuj earthquake: implications towards the northward spatial migration of the seismic activity in Kachchh, Gujarat. Journal of the Geological Society of India 74, 487–497
Maurya DM, Raj R, Chamyal LS (2000) History of tectonic evolution of Gujarat alluvial plains, western India during Quaternary: a review. Journal of Geological Society of India 55, 343–366
Maurya DM, Khonde N, Archana Das, Chowksey V, Chamyal LS (2013) Subsurface sediment characteristics of the Great Rann of Kachchh from prelimininary textural analysis of two continuous cores. Current Science 104, 1071–1077
Maurya DM, Tiwari M, Rajawat AS, Kumar H, Khonde N, Chamyal LS (2016) Geomorphic characterization of the Banni Plain, Kachchh, using orbital imaging radar (RISAT 1C) and optical remote sensing data. In: Recent studies on the Geology of Kachchh, Thakkar MG (ed), Geological Society of India Special Publication No. 6, pp. 168–178
Maurya DM, Chowksey V, Patidar AK, Chamyal LS (2017) A review and new data on neotectonic evolution of active faults in the Kachchh Basin, Western India: legacy of post-Deccan Trap tectonic inversion. Geological Society, London, Special Publications 445, 237–268
Merh SS (2005) The Great Rann of Kachchh: perceptions of a field geologist. Geological Society of India 65, 9–25
Merh SS, Patel PP (1988) Quaternary geology and geomorphology of the Ranns of Kutch. In: Proceedings seminar on recent quaternary studies in India. M.S. University, Baroda, India, pp 377–393

Mukherjee S (2019) Introduction to "Tectonics and Structural Geology: Indian Context". In: Mukherjee S (ed) Tectonics and structural geology: Indian context. Springer International Publishing AG, Cham, pp 1–5. ISBN: 978-3-319-99340-9

Niviere B, Marquis G (2000) Evolution ofterrave risers along the upper Rhine Graben inferred from morphologic dating methods: evidence of climatic and tectonic forcing. Geophysical Journal International 141, 577–594

Nash DB (1984) Morphologic dating of alluvial terrace scarps and fault scarps near West Yellowstone, Montana. Geological Society of America Bulletin 95, 1413–1424

Nash DB (1987) Reevaluation of the linear-diffusion model for morphologic dating of scarps. In: Proceedings of conference XXXIX directions in paleoseismology, open-file report, pp 87–673

Oldham RD (1926) The Cutch Earthquake of 16th June 1819 with a revision of the great earthquake of 12th June 1897. Memoir of Geological Survey of India 46, 71–146

Quittmeyer RC, Jacob KH (1979) Historical and modern seismicity of Pakistan, Afghanistan, northern and southern Iran. Bulletin of Seismological Society of America 69, 773–823

Patidar AK (2010) Neotectonic studies in southern mainland Kachchh using GPR with special reference to Katrol Hill Fault, Ph.D. thesis, The M S University of Baroda, Vadodara, India, 163p. Available online at www.shodhganga.com

Rajendran K, Rajendran CP (1999) Seismogenesis in the stable continental interiors: an appraisal based on two examples from India. Tectonophysics 305, 355–370

Rajendran CP, Rajendran K (2001) Characteristics of deformation and past seismicity associated with the 1819 Kutch Earthquakes, northwestern India. Bulletin of Seismological Society of America 91, 407–426

Rastogi BK, Kumar S, Aggrawal SK, Mohan K, Rao N, Rao NP, Kothyari GC (2013) The October 20, 2011 Mw 5.1 Talala earthquake in the stable continental region of India. Natural Hazards. 65, 1197–1216

Roy B, Merh SS (1982) The Great Rann of Kutch: intriguing Quaternary terrain, in Recent Researches in Geology, Series 9, Hindustan Publication Company, Delhi, pp. 100–108

Summerfield MA (1991) Global Geomorphology: an introduction to the study of landforms, Essex: Longman, 537p

Thakkar MG, Ngangom M, Thakker PS, Juyal N (2012) Terrain response to the 1819 Allah Bund earthquake in western Great Rann of Kachchh, Gujarat, India. Current Science 103, 208–212

To A, Bürgmann R, Pollitz F (2004) Postseismic deformation and stress changes following the 1819 Rann of Kachchh, India earthquake: was the 2001 Bhuj earthquake a triggered event? Geophysical Research Letters 31(13)

Valdiya KS (2002) Saraswati: the river that disappeared, Universities Press, Hyderabad, 116pp

Wallace RE (1977) Profiles and ages of young fault scarps, north-central Nevada. Geological Society of America Bulletin 88, 1267–1281

Wynne AB (1872) Memoir on the geology of Kutch. Memoir of Geological Survey of India 9, 1–294

Interplay Between Tectonics & Eustacy in a Proterozoic Epicratonic, Polyhistory Basin, North Dharwar Craton

Shilpa Patil Pillai and Vivek S. Kale

1 Introduction

Modern developments in sedimentary basin analysis (e.g. Einsele 2000; Miall 2000; Allen and Allen 2013) have accepted the role of Wilson cycle in the formation, growth and termination of sedimentary basins. The tectonic classifications of sedimentary basins (Dickinson 1974; Kingston et al. 1983; Allen et al. 2015) reflect this perception. The sediment fill of a basin manifests the interplay of sediment supply, eustacy and subsidence. Climate, topography and petrographic constitution of the source region control the sediment supply. The geometry and volume of the fill depends upon the available accommodation space, moderated by tectonics and eustacy. In Precambrian basins, their deformation and metamorphism hampers the elucidation of the contents, distribution patterns and sequences of the basin fills. Sparse age controls in Precambrian sediments when compared to fossiliferous Phanerozoic sediments further hampers their accurate elaboration (Eriksson et al. 2001). The Indian Peninsular Shield amalgamated into a singular continental block towards the end of the Archean (2700–2500 Ma) with a series of K-rich granitic batholiths representing the terminal events of this unification (Naqvi and Rogers 1987; Ramakrishnan and Vaidyanadhan 2010; Meert et al. 2010). Two contrasting regimes punctuate its ensuing history (Kale 1995). The narrow, linear metamorphosed Early and Middle Proterozoic Mobile Belts with compressive tectonic history (Radhakrishna and Naqvi, 1986; Ramakrishnan and Vaidyanadhan 2010) are juxtaposed against the wide, extensional epicratonic platform basins that display

S. Patil Pillai
Department of Environmental Sciences, Savitribai Phule Pune University, Ganeshkhind, Pune 411007, India

V. S. Kale (✉)
Advanced Center of Water Resources Development and Management (ACWADAM), Kshipra Society, Karvenagar, Pune 411052, India
e-mail: dr.vivekale@gmail.com

S. Mukherjee (ed.), *Tectonics and Structural Geology: Indian Context*, Springer Geology, https://doi.org/10.1007/978-3-319-99341-6_4

a paucity of pervasive metamorphism and/or deformation (Fig. 1). The latter are known as the 'Purana' basins (Radhakrishna 1987; Kale and Phansalkar 1991). They contain thick sequences of shallow marine sediments deposited in discrete basins which are in tectonic contacts with the mobile belts on one side (Kale 2016). The Kaladgi and Bhima basins on the northern exposed edge of the Dharwar craton are the exceptional Purana basins that do not have any Proterozoic mobile belts adjoining them. The Kaladgi basin is additionally unique amongst the Purana basins in that it has suffered stronger deformation along the median sectors of the basin, unlike all others which have deformation restricted to their margins (Kale 1991).

The three Purana basins surrounding the Dharwar craton (Fig. 1) display diverse modes of evolution (see Kale and Phansalkar 1991; Kale 1991, 2016; Meert et al. 2010; Allen et al. 2015). The Cuddapah basin has a record of its early growth as an

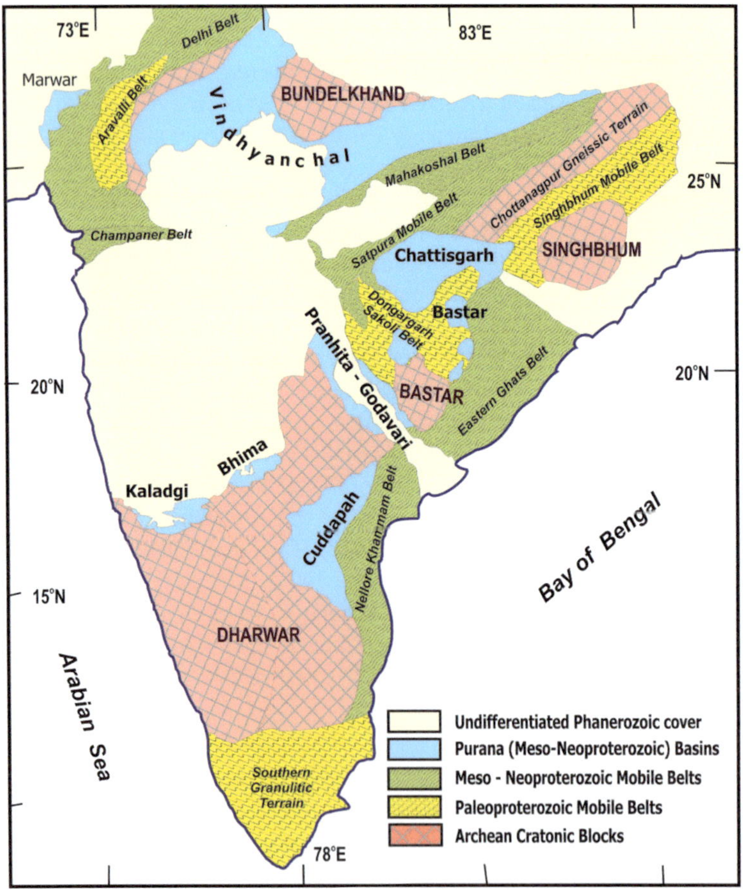

Fig. 1 Precambrian terrains of Peninsular India (modified after Kale 1995; Sharma 2009; Ramakrishnan and Vaidyanadhan 2010)

extensional basin attended by basic volcanism; that eventually gave way to a retro-arc basin with overthrusting of the Eastern Ghats Mobile Belt (including the Nellore belt) from the east (Matin 2015 and citations therein). The Bhima basin in the north is a pull-apart basin (Kale and Peshwa 1995). The Kaladgi basin on the north-west shows events of initial extension (with minor transtensional component), compressive deformation and late extension (Dey 2015; Patil Pillai et al. 2018). The different margins of the Dharwar craton have obviously suffered diverse tectonics and basin growth histories during the Proterozoic.

The Kaladgi basin hosts commercially exploited limestone reserves and asbestos mines located in its basement (Jayaprakash 2007). It is a "Category IV" petroliferous basin (Pratap et al. 1999; Biswas 2008; Kalpana et al. 2010). Uranium mineralization has been recorded from various locations within it (Dey et al. 2008; Sridhar et al. 2014). Our group has conducted extensive field studies of these sediments across the last couple of decades; supplemented by petrographic laboratory studies. We enumerate the contents and architecture of this unique Purana basin: and demonstrate signatures of the tectono-eustatic interplay during its evolution.

2 Regional Geology

2.1 Location

The Kaladgi basin extends between the 73°E and 76°E longitudes and 15° 30′N to 17°N latitudes, from the Konkan coastal tract in Sindhudurg district; across the Western Ghats and the Deccan Plateau in Kolhapur district of Maharashtra state into the Belgaum, Bagalkot and Bijapur districts of northern Karnataka state. Isolated outliers of these sediments are scattered south of the main basin around Saundatti and Gajendragad. Inliers within Deccan Traps occur near Kallamavadi and near Jamkhandi to the north of the main basin. The westernmost exposures are found as patchy inliers (within the Deccan Traps) in the Konkan coastal strip, around Phonda and Malvan. They are collectively referred to as the Konkan Kaladgis (Sarkar and Soman 1985; Deshpande and Pitale 2014). These disconnected outcrops show that the main basin extended far beyond its currently exposed limits. The present exposures of about 8000 km^2 represent only a fraction of the span of the basin, that can be interpreted to occupy more than 30,000 km^2.

The sediments in this basin are subdivided into the Bagalkot and Badami Groups (Viswanathiah 1979) separated by an angular and erosional unconformity. The Bagalkot Group occupies an irregular ovoid shaped basin in the eastern parts (Inset Fig. 2). The Badami Group covers more than 300 km in an E-W direction.

It is significant that while the Konkan Kaladgis are exposed at elevations of less than 200 m above mean sea level, none of the remaining exposures (east of Kallamavadi) occur at elevations below 550 m. This elevation difference may the

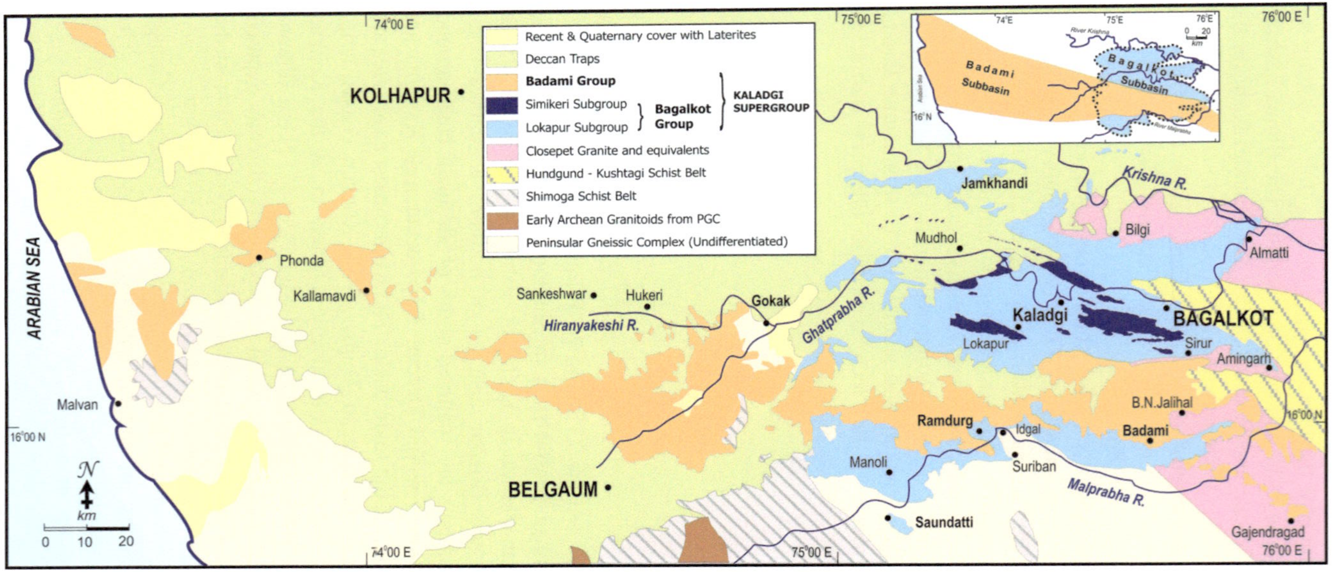

Fig. 2 Regional geological map of the Kaladgi basin and its surrounding areas (compiled from Foote 1876; GSI 1981; Jayaprakash 2007). Inset is a sketch of the geographic spread of the exposures of the older Bagalkot and younger Badami Groups showing the two subbasins that developed at different times

Table 1 Lithostratigraphy of the Kaladgi basin (after Kale et al. 1999; Kale and Patil Pillai 2011)

Late Cretaceous to Paleogene **Deccan Traps** (Trappean basalts, infra- & inter-Trappean beds)			
Unconformity			
Kaladgi Supergroup	*Neoproterozoic* **Badami Group**		• Konkankoppa Limestone Formation • Halkurki Shale Formation • Cave Temple Arenite Formation
	Angular Unconformity		
	Mesoproterozoic **Bagalkot Group**	**(*) Mallapur Intrusives**	
		Simikeri Subgroup	• Lakshanhatti Dolomite Formation • Niralkeri Chertbreccia Formation • Arlikatti Argillite Formation • Muchkundi Quartzite Formation
		Disconformity	
		Lokapur Subgroup	• Petlur Carbonate Formation • Mahakut Chertbreccia Formation • Yadhalli Argillite Formation • Saundatti Quartzite Formation
Nonconformity			
Paleoproterozoic **Mafic dykes** *Archaean* **Closepet Granite, Dharwar Supergroup & Peninsular Gneisses**			

The Mallapur Intrusive (sensu: Jayaprakash et al. 1987) should be separated from the Simikeri Subgroup as the intrusive bodies post-date the Simikeri sedimentation and early deformation (Patil Pillai et al. 2018) and is hence shown with an asterisk before it

result of the downfaulting of the coastal strip prior to the Late Cretaceous—Early Paleogene eruption of the Deccan Traps since the trappean basalts do not display any evidence of faulting of this magnitude. It is likely that this elevation difference is an artifact of the pre-Trappean downfaulting related to the rifting and separation of the Indian plate from Madagascar—Syechellis in the Late Cretaceous, but it could also be a much earlier event. The significance of this is that the Kaladgi sediments have been subjected to Phanerozoic upheavals along with the rest of the Dharwar craton. Its local implications could be several, when trying to correlate the terrain evolution of this basin.

Archaean cratonisation, Proterozoic sedimentation, and the Cretaceous-Tertiary Deccan volcanism have contributed to the overall geology in this region that was eventually subjected to emergent erosion, fluvio-colluvial sedimentation and localised lateritization during Quaternary times. Table 1 summarises the sequence of strata exposed in the Kaladgi basin and its surroundings.

2.2 Basement

The basement for the Kaladgi Supegroup consists of Archean to Early Paleoproterozoic rocks of the Dharwar craton (Radhakrishna and Ramakrishnan 1988; Ramakrishnan and Vaidyanadhan 2010; Valdiya 2016). They are collectively termed as the "basement" or "Archean basement" in the ensuing text. These sequences not only served as the basement for the Kaladgi sediments, but also have acted as the source for their detrital sediments (Sambasiva Rao et al. 1999; Dey et al. 2008, 2009).

This basement consists of a typical cratonic assemblage of gneisses, schist belts and intrusive granitic batholiths (Fig. 2). The Peninsular Gneissic Complex is composed of amphibolite facies tonalitic-trodndhjemitic-granodioritic (TTG) gneisses. The NNW-SSE Shimoga Schist belt exposed southwest of the basin is composed of genisses, meta-conglomerates, phyllites and banded iron formations (BIFs). The Hundgund-Kushtagi Schist Belt forms the basement in the eastern parts of the Kaladgi basin. It is composed of BIFs, basalt, minor ultramafic and metasediments engulfed by the Peninsular Gneisses and intruded by the Clospet Granites. The 2600 Ma old potassic granites of Closepet Granites and equivalent rocks constitutea NW-SE trending belt. Towards the northern parts of the Dharwar craton (in and around the Kaladgi basin), the basement trends swirl from NW-SE to almost E-W. Basic dykes intrude this mixed basement (Ramakrishnan and Vaidyanadhan 2010). Some of the dykes in it may be as young as 1800 Ma (French et al. 2008; Meert et al. 2011).

2.3 Kaladgi—Basement Unconformity

The pre-Kaladgi basement profile was essentially a peneplained one with local undulations, such as one below the Chitrabhanukot Dome, that emerge from the geophysical data (Ramakrishna and Chayanulu 1988; Mallick et al. 2012). Deeply weathered basement (sometimes with a paleosol horizon on top) is exposed immediately below the sedimentary cover, at several locations along the northern and southern rims of the basin.

Linear ridges of quartzites present on the eastern, northern and southern rims of the main basin host the basal terrigenous clastics of the Bagalkot Group resting on the basement. This contact between the basement and the capping sediments is an angular and erosional non-conformity (Fig. 3a). Shear zones and faults that affect both the Kaladgi sediments and their basement are exposed at places where the contact between them is a structural one (see: Fig. 8). The exposures east of Bilgi, between Idgal and Suriban, southeast of Ramdurg (Fig. 3b) and near Saundatti are excellent examples of such structural contacts.

The subhorizontal Badami sediments rest on the dipping Bagalkot Group sediments with an angular and erosional unconformity in the eastern parts of the

Fig. 3 **a** Exposure of the Badami Group (on the top with subhorizontal disposition) resting unconformably upon the steeply dipping Bagalkot Group sediments (appearing as barren linear ridges) with a nonconformable contact with the basement at B. N. Jalihal looking eastwards. The location is about 3 km NE of Badami. **b** Steeply dipping Saundatti Quartzite exposing the sheared contact with the basement (on the right hand side—not in the picture). The photo is looking northeastwards at a location about 4 km east of Ramdurg. Author standing is 1.7 m tall. **c** Erosional and angular unconformity between the northwards dipping Manoli Argillite (Bagalkot Group) and the capping Cave Temple Arenite (Badami Group) at Torgal Tanda located about 3 km south of Ramdurg. The conglomerate is 7 m thick. **d** Thin cap of subhorizontal Cave Temple Arenites resting directly on the Archaean Basement at the Gokak hill forming flat-topped mesas

Kaladgi basin (Fig. 3a, c). This was first recorded at B. N. Jalihal (northeast of Badami) and Torgal Tanda (south of Ramdurg) by Viswanathiah (1968). In the western parts, the Badami sediments are exposed resting directly upon the basement without the intervening Bagalkot Group sediments, forming flat topped mesas and buttes (Fig. 3d).

2.4 *Phanerozoic Cover*

Sediments in the Kaladgi basin are capped by the Deccan Trap basalts, which conceal large tracts of the basin, making it difficult to establish their lateral continuity particularly in the western parts of the basin (Fig. 2). The Deccan Trap basalts were erupted close to the Cretaceous—Tertiary Boundary and manifest one of the largest continental flood basalt provinces in the world (Renne et al. 2015). At places near Badami, Belgaum (where it is quarried) and along the western Konkan coastal tract Late Tertiary—Quaternary lateritic caps are developed on top of the older strata (Fig. 2). In some of the river valleys Quaternary fluvial sedimentary deposits are developed capping the Kaladgi sediments. All of them contribute to masking of the exposures of the Kaladgi rocks and making the lateral correlation difficult.

3 Kaladgi Supergroup

3.1 Stratigraphy

The early description of the geological successions in this region by Captain Newbold in the mid nineteenth century were compiled with additional inputs by Foote (1876). He classified this sedimentary sequence into six subunits ascribed to the "***Upper*** and ***Lower Kaladgi Series***" that remained unquestioned for almost a century. Recognition of the unconformity separating the undeformed succession of sandstones, shales and limestones ('Unit 3' of Foote 1876) and the deformed lower sediments around B. N. Jalihal and Torgal Tanda by Viswanthiah (1968) led to the recognition of the two Groups (Badami and Bagalkot) from this basin. Jayaprakash et al. (1987) identified three cycles of sedimentation within the basin and further refined the stratigraphy by dividing the deformed Bagalkot Group into two subgroups, namely the Lokapur and Simikeri separated by a basin-wide disconformity. This was further elaborated upon by Jayaprakash (2007).

The spread of the exposures of the two constituent Groups (after discounting the younger cover) shows that they were deposited in slightly different geographical realms (Fig. 4). They are recognised as subbasins of the Kaladgi basin, and can be justified to be sedimentologically, structurally and temporaly different from each other. They overlap in the eastern part of this basin, but the older Bagalkot Group does not extend in the western areas that host the Badami Group (Fig. 2 Inset).

Jayaprakash et al. (1987) recognised 21 members assigned to 8 formations in the Bagalkot Group and 6 members from 2 formations in the Badami Group. The

Fig. 4 **a** Map of the Bagalkot subbasin depicting the exposures of various stratigraphic units. ▶ Chitrabhanukot Dolomite and Chikshellikeri Limestones are Members of the Petlur Carbonate Formation. The locations of the stratigraphic logs (Fig. 6a) are given. Note the outline of this basin marked in this map is traced by connecting the outermost exposures of the Bagalkot Group as exposed today. The detached exposures south of Manoli are separated from the others by structural contacts and hence shown separately. They comprise of Saundatti Quartzite. They indicate a larger southward expanse of this subbasin, which is detached by structural disruptions or has been disconnected due to subsequent erosion. **b** Map of the Badami subbasin depicting the distribution of various stratigraphic units. Locations of the measured logs (Fig. 6b) are given for reference. West of the limits of the Bagalkot subbasin, the Badami sediments rest directly on the Archean basement complex

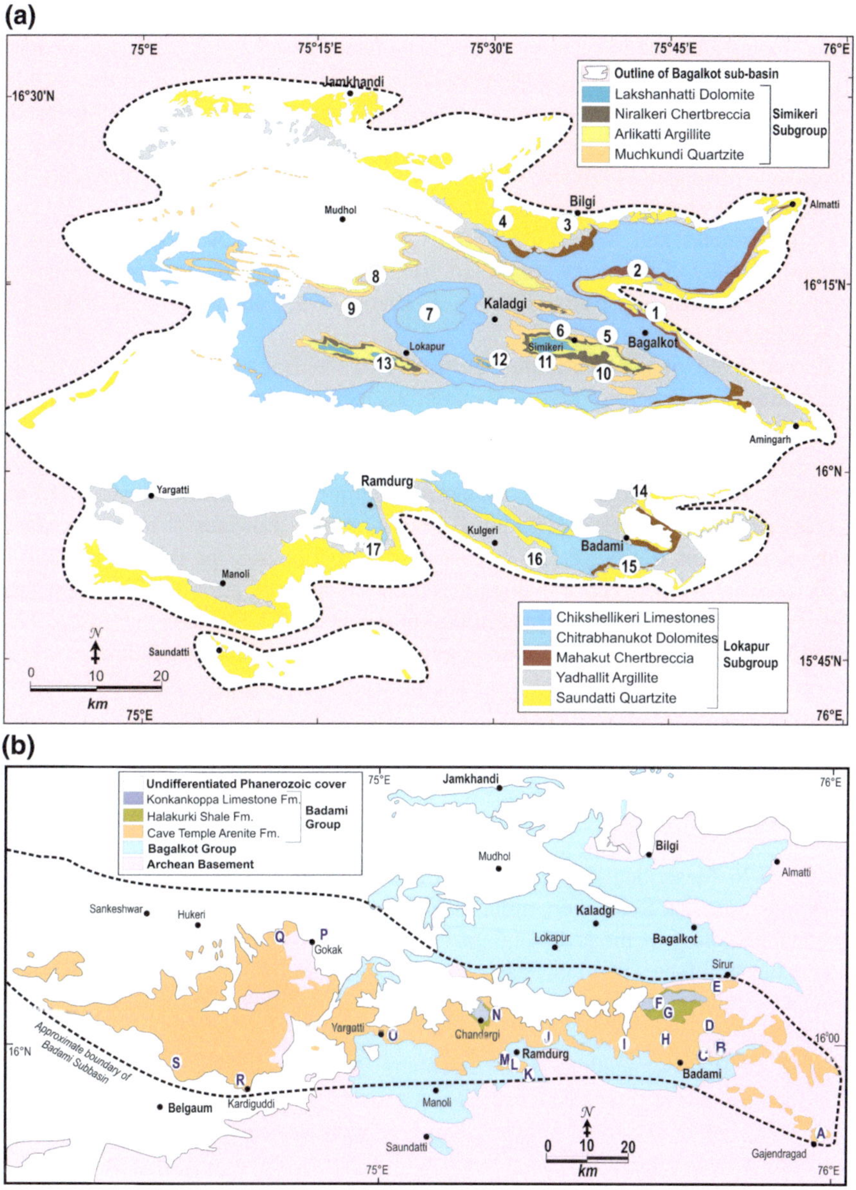

Lokapur and Simikeri Subgroups are characterised by basal quartzitic sandstones with minor rudaceous component. Similarly, the Badami Group is floored by sandstone—conglomerate beds. Lithological boundaries between horizons are generally gradational and interfingered in nature. Upward fining arenites pass into the argillites. The siliceous and ferruginous argillites grade into calcareous argillites and eventually into the carbonates.

These horizons display a lateral facies change into the adjoining lithology. This lateral facies changes have been earlier elaborated for a part of the basin by Kale et al. (1996) and Patil Pillai (2004). Consequently there is a repetition of the argillite—carbonate horizons encountered along several traverses across the basin. Jayaprakash et al. (1987) and Jayaprakash (2007) recorded these as a number of 'members' that they clubbed into 'formations'. This is not valid within the framework of standard stratigraphic practices and Walther's Law of facies succession (Brenner and McHargue 1988; Catuneanu 2006). A process responsive lithostratigraphy (Table 1) of simplified, lithology-based, mappable formations was erected for the Kaladgi basin (Kale et al. 1996, 1999; Patil Pillai 2004; Kale and Patil Pillai 2011).

Direct onlap of the quartzitic sandstones of the Bagalkot Group without the intervening conglomerates is seen at numerous locations located along the structural contact with the basement. This indicates landward transgressive flooding beyond the structural boundaries of the basin. While the Lokapur Subgroup is spread all across the Bagalkot subbasin, the Simikeri Subgroup is restricted to the cores of the synclines (Figs. 2 and 4). The Badami Group does not show any structural contacts with its basement. In the western exposures of this subbasin, the Bagalkot Group is not present at all between the Badami sediments and the Archean basement (see Fig. 2). At no place does the Badami Group rest upon the Simikeri Subgroup. This could be an indication of the limited development of the Simikeri Subroup, which has no representatives south of the Badami subbasin.

3.2 Constitutents

Foote (1876) described the Kaladgi sediments within the framework of the contemporary formats. Since then, more than a century of research has been invested in the understanding of the sediments of this basin. Viswanathiah (1979) recognized four major lithofacies from this basin, namely the sandstones (/quartzites), shales (/phyllites), carbonates (limestones and dolomites) and chertbreccias. The discovery of stromatolites from the older carbonates (Viswanathiah and Chandrasekhara Gowda 1970; Viswanathiah and Sreedhara Murthy 1979) was interpreted as additional evidence of shallow marine environments.

Jayaprakash et al. (1987) showed that there were three cycles of shallow marine epicratonic sedimentation within the basin, yielding three successions separated from each other by unconformable contacts. Kale et al. (1996) further refined this view using lithofacies classification to the southern parts of the basin (see: Figs. 7 and 8 of Kale et al. 1996) and later to the basin as a whole (Kale et al. 1999; Kale and Patil Pillai 2011). A summary of the petrological characters of these facies based on our studies is given below.

The Bagalkot Group is characterized by two sequences comprising of four lithofacies, namely (a) detrital siliciclastic arenites and rudites (sandstones and conglomerates), (b) argillites (siltstones, shales, mudstones and phyllites),

(c) chertbreccias and (d) carbonates (limestones and dolomites/dolostones). They occur as concentric stripes with an essentially E-W to ESE-WNW trend, swirling parallel to the rim of the basin. The clastics occur closer to the basin margin and others successively inwards towards the basin axis. In the Badami Group, the arenaceous lithofacies dominates significantly, and the argillites and carbonates are restricted to a narrow linear patch in the northern part of the eastern sector of this subbasin (Fig. 4b). Chertbreccia is absent in the Badami succession.

3.2.1 Detrital Siliciclastics

Cap the basal unconformities. In case of the Lokapur Subgroup (=Saundatti Quartzite), they rest on the weathered basement; while those from the Simikeri Subgroup (=Muchkundi Quartzite) rest on the deformed sediments of the Lokapur Subgroup. They occur as ridges that stand out in the low-lying terrain surrounding them. This facies from the Badami Group (=Cave Temple Arenite) rests either on the basement or on the deformed sediments of the Bagalkot Group, as gently undulating table-lands.

Conglomerates occuring immediately above the basement are several meters thick. Thin, sandy cross-stratified interbeds are common in them. Further upwards in the sequence, they are present as thin sheets or thicker lensoid beds with poor internal stratification (Fig. 2c) interbedded within the sandstones/quartzites.

The conglomerates are generally polymictic, matrix supported, poorly sorted rudites. The clasts in the Bagalkot conglomerates are dominantly composed of quartzitic, vein quartz, granitoid, schists and gneissic fragments embedded in a clastic matrix which is often silicified (Fig. 5a). The Badami conglomerates contain clasts derived from the Bagalkot sediments as well.

The Bagalkot sandstones display planar bedding (Fig. 5b) separated by thin mud or silt partings, with low angle internal trough cross-stratification. Low angle cross-laminations, climbing ripple laminations and widespread wave ripple marks on bedding planes (Fig. 5c) suggest a coastal shoreface regime of deposition in a variety of marginal to shallow marine environments such as strandplains, beaches, deltas, estuaries, off-shore sand bars, and beaches that occasionally were influenced by tidal currents (Kale et al. 1996; Jayaprakash 2007). Some of these beds, notably from the exposures around Bilgi and Jamkhindi on the northern parts of the basin are interpreted to be deposited under fluvial influence in deltaic/esturine environments (Bose et al. 2008). Some of the coarser detritus (rudites) may be lag deposits or gravelly shoreline rim deposits on the landward side of the basin, representing inception of the transgression in it. Their provenance is the Dharwarian basement from where they were transported in a fluvial medium (Dey et al. 2009; Mukhopadhyay et al. 2014). Palaeocurrent studies (Jayaprakash et al. 1987; Stephen George 1999) have indicated a centripetal pattern. The northern exposures display southward directed palaeocurrents and the southern exposures in the Ramdurg–Saundatti tract have northerly palaeocurrent directions. Change in the orientations of the current ripples in successive beds (Fig. 5c) suggest limited tidal

(a)
(b)
24 01 2015
(c)
(d)
(e)
(f)
(g)
(g)
25 01 2015

◂**Fig. 5** **a** Thickly bedded conglomerates with parallel stratified interlayers of sandstones occurring at the base of the Badami Group (at Gokak falls). They grade upward into the sandstone beds. The photographed section is about 5 m high. **b** Parallel bedded quartzitic sandstones of the Saundatti Quartzite exposed in a quarry face near Bilgi. Beds dip gently towards the south (right of photo). The quarry face is about 10 m high. **c** Current ripple marks on the bedding plane of Muchkundi Quartzite exposed near the Muchkundi reservoir. Note the change in ripple axis orientations in successive beds. **d** Trough cross bedded sandstones of Badami Group (exposed around Badami town). The geometry of troughs, indicate small shifting channels similar to those in braided channels of mature river systems close to their mouth (in estuaries/deltas). **e** Lenticular beds of Yadhalli Argillite exposed near Manoli. Note the gently wavy nature of the bedding in the upper part of the photograph. **f** Thin parallel bedding with alternations of grey (siliceous) and purple (ferruginous) shales of Halkurki Formation exposed near Halkurki. They are extremely friable and weather easily. Hammer for scale is 34 cm long. **g** Alternate laminations of calcareous (buff colored) and pure crystalline limestones (dark grey colored) from the Chikshellikeri Limestone exposed near Gaddankeri. Note the climbing ripple laminated beds interspersed within the parallel laminated bedding. They indicate reworking of the lime-mud under influence of currents. Pinching and swelling of the lamina thicknesses is due to uneven compaction. **h** Variegated parallel stratified Kokankoppa Limestone with an interbedded limestone breccia horizon. The clasts of the breccia are embedded in an impure calcareous muddy matrix. Such intraformational breccia horizons testify to intermittent emergence (shallowing) or may manifest synsedimentary deformation

influence during their deposition. The sandstones from the Bagalkot Group have been altered to glassy orthoquartzites due to diagentic recrystallisation and further during deformation.

The Badami sandstones are dominantly trough cross-bedded (Fig. 5d) with occassional birectional and parallel stratification. This suggests their deposition in braided fluvial to transitional marine environments with minor tidal influence. Slope-base conglomerates and the sandbars, over-bank mud drapes on hummocky sand beds and upward fining beds that are interrupted by gravel beds or local erosional surfaces (overlain by coarser sands) are often seen. These characters and the stratal patterns of dominantly wide amplitude trough-cross bedded channels suggest the deposition in braided channels. They are therefore attributed to a wide, mature braided river system that drained into a shore-face environment. Few of these sandstones (e.g. those exposed north of Ramdurg and northwest of Badami) display tidal silty interlayers. Their stratification with low amplitude shallow and wide troughs stacked one above the other (Fig. 5d) is also suggestive of their being estuarine or deltaic channel deposits.

3.2.2 Argillites

Siltstones, shales and mudstones represent this facies. The Bagalkot Group also contains horizons of phyllitic shales (with chlorite neocrystallisation) having well-developed slaty cleavages. This is the result of the localised, low grade metamorphism in the strongly deformed northern sector of the Bagalkot subbasin.

The argillites shows basinward progressive compostional gradation from siliceous near the basin margins to ferruginous to calcareous in the centre of the basin. The argillites occur as flat and laterally extensive facies that occupy vast tracts within the basin, with subparallel to parallel stratification. They display a variety of primary sedimentary structures such as flaser and lenticular bedding (Fig. 5e), discontinuous or lensoid bedding, mud cracks, clay gals, rip up clasts, raindrop impressions etc. These features testify to their deposition in a tidal regime which was intermittently flooded. The occurrence of manganese nodules within the Arlikatti Argillite near Niralkeri village reported by Patil Pillai et al. (1999) indicates the possibility of a locally developed condensed sequences within the basin. The argillaceous sediments from the Bagalkot Group (Yadhalli Argillite and Arlikatti Argillite) are products of shallow intertidal off-shore flats, that suffered phases of emergent and submergent conditions.

The shales of Badami subbasin are dominantly thinly stratified (Fig. 5f) and are ocassionally glauconitic. This indicates their deposition in transitional marine settings. The calcareous shales (generally occurring at the top of the argillite beds) progressively grade into the impure limestones of the carbonate facies. Distribution and sedimentological characters suggest that they (Halkurki Shale) were deposited in a lagoonal environment.

3.2.3 Chertbreccias

They occur only in the Bagalkot Group but are not present in the Badami Group. Two separate horizons are present in the Lokapur and Simikeri Subgroups (Mahakut and Niralkeri Chertbreccia Formations respectively). In field exposures they can be easily misidentified as fault breccias, but are unequivocally interbedded horizons within the rest of the Bagalkot sediments. They have been deformed coherently with the other enveloping sediments of this Group. They are interpreted as washed fault-generated debris produced during the synsedimentary tectonic movements along the basinal faults and subsequently silicified (Kale and Patil Pillai 2011).

The chertbreccia and argillaceous beds display a mutually exclusive development in the Bagalkot sequence. In the sectors of the basin that suffered synsedimentary tectonism, where chertbreccia deposits developed, the extent of the mudflats was curtailed. This led to accumulation of thinner deposits of the argillaceous sediments. In other sectors such as in the southern sector around Manoli and west of Badami, where the basin floor subsided in a more gentle continuum, wide tidal mudflats yielded thicker argillaceous deposits (see: Kale et al. 1996).

3.2.4 Carbonates

This facies is represented by limestones and dolomites, with the latter being restricted to the Bagalkot Group. The limestones display a variety of colours (pink, white, green and grey) attributed to the impurities in them. The limestones of Bagalkot Group display flat, parallel-laminated alternations of impure lime muds and pure crystalline limestone. Occurrence of climbing ripple laminations and cross-stratified lenses (Fig. 5g) indicate that they were deposited in a very shallow shelf affected by currents. The dolomites or dolostones of the Bagalkot Group display tabular parallel-laminated stratification. They are often diagenetically silicified. They contain wavy algal-laminated dolomites with stromatolitic colonies (Viswanathiah and Sreedhara Murthy 1979; Sharma et al. 1998) indicating water depths of less than 20 m and a warm equitable climate. These carbonates are interpreted to be deposited in a carbonate flat regime with occasional interruptions of storm surges manifested as thin sandy lenses. Occurrence of intraformational limestone breccia horizons (Kale et al. 1998) in them is indicative of their episodic emergence or the manifestation of synsedimentary deformational events during their accumulation.

The flaggy limestones of Badami Group are varigated (Fig. 5h) and contain thin lensoid intercalations of siliciclastics. They display wave ripples, clay galls, rip up clasts, tidal couplet laminae suggesting a tidally influenced depositional setting. Presence of glauconites in the sandy lenses suggests locally developed reducing environments during their deposition.

3.2.5 Mafic Bodies

Basic dykes and quartz veins intruding discordantly into the sediments have been recorded from the northern folded sector of the Bagalkot subbasin (Foote 1876; Jayaprakash et al. 1987; Jayaprakash 2007). They are collectively named as Mallapur Intrusives Formation of the Simikeri Subgroup. Their occurrence along the axial planes of folds in the Simikeri Subgroup indicates that they were emplaced after the cessation of sedimentation. Therefore they are separated as an independent formation post-dating the Simikeri Subgroup as listed in Table 1 (Patil Pillai et al. 2018).

3.3 Lithologs

Besides the detailed lithological mapping (Fig. 4), logs of stratigraphic sections were reconstructed based on sections measured during field studies (Fig. 6). They include the sedimentary structures, stratal patterns and thicknesses of the various lithotypes.

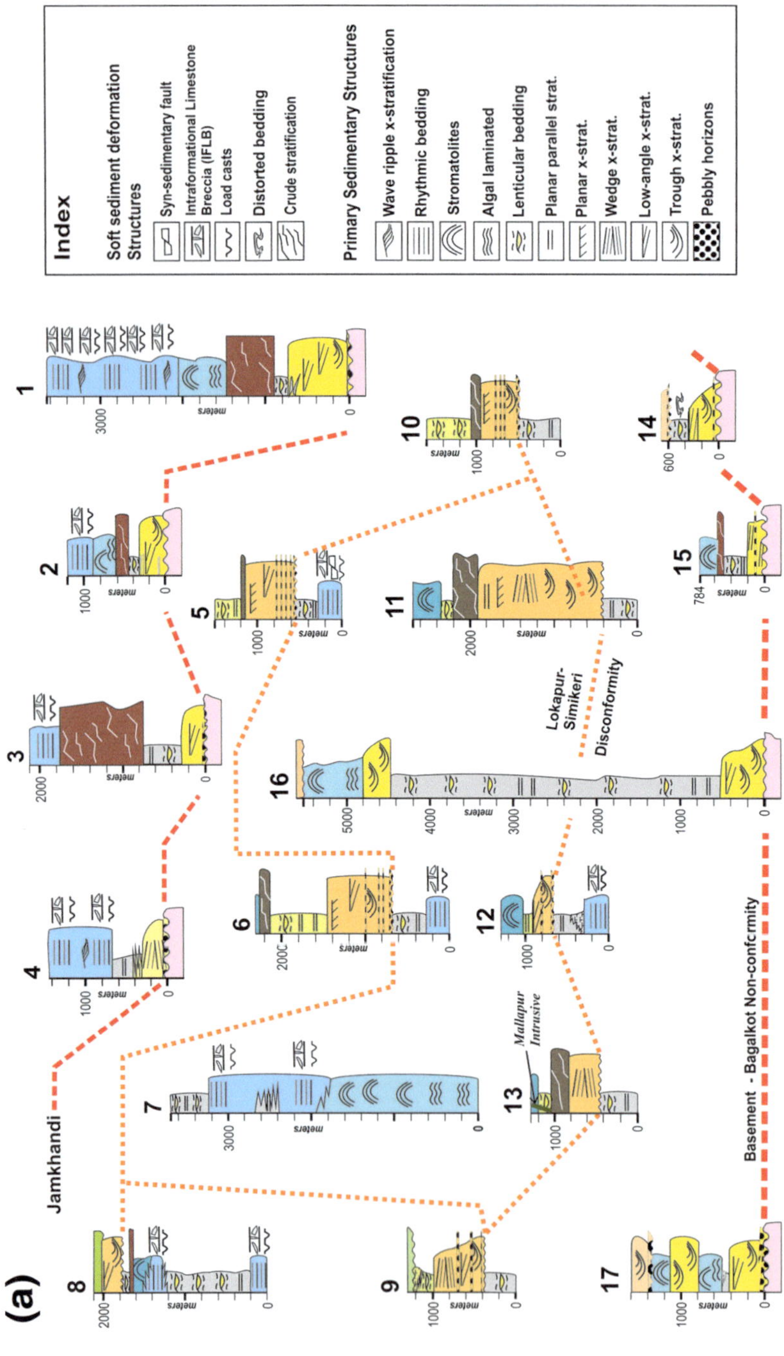
(a)
Index
Soft sediment deformation Structures
Syn-sedimentary fault
Intraformational Limestone Breccia (IFLB)
Load casts
Distorted bedding
Crude stratification
Primary Sedimentary Structures
Wave ripple x-stratification
Rhythmic bedding
Stromatolites
Algal laminated
Lenticular bedding
Planar parallel strat.
Planar x-strat.
Wedge x-strat.
Low-angle x-strat.
Trough x-strat.
Pebbly horizons
Jamkhandi
Lokapur-Simikeri Disconformity
Mallapur Intrusive
Basement - Bagalkot Non-conformity
meters

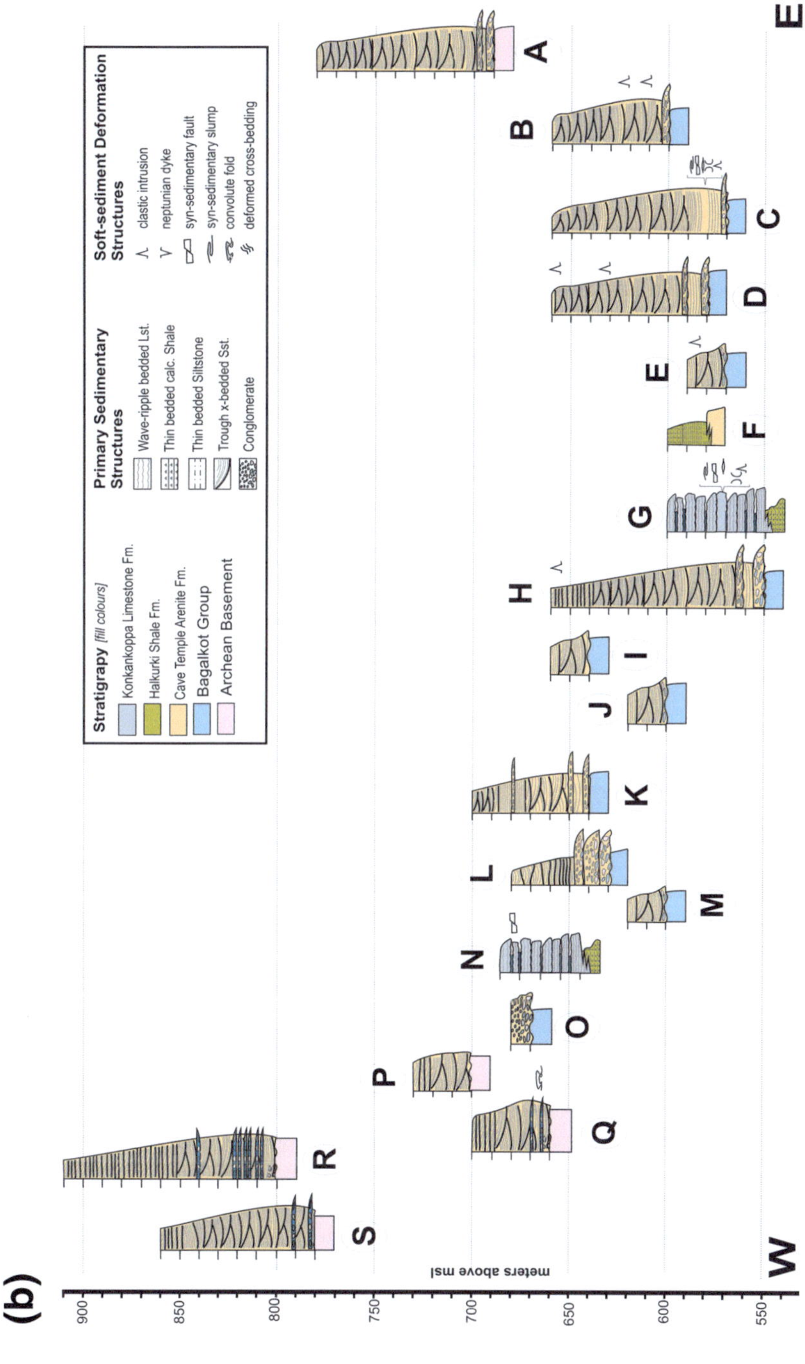

Fig. 6 (continued)

◂**Fig. 6** **a** Measured stratigraphic lithologs reconstructed from vertical sections/traverses across different exposures of the Bagalkot Group. The locations of the logs are shown in Fig. 4a. The lithofacies encountered are colored as per the lithological map of the Bagalkot subbasin given in the inset, with significant sedimentary structures and bedding patterns superimposed on them. The logs are broadly arranged west to east (left to right successively) in a north-up sequence, with reference to their base. Indicative correlations of the stratigraphic boundaries are shown with dashed red lines. These logs show (a) the interfingering nature of the lithofacies leading to recurrence of the same lithofacies in the same sequence; and (b) paucity of chertbreccias in sections where the argillites are prolific and vice versa as discussed in the text. **b** Measured stratigraphic lithologs reconstructed from vertical sections/traverses across different exposures of the Badami Group. Locations of the logs are given in Fig. 4b. The logs are arranged west to east (left to right successively) and plotted against the elevations above msl rather than in terms of absolute thickness as in case of the Bagalkot logs (**a**). This is enabled by the fact that barring low (<5°) rolling dips, the Badami sediments occur in a subhorizontal disposition indicating that they have not suffered any post-depositional deformation

Petrological studies of the strata from these sections, to establish their textural and composition are embedded in these logs. These logs provide a synoptic view of the various lithotypes occuring in different sections. In conjunction with the geological map they provide a clear understanding of the vertical and lateral distribution of various horizons from the Kaladgi basin.

3.4 Subsidence Analysis

A combination of supply (types and rates) and accommodation space (as a function of relative subsidence, driven by eustacy and tectonics) defines the sedimentation patterns and contents in a sedimentary basin (Catuneanu et al. 2005; Allen and Allen 2013). The motifs of the sediment piles that accumulate at different locations within a basin manifest the patterns of relative subsidence in those parts. The reconstruction of successive stages of sediment accumulation within a basin plotted against time enables understand the relative subsidence of the basin floor at different stages of its evolution (Gallagher and Lambeck 1989).

This analysis requires primary data of (a) the lithotypes and their compaction and diagenetic history, (b) estimated paleobathymetry of the different types of sediments, (c) stratigraphic (plotted as vertical) succession of the constituent sediments, (d) the ages of the different strata within the stratigraphic succession; and (e) backstripping that enables quantify the removal or loss of sediments that were accumulated but are not preserved (Sclater and Christie 1980; Bond and Kominz 1984).

In case of the Phanerozoic successions, in which accurate ages of the strata are provided by their fossil assemblages, such subsidence analysis has led to the development of a robust global eustatic sea-level curve (e.g. Vail—EXXON curve —Vail et al. 1977; Haq et al. 1987; Miller et al. 2005). However, this is not possible in the unfossiliferrous Precambrian sequences where age data is sparse

(Catuneanu et al. 2005). In such cases, accurate determination of the net rate of subsidence of the basin floor cannot be plotted and needs to be estimated using other proxies. Kale (1991) used these principles to plot the 'sediment accumulation curves' ('SAC's) of the Purana basins, using the total accumulated stratigraphic thickness in the basin as a proxy of 100% time of the basin evolution. The maximum stratigraphic thickness of each stratigraphic unit computed as a percentile value of the total thickness represents the relative time taken by it in the basin. This yields a relative (rather than absolute—as in case of the Phanerozoic) scale for the basin. We have used our observations (as enumerated in Figs. 4 and 6) and the maximum thicknesses of strata as defined by Jayaprakash (2007) in these computations. The cumulative uncompacted maximum stratigraphic thickness for the Bagalkot Group is 13,717 m and for the Badami Group is 1846 m. These values match well with the geophysically determined maximum depth to basement (e.g. Ramakrishna and Chayanulu 1988; Kale 1991) of the Kaladgi basin.

3.4.1 Sediment Accumulation Curves

This analysis uses Cant's (1989) equation that derives a relationship between eustacy (E), subsidence (Sub), sediment accumulation (Sed) and the depth of the depositional interface (D). It is postulated that the change in water depth (ΔD) during the deposition of a single uninterrupted stratum with thickness (ΔSed), in a basin where space is created due to subsidence (tectonic or gravity driven) of the basin floor (ΔSub) and eustatic sea-level change (ΔE) will be as follows:

$$\Delta Sub + \Delta E - \Delta Sed = \Delta D \tag{1}$$

If the combined effects of subsidence and eustacy are clubbed together into a single parameter (ΔDI), this equation can be simplified into the following form:

$$\Delta DI - \Delta Sed = \Delta D \tag{2}$$

In other words, the sum of sediment thickness and its depositional depth of each strata gives the net change in the depth of the depositional interface due to subsidence and eustacy:

$$\Delta DI = \Delta Sed + \Delta D \tag{3}$$

It is assumed that during the deposition of a single lithotype, the sediment accumulation (ΔSed) kept pace with the net subsidence (ΔDI) and that the change in this resulted in the succeeding lithotype to be deposited in succession. The uncompacted thickness of each stratum in a single stratigraphic section is computed to generate the progressive accumulation, computed as the 'cumulative ΔDI'. Plotting the successive values of ΔDI against time (as determined by the age of the sediments in a sequence) provides a glimpse of the nature of changes in the

Table 2 Listing of compaction factors and depositional depths used in the subsidence analysis of the Kaladgi sediments

Lithology	Value
Compaction factor	
1. Highly compacted and recrystallized sandstones (quartzites)	0.76
2. Well-indurated, cemented sandstones	0.86
3. Siltstones	0.53
4. Mudstones and shales	0.41
5. Carbonates (without significant recrystallisation)	0.75
6. Carbonates (recrystallized)	0.69
Depositional depth in meters	
1. Carbonates (unaffected by influence of surface waves)	-7
2. Carbonates (deposited within wave-influence depths)	-5
3. Tidal muds and silts (=shales)	-3
4. Supratidal and beach sandstones	+3
5. Conglomerates	+7
6. Chertbreccias (reworked tectonically generated debris)	0

progressive changes in the accommodation space available in a basin. Such plots are called Sediment accumulation curves (SACs).

For ancient sediments, the stratigraphic thickness represents the compacted thickness of the preserved sediments. The compaction curves by Sclater and Christie (1980) provide ranges of 0.76–0.86 (sandstones), 0.41–0.53 (siltstones and shales) and 0.69–0.75 (carbonates) for different overburden depths. Subsequent studies of sediment compaction (Baldwin and Butler 1985; Audet and Fowler 1992; Bjørlykke 2014) have also shown that authigenic recrystallisation, cementation and diagenetic changes can affect the compacted thicknesses of sediments. Taking these factors into account, we have used the estimated values of compaction factors for the different lithotypes occurring in the Kaladgi basin as given in Table 2.

The optimum depositional depths for these lithotypes estimated assuming the value of zero for the contemporary mean sea level are also listed in Table 2. Positive values indicate deposition above the mean sea-level while negative values indicate submergent accumulation of the sediments.

The uncompacted thickness of each stratum in a single stratigraphic section is successively added to generate the progressive accumulation, computed as the 'cumulative ΔDI'. The percentile ratio of the uncompacted stratal thickness with the maximum Group-wise thicknesses gives a proxy of the time taken by that sequence to accumulate. The locations of the SACs for the two constituent Groups (Fig. 7) are the same (in numbers) as the stratigraphic sections (Figs. 4 and 6) that were used to plot them.

The overall flatness of the Badami SACs in comparison to those of the Bagalkot Group indicates a relatively slower accumulation in a gradually subsiding basin floor of the former. This indicates a paucity of tectonic influence in the subsidence

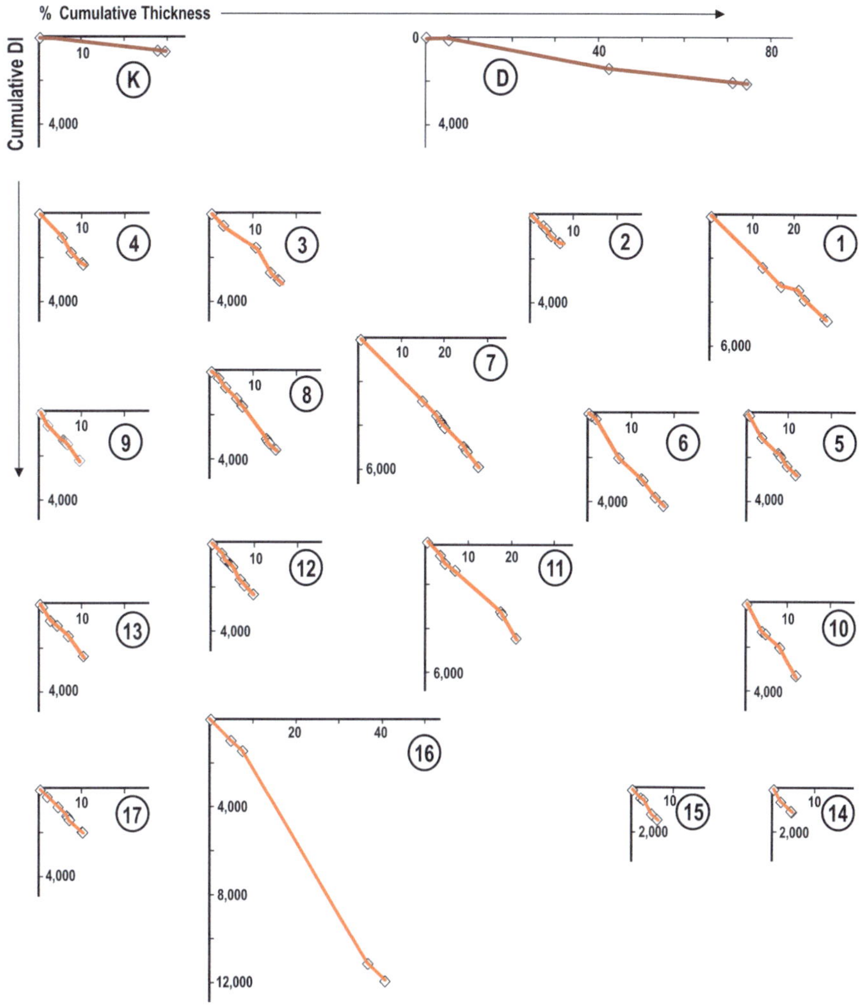

Fig. 7 SACs of the Kaladgi Supergroup. The curves are numbered as per the stratigraphic section (Fig. 6a, b) used to plot them. For the Bagalkot Group rows are arranged from east to west (right to left) respectively to give a perception of the along-strike changes in subsidence patterns. The rows are arranged north to south (top-bottom) respectively. For the Badami Group only two representative SACs are given

history of the Badami Group. The steeper Bagalkot SACs suggest a faster subsidence, which must have been enabled by tectonic deepening besides the eustatic changes in the sea-level through its evolutionary history.

The SACs of the section numbers 2, 7, 9, 14, 15, 16 displays a consistent gradient with little or no deflection; pointing to a rapid subsidence with a fairly consistent rate. These sections are not located close to any of the basement shears. On the other hand, the SACs of section numbers 1, 3, 6, 10, 11 that are located near the basement shears display intervening deflections. These deflections reflect modifications in the relative subsidence rates due to synsedimentary tectonic events. Kale et al. (1998) showed the record of increasing frequency of the intraformational limestone breccia horizons in the Gaddankeri sector indicating enhanced tectonic modifications of the basin floor. Similar observations recorded by us in the other sectors in the form of SSDs in the measured sections (Fig. 6) reinforce this interpretation.

3.5 Depositional System

The broad picture that emerges is of a continent-margin shallow platform complex with segments of continental deposits. The geographic spread of the basin, with the basement exposed on all sides shows that it was in all probability a land-locked basin with sediments supplied from all sides and deposited in it. Our data supports the interpretation (Kale et al. 1996; Jayaprakash 2007) of the Kaladgi sediments being deposited in transitional to shallow-marine regimes with local gentle depositional dips of 2°–5°. The detrital siliciclastics (quartzitic sandstones and conglomerates) were deposited in near-shore environments. Carbonate mudflats that yielded the limestones and dolomites were further away from the shoreline, on a shelf with shallow depths within the wave-base. The intervening mudflats yielded the silicious and ferrugenous argillites on the landward side and calcarous mudstones on the open sea-side. The bedding plane markings, mud cracks, rip up clasts and thin lensoid bedding evidence their accumulation under tidal influence with alternate pulses of flooding and subaerial emergence.

Intricate sediment patterns in some sectors of the Bagalkot subbasin (particularly the parts between Mudhol–Lokapur–Bagalkot) as depicted in the map (Fig. 4a) and evident in the stratigraphic logs (Fig. 6a) suggest that the gently sloping profile of the basin floor in these sectors was subjected to deviations, leading to uneven relative subsidence. Pulses of temporary (and localised) shallowing or deepening in excess of that expected during a normal trangressive sequence are reflected by the alternations of shales and carbonates in these sectors. Some of these deviations could be artifacts of the original basin floor geometry, such as basement highs and lows. Deviations in the sediment supply rates may also have contributed to the diversity in the dispersal (and consequently depositional) patterns of the sediments. Some of these variations are interpreted to be consequences of synsedimentary instability of the basin floor, created by tectonic events. The distribution of the intraformational breccias in Petlur Carbonate, chertbreccias and seismites in other horizons (Kale et al. 1998; Kale and Patil Pillai 2011; Patil Pillai and Kale 2011) reinforce this possibility.

4 Structural Configuration

For understanding the evolution of the basin, it is necessary to try and separate the diverse influences (basement undulations, supply rate variations, sea-level changes and tectonics) and their imprints in the sedimentary record. The primary requirement for enabling this was to establish the three-dimensional geometry of various horizons. This is relatively straight forward in case of the subhorizontal Badami Group. However, the tightly folded sequence of the Bagalkot Group does not permit the lateral continuity of the lithostratigraphic units to be established without deciphering the structural patterns and 'unfolding' the strata.

The deformation patterns in the basin were earlier documented by Awati and Kalaswad (1978), Nair and Raju (1987), Jayaprakash et al. (1987), Gokhale and Pujar (1989), Jayaprakash (2007), Mukherjee et al. (2016), Jadhav and Kshirsagar (2017) amongst others. Traverse mapping to validate earlier observations and field observations on the orientations of the bedding planes, fractures and phyllitic cleavage (in case of the argillites) have been used to understand the sense of movement in different folds and faults in the basin.

Figure 8 compiles these observations in the Bagalkot subbasin that has undergone significant tectonic deformation. It is apparently divided into the northern and southern sectors (Figs. 4a and 8) that are structurally different from each other, separated by younger cover. The northern sector hosts tight eastwards or westwards plunging folds. The southern sector exposes monoclinal, northward dipping strata from the Lokapur Subgroup (Fig. 8). This stronger deformation of the Bagalkot Group in the central parts than that along its fringes is akin to mobile belts. This contrasts to that recorded from the other Purana basins, which display deformed margins and relatively undeformed mid-basinal sectors.

The narrow linear Badami subbasin (Inset Figs. 2 and 4b) is almost undeformed. The Badami Group is characterized by subhorizontal dips and small local faults showing dragging along them. No other structural deformation is recorded from this Group.

4.1 *Regional Faults/Shears*

The Kaladgi basin displays several sets of inherent as well as superimposed faults (with evidence of brittle deformation and displacement) and shears (zones of quasi-ductile to ductile deformation). The sense of slip along them is recorded based on slickenside surfaces, drag of the bedding and joint patterns along them. They dominantly trend E-W to ESE-WNW with gentle swirls along their lengths and occur close to the margins of the basin.

The Sirur Shear zone extends for >100 km across the central part of the basin and further eastwards into the basement terrain. It displays a minor right-lateral sense of movement and a larger dip-slip component with downthrows towards

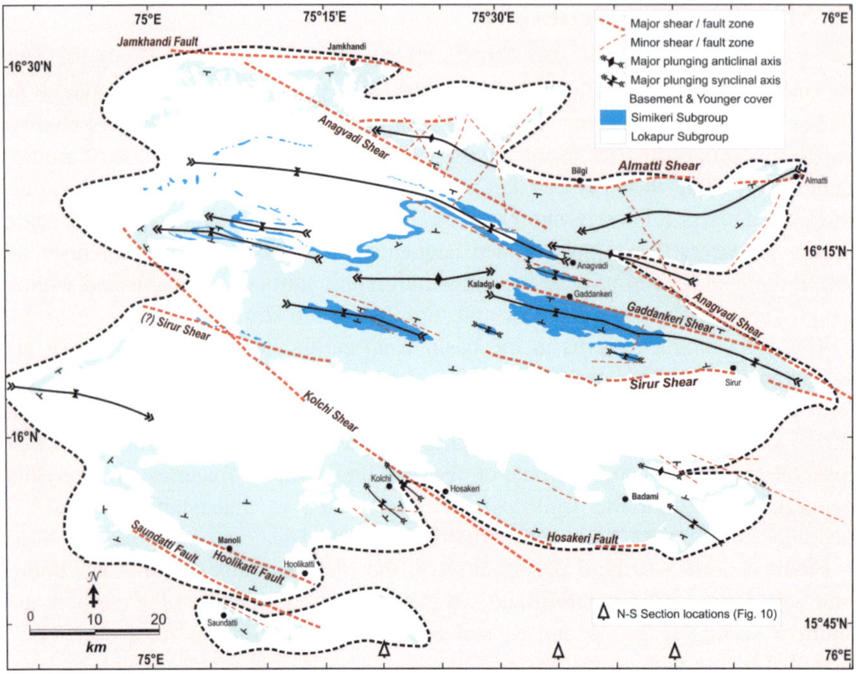

Fig. 8 Structural map of the Bagalkot subbasin showing the major folds and faults. The dotted outline marks the boundary of this subbasin that is surrounded on all sides by basement rocks. Regional shear zones such as the Kolchi and Anagwadi Shears have been recorded in geophysical studies as crustal scale lineaments. The Sirur Shear separates the northern folded sector and the southern monoclinal (northward dipping) sector of this subbasin

north. This shear zone (Fig. 8) is largely concealed below younger cover (of the Badami Group and Deccan Traps). It occurs along the boundary between the northern and southern sectors of the Bagalkot subbasin. It also marks the northern edge of the Badami subbasin in the eastern sector (Fig. 4b). This suggests that it may have suffered a reversal of sense of movement, yielding southward downthrows during the Badami sedimentation. Most of the other E-W trending faults display dominantly dip-slip displacements with basin-ward downthrows, and negligible to no strike-slip components.

The NW-SE trending sinistral Anagvadi and Kolchi Shears extend into the basement. The imprints of the former are exposed across the Bagalkot subbasin. The northward continuity of the Kolchi Shear is concealed below younger cover and is hence shown with a thinner line in Fig. 8. Both correspond to megascale geophysical lineaments of Peninsular India (Sridhar et al. 2017; Rajaram et al. 2016) that extend far beyond the bounds of the Kaladgi basin.

Minor faults have been mapped in several parts of the basin with NNE-SSW, N-S, NNW-SSE trends (Fig. 8). They appear related to folding as sympathetic fractures that show brittle deformation when transecting the quartzitic and conglomeratic horizons, while the argillites and carbonates suffered ductile dragging coherent with the sense of movement (see Mukherjee 2014).

The synsedimentary deformation from the Bagalkot Group is oriented parallel to the adjoining shears and faults (Kale et al. 1998; Kale and Patil Pillai 2011; Patil Pillai and Kale 2011). This indicates that they may be inherited structures from the basement that were reactivated during and after the sedimentation in the Kaladgi basin. Faulted contact of the basement and cover sediments when traced laterally, assumes the nature of an unconformable relation, with the trace of the fault concealed below the sedimentary cover. This suggests that although faulting played an active role in the basin subsidence, transgressive onlap of sediments beyond the faults did take place. The combination of the NW-SE trending regional shears and the E-W trending Sirur Shear suggest that basinal growth took place in a transtensional regime.

Several cross-faults that display quasi-ductile deformation in the argillaceous and carbonate sediments cross the quarzitic ridges as brittle faults lined with fault breccia. This aspect of competency contrast provides a better understanding of the stress distribution during deformation and provides clues to the sense of slip along them.

4.2 Folds

Folds in the Bagalkot subbasin are easily mappable from aerial photographs and satellite imageries (Awati and Kalaswad 1978; Nair and Raju 1987) due to the rims of the quartzite ridges along them (Fig. 9). These folds are large amplitude, small wavelength 'close folds' with inter-limb angles of less than 30°. They display plunging fold axes and curviplanar axial planes (as shown in Figs. 8 and 9).

This has been interpreted as a weaker event of cross folding (Mukherjee et al. 2016; Jadhav and Kshirsagar 2017). Alternatively, the fold geometry may have also been influenced by inherited structures such as the basement shears and paleotopographic undulations, yielding a perception of a second phase of coaxial folding.

Incipient greenschist facies metamorphism is manifested as development of chlorite, sericite, albite (Govinda Rajulu and Chandrasekhar Gowda 1972) and phyllitic cleavages in argillaceous sediments; and uralitization of pyroxenes and albitisation of plagioclase in the Mallapur Intrusives (Patil Pillai et al. 2018). It is pronounced at locations where the secondary faults cut across them and along fold axes, but becomes indistinct away from the faults and on the fold limbs.

4.3 Synthesis

Three N-S geological cross-sections are reconstructed to depict the complex nature of deformation and the stratigraphic relationships between various horizons in the

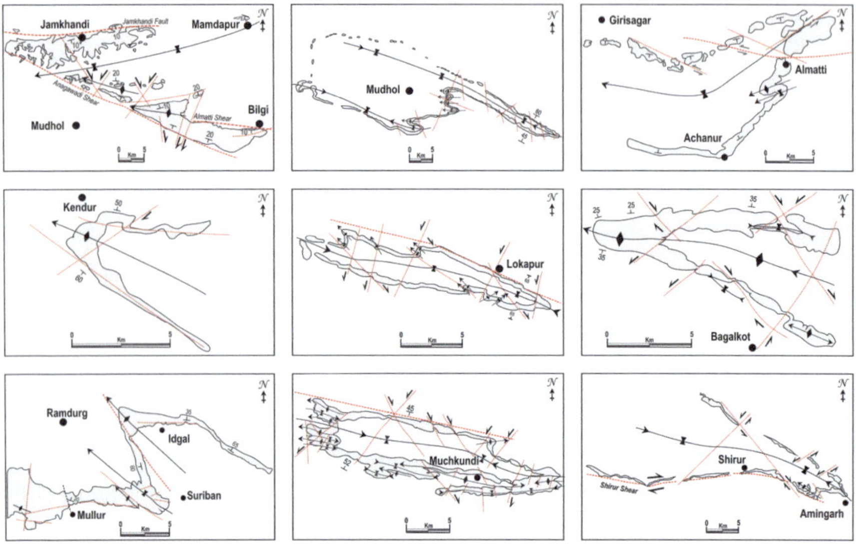

Fig. 9 Sketch maps of major folds in the Bagalkot subbasin. The fold axes are shown with their plunge direction and anticlinal/synclinal symbol. Faults are shown with thin dotted red lines. Strike-slip faults are depicted with the sense of slip along them. Faults with dominant dip-slip component are shown without any marking of sense of movement along them

Kaladgi basin (Fig. 10). Besides enabling understanding of the lateral facies changes, they also show the relationships between various folds and faults in the deformed sector of the Bagalkot subbasin.

The E-W and NW-SE trending regional shears in the basement continue as faults affecting the sediments. The parallel orientations of the synsedimentary deformational structures to them (Patil Pillai and Kale 2011) indicate that they were active during the sedimentation in the basin (D_0) and have influenced the basin geometry. The sediment distribution patterns as lithofacies stripes parallel to these basement structures reinforces this interpretation. They controlled its deepening by subsidence of the continent-margin shallow marine platform.

The pattern of the SACs (Fig. 7), the shore-line—shelf depositional regimes of the sediments and their stripping patterns collectively indicate that extensional subsidence was responsible for the growth of the basin. Presence of synsedimentary extensional faults as documented by Patil Pillai and Kale (2011; Fig. 5) in the lower sequences of the Kaladgi basin and the onlap of sediments across bounding faults are additional evidences in support of multiple transgressions that occurred in a progressively deepening basin. In absence of systematic geophysical or drilling data, it is difficult to validate the depth penetration of the structures recognized during surface mapping as depicted in the cross-sections (Fig. 10).

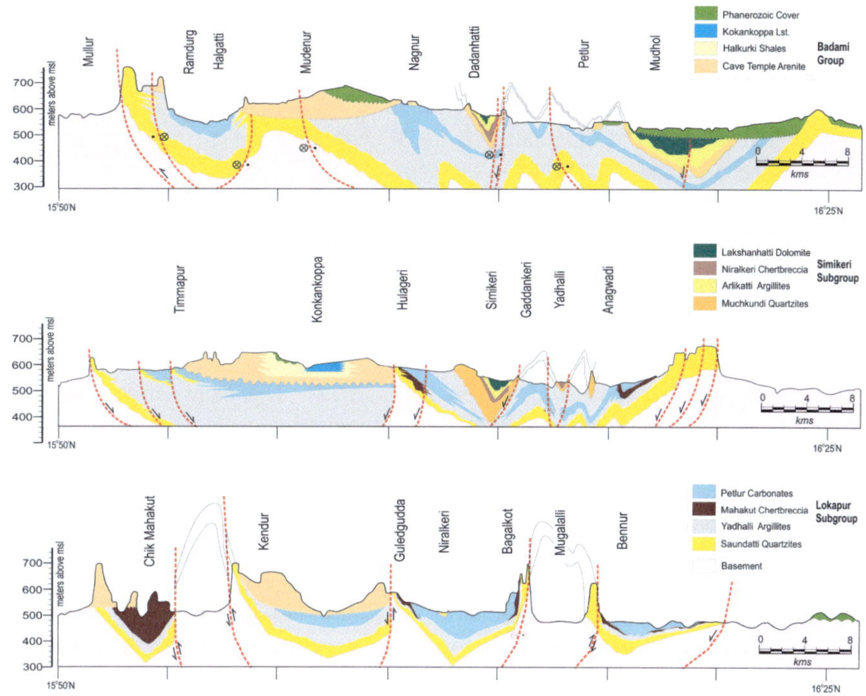

Fig. 10 Sketches of N-S geological cross sections across the eastern part of the Kaladgi basin at locations given in Fig. 9. The sections are N-S sections along the longitudes of 75° 20′E, 75° 35′E and 75 45′E from top to bottom respectively. Note that the subsurface extension of the dips of beds and of the fault/shear zones have limited validation from field observation

These features suggest a dominantly extensional regime during the growth of this basin, with minor but significant strike-slip component. We therefore conclude that the inception and growth of the Bagalkot subbasin occurred in a transtensional tectonic regime.

The parallel orientation of fold axes with these basement structures suggests their control on the deformational history of the sediments at a later date (D_1). Jadhav (1987) and Jadhav and Phadke (1989) interpreted the large amplitude, WNW-ESE trending sinusoidal folds in the multilayered succession of Bagalkot Group to form by buckling under relatively homogenous stress conditions. Based on the fold orientations and field data, we infer that the initial buckling was caused by transpressional stresses active in NE-SW and NW-SE directions during the D_1 event. As the strain gradually increased, sinistral as well as dextral verging cross-folds developed along the limbs of these folds, accompanied by transverse fractures and faults during a later event (D_2). This interpretation of reactivated basement shears governing the deformation history is more plausible than the gravity gliding model proposed by Mukherjee et al. (2016).

The Badami sediments were deposited on the eroded surface of the deformed Bagalkot Group (D_3). These younger sediments do not display significant deformation except for some local faults. Its depositional characters dominated by continental deposition suggest that the younger Badami subbasin was nested in the uplifted and eroded fold-mountains of the Bagalkot Group in the north and the ridges of the Saundatti Quartzites in the south.

5 Basin Analysis

This basin had developed on the continental crust of the Dharwar craton (Kale 1991; Basu and Bickford 2015; Allen et al. 2015). The Dharwarian basement served as the source of its sediments. The geometric relationship between the regional basement shears that were active during the sedimentation and the sedimentation patterns of the Bagalkot Group indicates that movements along them facilitated the inception and growth of the basin. The tectonic influence on the evolutionary history of this basin therefore precludes the possibility of this being an intracratonic sag as suggested earlier (Jayaprakash et al. 1987; Jayaprakash 2007; Dey et al. 2009; Dey 2015). A different model of the evolution of the Kaladgi basin emerges based on the evidence and analyses documented above.

5.1 Phases of Basin Evolution

While integrating these observations, the ages of different events (Table 3) inferred by Patil Pillai et al. (2018) are used. It is necessary to recognize that the ranges of the ages inferred in the Table 3 are not the durations of the sedimentation. An average sediment accumulation rate of 0.1 mm/year yields a duration of $\sim$130 million years to the Bagalkot Group and $\sim$19 million years for the Badami Group. Modern accumulation rates in deltas range between 1 mm/year (=1 m/k year) and 10 cm/year (=0.1 km/k year). They are slightly different, but in the same magnitude range for shelf seas (Saddler 1999; Sommerfield 2006). Using analogs of Phanerozoic sediments from the near-shore systems, the actual durations of sedimentation in the Kaladgi basin are more likely to be a fraction of the estimates based on average sedimentation rates (Kale 2015, 2016). This adds to certain uncertainties when tagging the ages of the events in this basin.

5.1.1 Early History

Given the ages of the Mesoproterozoic dyke swarms, occurring in the adjoining areas (Kumar et al. 2015), it is evident that the Kaladgi sedimentation cannot be older than 1800 Ma (see also Basu and Bickford 2015). This cratonic block of the

Table 3 Chronology of events in the Kaladgi basin modified after Patil Pillai et al. (2018)

Subbasin	Sedimentological events	Tectono-thermal events	*Estimated age range*/measured age (Ma)
Badami	Continental sedimentation with localised marine transgression	Extension (**D_3**)	~900–800
	Emergence and erosion of folded strata (+basement)	Hiatal break	1100–900
Bagalkot	Post depositional events	Low grade metamorphism and deformation (**$D_{2'}$**)	1150–1100
		Mafic intrusion	1154 ± 4
		Folding and faulting (**D_2**)	
	Basin deepening with synsedimentary intra-basinal tectonics	Extension	~1200
	Local intra-basinal reorganisation (?)	Intra-basinal tectonics with co-axial deformation (**D_1**)	>1200
	Peak transgression	Extension and subsidence (**D_0**)	<1300 + 50
	Basin inception		~1400 + 50
Basement	Emergence and peneplanation	Continental regime	>1600 Ma

northern Dhawar craton was subjected to an emergent period of continental weathering and erosion leading to a peneplained topography with gentle gradients following the event of dyke intrusion. This is consistent with the period of stability of supercontinental assembly (between 1800 and 1500 Ma) predicted by Condie et al. (2015).

The reorganization of continental blocks and assembly of Rodinia after the breakup of Nuna started around 1400 Ma (Cawood and Hawkesworth 2014; Condie et al. 2015). Similar coeval extensional regimes and cratonic rifting are recorded from the other Purana basins of Peninsular India (Kale 2016). It is speculated that the primary inception of this basin on the northern edge of the Dharwar craton may be a part of this global tectono-thermal event. Transtensional reactivation of some of the basement shears, like the Sirur Shear, enabled tectonically driven subsidence in the Bagalkot subbasin. It is generally accepted that during the existence of the Mesoproterozoic supercontinents, due to their low topographic gradient, even small sea-level increase could lead to flooding of large areas creating wide-spread platform basins (Eriksson et al. 2006; Bradley 2011), similar to the Kaladgi basin.

5.1.2 Growth

The growth of the Kaladgi basin between 1400–1300 Ma is consistent in the global context of marginal break-up of the supercontinents. The upper crustal regime would normally be expected to fail (under extensional stress) along brittle fault planes. The brittle upper crustal levels of the continental crust would essentially display failure during extension with fault complexes analogous to those found in present day continent margin rifts (Gibbs 1984; Allen and Allen 2013) as postulated in the model in Fig. 11a. The cross-sections across the Kaladgi basin (Fig. 10), and fault-plane orientations, can be modeled into an analogous listric system.

A postulated listric geometry of the main faults/shear zones that governed the growth of this basin is endorsed by the fact that at several locations, mesoscopic listric and antithetic fault complexes of meter-scale dimensions are seen (Fig. 11b–e). The field exposures not only endorse the basin-scale listric geometry of the growth fault complex, but also show it's replicability at smaller scales. The maximum recorded vertical thickness of sediments (of ~5000–6500 m) suggests that the depth of the detachment plane was located at those depths.

Given the element of horizontal movements recorded from various associated faults (see Figs. 8 and 10); the extension was associated with a degree of horizontal shear. It is therefore interpreted that the early growth of the Kaladgi basin was in a transtensional regime rather than purely extension driven by gravity as previously modeled by Mukherjee et al. (2016). The essentially extensional, fault-driven subsidence (=D_0) of the floor of the Bagalkot subbasin enabled accumulation of thick sediments in it.

A minor, localized event of basin floor reorganization led to the intrabasinal unconformity between the Lokapur and Simikeri Subgroups in the northern sector of this subbasin (Kale et al. 1998; Jayaprakash 2007). The Simikeri sediments are confined within the doubly plunging synclinorial areas. These areas are closely aligned with the Anagvadi, Sirur and Gaddankeri Shears (see Fig. 8). We infer that the localized restructuring of the basin floor in these narrow parts of the Bagalkot subbasin may be linked to reactivation of these shears (D_1). The incremental occurrence of intraformational limestone breccia horizons towards the Gaddankeri Shear (Kale et al. 1998) endorses it.

Sedimentation in other parts of the basin appears to be uninterrupted by this localized event (D_1). It may have also produced the first generation folds in the central parts of this subbasin, leading to the exposure of some of the strata to weathering and erosion. The emergent strata of the Lokapur Subgroup have contributed sediments to the Simikeri Subgroup as evident in basal conglomerates from the Muchkundi Quartzites. This event reflects as kinks in the SACs that cover both the subgroups from the central parts of the basin (e.g. nos. 6 and 10 in Fig. 7). Extension, consequent relative deepening of the basin and sedimentation continued after that for a relatively shorter duration, during which the thinner Simikeri Subgroup was deposited. Based on the chronological data available (Patil Pillai et al. 2018) and the assumption of sedimentation rates ranging between ~0.1 to 1.0 mm/year; the event D1 is estimated to have occurred slightly before 1200 Ma.

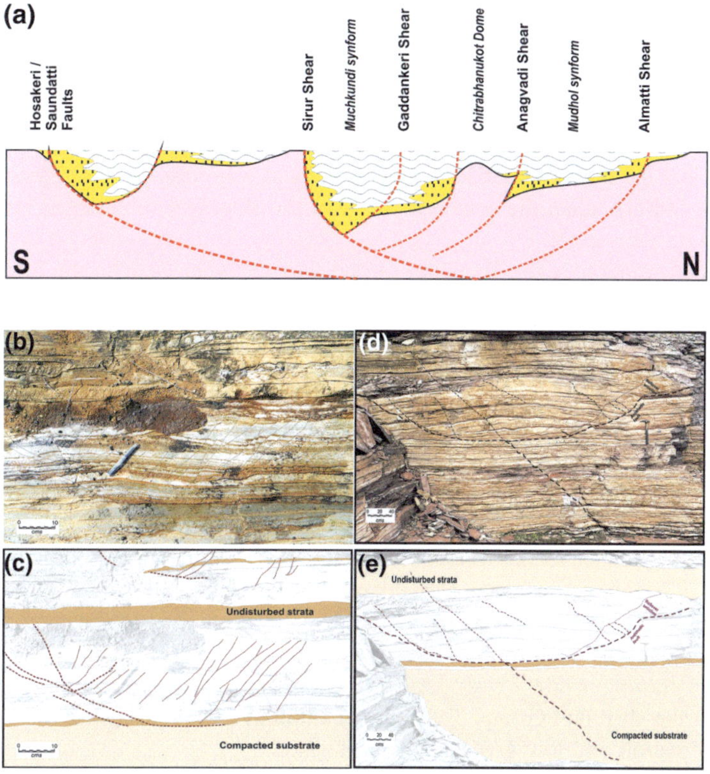

Fig. 11 **a** Schematic N-S model of the fault/shear framework of the Bagalkot subbasin based on models of continental rifting with a listric geometry predicted by Bott and Mithen Bott and Mithen 1983; Dula 1991). The inferred listric geometry is based on field evidence of movement along various faults. The postulated northward verging listric geometry is endorsed by the occurrence of similar mesoscopic arrays of faults, whose examples are given below (**b** and **d**). **b** Field photograph of subvertical dipping Chikshellikeri Limestone near Gaddankeri. Uneven compaction of the alternating impure and pure limestone strata have produced several synsedimentary deformation features such as compaction lobes, flame structures and pinching and swelling of the parallel bedded strata. Note the small-scale fault system transecting it, which has not affected the overlying (top of the photo) strata indicating its intraformational nature. **c** Sketch of the photograph in **b**. Note two sets of small listric faults and antithetic faults terminating against the substrate of undisturbed strata can be marked in this exposure. The lack of penetration into the underlying strata and termination against the overlying bed confirms their intraformational status. **d** Field photograph of subhorizontal Kokankoppa Limestone at Chandargi village (see Fig. 4b). Weakly stylolitic parallel bedded limestones are interspersed with thin muddy impure beds resulting in a flaggy appearance to them. A small fault system has affected only the strata in between; that shows a 'saucer-like geometry' analogous to listric fault systems. **e** Sketch of the photograph in **d**. Their intraformational nature of the listric array of faults is established by the lack of penetration into the underlying strata and termination against the overlying bed. Typical antithetic faults and roll-over anticline is observed on the northern (right) side of the picture as marked

5.1.3 Deformation

The Bagalkot sediments were folded and faulted during the event recognized as D_2. The sub-parallel alignment of the fold-axes with the NW-SE to WNW-ESE regional shear zones indicates that this event was related to another pulse of activity along them or that these zones channelized N-S oriented compressive stress systems. The D_2 event did not affect the southern sector of the Bagalkot subbasin as much as it did the northern sector.

The deformation of the Bagalkot Group was interspersed with the intrusion of dykes (=Mallapur Intrusives) along the D_2 fold axes. The dykes are dated at 1154 ± 4 Ma by $^{40}Ar/^{39}Ar$ method (Patil Pillai et al. 2018). The estimated age (~1200 Ma: based on the dating of a dyke (Mallapur Intrusives) that is emplaced along the D2 fold axes) of this event corresponds well with the period of super-continental reassembly leading to the development of Rodinia (Condie and Kroner 2013; Condie et al. 2015).

Further deformation and low grade greenschist facies metamorphism of the sediments along with the intrusives is recognized as the event D_2', since there is insufficient evidence to say whether it represents the peak of the deformation event (D_2) or is a separate episode altogether. Cross faulting of the earlier folds and co-axial second generation shear folds observed here are likely to have evolved at a later stage than the main folding. Jadhav and Phadke (1989) recognize two separate events of folding while Jayaprakash (2007) records only a single deformational event for the Bagalkot Group. The sedimentation in this subbasin ceased following the deformation and low grade metamorphism.

5.1.4 Hiatal Break and Younger Cycle

The subsequent period of quiescence coincides with the period of stability of the Rodinia supercontinent estimated between ~1100 and 900 Ma. This period was marked by the emergence and erosion of the folded hills of the Bagalkot Group along with its basement. The deposition in the Badami subbasin is essentially a continental sedimentation with very little component of the marine interface. The presence of ichnofossils from the Badami sediments (Kulkarni and Borkar 1997) confirm their Neoproterozoic age. Given that some parts of the Badami Group rest directly upon the basement, suggests a significant removal of the preceding Bagalkot sediments prior to the inception of these younger sediments. We therefore project the age of the extensional growth (D3) of the Badami subbasin to be between 800 and 900 Ma corresponding to the age of the early breakup of Rodinia before its reorganization into the Gondwanaland assembly (Cawood and Hawkesworth 2014).

The thickly-bedded, poorly-sorted conglomerates at the base of the Badami Group are analogous to debris-wash deposits commonly found at the base of slopes in a continental regime. Fluvial influence in the form of braided channel sands is overprinted on the sandstones of the Badami Group. Taken together with the narrow linear expanse of the Badami subbasin, a bulk of these sediments are interpreted as continental fluvial deposits derived from the flanking hill ranges of the folded Bagalkot Group and its basement. They are eventually overlain on the eastern side by the mudflat siltstones and flaggy impure limestones. This indicates that a transgressive event on a shallow shelf occurred during this sequence of sedimentation. The adjoining sandstones in this tract between Yargatti–Chandargi (on the west) and Badami (in the east) show evidence of deposition in a near-shore environment. It is concluded that the Badami sedimentation occurred in a nested basin, attended by limited extension and subsidence. The subsidence curves (Fig. 7b) indicate this to be a very slow subsidence that was overtaken by a marine transgression during its terminal phase.

5.2 *Relative Sea-Level*

All configurations of the assembly of Proterozoic supercontinents (Rogers and Santosh 2004; Piper 2013, 2018) show the Indian peninsula to be located close to the edges of the successive supercontinents. The Proterozoic eustacy is primarily driven by gradually reducing crustal growth rates, large continental assemblies (that depress the sea-level) and consequent evolution of large continental freeboards, enabling flooding of large areas of craton-margins (Eriksson et al. 2001, 2006; Bradley 2011). This led to platforms that promoted the spread of cynobacterial mats (stromatolites and microbial mat deposits) along with mudflats where argillaceous and carbonate sediments accumulated in thick sequences. The proximity to the oceanic waters (due to its position on the edge of supercontinental assemblies) enabled flooding in the Purana basins as also in the Kaladgi basin.

Taking the depositional environments and geographical interrelations into account, the relative sea-level curve (Fig. 12) indicates that eustatic fluctuations contributed to the evolution of the basin. There have been at least two major flooding events in the history of the Kaladgi basin, one during the Mesoproterozoic and the second during the Neoproterozoic.

The transgressive events were closely interlinked with the tectonic subsidence in the Bagalkot subbasin. The tectonic influence on evolution of the younger Badami subbasin appears to be milder. The tectonic subsidence has left its imprints in the form of growth faults, seismites, synsedimentary deformational structures and fining upwards sequence. The tectonic subsidence, as discussed above, was essentially extensional with minor strike-slip component for the Bagalkot subbasin. It was purely extensional during the Badami sedimentation.

This compiled data of sedimentary environments, subsidence, deformation and stratigraphic succession of the Kaladgi Supergroup enables a more robust analysis of its basinal history. It is evident that this interpretation matches well with recent understanding of Proterozoic systems (e.g. Condie and Kroner 2008, 2013; Cawood and Hawkesworth 2014) and the supercontinental and climatic cycles (Evans 2013;

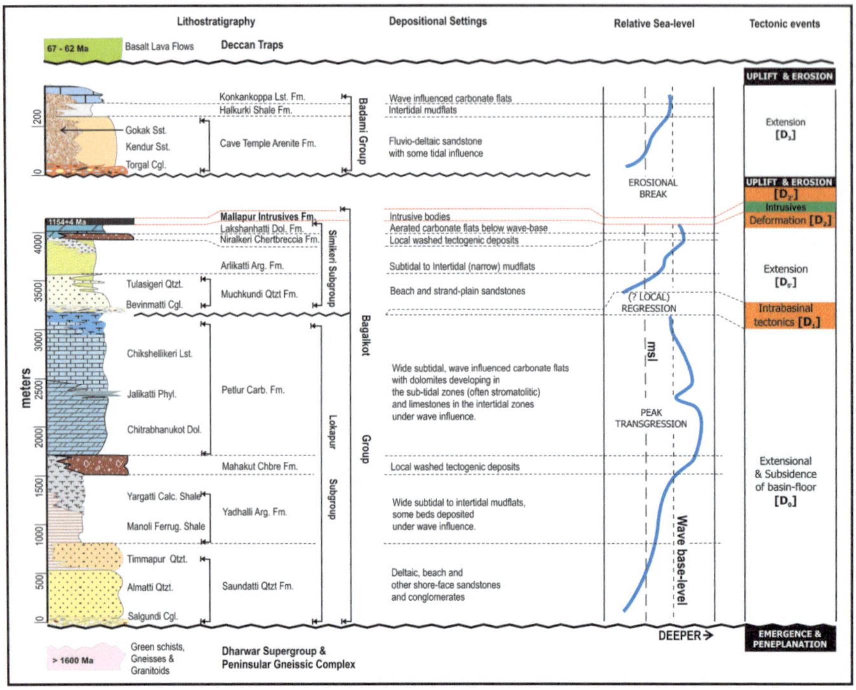

Fig. 12 Depositional environments of the successive lithostratigraphic units from the Kaladgi Supergroup. The relative sea-level curve plotted is based on the depth of deposition of corresponding lithologies and their mutual interrelations (adjoining/overlapping) across different parts of the basin. The tectonic events interpreted in this work are depicted in the next column to demonstrate the relative temporal relation between the tectonic events and sea-level fluctuations as inferred by us

Piper 2013; Rooney et al. 2015; Condie et al. 2015) that controlled Proterozoic sedimentary basin evolution (Eriksson et al. 2006; Allen et al. 2015).

5.3 *Basin Classification*

The Kaladgi basin was described as an intracratonic sag/rift basin by earlier workers (Jayaprakash 2007; Dey 2015). Using the classification parameters of sedimentary basins (see Allen et al. 2015), the Kaladgi basin qualifies as an intracratonic/ continent interior basin, bounded on all sides by the basement rocks of the Dharwar craton. Miall et al. (2015) consider this to be a 'fault-bounded basin related to far field tectonism', given that there is no contemporary tectonic zone in its vicinity. Our work shows that the two constituent Groups in this basin display different tectonic styles. A unified classification of this basin is therefore not appropriate. It must therefore be recognized as a polyhistory basin (sensu Kingston et al. 1983).

Although reactivation of pre-existing shear zones appear to have played an important role in its evolution, the fault-bounded Bagalkot subbasin shows typical geometry of a rifted continental crust with listric faults, suggesting an extensional origin. Its overall geometry and sediment distribution is consistent with typical Precambrian intracratonic fault-bounded sag basins (Gibbs 1984; Pirajno and Santosh 2015; Roberts et al. 2015). On the other hand, the Badami subbasin is best ascribed to a 'successor basin' (post-tectonic intermontane basin: sensu Allen et al. 2015) that ensued after the tectonic deformation of the Bagalkot Group.

Its tectonic history bears a close temporal relationship with the break-up and reassembly of the Proterozoic supercontinental cycles; but lacks any tectonic province or oceanic crustal interface in its proximity. It is therefore concluded that the Kaladgi basin is an intracratonic polyhistory basin and its origin and growth was driven by reactivation of basement shear zones.

6 Conclusions

The Kaladgi Supergroup is comparable to the large platform sediments accumulated in the Purana basins of Peninsular India, but differs from them in the nature and intensity of its deformation. Its position on the northern edge of the Dharwar craton, lack of proximity to any orogenic belt or large intrusive complex; pose difficulties in ascertaining its origin and growth.

Our studies conclude that the Bagalkot Group was deposited during a eustatic sea-level rise and transgressive incursions on the cratonic platform, in close association with basement tectonics resulting from reactivation of the basement structures. Activity along synsedimentary growth faults modified the geometry and geography of the depositional interface in different sectors of the Bagalkot subbasin (Fig. 11). We recognize cycles of extensional subsidence interspersed with folding and faulting during the history of this basin (Fig. 12). Our work shows that there have been at least two clear transgressive events (and a possible third one that was either a part of the first or an independent one) during the Mesoproterozoic and Neoproterozoic times. Whether they are contemporaneous with the transgressive events in the Cuddapah and Bhima basins surrounding the Dharwar craton (Kale 2016) and how this compares in the global framework of the supercontinental assemblies and breakups during the Proterozoic remains in the realm of future studies.

We conclude that the Kaladgi basin provides a rare glimpse of how eustacy and tectonics closely worked in generating this Proterozoic polyhistory basin. The ages of the constituent sediments are as yet poorly constrained and remain an area requiring focused attention. Similarly, a detailed isotopic profiling of the sediments in the basin may not only help understand its provenance character, but actually may enable decipher the unroofing history of the emergent landmass of the Dharwar craton during its evolution.

Acknowledgements This work is the result of more than 3 decades of research in the Kaladgi basin that was funded by CSIR, DST and AMD. Different students and colleagues at different times have helped in the data collection and we acknowledge their contributions to the work. We are thankful to Prof. V. V. Peshwa, Prof. T. K. Biswal, Prof. V. G. Phansalkar, Dr. Amogh Chitrao, Dr. Anand Kale, Dr. Himanshu Kulkarni and Dr. Pradeep Jadhav for their suggestions and discussions on this work. We thank Prof. Soumyajit Mukherjee for inviting us to contribute this chapter and his patient support and encouragement. The comments and suggestions from the reviewers helped improve the manuscript significantly.

References

Allen PA, Allen JR (2013) Basin analysis—principles and applications, 3rd edn. Wiley-Blackwell, 549pp. ISBN: 978-0-470-67377-5
Allen PA, Eriksson PG, Alkmim FF, Betts PG, Catunaeunu O, Mazumdar R, Meng Q, Yong GM (2015) Precambrian basins of India: stratigraphic and tectonic context. In: Mazumdar R, Eriksson PG (eds) Precambrian basins of India: stratigraphic and tectonic perspective, vol 43. Geological Society, London, Special Publications, pp 5–28
Audet DM, Fowler AC (1992) A mathematical model for compaction in sedimentary basins. Geophysical Journal International 110, 577–590
Awati AB, Kalaswad S (1978) Structure of the Kaladgis around Yadwad, Belgaum district (A study based on Landsat-1). Bulletin of Earth Science 6, 43–47
Baldwin B, Butler CO (1985) Compaction curves. AAPG Bulletin 69, 622–626
Basu A, Bickford ME (2015) An alternate perspective on the opening and closing of the intracratonic Purana Basins in Peninsular India. Journal of Geological Society of India 85, 5–25
Biswas SK (2008) Petroliferous basins of India—fifty years' history and perspective. Memoir of Geological Society of India 66, 159–201
Bjørlykke K (2014) Relationships between depositional environments, burial history and rock properties. Some principal aspects of diagenetic process in sedimentary basins. Sedimentary Geology 301, 1–14
Bond GC, Kominz MA (1984) Construction of tectonic subsidence curves for the early Palaeozoic miogeocline, Southern Canadian Rocky mountains: implications for subsidence mechanisms, age of breakup, and crustal thinning. Geological Society of America Bulletin 95, 155–173
Bose PK, Sarkar S, Mukhopadhyay S, Saha B, Eriksson P (2008) Precambrian basin-margin fan deposits. Mesoproterozoic Bagalkot Group, India. Precambrian Research 162, 264–283
Bott MHP, Mithen DP (1983) Mechanisms of graben formation: the Wedge-subsidence Hypothesis. Tectonophysics 94, 11–22
Bradley DC (2011) Secular trends in the geological record and supercontinental cycles. Earth Science Reviews 108, 16–33
Brenner RL, McHargue TR (1988) Integrative stratigraphy—concepts and applications. Prentice Hall, New Jersey, p 419
Cant DJ (1989) Simple equations of sedimentation: applications to sequence stratigraphy. Basin Research 2, 73–81
Catuneanu O (2006) Principles of sequence stratigraphy. Elsevier, Amsterdam, p 375
Catuneanu O, Martins-Neto MA, Eriksson PG (2005) Precambrian sequence stratigraphy. Sedimentary Geology 176, 67–95
Cawood PA, Hawkesworth CJ (2014) Earth's middle age. Geology 42, 503–506
Condie KC, Kroner A (2008) When did plate tectonics begin? evidence from the geologic record. Geological Society of America, Special Papers 440, 281–291

Condie KC, Kroner A (2013) The building blocks of continental crust: evidence from a major change in tectonic setting of continental growth at the end of the Archean. Gondwana Research 23, 394–402

Condie KC, Pisarevsky SA, Korenga J, Gardoll S (2015) Is the rate of supercontinental assembly changing with time? Precambrian Research 259, 278–289

Deshpande GG, Pitale UL (2014) Geology of Maharashtra, 2nd edn. Geological Society of India, Bangalore, p 266

Dey S (2015) Geological history of the Kaladgi–Badami and Bhima basins, south India: sedimentation in a Proterozoic intracratonic setup. In: Mazumder R, Eriksson PG (eds) Precambrian basins of India: stratigraphic and tectonic context. Geological Society, London, Memoir, 43, 283–296

Dey S, Rai AK, Chaki A (2008) Widespread arkose along the northern margin of the Proterozoic Kaladgi basin, Karnataka: product of uplifted granitic source of K-metasomatism? Journal of Geological Society of India 71, 79–88

Dey S, Rai AK, Chaki A (2009) Palaeoweathering, composition and tectonics of provenance of the Proterozoic Intracratonic Kaladgi–Badami basin, Karnataka, southern India: evidence from sandstone petrography and geochemistry. Journal of Asian Earth Sciences 34, 703–715

Dickinson WR (1974) Plate tectonics and sedimentation. In: WR Dickinson WR (ed) Tectonics and sedimentation: Society of Economic Paleontologists and Mineralogists Special Publication 22, 1–27

Dula WF Jr (1991) Geometric models of listric normal faults and rollover folds. AAPG Bulletin 75, 1609–1625

Einsele G (2000) Sedimentary basins: evolution, facies and sedimentary budget. Springer-Verlag, Berlin, p 792

Eriksson PG, Martins-Neto MA, Nelson DR, Aspler LB, Chiarenzelli JR, Catuneanu O, Sarkar S, Altermann W, de W Rautenback CJ. (2001) An introduction to Precambrian basins: their characters and genesis. Sedimentary Geology 141–142:1–35

Eriksson PG, Mazumder R, Catuneanu O, Bumby AJ, Ilondo BO (2006) Precambrian continental freeboard and geological evolution: a time perspective. Earth Science Reviews 79, 165–204

Evans DAD (2013) Reconstructing pre-pangean supercontinents. GSA. Bulletin 125, 1735–1751

Foote RB (1876) Geological features of the south Mharatta country and adjoining districts. Memoir of Geological Survey of India 12, 70–138

French JE, Heaman LM, Chacko T, Srivastava RK (2008) 1891–1883 Ma southern Bastar–Cuddapah mafic igneous events, India: a newly recognized large igneous province. Precambrian Research 160, 308–322

Gallagher K, Lambeck K (1989) Subsidence, sedimentation and sea-level changes in the Eromanga Basin, Australia. Basin Research 2, 115–131

Geological Survey of India (1981) Geological and mineral map of Karnataka and Goa on 1:0.5 million scale. GSI, Kolkatta

Gibbs AD (1984) Structural evolution of extensional basin margins. Journal of Geological Society, London 141, 609–620

Gokhale NW, Pujar GS (1989) Bedding plane faults in the Kaladgi rocks, Basidoni, Belgaum district, Karnataka state. Current Science 58, 1088–1089

Govinda Rajulu BV, Chandrasekhar Gowda MJ (1972) Albitised slates from the Upper Kaladgi (Precambrian) formations, Lokapur, Bijapur district, Mysore state. Journal of Geological Society of India 13, 247–261

Haq BU, Hardenbol J, Vail PR (1987) Chronology of fluctuating sea levels since the Triassic (250 million years ago to present). Science 235, 1156–1167

Jadhav PB (1987) Microstructures of quartzites of Yadwad–Lokapur–Bagalkot area, Bijapur district, North Karanataka. Unpubl. Ph.D. thesis, Pune University, 204pp

Jadhav PB, Kshirsagar LK (2017) Analysis of folding in the Proterozoic sedimentary sequence of bagalkot group exposed around Yadwad, N Karnataka, India: a study based on remote sensing techniques and field investigations. Journal of Indian Society of Remote Sensing 45, 247–258

Jadhav PB, Phadke AV (1989) Coaxial deformation of Proterozoic Kaladgi Quartzites of Karnataka craton, India. 28th International Geologial Congress, Washington 2, 111
Jayaprakash AV (2007) Purana basins of Karnataka. Memoir of Geological Survey of India, 129pp
Jayaprakash AV, Sundaram V, Hans SK, Mishra RN (1987) Geology of the Kaladgi-Badami basin, Karnataka. Memoir of Geological Society of India 6, 201–225
Kale VS (1991) Constraints on the evolution of the Purana basins of Peninsular India. Journal of Geological Society of India 38, 231–252
Kale VS (1995) Association of the Purana basins and the Middle Proterozoic Mobile belts in Peninsular India: implications on targeting Uranium deposits. Exploration and Research for Atomic Minerals 8, 95–110
Kale VS (2015) Discussion on an alternate perspective on the opening and closing of the Intracratonic Purana basins in Peninsular India. Journal of Geological Society of India 85, 627–631
Kale VS (2016) Proterozoic basins of Peninsular India: status within the global Proterozoic systems. Proceedings of the Indian National Science Academy 82, 461–478
Kale VS, Patil Pillai S (2011) A Reinterpretation of two Chertbreccias from the Proterozoic basins of India. Journal of Geological Society of India 78, 429–445
Kale VS, Peshwa VV (1995) Bhima basin. Geological Society of India, Bangalore, p 142
Kale VS, Phansalkar VG (1991) Purana basins of Peninsular India: a review. Basin Research 3, 1–36
Kale VS, Ghunkikar V, Paul Thomas P, Peshwa VV (1996) Macrofacies architecture of the first transgressive suite along the southern margin of the Kaladgi Basin. Journal of Geological Society of India 48, 75–92
Kale VS, Nair S, Patil S (1998) Testimony of intraformational limestone breccias on Lokapur-Simikeri disconformity, Kaladgi Basin. Journal of Geological Society of India 51, 43–48
Kale VS, Patil Pillai S, Nair Pillai S, Phansalkar VG, Peshwa VV (1999) A process-responsive litho-stratigraphic classification of the Kaladgi Basin, Karnataka. Abstract field workshop on integrated evaluation of the Kaladgi & Bhima Basins. Geolgical Society of India, Bangalore, 4–7
Kalpana MS, Patil DJ, Dayal AM, Raju SV (2010) Near surface manifestation of hydrocarbons in Proterozoic Bhima and Kaladgi basins: implications to hydrocarbon resource potential. Journal of Geological Society of India 76, 548–556
Kingston DR, Dishroom CP, Williams PA (1983) Global basin classification system. American Association of Petroleum Geologists 67, 2175–2193
Kulkarni KG, Borkar VD (1997) On occurrence of *Cochlichnus* in the Proterozoic rocks of Kaladgi basin. Gondwana Geological Magazine 12, 55–59
Kumar A, Parasharumulu V, Nagaraju E (2015) A 2082 Ma radiating dyke swarm in the Eastern Dharwar craton, southern India and its implications to Cuddapah basin formation. Precambrian Research 266, 490–505
Mallick K, Vasanthi A, Sharma KK (2012) Kaladgi–Badami Basin, Karnataka–Maharashtra, India: In: Bouguer gravity regional and residual separation: application to geology and environment, Capital Publishing Co. (Springer), New Delhi, pp 60–67
Matin A (2015) Tectonics of the Cuddapah basin and a model of its evolution: a review. In: Mazumder R, Eriksson PG (eds) Precambrian basins of India: stratigraphic and tectonic context, vol 43. Geological Society of London Memoir, pp 231–254
Meert JG, Pandit MK, Pradhan VR, Banks J, Sirianni R, Stroud M, Newstead B, Gifford JOUR (2010) Precambrian crustal evolution of Peninsular India: A 3.0 billion year odyssey. Journal of Asian Earth Sciences 39, 483–515
Meert JG, Pandit MK, Pradhan VR, Kamenov G (2011) Preliminary report on the paleomagnetism of 1.88 Ga dykes from the Bastar and Dharwar cratons, Peninsular India. Gondwana Research 20, 335–343
Miall AD (2000) Principles of sedimentary basin analysis, 3rd edn. Springer-Verlag, Berlin, 616pp

Miall AD, Catunenu O, Eiksson PG, Mazumder R (2015) A brief synthesis of Indian Precambrian Basins: classification and genesis of basin fills. In: Mazumder R, Eriksson PG (eds) Precambrian basins of India: stratigraphic and tectonic context, vol 43. Geological Society of London Memoir, pp 339–347

Miller KG, Kominz MA, Browning JV, Wright JD, Mountain GS, Katz ME, Sugarman PJ, Cramer BS, Christie-Blick N, Pekar SF (2005) The Phanerozoic record of global sea-level change. Science 310, 1293–1298

Mukherjee S (2014) Atlas of shear zone structures in meso-scale. Springer, Cham, p 124

Mukherjee MK, Das S, Modak K (2016) Basement-cover structural relationships in the Kaladgi Basin, southwestern India: indications towards a Mesoproterozoic gravity gliding of the cover along a detached unconformity. Precambrian Research 281, 495–520

Mukhopadhyay S, Choudhuri A, Samanta P, Sarkar S, Bose PK (2014) Were the hydraulic parameters of Precambrian rivers different? Journal of Asian Earth Sciences 91, 289–297

Nair MM, Raju AV (1987) A remote sensing approach to basinal mapping of the Kaladgi and Badami sediments of Karnataka state. In: Radhakrishna BP (ed) Purana Basins of Peninsular India (Middle to Late Proterozoic), vol 6. Memoir of Geological Society of India, pp 375–381

Naqvi SM, Rogers JJW (1987) Precambrian Geology of India Oxford Monographs Geology & Geophysics. Clarendon Oxford Press, Oxford, p 223

Patil Pillai S (2004) Testimony of sedimentation and structural patterns of the Bagalkot-Simikeri area on the evolution of the Proterozoic Kaladgi Basin. Unpublished Ph.D. thesis, Pune University, Pune

Patil Pillai S, Kale VS (2011) Seismites in the Lokapur subgroup of the Proterozoic Kaladgi Basin, South India: A testimony to syn-sedimentary tectonism. Sedimentary Geology 240, 1–13

Patil Pillai S, Peshwa VV, Nair S, Sharma M, Shukla M, Kale VS (1999) Occurrence of a manganese-bearing horizon in the Kaladgi basin. Journal of Geological Society of India 53, 201–204

Patil Pillai S, Pande K, Kale VS (2018) Implications of new 40Ar/39Ar age of Mallapur Intrusives on the chronology and evolution of the Kaladgi Basin, Dharwar Craton, India. Journal of Earth System Science 127, 32, p 18

Piper JDA (2013) A planetary perspective on Earth's evolution: lid tectonics before plate tectonics. Tectonophysics 589, 44–56

Piper JDA (2018) Dominant lid tectonics behavior of continental lithosphere in Precambrian times: paleomagnetism confirms prolonged quasi-integrity and absence of supercontinental cycles. Geoscience Frontiers 9, 61–89

Pirajno F, Santosh M (2015) Mantle plumes, supercontinents, intracontinental rifting and mineral systems. Precambrian Research 259, 243–261

Pratap M, Sharma R, Das SK, Tripathi B (1999) Tectono-stratigraphic styles and Hydrocarbon prospectivity of Bhima-Kaladgi Basins vis-à-vis Vindhyan and other coeval Proterozoic Basins. Abstract Field workshop on Integrated evaluation of the Kaladgi & Bhima Basins. Geological Society of India, Bangalore, 61–63

Radhakrishna BP (1987) Introduction: Purana Basins of Peninsular India (Middle to Late Proterozoic). Memoir of Geological Society of India 6, i–xv

Radhakrishna BP, Naqvi SM (1986) Precambrian continental crust of India and its evolution. Journal of Geology, 94, 145–166

Radhakrishna BP, Ramakrishnan M (1988) Archaean_Proterozoic Boundary in India. Journal of Geological Society of India 32, 263–278

Rajaram M, Anand SP, Erram VC, Shinde BN (2016) Insight into the structure below the Deccan Trap covered region of Maharashtra, India from geopotential data. In: Mukherjee S, Misra AA, Calvès G, Nemčok M (eds) Tectonics of the Deccan Large Igneous Province. Geological Society, London, Special Publications, 445, 219–236

Ramakrishna TS, Chayanulu AYSR (1988) A geophysical reappraisal of the Purana basins of India. Journal of Geological Society of India 32, 48–60

Ramakrishnan M, Vaidyanadhan R (2010) Geology of India, vol 1. Geological Society of India, Bangalore, p 556

Renne PR, Sprain CJ, Richards MA, Self S, Vanderkluysen L, Pande K (2015) State shift in Deccan volcanism at the Cretaceous-Paleogene boundary possibly induced by impact. Science 350, 76–78
Roberts NMW, Slagstad T, Viola G (2015) The structural, metamorphic and magmatic evolution of Mesoproterozoic orogens. Precambrian Research 265, 1–9
Rogers JJW, Santosh M (2004) Continents and Supercontinents. Oxford University Press, New York, p 292
Rooney AD, Strauss JV, Brandon AD, Macdonald FA (2015) A Cryogenian chronology: two long-lasting synchronous Neoproterozoic glaciations. Geology 43, 459–462
Saddler PM (1999) The influence of hiatuses on sediment accumulation rates. GeoResearch Forum 5, 15–40
Sambasiva Rao VV, Sreenivas B, Balaram V, Govil PK, Srinivasan R (1999) The nature of the Archean upper crust as revealed by the geochemistry of the Proterozoic shales of the Kaladgi basin, Karnataka, South India. Precambrian Research 98, 53–65
Sarkar PK, Soman GR (1985) Heavy mineral assemblages from the Konkan Kaladgi sediments. In: Bhaskara Rao B (ed) Proceedings of 5th Indian Geological Congress, IIT, Bombay, pp 19–25
Sclater JG, Christie PAF (1980) Continental stretching: an explanation of the post-Mid Cretaceous subsidence of the central North Sea Basin. Journal of Geophysical Research 85(B7), 3711–3739
Sharma RS (2009) Cratons and fold belts of India. Springer-Verlag, Berlin, p 304
Sharma M, Nair S, Patil S, Shukla M, Kale VS (1998) Occurrence of tiny digitate stromatolite [Yelma Digitata Grey, 1984], Yargatti Formation, Bagalkot Group, Kaladgi Basin, Karnataka, India. Current Science 75, 360–365
Sommerfield CK (2006) On sediment accumulation rates and stratigraphic completeness: lessons from Holocene ocean margins. Continental Shelf Research 26, 2225–2240
Sridhar M, Chaturvedi AK, Rai AK (2014) Locating new Uranium occurrences by integrated weighted analysis in Kaladgi basin, Karanataka. Journal of the Geological Society of India 84, 509–512
Sridhar M, Markandeyulu A, Chaturvedi AK (2017) Mapping subtrappean sediments and delineating structure with the aid of heliborne time domain electromagnetics: Case study from Kaladgi basin, Karanataka. Journal of Applied Geophysics 136, 9–18
Stephen George T (1999) Sedimentology of the Kaladgi Basin. Abstract volume. Field workshop on integrated Evaluation of the Kaladgi & Bhima Basins. Geological Society of India, Bangalore, 13–17
Vail PR, Mitchum RM Jr, Thompson SIII (1977) Seismic stratigraphy and global changes of sea level, part 3: relative changes of sea level from coastal onlap. In: Payton CE (ed.) Seismic stratigraphy-Applications to hydrocarbon exploration: American Association of Petroleum Geologists Memoir 26, 63–81
Valdiya KS (2016) The making of India: geodynamic evolution, 2nd edn. Springer International, Switzerland, 924pp
Viswanathiah MN (1968) Badami series: a new post-Kaladgi formation of Karnatak state. Bulletin of Geological Society of India 5, 94–97
Viswanathiah MN (1979) Lithostratigraphy of the Kaladgi and Badami Groups, Karnataka. Indian Mineralogist 18, 122–132
Viswanathiah MN, Chandrasekhara Gowda MN (1970) Algal stromatolites from Kaladgi (Precambrian) formations near Alagundi, Bijapur district, Mysore state. Journal of Geological Society of India 11, 378–385
Viswanathiah MN, Sreedhara Murthy TR (1979) Algal stromatolites from Kaladgi Group around Bilgi, Bijapur district, Karnataka. Journal of Geological Society of India 20, 1–6

NE-SW Strike-Slip Fault in the Granitoid from the Margin of the South East Dharwar Craton, Degloor, Nanded District, Maharashtra, India

Md. Babar, R. D. Kaplay, Soumyajit Mukherjee, Souradeep Mahato and Chandrakant Gurav

1 Introduction

Present study emphasizes deformations in the basement granite from the 'South East Deccan Volcanic Province' (SEDVP) around Degloor/Diglur. W dipping thrusts (Kaplay et al. 2013), steep normal faults, fault planes dipping towards S, from basalts and deformations from Quaternary deposits (Kaplay et al. 2017a) were reported from the area around Nanded. Similarly, Kaplay et al. (2017b) studied the margin of SEDVP around Kinwat lineament, Maharashtra, India, (northern part of Nanded district). The E-W strike-slip faults (with maximum net-slip of 24 cm) were identified in the basement granite from the Kinwat region. The zone was designated as the 'Western Boundary East Dharwar Craton Strike-Slip Zone' (WBEDCSZ). This prompted us to study the structural aspects of another margin of SEDVP with basement granite near the Degloor region. The study area (Fig. 1) is located ~135 km SSW of Kaddam fault, the area from where strike-slip faults were reported from the basement granite (Sangode et al. 2013).

Basement structures affected Deccan tectonics. For example, ~N-S Dharwar trend in the basement was inherited as weak planes in Deccan trap in subsequent strike-slip faulting around Mumbai (Misra et al. 2014).

Md. Babar · C. Gurav
Department of Geology, Dnyanopasak College, Parbhani 431401
Maharashtra, India

R. D. Kaplay
School of Earth Sciences, S.R.T.M. University, Nanded 431606
Maharashtra, India

S. Mukherjee (✉) · S. Mahato
Department of Earth Sciences, Indian Institute of Technology Bombay,
Powai, Mumbai 400076, Maharashtra, India
e-mail: soumyajitm@gmail.com

S. Mukherjee (ed.), *Tectonics and Structural Geology: Indian Context*,
Springer Geology, https://doi.org/10.1007/978-3-319-99341-6_5

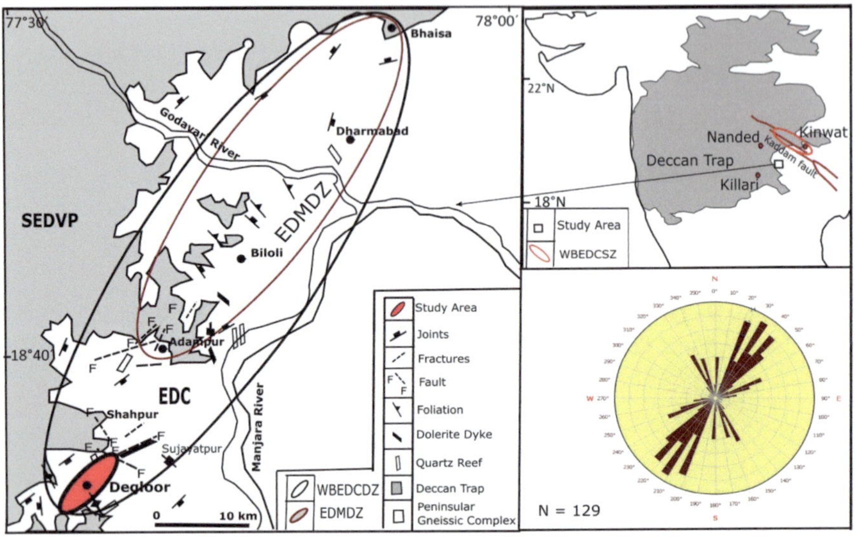

Fig. 1 Location map showing deformation zones along SEDVP contact with EDC. SEDVP: South East Deccan Volcanic Province, EDC: East Dharwar Craton, EDMDZ: East Dharwar Margin Deformation Zone, and WBEDCDZ: 'Western Boundary East Dharwar Craton Deformed Zone'. Right corner inset: Rose diagram- dominant trend of the strike-slip faults from the regions 1–5. Data population: N = 129. (Geology after Banerjee et al. 2012)

The Degloor region is characterized by westernmost part of East Dharwar Craton with basement granite. The Deccan trap exposes about ~10 km from the Degloor granite. The Degloor region consists of Palaeoproterozoic high K-granitoids (Banerjee et al. 1993; Wesankear and Patil 2000). These rocks are also referred to as the Nanded granitoids (NG) by Banerjee et al. (2012). They are the continuation of 'Peninsular Gneissic Complex' (PNG) along the Deccan Traps boundary (SEDVP) (Banerjee et al. 2012). These granites are coarse-grained pink and gray with porphyritic texture. Quartzo(-feldspathic) veins and pegmatite dykes intrude them profusely. The dark coloured basic enclaves are also commonly associated with these granites. At places WNW trending mafic dykes also exposed. The granites exposed as detached mounds/along the course of the river. These granites are part of the reactivated and remobilized EDC (GSI 1979, 2001).

Banerjee et al. (2012) report that these granitoids demonstrate well developed mostly NE and NW trending joints and local foliations. These authors also report (i) minor slip in pink feldspar bands near Sagroli- ~24 km NE of Degloor; and (ii) several minor faults close to the 'South East Deccan Volcanic Province-East Dharwar Cratorn margin, and minor folding within the EDC. However, Banerjee et al. (2012) did not present structural details. On the other hand we worked on this basement granite for detail structural geology. The methodology of the present study have been data collection through fieldwork, and their interpretation.

2 Structures Observed in Degloor—The Present Study

Reverse faults, normal faults (step faults), strike-slip faults, brittle shear along the boundary of veins, minor folds and boudins are observed in the basement rock. In Fig. 2, a thin quartz vein is faulted in normal manner at as many as 12 locations. These faults are (sub) parallel to each other and constitute a step-like array with small throw. The fault plane dips 80° due W. The slip of fault plain varies from 1.6 to 6.5 cm.

Brittle strike-slip faults are numerous in the study area. In some cases, single veins are strike-slip faulted (Fig. 3a). In Fig. 3b, a pegmatite vein is dextrally faulted. In Fig. 3c, slightly folded pegmatite vein is slipped possibly with a "reverse drag" (Mukherjee and Koyi 2009). A medium size pegmatite vein (A) in granitic country rock (B), at the contact of basic dyke (C) shows reverse slip but no drag close to the fault plane (Fig. 3d). A pegmatite vein is strike-slip sinistrally faulted twice in two directions. Sigmoid P-planes decode the former slip, and sharp offset the later one (Fig. 3e). Brittle shear fractures and fault planes were also observed; Y-planes and the fault planes in Fig. 4a–d and also in Fig. 5 trend NE, showing both sinistral and dextral slip. A thin vein, observed on a horizontal plane, is reverse dragged only at one side of the fault plane. At the other side, no drag is noted (Fig. 3f). The slip is 17 cm and fault strikes ~NE. A rather irregular basic intrusion in granitic rock shows faulting. This is a strike-slip fault (Fig. 6a).

In other cases, single veins are multiply faulted (Fig. 6b). Fault 'A' strikes N20° E and slips 9.0 cm; 'B' strikes N50°E and slips 5.0 cm; 'C' strikes N75°W and slips 24 cm; and 'D' strikes N80°E and slips 28 cm. In Fig. 6c, the quartzo-feldspathic vein is dextral faulted at three places. The en-echelon joints

Fig. 2 Step faulting in quartz vein (12 faults in vertical section). We tend to think them as a manifestation of a major fault, since our observations repeat

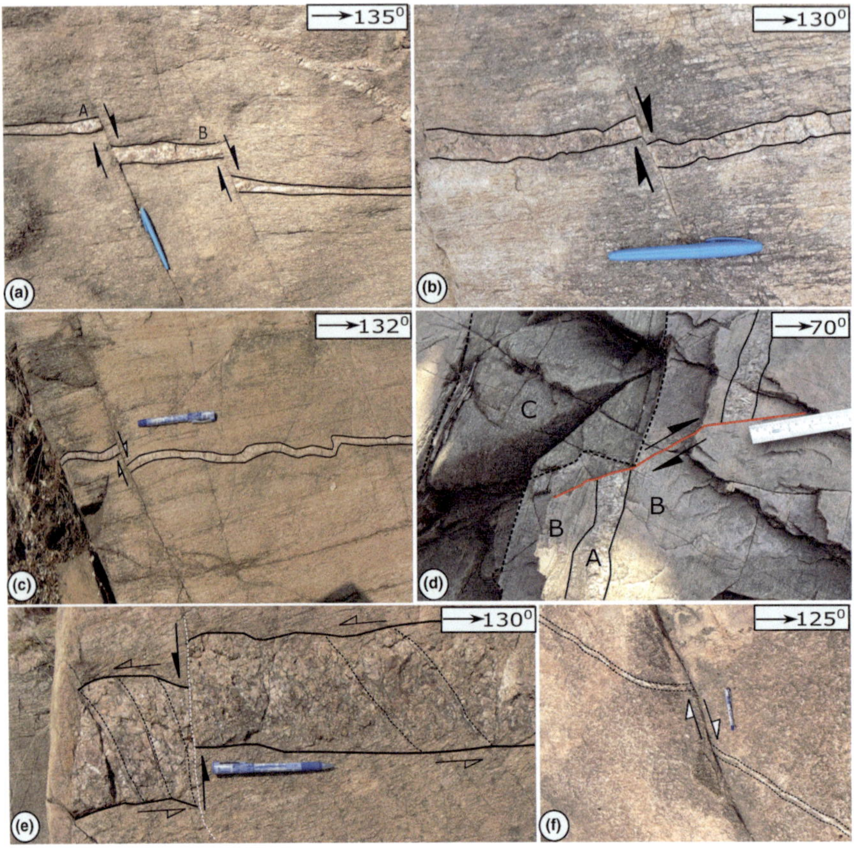

Fig. 3 **a** Normal step faults exposed on steeply inclined exposure, at 'A' and 'B' dip towards E at 25°. Net slip: 5.5 cm both at 'A' and 'B'. **b** Dextral sheared pegmatite vein, 2.1 cm net slip along NE trending fault plane. **c** Joints acted locally as brittle-ductile fault planes (strike N20°E, 3.2 cm slip), at horizontal outcrop. **d** Dextral brittle slip. Net slip 12 cm; N45°E (horizontal exposure). **e** Sinistral shear (at horizontal outcrop). Strike of the fault is N20°E; net slip is 5.0 cm. **f** A thin vein shows bending before it is deformed in an brittle manner (horizontal exposure)

locally acted as fault planes. In Fig. 6d as well, the quartzo-feldspathic vein faulted along joint planes at three places both dextrally and sinsitrally.

In Fig. 7a, the vein is faulted at four places. At 'A' the sense of movement is sinistral, while at 'B', 'C' and 'D' the sense is dextral. At 'B' the vein has not lost the continuity indicating local ductile deformation. At 'A', the vein is bent close to the fault plane indicating drag.

A quartz vein (Fig. 7b) slips at three places along parallel set of N6°W-S6°E trending faults. The displacement at 'A' is 7 cm, at 'B' it is 4.2 cm and at 'C' it is 4 cm. In Fig. 7c, curved joints act as fault planes slipping the vein at five places. One more slip is noted at 'F'. Some of these fault planes are curved. In some of the cases, one of the veins is strike-slip faulted (Fig. 7d) while the other parallel vein is

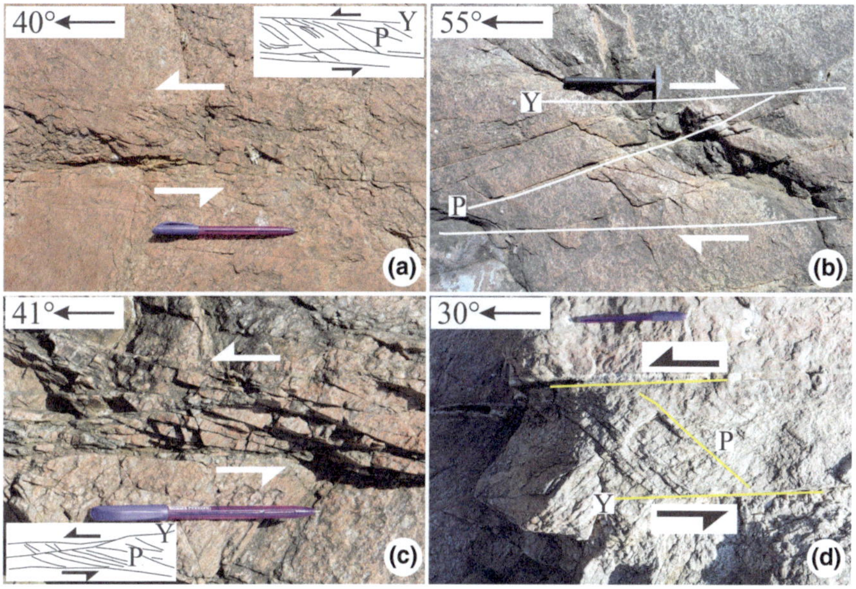

Fig. 4 Brittle strike slip shear observed on horizontal exposures. Y-plane trends ~ NE. **a**, **c**, **d** Sinistral slip. **b** Dextral slip. Additionally: **d** secondary quartz along Y-plane

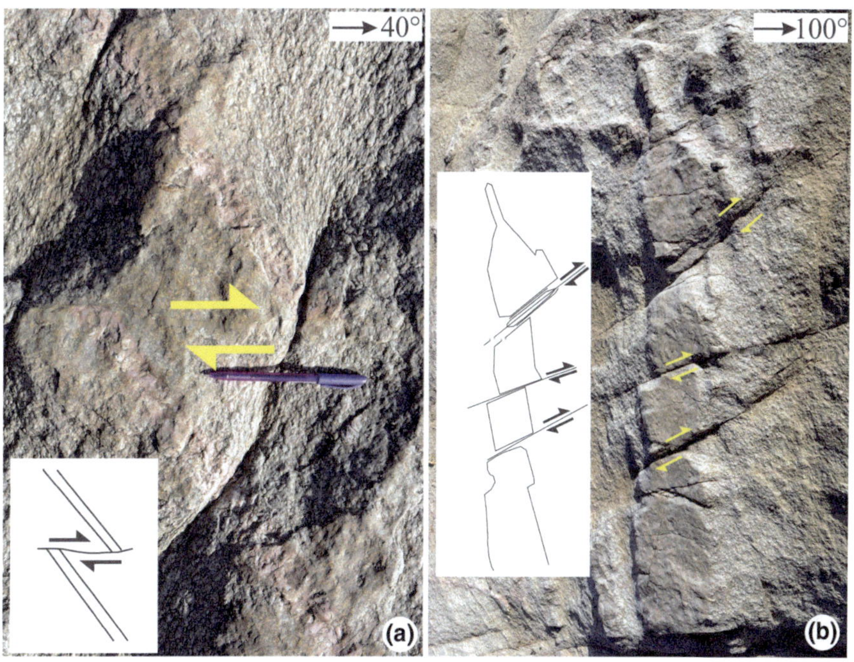

Fig. 5 **a** Quartzo-feldspathic vein dextral slipped along N40°E striking fault. **b** Three fault planes marked on the diagram are vertical, striking N69°E, N85°E and N54°E from bottom to top of the image also show dextral slip

Fig. 6 **a** Basic intrusion faulted with 2.2 cm slip. Fault strikes NE. **b** Multiple faults (at horizontal outcrop). **c** Joints acting as fault planes, at horizontal outcrop. Fault strikes N65°E at 'A' with 3 cm slip, at 'B': strike: N65°E, slip: 2.0 cm; at 'C', strike: N65°E, slip: 3.5 cm. **d** 'A' and 'B' are the joints which acted as fault planes in displacing pegmatite vein (horizontal outcrop). At 'A' the fault strikes N48°E, with 4.0 cm slip, at 'C', strike: N78°W, slip: 9.0 cm

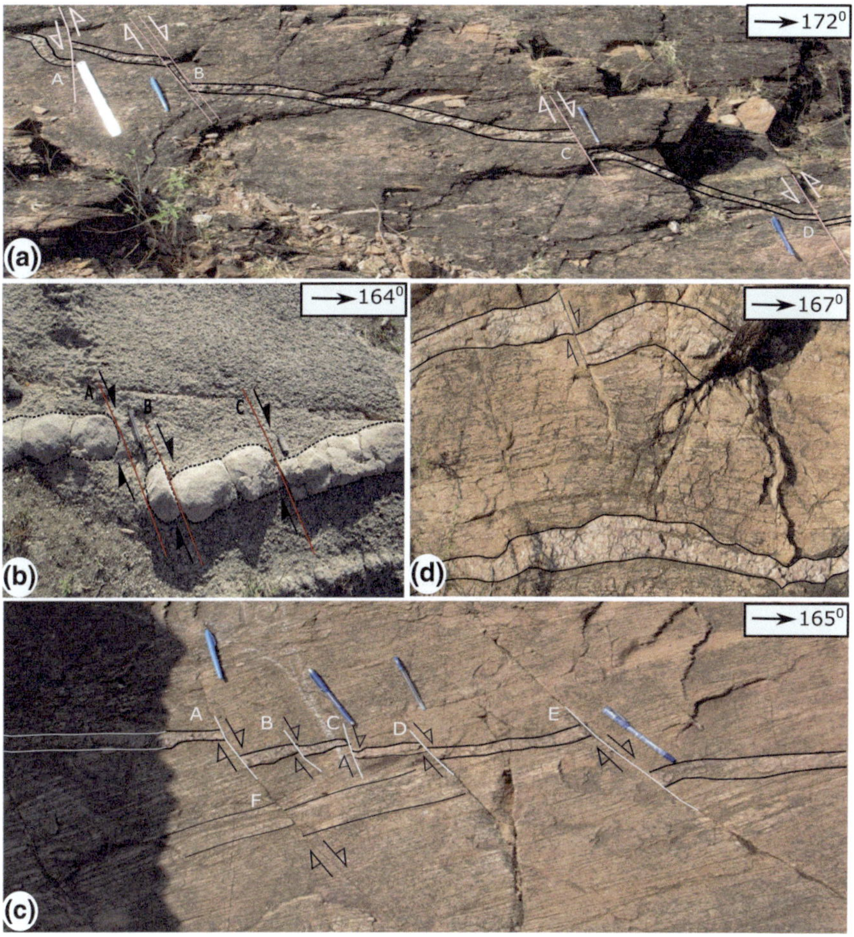

Fig. 7 **a** Vein displaced at four places, at horizontal outcrop. Strike of the fault at 'A': N30°E, slip: 7.0 cm, at 'B' strike: N30°E, slip: 10 cm, at 'C' strike: N65°E, slip: 6.5 cm; at 'D' strike: N50°E, slip: 13 cm. **b** Quartz vein faulted at three places: 'A', 'B' and 'C'. Sense of slip: dextral. **c** Curvilinear en-echelon joints acting as fault planes, at horizontal outcrop. Slip at 'A' is 8.0 cm, at 'B' 0.2 cm, at 'C' 4 cm, at 'D' 2 cm, at 'E' 14 cm; and at 'F': 3.8 cm. The strike all these small faults is N35°E. **d** At 'A' the vein shows deformation with 8 cm strike slip. Fault strikes N60°E. At 'B' the vein is unaffected by faulting and is open folded (at horizontal outcrop)

undeformed. In other words, in meso-scale one can see termination of fault planes. In Fig. 8a, the topmost vein (A) faults at four places (locations 'D' to 'G') along parallel en-echelon joints while the lower vein ('B') is not. The host rock ('C') above and below the vein 'A' is also faulted. In some other cases, parallel veins faulted (Fig. 8b). Here, left vein shears sinistrally and the right vein oppositely and dextrally. In Fig. 8c, however, both the veins shear dextrally.

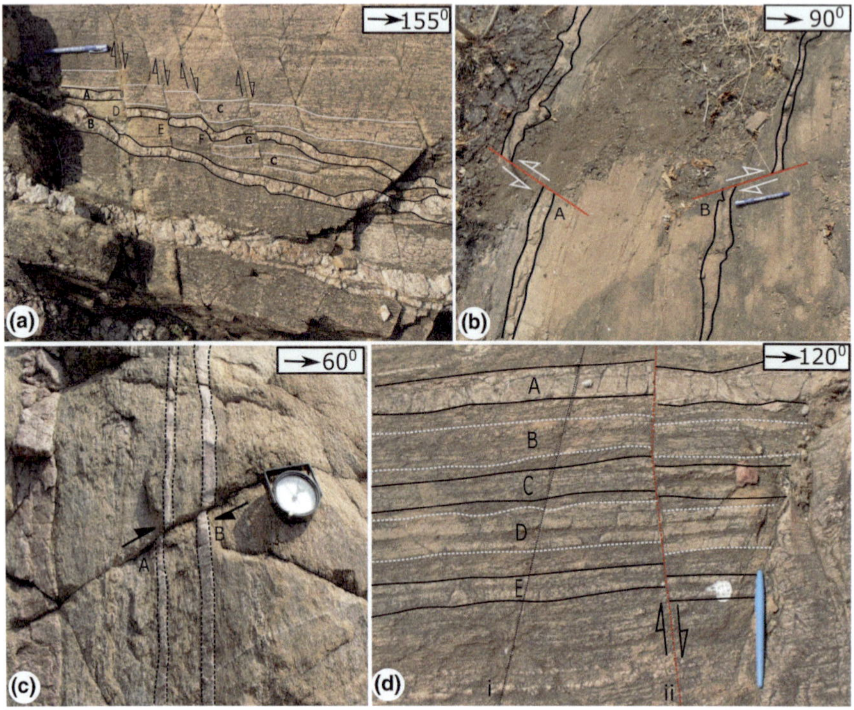

Fig. 8 **a** En-echelon joints (marked as 'D', 'E', 'F' and 'G') displaced the vein in brittle and ductile-brittle manner (at horizontal outcrop). At 'D' slip: 2.0 cm, at 'E': 0.6 cm, at 'F' the vein is ductile deformed, at 'G' slip is 0.5 cm. These faults strike ~NE. **b** brittle deformation (at horizontal outcrop). Strike of the fault at 'A' is N40°W with 26 cm slip; at 'B' strike: N70°E, slip 14.0 cm. **c** Two parallel veins brittle deformed. At 'a' 2 cm strike slip, at 'B' 1.5 cm slip. Fault strikes N40°E. **d** Multilayered faulting

Several layers in the rock got faulted (Fig. 8d). Two conjugate joints, i and ii, cut through the layers. Joint i does not slip the layer, however, joint ii does dextrally. Therefore joint ii can be called more accurately a fault. The strike of the fault is N30°E. The displacement for layer 'A' is 0.8 cm, for 'B' 0.9 cm, for 'C' 1.1 cm, for 'D' 1.2 cm, and for 'E' 1.4 cm. Slip increases locally towards SW. Commonly fault planes demonstrate maximum slip near their central part and at the margins minimum or zero (review in Mukherjee 2014a). Figure 8d is an exception.

Conjugate strike-slip faults at ~30° are observed in this study at places (Figs. 6d, 9a, b and 10). Fault 'A' strikes N50°E and slips 7 cm. Fault 'B' strikes N80°E and slips 6 cm. Spectacular conjugate multilayer strike-slip faulting is observed in the ancient granitoid rock (Fig. 10). On a ~N striking fault, brittle-ductile slip of 3 cm is noted at 'A', 5 cm at 'B', and 1.2 cm at 'C'. On a

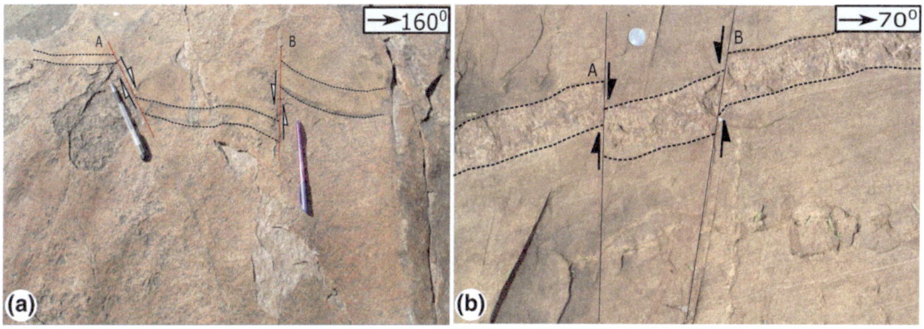

Fig. 9 **a**, **b** Conjugate strike-slip faults

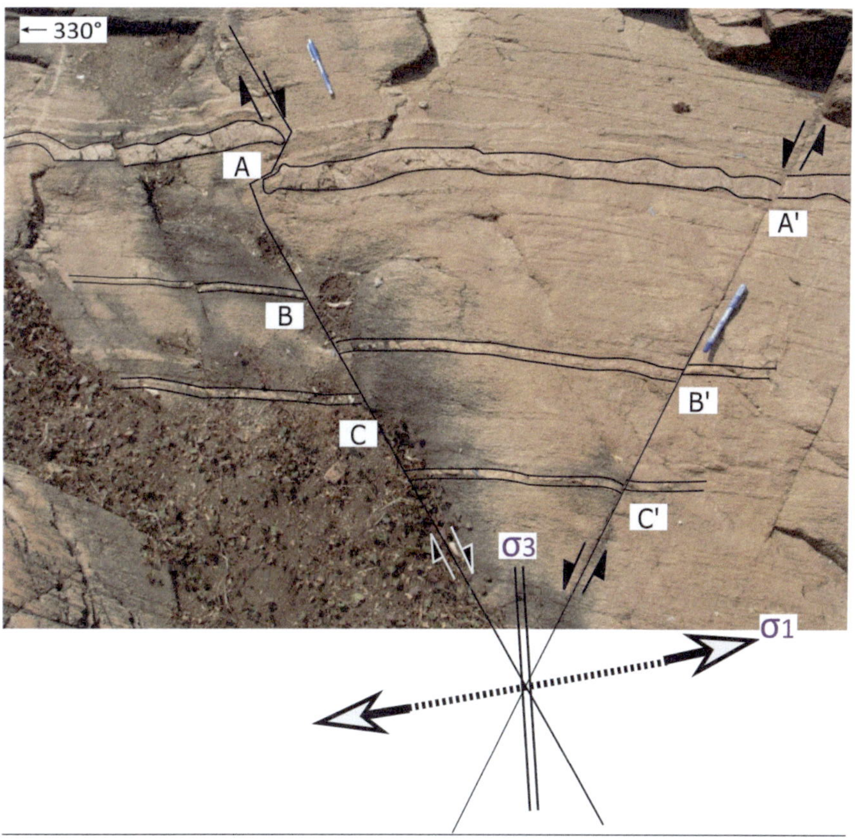

Fig. 10 Multilayer strike-slip faults (at horizontal outcrop)

N30°E striking plane, slip at 'A/' is 1 cm, at 'B/' is 1.1 cm, and at 'C/' is 1 cm. The thickest pegmatite layer is sheared sinistrally. Note fracture patterns inside the vein. That shear acted prior to the conjugate faulting. The left hand side geographic direction in this snap is N30°W/330°. A pegmatite vein faults along two sub-parallel planes. Since the angle between the two fault planes is rather low, they are not conjugate faults. Fault 'A' strikes N25°W and slips 2 cm. Fault 'B' strikes N6°W and slips 4 cm.

'Intrusion-in-intrusion' feature is ductile sheared (Fig. 11). Basic intrusion (B), with quartz vein (C) within, inside host granitic rock (A), is ductile sheared along N20°W-S20°E. Both the basic intrusion and the quartz vein thin significantly inside the shear zone (Fig. 11). Outside and inside the shear zone, 'B' is 22 and 6 cm thick, respectively. Likewise, 'C' is 6 and 1 cm thick.

Though major/regional folds are absent, small (ptygmatic) folds are found in vertical and horizontal sections. Synclinal type of fold is observed at one place (Fig. 12a). The axial plane is slightly inclined. A folded structure, at the horizontal outcrop, is exposed in granitic gneiss (Fig. 12b).

Folding in quartzo-feldspathic vein/pegmatite veins is also a common feature observed in the study area (Fig. 12c). At many places, pegmatite veins intruded in host granitoids is ptygmatic folded (Fig. 12d; also see Mukherjee et al. 2015). Drawing tectonic inferences from ptygmatic folding would be difficult because of their irregular and sometimes inconsistent geometries. 'S' and M-type folding in

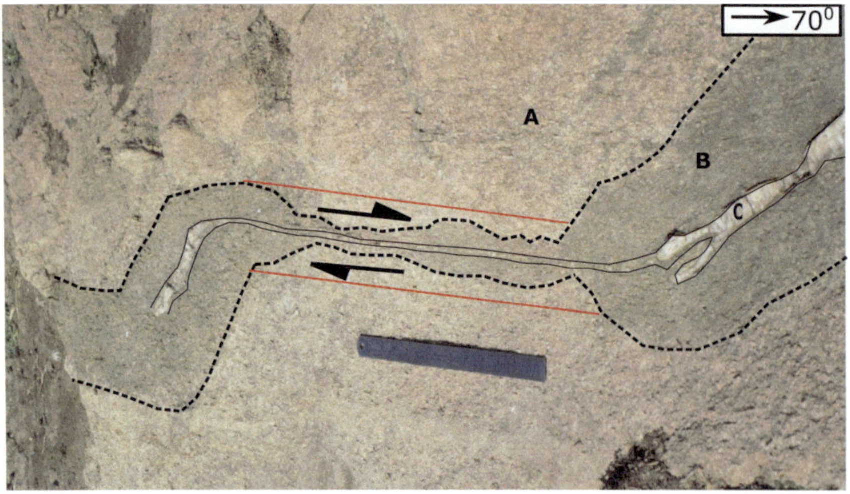

Fig. 11 Quartz vein (C) within basic intrusion (B) in host granitoid (A) ductile deformed. Shear zone is marked by red lines (at horizontal outcrop)

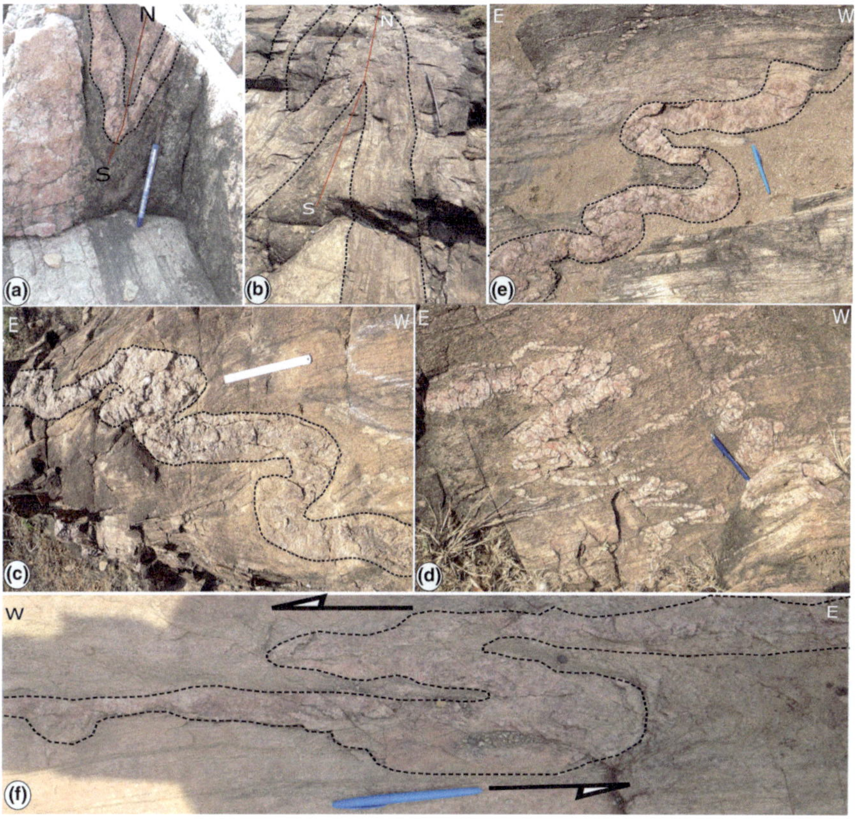

Fig. 12 **a** Asymmetric folds at vertical outcrop. **b** Folding in granitic gneiss (horizontal outcrop). Axial plane trends N. **c** Folding in pegmatite vein (horizontal outcrop). **d** Ptygmyatic folding. (E-W trending photo). **e** Pegmatite vein folded (W-E trending photo). **f** Folding in pegmatite vein (horizontal outcrop)

pegmatite dyke in granitoid (Figs. 12e, f and 13a, b) do exist though we do not decipher any regional folding. 'Z'-type tight fold with hinge thicker than the limbs is observed in one of the pegmatite veins. Multiple (faint) faults also occur in the close vicinity (Fig. 13c). Pinch and swell structures of pegmatite layer indicates local brittle-ductile NW-SE extension. Not all swells resemble geometrically, while few are sub-elliptical.

There are two major trends of vertical joint planes observed. Trend-1 varies N10° to 340°NW and trend-2 260°SW to 280°NW (Fig. 14).

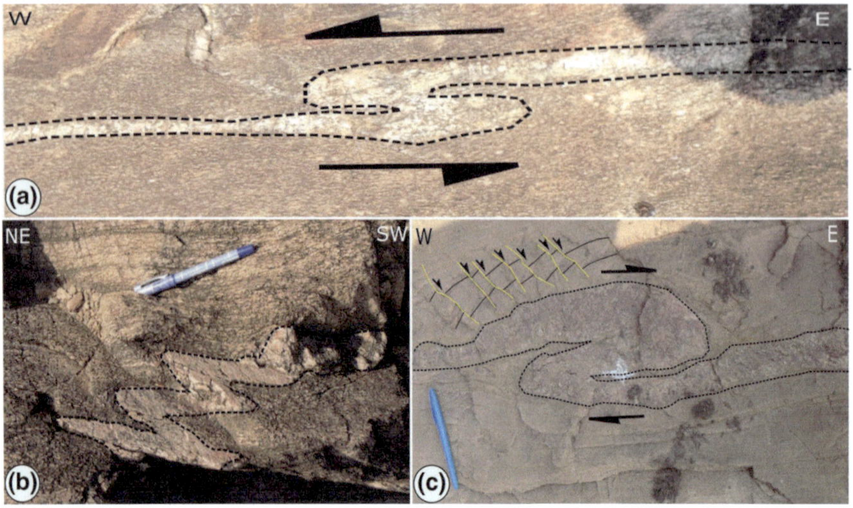

Fig. 13 **a** Folding in pegmatite vein. **b** 'M' type of folded pegmatite vein. **c** 'Z' type folding with multiple faulting in its close vicinity (W-E trending photo)

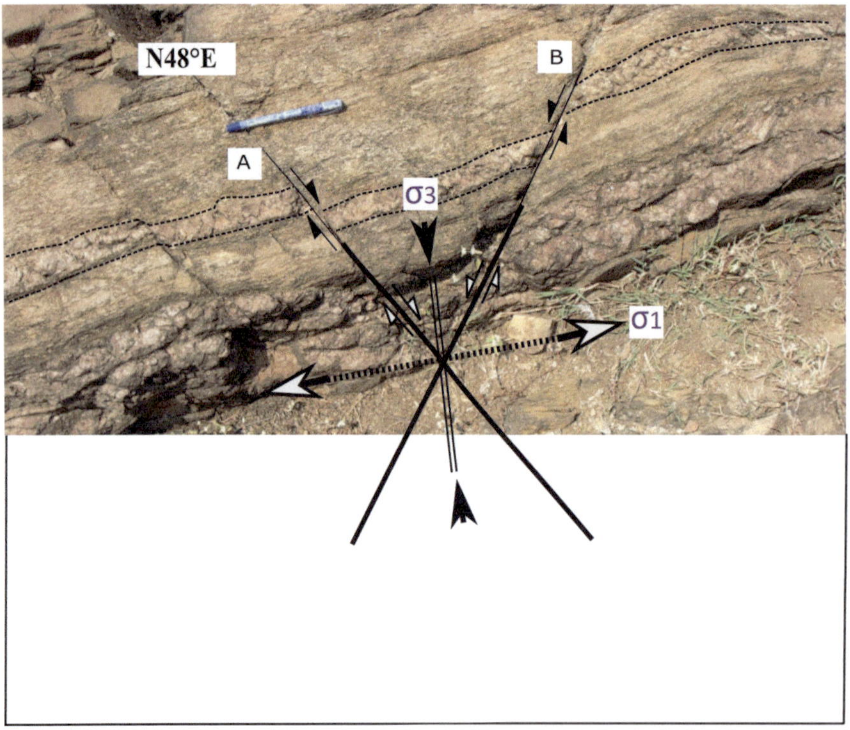

Fig. 14 Mechanism of strike-slip conjugate faulting. Strike of the fault is N50°E. Red dash lines: actual fault lines, black solid lines intersecting shear axes. σ_1: short solid black arrow, σ3, compressive stress axis: open arrow

3 Discussions

The faults/fractures/shear zones, in the Degloor region, are represented by a number of N to NNE, NE, NE and E bound lineaments (Banerjee et al. 2008). One NE trending fault is reported in between Shahapur and Sujayatpur, 9 km NE to the present study area and a shear zone near Madnoor. However, there has been no report of faults from the study area by the previous workers.

We document 129 small faults (Table 1) out of which 116 (~90%) are strike-slip, 12 (~9%) are normal faults, and only one (1%) is a reverse fault from the study area (Fig. 1). From these faults, 72% are of the strike-slip faults trend NE. 80% of the total strike-slip faults shear dextrally. Net slip of all the strike slip faults, where decipherable, varies 0.5–82 cm.

Conjugate or 'X' type joints at ~30°, devoid of any slip along them, is also observed. Conjugate joins were not observed in earlier studies from (other parts of) Deccan Traps such as Devey and Lightfoot (1986), Mitchell and Widdowson (1991), Babar et al. (2017), Bhave et al (2017), Kaplay et al. (2017a, b), Misra and Mukherjee (Misra and Mukherjee 2017) and Mukherjee et al. (2017). Most prominent sets of joints strike N50°E at 'A' and N80°E. These trends mostly match with the conjugate faults reported in Fig. 6d (N48°E), Fig. 9a (N50°E), Fig. 9b (N50°E), and Fig. 10 (N30°E). These joints played crucial role in displacing the veins in the region in a strike-slip manner. Similar types of joints have displaced various veins at other outcrops (Figs. 9a, b and 10). The angle between these two joints is 30°.

This work finds that the deformation style observed in the basement granite near Degloor, close to the contact of SEDVP with E Dharwar Craton, is dominantly strike-slip. This suggests that the region belongs to a tectonic regime with intermediate stress axis σ_2 vertical (in the past). Fault planes most dominantly trends NE.

Numerous conjugate strike-slip faults exist in the study area (Figs. 6b, d, 8b, 9a, b and 10). At some places, conjugate shear develop within the strike slip fault system (Fig. 14). The state of stress in such type of faulting consists of two horizontal principal stresses that are compressional push (σ_1; Fig. 14) in one direction and tensional pull (σ_3) in the perpendicular direction. The σ_1 direction bisects the acute angle, and the σ_3 the obtuse angle between the two fault planes that constitute the single conjugate set. The mechanism of faulting is shown in Fig. 14.

We refer the study area as the 'Degloor Strike-slip Shear Zone'. Note that strike-slip shear is reported from the basement granite nearer to the margin of SEDVP with Dharwar Craton at Kinwat (Kaplay et al. 2017a, b), which is ~135 km NNE of the present study area. The strike-slip faults, with movement parallel to the NE trend of Kaddam fault, are also reported close to Kaddam lineament (Sangode et al. 2013). Thus the Kinwat-Degloor-Kaddam triangular zone, so far explored, shows strike-slip tectonics. However, the deformation style, ~NE-SW at Degloor, does not match with the trend of those at northern margin of SEDVP with Dharwar craton near Kinwat (~E-W) and at Kaddam (~NW-SE),

Table 1 Deformation types and characters

Structure No.	Deformation regime	Fault type	Shear sense	Fault trend	Net slip (cm)
F1	Brittle	Strike slip	Dextral	N43°E	3.5
F2	Brittle	Strike slip	Dextral	N56°E	3.0
F3	Brittle	Strike slip	Dextral	N70° E	4.0
F4	Brittle	Strike slip	Dextral	N48°E	3.5
F5	Brittle	Strike slip	Dextral	N25°E	3.2
F6	Brittle	Strike slip	Dextral	N20°W	2.5
F7	Brittle	Strike slip	Dextral	N20°W	2.4
F8	Brittle	Strike slip	Dextral	N20°W	1.0
F9	Brittle	Strike slip	Dextral	N20°W	0.7
F10	Brittle	Strike slip	Dextral	N26°W	0.5
F11	Brittle	Strike slip	Dextral	N25°W	1.3
F12	Brittle	Strike slip	Dextral	N25°W	0.4
F13	Brittle	Strike slip	Dextral	N27°W	0.7
F14	Brittle	Strike slip	Sinistral	N25°W	1.6
F15	Brittle	Strike slip	Dextral	N26°W	0.7
F16	Brittle	Strike slip	Dextral	N28°W	1.8
F17	Brittle	Strike slip	Dextral	N28°W	1.7
F18	Brittle	Strike slip	Dextral	N28°W	1.5
F19	Brittle	Strike slip	Dextral	N28°W	2.2
F20	Brittle	Strike slip	Dextral	N25°E	2.0
F21	Brittle	Strike slip	Dextral	N43°E	3.5
F22	Brittle	Strike slip	Dextral	N41°E	3.0
F23	Brittle	Strike slip	Dextral	N10°W	3.4
F24	Brittle	Strike slip	Dextral	N14°W	0.8
F25	Brittle	Strike slip	Dextral	N65°E	4.0
F26	Brittle	Strike slip	Sinistral	E-W	2.7
F27	Ductile/Brittle	Strike slip	Dextral	N47°E	0.6
F28	Ductile/Brittle	Strike slip	Sinistral	E-W	2.8
F29	Brittle	Strike slip	Dextral	N78°E	1.8
F30	Brittle	Strike slip	Dextral	N55°W	4.0
F31	Brittle	Strike slip	Dextral	N28°E	13
F32	Brittle	Strike slip	Dextral	N26°E	1.1
F33	Brittle	Strike slip	Dextral	N45°E	12
F34	Brittle	Strike slip	Dextral	N24°E	07
F35	Brittle	Strike slip	Dextral	N05°W	05
F36	Brittle	Strike slip	Dextral	N65°E	26
F37	Brittle	Strike slip	Dextral	N65°E	28
F38	Brittle	Strike slip	Dextral	N23°E	03
F39	Brittle	Strike slip	Dextral	N23°E	03
F40	Brittle	Strike slip	Sinistral	N70°E	5.7

(continued)

Table 1 (continued)

Structure No.	Deformation regime	Fault type	Shear sense	Fault trend	Net slip (cm)
F41	Brittle	Strike slip	Sinistral	N70°E	1.1
F42	Brittle	Strike slip	Sinistral	N60°E	08
F43	Brittle	Strike slip	Dextral	N65°E	03
F44	Brittle	Strike slip	Dextral	N65°E	02
F45	Brittle	Strike slip	Dextral	N65°E	3.5
F46	Brittle	Strike slip	Dextral	N45°E	09
F47	Brittle	Strike slip	Dextral	N85°E	3.0
F48	Brittle	Strike slip	Dextral	N85°E	3.5
F49	Brittle	Strike slip	Dextral	N20°E	7.5
F50	Brittle	Strike slip	Sinistral	N35°W	03
F51	Brittle	Strike slip	Sinistral	N40°E	04
F52	Brittle	Strike slip	Dextral	N20°E	3.2
F53	Brittle	Strike slip	Dextral	N47°E	02
F54	Brittle	Strike slip	Dextral	N48°E	04
F55	Brittle	Strike slip	Sinistral	N80°W	09
F56	Brittle	Strike slip	Dextral	N30°E	07
F57	Brittle	Strike slip	Dextral	N30°E	10
F58	Brittle	Strike slip	Dextral	N65°E	6.5
F59	Brittle	Strike slip	Dextral	N50°E	13
F60	Brittle	Strike slip	Dextral	N45°E	14
F61	Brittle	Strike slip	Dextral	N70°E	26
F62	Brittle	Strike slip	Sinistral	N40°W	09
F63	Brittle	Normal	–	–	12.5
F64	Brittle	Strike slip	Dextral	N20°E	02
F65	Brittle	Strike slip	Dextral	N50°E	02
F66	Brittle	Strike slip	Dextral	N50°E	05
F67	Brittle	Strike slip	Sinistral	N75°W	24
F68	Brittle	Strike slip	Sinistral	N80°E	28
F69	Brittle	Strike slip	Sinistral	N39°E	02
F70	Brittle	Strike slip	Dextral	N37°E	05
F71	Brittle	Strike slip	Dextral	N40°E	2.2
F72	Brittle	Strike slip	Dextral	N40°E	6.0
F73	Brittle	Strike slip	Sinistral	N40°E	1.0
F74	Brittle	Strike slip	Sinistral	N40°E	1.5
F75	Brittle	Strike slip	Dextral	N20°E	18
F76	Brittle	Strike slip	Dextral	N45°E	11
F77	Brittle	Strike slip	Dextral	N45°E	08
F78	Brittle	Strike slip	Dextral	N43°E	6.5
F79	Brittle	Normal	–	–	4.5
F80	Brittle	Normal	–	–	1.6

(continued)

Table 1 (continued)

Structure No.	Deformation regime	Fault type	Shear sense	Fault trend	Net slip (cm)
F81	Brittle	Normal	–	–	2.0
F82	Brittle	Normal	–	–	6.5
F83	Brittle	Normal	–	–	3.0
F84	Brittle	Normal	–	–	6.0
F85	Brittle	Normal	–	–	4.0
F86	Brittle	Normal	–	–	1.6
F87	Brittle	Normal	–	–	2.0
F88	Brittle	Normal	–	–	4.0
F89	Brittle	Normal	–	–	3.5
F90	Brittle	Strike slip	Dextral	N35°E	08
F91	Brittle	Strike slip	Dextral	N35°E	04
F92	Brittle	Strike slip	Dextral	N35°E	02
F93	Brittle	Strike slip	Dextral	N35°E	14
F94	Brittle	Strike slip	Sinistral	N80°E	12
F95	Brittle	Strike slip	Sinistral	N80°E	19
F96	Brittle	Strike slip	Sinistral	N20°E	15
F97	Brittle	Strike slip	Sinistral	N20°E	22
F98	Brittle	Strike slip	Dextral	N20°E	22
F99	Brittle	Strike slip	Dextral	N24°E	04
F100	Brittle	Strike slip	Dextral	N24°E	04
F101	Brittle	Strike slip	Dextral	N30°E	12
F102	Brittle	Strike slip	Dextral	N30°E	11
F103	Brittle	Strike slip	Dextral	N30°E	06
F104	Brittle	Strike slip	Sinistral	E-W	07
F105	Brittle	Strike slip	Dextral	N35°E	6.5
F106	Brittle	Strike slip	Dextral	N35°E	09
F107	Brittle	Strike slip	Dextral	N26°E	05
F108	Brittle	Strike slip	Sinistral	N20°E	05
F109	Brittle	Thrust	–	–	13
F110	Brittle	Strike slip	Dextral	N40°E	1.5
F111	Brittle	Strike slip	Dextral	N40°E	2.0
F112	Brittle	Strike slip	Dextral	N05°W	10
F113	Brittle	Strike slip	Dextral	N20°W	1.0
F114	Brittle	Strike slip	Dextral	N60°E	07
F115	Brittle	Strike slip	Sinistral	N30°E	06
F116	Brittle	Strike slip	Sinistral	N20°W	82
F117	Brittle	Strike slip	Dextral	N40°W	10
F118	Brittle	Strike slip	Dextral	E-W	05
F119	Brittle	Strike slip	Dextral	E-W	05
F120	Brittle	Strike slip	Dextral	E-W	03

(continued)

Table 1 (continued)

Structure No.	Deformation regime	Fault type	Shear sense	Fault trend	Net slip (cm)
F121	Brittle	Strike slip	Dextral	N25°E	04
F122	Brittle	Strike slip	Dextral	N40°E	05
F123	Brittle	Strike slip	Dextral	N30°E	2.5
F124	Brittle	Strike slip	Dextral	N30°E	07
F125	Brittle	Strike slip	Dextral	N30°E	4.0
F126	Brittle	Strike slip	Dextral	N30°E	2.0
F127	Brittle	Strike slip	Dextral	N30°E	1.5
F128	Brittle	Strike slip	Dextral	N30°E	1.0
F129	Brittle	Strike slip	Dextral	N30°E	3.5

which is located along the SE of Kinwat region. This NE-SW trend of strike-slip faults from Dharwar Craton near Degloor mismatches with the trend of the N-S trending strike-slip faults near Mumbai reported by Misra et al (2014) (also see Misra and Mukherjee 2015). The easternmost fault reported at Malshej Ghat (Misra et al. 2014) is ~500 km west of Degloor. These faults cannot be related with the study area presumably because they are far away from the study area.

The Degloor region is bound at SW by seismically active Killari region, NW by the present day microseismically active Nanded and NE by seismic event nearer to Kaddam fault/lineament reported in 2015 (Kaplay et al. 2017a, b). The study area locates exactly at the edge of one of the side of a seismically active triangle i.e., Killari-Kaddam-Nanded. Killari seismically activated since 1993, Nanded since 2006 and a small-scale seismicity is reported recently in 2015 nearer to Kaddam fault/lineament (Kaplay et al. 2017a, b). So far no (micro)seismic events are reported from Degloor. The region Killari and Nanded in Maharashtra and Kaddam in the Indian state Telangana lies in the zone III of Indian map of seismic zonation (Subhadra et al. 2015). Killari is a typical example of intracratonic seismicity and a 'slow-deforming non-rifted zone' (Rajendran 2016). Nanded and Kaddam too exemplify intracratonic microseismicity. The studies in Killari region suggested that Killari had experienced similar earthquakes in past (Rajendran et al. 1996). The Nanded region also experienced earthquakes in 1942 (review in Valdiya 2015). Killari and Nanded seismicity occurred in the SEDVP, which was previously considered tectonically stable. The Kaddam event occurred in granitic province.

The E-W strike-slip deformation zone is reported at the SEDVP contact around Kinwat (Kaplay et al. 2017a). A stretch of 60 km (Adampur-Bhaisa) SW of Kinwat along SEDVP contact has been recently designated as the "East Dharwar Margin Deformation Zone" (EDMDZ; (Kaplay et al., submitted). The reporting of strike-slip faults from Degloor, 15 km SW of Adampur-Bhaisa stretch, is marked as the EDMDZ in Fig. 1 confirms that the stretch EDMDZ continued further SW up to Degloor and hence this zone is designated as 'Western Boundary East Dharwar Craton Deformed Zone'(WBEDCDZ) which includes entire EDMDZ. The zone might continue SW right down up to Killari and also towards NE from Bhisa to Kinwat, but is to be checked by fieldwork. The study area is bound towards SE by seismically active Killari, towards NW by microseismically active Nanded and towards NE by recently reported microseismic event near Kaddam.

4 Conclusions

A. Deformation style, ~NE-SW structures, at Degloor does not match with those at the South East Deccan Volcanic Province (SEDVP), with Dharwar craton at Kinwat with ~E-W trend and the NW-SE strike-slip faults at Kaddam. Therefore, the stretch of ~105 km from Degloor to Bhainsa is designated separately as the 'Western Boundary East Dharwar Craton Deformed Zone' (WBEDCDZ).

B. Orientation of stress axes deciphered from this study: Fig. 6d: σ_3 (N83°E), σ_1 (N8°W/352°). Figure 9a: σ_3 (N65°E), σ_1 (N25°W/335°). Figure 9b: σ_3 (N20° W/340°), σ_1 (N66°E/66°). Figure 10: σ_3 (N°55E/°), σ_1 (N40°W/320°). This shows the extent of local variation of stress regime. This work presents the first detail information on orientation of stress axes from the remobilized EDC. Future structural geological works in the same line can provide comparison of how stress pattern had varied in the past. A statistical model highlighting the probability of earthquake occurrence can better justify the suggestion of installing a seismic network along the SEDVP.

Acknowledgements D. Kaplay thanks the School of Earth Sciences, S.R.T.M. University. Dr. Babar thanks the Principal, Dnyanopasak College, Parbhani, Maharashtra for support and encouragement. S. Mukherjee was supported by the CPDA grant of IIT Bombay. A research sabbatical provided by IIT Bombay helped him to finish this work. Fieldwork by S. Mahato was supported partially by IIT Bombay. The anonymous reviewers provided several positive inputs in two rounds. Mukherjee (2019) summarizes this work.

References

Babar MD, Kaplay RD, Mukherjee S and Kulkarni PS (2017) Evidences of deformation of dykes from Central Deccan Volcanic Province, Aurangabad, Maharashtra, India. In: Mukherjee S, Misra AA, Calvès G, Nemčok M (eds) Tectonics of the Deccan large igneous province. Geological Society, London, pp 337–353 (Special Publications 445)

Banerjee R, Veena K, Pandey BK, Parthasarathy TN (1993) Rb-Sr geochronology of the radioactive granites of Nanded area, Maharashtra, India. Journal of Atomic Mineral Science 1, 111–117

Banerjee R, Jain SK, Shivkumar K (2008) Geochemistry and petrogenesis of uraninite bearing granitoids and radioactive phosphatic cherty cataclasite of Thadisaoli Area, Nanded District, Maharashtra. Memoir of Geological Society of India 73, 55–84

Banerjee R, Shivkumar K, Maithani PB (2012) Major and trace element geochemistry of palaeoproterozoic nanded granitoids of Diglur-Biloli-Dharmabad Sector, Nanded District, Maharashtra: geodynamic and petrogenetic implications. Journal of Applied Geochemistry 26–58

Bhave N, Siddique MI, Desai J, Pillai SP, Dole G, Kulkarni H, Kale VS (2017) Faulting in Deccan traps in the vicinity of Koyna-Warna Seismic Zone. Journal of Geological Society of India 90, 748–751

Devey CW, Lightfoot PC (1986) Volcanological and tectonic control of stratigraphy and structure in the western Deccan Traps. Bulletin of Volcanology 48, 195–207

Geological Survey of India (1979) Geological map of Diglur Quadrangle, Andhra Pradesh—Karnataka—Maharashtra, on 1:253,440 scale

Geological Survey of India (2001) District resource map of Nanded District, Maharashtra, on 1:300,000 scale with Explanatory brochure

Kaplay RD, Babar MD, Mukherjee S, Wable D, Pisal K. (Submitted) Granitic rocks underlying Deccan trap along the margin of east Dharwar craton, Muntyal (Maharashtra)—Bhaisa (Telengana), India, general description and deformation. Proceedings of National Academy of Sciences, India Section A: Physical Sciences

Kaplay RD, Vijay Kumar T, Sawant R (2013) Field evidence of deformations in Deccan Basalt in micro-Seismically active Nanded, Maharashtra. Current Science 105, 1051–1052

Kaplay RD, Babar MD, Mukherjee S, Kumar TV (2017a) Morphotectonic expression of geological structures in eastern part of south east Deccan volcanic province (around Nanded, Maharashtra, India); In: Mukherjee S, Misra AA, Calvès G, Nemčok M (eds) Tectonics of the Deccan large igneous province. Geological Society of London, pp 317–334 (Special Publications 445)

Kaplay RD, Vijay Kumar T, Mukherjee S, Wesanekar PR, Babar MD, Chavan S (2017b) E-W strike slip shearing of Kinwat Granitoid at South East Deccan Volcanic Province, Kinwat, Maharashtra, India. Journal of Earth System of Science 126, 71

Misra AA, Mukherjee S (2015) Tectonic inheritance in continental rifts and passive margins. Springer Briefs in Earth Sciences

Misra AA, Mukherjee S (2017) Dyke-brittle shear relationships in the Western Deccan Strike Slip Zone around Mumbai (Maharashtra, India); In: Mukherjee S, Misra AA, Calvès G, Nemčok M (eds) Tectonics of the Deccan large igneous province. Geological Society, London, pp 269–295 (Special Publications 445)

Misra AA, Bhattacharaya G, Mukherjee S, Bose N (2014) Near N–S paleo-extension in the western Deccan region, India: does it link strike-slip tectonics with India–Seychelles rifting? International Journal of Earth Sciences 103, 1645–1680

Mitchell C, Widdowson M (1991) A geological map of the southern Deccan Traps, India and its structural implications. Journal of Geological Society 148, 495–505

Mukherjee S (2014) Atlas of shear zone structures in Meso-scale. Springer Geology, pp 71–75

Mukherjee S (2019) Introduction to "Tectonics and Structural Geology: Indian Context". In: Mukherjee S (ed) Tectonics and structural geology: Indian context. Springer International Publishing AG, Cham, pp 1–5. ISBN: 978-3-319-99340-9

Mukherjee S, Koyi HA (2009) Flanking microstructures. Geological Magazine 146, 517–526

Mukherjee S, Punekar J, Mahadani T, Mukherjee R (2015) A review on intrafolial folds and their morphologies from the detachments of the western Indian Higher Himalaya. In: Mukherjee S, Mulchrone KF (eds) Ductile shear zones: from micro- to macro-scales. Wiley Blackwell, London, pp 182–205

Mukherjee S, Misra AA, Calvès G, Nemčok M (2017) Tectonics of the Deccan Large igneous province: an introduction; In: Mukherjee S, Misra AA, Calvès G, Nemčok M. (eds) Tectonics of the Deccan large igneous province. Geological society of London, pp 1–9 (Special Publications 445)

Rajendran CP (2016) Paleoseismology: 2010–2016. Proceedings of Indian National Science Academy 82, 695–704

Rajendran CP, Ranjedran K, John B (1996) The 1993 Killari (Latur), Central India, Earthquake: an example of fault reactivation in the Precambrian Crust. Geology 24, 651–654

Sangode SJ, Meshram DC, Kulkarni YR, Gudadhe SS, Malpe DB, Herlekar MA (2013) Neotectonic response of the Godavari and Kaddam Rivers in Andhra Pradesh, India: implications to quaternary reactivation of old fracture system. Journal Geological Society of India 81, 459–471

Subhadra N Padhy S, Prabhakara PP, Seshunarayana T (2015) Site-specific ground motion simulation and seismic response analysis for microzonation of Nanded City, India. Natural Hazards 78. https://doi.org/10.1007/s11069-015-1749-z

Valdiya KS (2015) The making of India: geodynamic evolution. Springer. ISBN 978-3-319-25029-8

Wesanekar PR, Patil RR (2000) Rb-Sr dating of pink Granites of Deglur, Nanded district, Maharashtra, India. In: Proceedings of National seminar on tectonomagmatism, geochemistry and metamorphism of Precambrian Terrain, pp 181–187

Synthesis of the Tectonic and Structural Elements of the Bengal Basin and Its Surroundings

Md. Sakawat Hossain, Md. Sharif Hossain Khan, Khalil R. Chowdhury and Rashed Abdullah

1 Introduction

The Bengal Basin is one of the largest peripheral collisional foreland basins in South Asia (Mukherjee et al. 2009; DeCelles 2012; DeCelles et al. 2014) consisting of Permo-carboniferous to Mesozoic and Tertiary deposits covered by the Recent alluvium. This is one of the thickest sedimentary basins of the world consisting of ~21 km thick Early Cretaceous–Holocene sedimentary succession (Curray 1991; Curray and Munasinghe 1991). Geographically, the Bengal Basin lies approximately between 20° 34′ to 26° 40′N and 87° 00′ to 92° 45′E with its major portion being constituted by Bangladesh and also covering parts of the Indian States of West Bengal, Tripura and Assam (Fig. 1). To the north, the basin passes into the Tertiary Shelf of the Assam Basin, and to the south, it continues as the Bengal Fan (Acharyya 2007). The Bengal Basin is bordered on the west by the Indian Shield and the Shillong Plateau to the north, the Indo-Burman Ranges to the east and the Bay of Bengal to the south (Uddin and Lundberg 2004).

The basin occupied dominantly by the Ganges-Brahmaputra-Meghna (GBM) delta named after the Ganges (Padma in Bangladesh), the Brahmaputra (Jamuna in Bangladesh) and the Meghna rivers, constituting the largest fluvio-deltaic to shallow marine sedimentary basin of the world. The Ganges enters the basin from the northwest, the Brahmaputra from the northeast, whereas the Meghna drains the Sylhet Trough and part of the Tripura hills before flowing into the Brahmaputra and finally into the Bay of Bengal. Cumulative sediment dispersal through the Bengal Basin into the Bay of Bengal per year is estimated to about 1.1 Gigatonne (GT), results the formation of the biggest submarine fan of the world, the

Md. S. Hossain (✉) · Md. S. H. Khan · K. R. Chowdhury · R. Abdullah
Department of Geological Sciences, Jahangirnagar University,
Dhaka 1342, Bangladesh
e-mail: sakawat@juniv.edu

S. Mukherjee (ed.), *Tectonics and Structural Geology: Indian Context*,
Springer Geology, https://doi.org/10.1007/978-3-319-99341-6_6

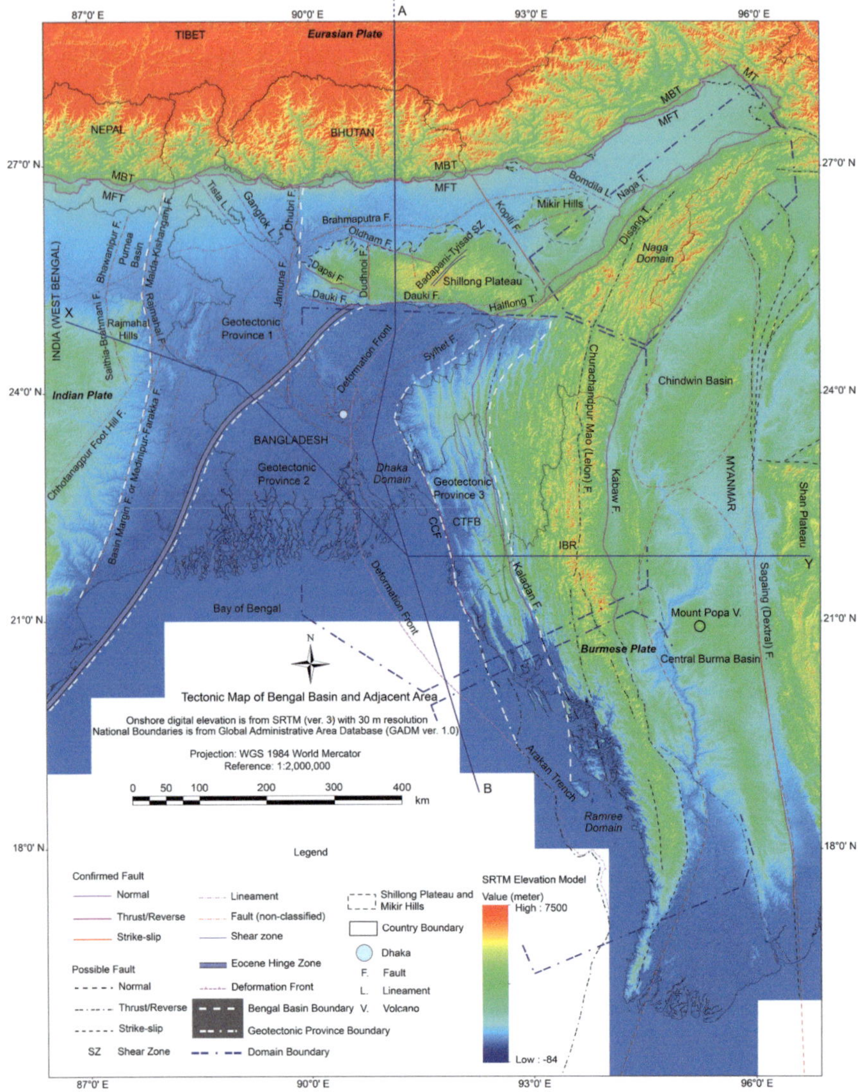

Fig. 1 Simplified map of the geotectonic provinces of the Bengal Basin and its surroundings. Solid blue lines indicate the cross-sections position. The boundaries of the three geotectonic provinces are taken from Bakhtine (1966), Guha (1978) and Matin et al. (1983), and the domain boundaries are taken from Wang et al. (2014). Faults and lineaments are mostly taken from Kayal (2008) and Wang et al. (2014). MBT: Main Boundary Thrust; MFT: Main Frontal Thrust; MT: Mishmi Thrust and T.: Thrust

Bengal Fan (Kuehl et al. 1989; Milliman et al. 1995; Curray et al. 2003; Goodbred et al. 2003; Syvitski et al. 2005; Curray 2014; Reitz et al. 2015). The GBM delta receives most of the sediments shed from the Himalaya and is subjected to tectonic influence from the Himalayan front and the Shillong Plateau uplift in the north and the accretionary fold belt in the east (Reitz et al. 2015). The Bengal Basin traps and accumulates less than half of the total sediment budget through flexural subsidence over a broad area, in addition to faulting, folding and localized compaction (Goodbred and Kuehl 1998, 1999). This accumulated sediment has caused the entire Bengal Basin margin to prograde more than 300 km since the Eocene (35 Ma) (Najman et al. 2008; Reitz et al. 2015).

The geotectonics of the eastern part of the Indian Plate is dominantly influenced by its collision with the Eurasian Plate (Mukherjee 2013a, 2015; Mukherjee et al. 2013, 2015) to the north and the Burmese Plate to the east, uplifting the Himalaya and the Indo-Burman Ranges (IBR) in the north and east, respectively (Alam et al. 2003; Vernant et al. 2014; Wang et al. 2014; Hossain et al. 2016). The Himalayan mountain range started to form in the Mid-Late Eocene (Garzanti et al. 1987; Klootwijk et al. 1992; Searle et al. 1997; Ali and Aitchison 2008; Baxter et al. 2016), whereas the IBR began to form approximately in the Middle Miocene (Acharyya 2007; Steckler et al. 2008). The Bengal Basin at the eastern part of the Indian Plate is tectonically active since the collision began (Eocene). Although the direction of subduction of the Indian Plate below the Eurasian Plate is ~N-S, whereas below the Burmese Plate, it is NE–SW (Goodbred et al. 2003; Reitz et al. 2015). In the north of the Bengal-Assam Basin, ~E trending Himalaya takes a southward turn towards Myanmar and connects with the IBR (Hossain et al. 2016). The geodetic measurements show that the motion of the Indian Plate relative to the Burmese Plate is ~36 mm/year (Socquet et al. 2006; Steckler et al. 2016) and the convergence between the Shillong Plateau and the Bengal Basin across the Dauki Fault varies from 3 mm/year in the west to >8 mm/year in the east (Socquet et al. 2006; Vernant et al. 2014). Plate boundary deformation along these collisional belts is broadly distributed over a series of reverse and strike-slip structures, including the Dauki and Haflong-Disang Faults to the north and Chittagong-Teknaf, Kaladan and Sagaing Faults to the east of the Bengal Basin, respectively (Hossain et al. 2014, 2016).

The Bengal Basin rests on lithosphere that is transitional (along the paleo-continental slope also known as the Eocene Hinge Zone) between thick, buoyant Indian continental lithosphere in the west and north and dense Indian oceanic lithosphere in the east (Bender 1983; Curray 1991; Reimann 1993; Uddin and Lundberg 2004; Singh et al. 2016). Along its northern edge, it is subjected to continent-continent collision whereas along eastern and southeastern edge, it is subjected to ocean-continent subduction. According to Ingersoll et al. (1995), due to continued oblique subduction of the Indian Plate beneath the Burmese Plate to the SE, the Bengal Basin turned into a Remnant Ocean Basin at the beginning of the Miocene. Due to eastward component of subduction of the oceanic part of the Indian Plate, the thick pile of sediments of the Bengal Basin has been deformed into a fold belt and a huge flat accretionary prism (Alam et al. 2003; Steckler et al. 2008). In general, by considering the overall regional tectonic setting, the Bengal Basin as a remnant ocean basin can be divided into three major geotectonic provinces: (i) the stable shelf to the northwest—passive to extensional cratonic margin in the west, (ii) the deeper foredeep basin at the center—remnant ocean basin, and (iii) fold belt to the east—the Chittagong Tripura Fold Belt (CTFB) (Bakhtine 1966; Alam 1972; Khan and Rahman 1992; Khan and Agarwal 1993; Reimann 1993; Shamsuddin and Abdullah 1997; Alam et al. 2003).

The northwest Stable Shelf of the Bengal Basin consists of an easterly dipping shelf that is separated from the Precambrian Indian Shield to the west by a prominent fault zone (Basin Margin Fault Zone: more than 350 km in length and width is variable in hundreds of metres) with dislocation and cataclasis (Raman et al. 1986; Matin and Misra 2009; Roy 2014). To the southeast it is bounded by the Eocene Hinge Zone that forms the paleo-continental slope (Sengupta 1966; Guha 1978; Bender 1983; Rahman et al. 1990a; Reimann 1993; Uddin and Lundberg 2004). Only a few studies have been performed on the basement rocks and their possible origin in the Stable Shelf part of the Bengal Basin (e.g., Ameen et al. 2007; Hossain et al. 2007, 2017; Ameen et al. 2016). The deeper Foredeep Basin lies in between the Eocene Hinge Zone to the west and CTFB to the east. The northern Boundary of the Foredeep Basin is marked by the Shillong Plateau consisting of Garo-Khasi-Janitia-Mikir Hills of Assam, India with $\sim$2000 m elevation. In contrast, immediate south of the plateau, the Foredeep Basin contains up to 20 km thick pile of sediments. The Shillong Plateau and the Foredeep Basin are separated by the E-W trending Dauki Fault (Bilham and England 2001; Biswas and Grasemann 2005a; Najman et al. 2016). This Foredeep Basin has been studied by Holtrop and Keizer (1970), Woodside (1983), Hiller and Elahi (1984), Murphy and Staff BOGMC (1988), Shamsuddin (1989), Johnson and Alam (1991), Mannan (2002), Alam et al. (2003), Rahman and Suzuki (2007), Najman et al. (2012), Farhaduzzaman et al. (2014) and Rahman and Worden (2016) to understand tectonic evolution of sedimentary basin architecture and provenance of the Neogene sediments.

Since 1914, the CTFB has been drawing attention to both structural and exploration geologists because of numerous earthquake epicenters and structural complexities including major thrust fault, and mud volcanoes (Bakhtine 1966; Hossain and Akhter 1983; Hossain 1985; Sikder 1998; Chowdhury et al. 2003, Steckler et al. 2008; Maurin and Rangin 2009a; Das et al. 2010; Wang et al. 2014; Steckler et al. 2016). Moreover, presence of several anticlinal folds, numerous faults and associated gas seepages attracted several national and international oil & gas companies to conduct seismic survey and drilling in the CTFB area (Shamsuddin and Abdullah 1997; Racey and Ridd 2015). Orogenic development and morphotectonic elements of the CTFB and IBR have been studied extensively by Bakhtine (1966), Gilbert (2001), Acharyya (2007), Richards et al. (2007), Bhattacharya et al. (2008), Das et al. (2011a) and Hossain et al. (2014). In addition, numerous investigations have been carried out in this area focusing mainly on 2D structural modeling (Sikder and Alam 2003; Mandal et al. 2004; Kabir and Hossain 2009; Maurin and Rangin 2009a; Najman et al. 2012; Abdullah et al. 2015), geochemistry and provenance studies (Najman et al. 2008; Dina et al. 2016; Rahman et al. 2017), active tectonics (Goodbred and Kuehl 1999; Khan et al. 2005; Wang et al. 2014; Khan et al. 2015), and earthquake and landslide hazard studies

(Ansary et al. 2000; Cummins 2007; Steckler et al. 2008; Wang et al. 2014; Steckler et al. 2016). Based on spatial deformation intensity and geometry of the folds, Bakhtine (1966) subdivided the CTFB into three zones: eastern highly compressed disturbed zone, middle asymmetric thrust faulted zone, and western quiet zone. Chowdhury (1970), Hossain and Akhter (1983), and Hossain (1985) performed single-fold geometric analysis on a few anticlines in the area. No comprehensive analysis of the fold geometric features has been carried out for a large population of folds till date. As a result, there are still many uncertainties concerning the original geometric shape as well as spatial distribution of the fold deformation intensity within the CTFB.

The first stratigraphic scheme of the Bengal Basin was originally proposed by Evans (1932) based purely on the exposures found in the CTFB and their simple long distance lithostratigrahic correlation with type sections in Assam. Over the years, several researchers have attempted to improve this scheme based on micro-paleontological studies (Ahmed 1968; Ismail 1978), palynological studies (e.g., Chowdhury 1982; Uddin and Ahmed 1989; Reimann 1993), and seismo-stratigraphic studies (Lietz and Kabir 1982; Salt et al. 1986; Lindsay et al. 1991; Alam et al. 2003), and consequently, abetted in solving some of the stratigraphic problems of the Bengal Basin. Three stratigraphic schemes for the three respective geotectonic provinces of the Bengal Basin have been proposed (Bakhtine 1966; Guha 1978; Khan 1991a, b; Alam et al. 2003) and these are: (i) Geotectonic Province 1—Stable Shelf (Zaher and Rahman 1980; Lindsay et al. 1991; Reimann 1993; Alam 1997; BOGMC 1997); (ii) Geotectonic Province 2—Central Deep Foredeep Basin (Hiller and Elahi 1988); and (iii) Geotectonic Province 3—CTFB (Fig. 1) (Gani and Alam 1999).

Deposition of sediments within the Bengal Basin is believed to be controlled by tectonic cycles or stages related to the interaction and collision patterns of major plates, and has taken place in five distinct phases (Alam et al. 2003): (1) Syn-rift stage: Permo-Carboniferous to Early Cretaceous (Sclater and Fisher 1974; Molnar and Tapponnier 1975; Uddin and Lundberg 2004); (2) Drifting stage: Cretaceous–Mid-Eocene (Banerji 1981; Ramana et al. 1994); (3) Early collision stage: Mid-Eocene–Early Miocene (Bender 1983; Uddin and Lundberg 1998, 2004); (4) Late Collision stage: Early Miocene–Mid-Pliocene (Ingersoll et al. 1995); and (5) Latest collision stage: Mid-Pliocene–Quaternary (Alam et al. 2003; Najman et al. 2016). However, proper correlation of the stratigraphic succession in the three tectonic provinces and with other adjacent foreland basins is yet to be done.

Based on the above discussion, it can be concluded that there is no comprehensive study that synthesizes the tectonic evolution and structural elements of the Bengal Basin as a whole including its adjacent areas. Therefore, this chapter is aimed at understanding the tectonic evolution, structural framework, and stratigraphic succession of the Bengal Basin. The work will not only enhance the understanding of present day structural and tectonic elements of the Bengal Basin but also help to comprehend their evolution with time.

2 Geological Boundary of the Bengal Basin

The Bengal Basin is bounded on all sides by different tectonics and structural controlled features which evolved at different geological times between the Late Cretaceous and Late Neogene, except in the south where the basin is open to the Bay of Bengal (Fig. 1) (Alam et al. 2003; Ali and Aitchison 2008; Roy and Chatterjee 2015; Baxter et al. 2016). The southeastern boundary of the Bengal Basin coincides with the eastern margin of the Chittagong Tripura Fold Belt (CTFB) bordering the western fringe of the Indo-Burman Ranges (IBR) (Acharyya 2007; Hossain et al. 2014; Wang et al. 2014). In the north, the basin is bounded by the Himalayan Foredeep, the Shillong Plateau and the Assam Basin (Roy and Chatterjee 2015; Najman et al. 2016). In the NNW corner, the Bengal Basin extends up to the Major/Main Frontal Thrust (MFT), which separates the Sub-Himalayan Zone from the Indo-Gangetic Alluvial Plain (Roy and Chatterjee 2015). The western and southwestern parts of the Bengal Basin consist of an easterly inclined shelf, separated from the Singhbhum Craton of the northeastern part of the Indian Shield (Acharyya 2003; Hossain et al. 2007, 2017; Kumar et al. 2017). The southern boundary of the basin is marked by the northern edge of the Bengal Fan (Fig. 2) (Curray et al. 2003). A generalized description of these bordered tectonic features is summarized below.

2.1 *Indo-Burman Ranges*

The Bengal Basin is flanked to the east by an approximately NNW–SSE trending fold belt known as Indo-Burman Ranges (IBR)/Paleogene Inner Burman Ranges that is separated from the CTFB/Neogene Outer Burman Ranges to the west by N-S striking Kaladan Fault (Najman et al. 2016). The IBR is a grand arcuate hill ranges which stretch from approximately 16–27.5°N originating from the collision of the Indian Plate with the Burmese Plate/Sunda Plates (Fig. 1). As a result of the collision, the oceanic part of the Indian Plate subducted below the Burmese Plate leading to the emergence of westward propagating accretionary wedge of sediments exhibiting an imbricate thrust system (Hossain et al. 2014) constituting the IBR. It evolves from a narrow-low hills in the south to a very broad-high range in the middle, and then again to a narrow-high range in the north. The IBR, which is bounded by the Kaladan Fault in the west and the Kabaw Fault in the east, is a distinct entity in the tectonic framework of the eastern Himalaya and SE Asia (Najman et al. 2016). It is thought to be the northern prolongation of the Indonesian island-arc linking with the Himalaya along the Tidding Suture Zone. In an east-west section, the IBR may be subdivided into two broad domains, a Western Belt and an Eastern Belt separated by the Carnian quartzose turbidites and flysch sediments (Acharyya 2007).

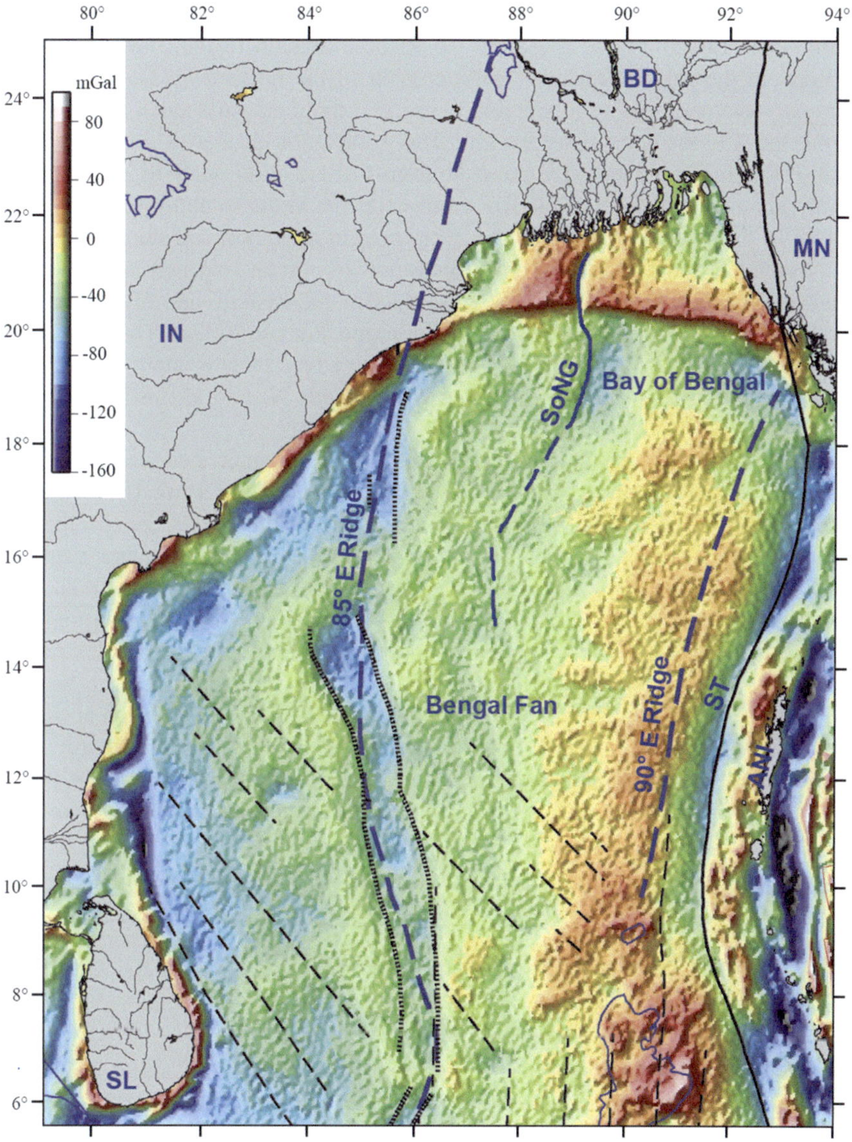

Fig. 2 Gravity mosaic of the Bengal Fan (Bay of Bengal) obtained from the satellite derived data (modified after Desa et al. 2013). Thick dashed lines mark the extent of the subsurface 85°E Ridge inferred using the satellite gravity mosaic. Blue dashed lines indicate the likely hotspots tracks of the 85°E and 90ºE Ridges. Fracture zones are marked by the thin dashed black lines. Thin blue line outlines the large igneous provinces. ANI: Andaman-Nicobar Islands; BD: Bangladesh; IN: India; MN: Myanmar; SL: Sri Lanka; SoNG: Swatch of No Ground; ST: Sunda Trench

Wang et al. (2014) divide the IBR and the CTFB into three domains along north-south direction, and these are the Ramree domain to the south, the Dhaka domain in the middle, and the Naga domain to the north (Fig. 1). The Ramree domain sustained oblique subduction to nearly orthogonal subduction and accretion along the 450 km long plate margin. This subduction produced a belt of deformation that increased not only the width from 170 km in the south to 250 km in the north, but also the height of the IBR from <1000 m in the south to >2000 m in the north. In the middle Dhaka domain, plate margin sustained collision, resulting in the formation and rapid westward propagation of a great fold belt and low angle subduction mega thrust within the sediments of the eastern side of the Ganges–Brahmaputra-Meghna (GBM) Delta (Maurin and Rangin 2009a). The total length of domain is ~500 km and the maximum width along 23°N is about 400 km. The height of the IBR in this domain increases towards N, and attains ~3000 m at 21.3°N.

The CTFB lies to the western margin of the Dhaka domain swing eastward to its northern end. The approximately east-west oriented Dauki Fault (Biswas et al. 2007; Hossain et al. 2016) marks the northern end of the Dhaka domain (beyond which CTFB is absent), and the beginning of the Naga domain. A major change in the collisional characteristic occurs at the beginning of the Naga domain approximately at 25°N. The Naga domain consisting of the Naga Hills (northern edge of the Burmese Plate) overrides the narrow continental shelf of the Assam Basin to the northeast (Wang et al. 2014). The significant geomorphologic difference between the Dhaka and the Naga domains seems to reflect the contrast in collision at shallow crustal level rather deep-seated ones. The length of the Naga domain is ~430 km and the width of the domain narrows from ~170 km in the southwest to ~90 km to the northeast, suggesting a northeastward depletion in total shortening across the hill range (Maurin and Rangin 2009a; Wang et al. 2014).

2.2 Assam Basin, Shillong Plateau and Himalayan Foredeep

The Bengal Basin in the northeast is bounded by the Shillong Plateau, which was uplifted during the Pliocene (Acharyya et al. 1986; Biswas et al. 2007), and separated the Assam Basin from the Bengal Basin at the juncture of the southwestern part of the Naga domain (Fig. 1). Approximately E-W stretching Shillong Plateau raised ~2000 m elevation as continental basement pop-up structure due to two seismically active steep reverse faults along its southern and northern edges. These two prominent faults are the E-W trending Dauki Fault in the south and the WNW-ESE trending Oldham Fault in the north (Fig. 1) (Bilham and England 2001; Biswas and Grasemann 2005a; Kayal et al. 2006; Kayal 2008). South-dipping Oldham Fault is thought to be the backthrust of the north dipping master Dauki Fault (Yin et al. 2010; see Mukherjee 2013b for backthrust kinematics in general, also see Bose and Mukherjee, submitted-a, b). The importance of the Oldham Fault is considered to be minor compared to the Dauki Fault for the plateau uplift (Biswas et al. 2007). However, another school of thought believes the Shillong Plateau to be a large anticlinorium floored by Archean and Proterozoic basement rocks with remnants of the passive margin strata on its steep southern face adjacent to the Surma basin (Singh et al. 2016).

The plateau is separated from the Mikir Massif to the east by the NW-SE trending Kopili Fault (also known as the Kopali Fracture Zone) (Evans 1964; Biswas and Grasemann 2005a; Biswas et al. 2007; Angelier and Baruah 2009). In the west, the plateau abuts against the Bengal Basin along a prominent fault, which continues northward up to the Himalayan front. Up to the point of inflection of the sharply bent Brahmaputra River course from westward to southward, the fault is known as the Jamuna Fault and to the further north as the Dhubri Fault (Desikachar 1974; Nandy 2001). According to Johnson and Alam (1991) and Uddin and Lundberg (1999), the paleo-Brahmaputra River flowed towards SW between the current day juncture of the Shillong Plateau and the Naga Domain (accretionary prism of the IBR) until the Miocene ended. The uplift of the Shillong Plateau as well as the westward encroachment and the final abutment of the CTFB accretionary prism (Neogene Outer IBR) against the already-uplifted Shillong Plateau led to a ~300 km westward shift of the Brahmaputra River to its present course during the Pliocene (Najman et al. 2012; Bracciali et al. 2015; Najman et al. 2016). Towards the east, the Shillong Plateau comprises the Garo, Khasi and Jaintia hills, which stretch for ~97 km in N-S and ~240 km in E-W (Morgan and McIntire 1959). The basement of the plateau is primarily composed of intensely stressed Precambrian and Early Paleozoic metamorphic and intrusive rocks overlain by the Cretaceous and Cenozoic sediments in its steep southern face as well as east and west margins (Wadia 1953; Nag et al. 2001; Biswas et al. 2007; Najman et al. 2016).

The Shillong Plateau presumably has had a major influence on strain partitioning and consequently tectonics of the Eastern Himalaya that influenced the seismic risk in the surrounding regions (Bilham and England 2001; Banerjee et al. 2008; Najman et al. 2016). According to Mitra et al. (2005) and Vernant et al. (2014), the southern edge of the Shillong Plateau is elevated by stresses arising from the master thrust (Dauki Fault), and the plateau is tilted northward at approximately 2°–4° as indicated by a 1–2 km increase in depth of the Moho between the northern edge of the plateau and the Himalayan Front. Under the Shillong Plateau and the Surma basin, the Moho has been observed at similar depths of ~37 km despite the contrast of craton vs. thick sedimentary basin. This possibly indicates the uplift of the overthrusting Shillong Plateau and the concurrent downward flexure of the Surma basin (Singh et al. 2016). The thickness of the crust of the Indian Plate beneath the Shillong Plateau is ~38–40 km (Singh et al. 2015, 2016). To the north and northwest, the Bengal Basin extends up to the Himalayan front where it is separated by the Main Frontal Thrust (MFT) or Himalayan Frontal Thrust (HFT) (Acharyya 1994; Ganguly 1997). The Assam Basin, situated in the north of the Shillong Plateau and northwest of the Naga domain, was connected to the

Bengal Basin up to the end of the Miocene. During the Pliocene, the upliftment of the Shillong Plateau separated the Assam Basin from the Bengal Basin except to the western part, across the N-S oriented Dhubri Fault.

2.3 *Indian Shield and Rajmahal Hills*

Peninsular India lying south of the Indo-Gangetic Alluvial Plain commonly referred to as the Indian Shield reportedly occupied a much wider area than the present triangular shaped region, made up of a diverse mosaic of igneous and metamorphic terrains that has undergone deformation and metamorphism (Roy 2014). These terrains, constituting the continental crust have attained tectonic stability since the Precambrian and are designated as cratons (Valdiya 2010, 2016). The cratons are flanked by fold belts, with or without a discernible suture or shear zone, suggesting that the cratons, as crustal blocks or micro-plates, moved against each other (Naqvi 2005). Alternatively, these cratons could be the result of fragmentation of a large craton that constituted the Indian Shield. Welded fold belts in between the neighbouring cratons indicate either rifting or splitting of cratons. The crustal blocks of Indian shield are divided into six cratons (Sharma 2010; Valdiya 2010, 2016). Among these cartons, northeastern Chhotanagpur Granite-Gneiss Complex (CGGC) and Singhbhum Craton (SC) together with Satpura Mobile Belt or Central Indian Tectonic Zone (CITZ) and Eastern Ghat Mobile Belt are lying along the western margin of the Bengal Basin (Misra 2006). The thickness of the crust of the Indian Plate near the western part of the Bengal Basin is ~38 km (Singh et al. 2015). The approximately N-S running Malda-Kishanganj Fault (Nandy 2001; Vaccari et al. 2011; Mohanty et al. 2013, 2014; Prasad and Pundir 2017) reaches up to the eastern and northeastern margins of the Rajmahal Hills which is considered as the northwestern margin of the Bengal Basin.

The Bengal Basin constitutes the eastern continuation of the Indo-Gangetic Alluvial Plain (IGAP), which separates the Extra Peninsular India to the north and the Peninsular India to the south (Oldham 1893; Roy 2014). The Indo-Gangetic plains are stretching across northern India between Punjab to the west and Assam to the east through Bangladesh. The Extra Peninsular India, which is the mountainous region formed by the Himalayan ranges and their extensions into the Baluchistan to the west and Myanmar to the east (Valdiya 2010).

To the north northwest, the Bengal Basin is separated from the Rajmahal Hills by the approximately N-S running Rajmahal Fault (Ghose et al. 2017). The Rajmahal Hills of West Bengal and Bihar is a fault bounded small tectonic element situated in the western edge of the Stable Shelf of the Bengal Basin (Fig. 1). The Early Cretaceous (100–118 Ma; McDougall and McElhinny 1970; Baksi 1995; Kent et al. 2002) Rajmahal Traps (RT), a sequence of tholeiitic basalts, crop out in the Rajmahal Hills (McDougall and McElhinny 1970). Geologically, the Rajmahal Hills is located at the juncture of the Singhbhum Craton (SC) towards west and the Rangpur Saddle of the Bengal Basin to the east (Bage et al. 2014; Singh et al. 2016).

Rajmahal Hills is bounded by the Rajmahal Fault and Saithia-Brahmani Fault in the east and the west sides, respectively (Ghose et al. 2017). The RT is the result of the Kerguelen Plume upwelling activity in the Early Cretaceous (Ghatak and Basu 2011, Roy and Chatterjee 2015), which resulted in a massive amount of mantle material covering the entire surface (Baksi et al. 1987).

According to Nandy (2001), Ghatak and Basu (2011), Sager et al. (2013) and Roy and Chatterjee (2015), the N–S running Jamuna Fault is coincident with the Ninety East Ridge along the path of the Kerguelen Plume in the Bay of Bengal as evidenced by the hotspot-related magma upwelling in the region. Roy and Chatterjee (2015) suggest that the Tista Fault, the Dauki Fault and N30°E–S30°W oriented Basin Margin Fault form a triple-point intersection of lineaments/fractures because of local doming up of the crust during upwelling of the mantle plume. Mantle upwelling extruded tholeiite lava through fractures over a wide region in and around the Bengal Basin. The Rajmahal Hills evolved along the then eastern continental margin of the Indian Plate, following rifting from the Gondwana. Over time, the upper part of this tectonic element subjected to extensional brittle deformation and formed graben structures (Singh et al. 2004). The basalts of the RT above the Rajmahal Hills are commonly vesicular that are filled with secondary minerals e.g., calcite, analsite, chalcedony, and agate (Ball 1877). The intertrappean beds are composed of sedimentary rocks like siltstone, claystone, and shale (Valdiya 2010). The central part carries the impression of up to 28 flows, which has been identified earlier by the Geological Survey of India (Pascoe 1975; Mukhopadhyay et al. 1986). On the other hand some other authors refer that the maximum exposed thickness of the lava is about 230 m or approximately ten flows (Kent et al. 1997). The Sylhet Traps are equivalent to the Rajmahal Traps, and are reported to be found in the northern edge of the Surma basin (Roy and Chatterjee 2015). Talwani et al. (2016) suggest that the Rajmahal and the Sylhet Traps are not separate eruptions as they are connected by a prominent magnetic doublet (Rahman et al. 1990a). According to them, both the Rajmahal and the Sylhet Traps are originated from same igneous activity.

2.4 Bengal Fan

The Bengal Fan (Fig. 2), which marks the southern end of the Bengal Basin and northeastern lobe of the Indian Ocean, floors the entire Bay of Bengal, forming the largest submarine fan in the world (Curray et al. 2003; Curray 2014). Continental slope of eastern Sri Lanka and India, southern Bangladesh, and the Sunda Trench extending from Myanmar to Andaman-Nicobar Islands marks the western, northern and eastern margins of the fan, respectively (Fig. 2). The southern end of the fan reaches up to the south distal end of the Bay of Bengal at $\sim$7°S. The fan was first apparently recognized by Dietz (1953), and delineated and named by Curray and Moore (1971). The length of the fan is $\sim$3000 km along N-S, the width is $\sim$1430 km along E-W, the area is $\sim 3 \times 10^6$ km^2 and the maximum sediment thickness of the fan is $\sim$16.5 km (Curray et al. 2003; Shanmugam 2016).

The development of the Bengal Fan is related to the creation of the Bay of Bengal, which was the direct result of the India–Asia collision and uplift of the Himalaya and thickening of the Tibetan Plateau. The initial Late Paleocene–Eocene collision of the Indian Plate with the subduction zone of the north side of the Tethys Ocean initiated a small fan in the northern Bay of Bengal, now lying beneath the younger deltaic and shelf sediments of the Bengal Basin (Curray and Moore 1974; Alam et al. 2003; Curray 2014). From the Eocene onward, collision continued resulting rapid clastic sedimentation in the Bengal Basin primarily from erosion of the high standing and rising Himalaya and the Tibetan Plateau, and formation of fan on top of this continental rise (Curray et al. 2003; Curray 2014). The GBM delta formed by the river Ganges, Brahmaputra, and Meghna gradually filled the Bengal Basin. The surplus sediment, which has passed through has been distributed across the entire Bay of Bengal to eventually form the world's biggest submarine fan. The fan gradually prograded southward during the Tertiary (Curray et al. 2003). The Swatch of No Ground (SoNG) is the main avenue that funneled down sediment from the Bengal Basin to the Bengal Fan (Curray et al. 2003). It is a shelf-incising aggrading canyon (Fournier et al. 2016) located 250 km $\sim$S of the confluence of the Ganges and the Brahmaputra rivers and only 30 km south from the coastal delta plain of Bangladesh.

Curray and Moore (1974) assumed that the formation of the modern Bengal Fan commenced during the Oligocene to the Early Miocene. Two lobes of the fan separated by the Ninety East Ridge were active until the Middle Pleistocene, namely the Bengal Fan proper in the west and Nicobar Fan in the east (Curray and Moore 1974). The SoNG, which was the main supplier of sediments to the Bengal Fan is largely cut off from its supply of sediment as the modern river sediments trapped on the Bengal shelf due to the Holocene transgression (Emmel and Curray 1985; Kuehl et al. 1989). The supply of sediments to the Nicobar Fan was cut off by the convergence of the Ninety East Ridge and the Sunda Trench during the Early Pleistocene. At present, only the axial area of the Sunda Trench is still receiving sediments from the fan in the Bay of Bengal (Curray and Moore 1974; Bender 1983). The surface features of the Bengal Fan are mainly active and abandoned turbidity current channels, and the Ninety East Ridge and Eighty Five East Ridge (Curray et al. 2003; Curray 2014).

3 Basin Geometry and Tectonic Evolution

The Bengal Basin, a complex collisional foreland basin, exhibits intense variability in Neogene sediment thickness that reflects a complicated depositional environment and tectonics (Curray et al. 2003; Uddin and Lundberg 2004; DeCelles 2012). With broad attributes of basin geometry, tectonic evolution and stratigraphy, the discussion begins with the basin geometry, followed by the tectonic evolution. Finally, the stratigraphy in context to the basin geometry and tectonic evolution has been discussed in the later part.

3.1 Basin Geometry

The Bengal Basin is a peripheral foreland basin formed due to continent–continent collision after the subduction of the oceanic part of the Indian Plate (Mukherjee et al. 2009). The basin geometry is complex and asymmetric having thinner, gently sloping sediment cover in the western and northern parts, which thicken in the basin-ward direction to the south and southeast (Curray et al. 2003; Uddin and Lundberg 2004). Gravity anomaly information indicates a considerable heterogeneity in different parts of the basin in terms of stratigraphy, sedimentation, tectonics and basin evolution (Uddin and Lundberg 2004). Considerable difference in the basement height results significant variations not only in sedimentary rock types but also in their individual thickness (Mukherjee et al. 2009). Based on geodynamic development, basement configuration and sedimentation pattern, the Bengal Basin can be divided into three distinct geotectonic provinces: (i) Stable Shelf or Geotectonic Province 1, (ii) Central Foredeep Basin or Geotectonic Province 2, and (iii) Folded Flank (CTFB) or Geotectonic Province 3.

The Stable Shelf or Geotectonic Province 1 has a sub-basin in the north that lies between the Main Frontal Thrust (MFT) to the north and the Rangpur Saddle (Bakhtine 1966) of the Garo–Rajmahal Gap, i.e. concealed Dinajpur Block (Ameen et al. 2007, 2016) to the south. Shallowest part of this sub-basin is in the south, where the thickness of the Neogene sediments overlying the Precambrian basement and the Phanerozoic Gondwana rift-basin rocks (Khan et al 1994) varies between 128 and 1160 m (Rahman 1987; Khan 1991a, b; Ameen et al. 2007). The southern part of the Stable Shelf that lies between the Indian Shield to the west and the Eocene Hinge Zone to the east, is a prominent fault zone with dislocation and cataclasis running along the N15°E–S15°W marked by the crowding of gravity contours, and shows the occurrence of laterite and lateritic soils, the protoliths (Matin and Misra 2009; Roy 2014; Roy and Chatterjee 2015). A number of half graben structures containing Permian coal deposits as well as numerous faults are present in the northeastern part of this sub-basin (Rabbani et al. 2000). The southernmost part of the Stable Shelf has a low southeasterly gradient, and is characterized by several down-to-basin growth faults. Sediments thicken uniformly from the west to the southeast (Khan and Agarwal 1993). In general, based on the concealed basement depth and gradient, the Foreland Shelf is divided into four different parts. From north to south, these are the Himalyan Foredeep, the Dinajpur Shelf, the Rangpur Saddle, and the Bogra Shelf (Sengupta 1966; Guha 1978; Khan and Rahman 1992; Khan and Agarwal 1993; Reimann 1993; Guha et al. 2010).

The Central Foredeep Basin or Geotectonic Province 2 basically comprises of three sub-basins between the Eocene shelf-break to the west and the Folded Flank of the Bengal Basin to the east characterized by a huge sediment thickness ranging between 10 and 22 km (Curray 1991). These sub-basins are the Surma basin to the north separated from the Faridpur Trough (a sub-basin) to the southwest by the Madhupur-Tripura High. On the otherhand, the Faridpur Trough is separated from the Hatia Trough (another sub-basin) to the east by the Barisal-Chandpur High

(Alam 1972; Das Gupta 1977; Guha 1978; Reimann 1993). In general, the Foredeep Basin shows small-amplitude, isometric or geographically equant gravity anomalies (Bakhtine 1966). The transitional zone of the continent–ocean crust of the Indian Plate occurs in this sub-basin between the Eocene shelf-break in the west and the Barisal–Chandpur Gravity High in the east (Shamsuddin and Abdullah 1997; Uddin and Lundberg 2004; Mukherjee et al. 2009).

The folded flank of the Bengal Basin or Geotectonic Province 3 comprises the eastern part of the Bengal Basin and is one of the active collisional orogenic belts in the world. Approximately NNW to SSE trending folds that have developed in the upper part of the thick deltaic sequence of the CTFB generally exhibit linear or elongated gravity anomalies with large amplitudes (Figs. 5 and 6) (Bakhtine 1966).

3.2 *Tectonic Evolution*

To understand the tectonic evolution of the Bengal Basin, it is necessary to have insight on the plate tectonic and paleogeographic reconstruction model of Gondwana to Asia. Therefore, for the convenience, plate tectonic and paleogeographic reconstruction model of Gondwana to Asia is discussed first and followed by a discussion on the tectonic evolution of the Bengal Basin.

3.2.1 Plate Tectonic and Paleogeographic Reconstruction Model

Although standard traditional models suggest that the India–Eurasia (Tibet part) collision event started 50–55 Ma ago (Rowley 1996; Hodges 2000; DeCelles et al. 2002; Zhu et al. 2005; Najman 2006), Ali and Aitchison (2005, 2008), Aitchison et al. (2007), Metcalfe (2013), Misra et al. (2015), Baxter et al. (2016) and Mukherjee et al. (2017) proposed new model with much younger age (Tables 1 and 2). Ali and Aitchison (2008) discourse the revised plate tectonic reconstructions model into two phases: (i) Break-up and dispersal of Gondwana—Middle Jurassic through end-Paleocene, and (ii) Convergence and collision of Indian Plate with the Eurasia Plate—Eocene. However, before discussing the revised plate tectonic reconstruction model, it is necessary to consider the palaeogeography of the Neotethys Ocean.

The commonly accepted standard models involved the present-day Indian Plate with some form of extension to the north as a passive margin, north of which lay oceanic crust of the Neotethys. As the Indian Plate moved towards the Eurasian Plate, the Neotethys was consumed beneath the Lhasa Block in a setting similar to the present-day Sunda Arc (Hall 2002). This palaeogeographic scenario is considered to be the ‘one ocean–two continent convergence model’. However, following seismic tomography studies of the mantle beneath the Indian Ocean–South Asia region, geological investigations of the India–Asia suture zone, and paleomagnetic and magnetic anomaly based India and Asia motion models, paleogeographic scenario of Neotethys are considered as the ‘two ocean–two continent

Table 1 Major tectonic events related to the Indian and Eurasian plate collision and subsequent development of the Bengal Basin

Geologic time	Age in Ma	Major tectonic events
Convergence and collision of the Indian plate with the Eurasian plate—Eocene		
Mid- to Late Pliocene	~3.5–2.5	The major thrust related uplift of the Shillong Plateau along the Dauki Fault in the south and the Oldham Fault in the north. Centre of the Bengal Basin started evolving as a foreland basin
Late Oligocene	~25	Remnant ocean basin took shape due to the collisional orogeny of the Barail–Cachar Hills at the northeastern corner of the Indian Plate
Eocene-Oligocene transition	~35	The continental crust of the Indian Plate impacted with the Tibet part of the Eurasian Plate, resulted in subduction of the northern part of the Indian Plate beneath the southern Tibet, therefore, initiated the continent–continent collision
Middle Eocene	~45	The northern edge of the Indian Plate was still at some distance south of the Lhasa Block, i.e. southern front of the Eurasian Plate
Early Eocene	~55	The collision of Indian Plate with a Neotethyan intra-oceanic arc (Dazhuqu Arc/Kohistan–Ladakh Arc), but not with the Eurasian Plate
Break-up and dispersal of the Gondwana supercontinent—middle Jurassic through Late Paleocene		
Late Paleocene	~55.9	Indian Plate appears to have reached its maximum level of isolation after being jettisoned by the Seychelles block
Cretaceous–Paleogene transition	~66	Indian Plate started breaking from the Seychelles
Late cretaceous	~90–85	India–Seychelles separated from the Madagascar
Mid-cretaceous	~118	The Kerguelen large igneous province began forming on the floor of the SE Indian Ocean, resulting the Rajmahal Trap eruptions, and provided a stepping-stone for migrations between the India–Seychelles–Madagascar and Australia–Antarctica for much of the Mid-Cretaceous
Mid-early cretaceous	~125	East Gondwana moved along the Davie Fracture Zone on a trail roughly parallel to the coast of the eastern Africa, which led to formation of true ocean floor (Somalia and Mozambique Basin)
Early cretaceous	~132	Australia–Antarctica began to drift away from the India–Seychelles–Madagascar with a fan-shaped expanse of ocean floor that separated the two elements by gradually widening to the east
Middle Jurassic	~170–175	Gondwana supercontinent started breaking following the rifting of South America–Africa (West Gondwana) from the India–Seychelles–Madagascar–Australia–Antarctica (East Gondwana)
Paleozoic—Mesozoic (Early—Middle Jurassic)	~180	Indian Plate presumably occupied a central location in the Gondwana supercontinent

Compiled from Smith and Hallam (1970), Norton and Sclater (1979), Ingersoll et al. (1995), Storey et al. (1995), Acton (1999), Torsvik et al. (2000), Bilham and England (2001), Biswas and Grasemann (2005a), Schettino and Scotese (2005), Rabinowitz and Woods (2006), Aitchison et al. (2007), Ali and Aitchison (2008), Yin et al. (2010), Metcalfe (2013), Misra et al. (2014, 2015), Misra and Mukherjee (2015), Baxter et al. (2016), Najman et al. (2016, 2017), Mukherjee et al. (2017)

Table 2 Major tectonic events related to the Indian and Burmese plate collision and subsequent development of the Folded Flank (CTFB) of Bengal Basin

Geologic time	Age in Ma	Major events
Late Pleistocene - Recent	~0.075	The fold has been progressively growing westward, and some of the folds in the westernmost part have been actively developing
Early Pleistocene	~2	Sediments in the upper part (~up to 5 km) within the major individual thrust started to deform above a de'collement level by combined thin-skinned and thick-skinned tectonic processes. Many of the folds in the westernmost part have been actively developing
Late Pliocene	~3.5–3	Oblique subduction of the Indian Plate beneath the Burmese Plate in an arc-trench setting developed accretionary prism and the CTFB started to develop within the upper parts of the thick deltaic sequence in the eastern margin of the Bengal Basin
Early Miocene	~20	Sediment contributions to the Bengal Basin began to arrive from the Indo-Burman Ranges (IBR) due to subduction of the Indian Plate beneath the Burmese Plate
Early Eocene	~45	The northeast corner of the Indian Plate making a glancing contact with the Sumatra Block, followed by the Burmese Plate

Compiled from Curray et al. (2003), Uddin and Lundberg (2004), Acharyya (2007), Ali and Aitchison (2008), Maurin and Rangin (2009a), Wang et al. (2014) and Najman et al. (2016)

convergence model' (Fig. 3) (Allégre et al. 1984; Abrajevitch et al. 2005; Aitchison et al. 2007; Ali and Aitchison 2008; Baxter et al. 2016). According to this model, during the Late Cretaceous–Cenozoic passage of the Indian Plate towards the Tibet part of the Eurasian Plate, the Neotethys comprises two oceanic plates separated by north-dipping subduction zone (Ali and Aitchison 2008). Considering the present-day island arcs as a guide, Ali and Aitchison (2008) and Baxter et al. (2016) suggest that there might have some volcanic islands (Neotethyan arc) spaced every 30–90 km apart present in the Neotethys ocean. If the 'two ocean–two continent convergence model' is correct, one might, therefore, predict that remnants of the two separate ocean slabs are present beneath the subduction system in the India–Eurasia collision belt in southern Tibet (Yarlung Tsangpo Suture Zone- Indus River) and northern India (Himalayan Foredeep) (Ali and Aitchison 2008).

Break-up and Dispersal of Gondwana—Middle Jurassic Through Late Paleocene

In standard plate tectonic reconstruction, the Indian plate presumably occupied a central location in the Gondwana supercontinent throughout the Paleozoic and much of the Mesozoic (~180 Ma), i.e. up to the beginning of the Middle Jurassic (Smith and Hallam 1970; Norton and Sclater 1979). In the Middle Jurassic (~170–175 Ma), this supercontinent started breaking following the rifting of South America–Africa

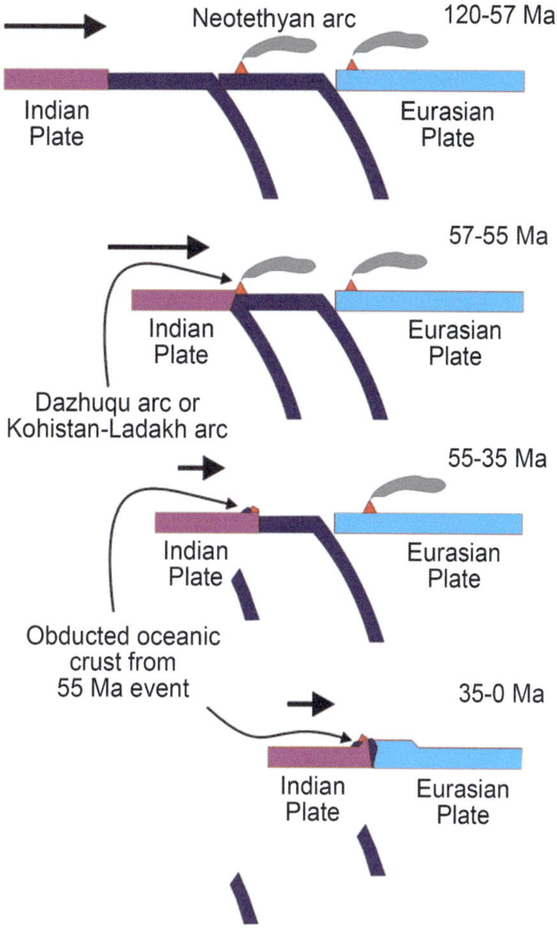

Fig. 3 Revised model describing the Late Cretaceous–Cenozoic convergence and collision of the Indian Plate with the Eurasian Plate. The model shows that the Neotethys comprises two oceanic plates separated by north-dipping subduction zone (modified after Ali and Aitchison 2008)

from India–Seychelles–Madagascar–Australia–Antarctica (Schettino and Scotese 2005). India–Seychelles–Madagascar–Australia–Antarctica together is known as the East Gondwana whereas South America–Africa is regarded as the West Gondwana. As the two parts of the Gondwana drifted over the next 50 Ma ($\sim$125 Ma), Indian Plate together with the rest of the East Gondwana moved along the Davie Fracture Zone on a trail roughly parallel to the coast of eastern Africa (Rabinowitz and Woods 2006). The East Gondwana migrated more than 400 km SSE relative to eastern Africa. This led to formation of true ocean floor (Somalia and Mozambique Basin) whose spreading ceased in the mid-Early Cretaceous and reached to its present-day separation of $\sim$430 km at about 116 Ma. On the other hand, Australia–Antarctica began to drift away from India–Seychelles–Madagascar for 11–12 m.y. in the Early Cretaceous ($\sim$132 Ma) with a fan-shaped expanse of ocean floor, separating the two elements by gradually widening to the east. The Kerguelen large igneous province began forming on the floor of the SE Indian Ocean at about 118 Ma, resulting the Rajmahal Trap eruptions. This could have provided a stepping-stone for migrations between India–Seychelles–Madagascar and Australia–Antarctica for much of the mid-Cretaceous. India–Seychelles–Madagascar was connected until the beginning of the Late Cretaceous (recent reviews in Misra et al. 2014, 2015; Misra and Mukherjee 2015; Mukherjee et al. 2017). In the Late Cretaceous (90–85 Ma), India–Seychelles separated from Madagascar (Storey et al. 1995; Torsvik et al. 2000; Aitchison et al. 2007). India started breaking from Seychelles around the Cretaceous–Paleogene boundary (~66 Ma), and India appears to have reached its maximum level of isolation after being jettisoned by Seychelles block at the end of the Paleocene (55.9 Ma). It is believed that the jettisoned Seychelles block coincided with the main Deccan Trap eruptions. The centre of the Indian Plate, which now sits on ~23.5°N (Tropic of Cancer), was at ~30°S in the Late Cretaceous, some 6000 km to the south from where it is today (Ali and Aitchison 2008).

Convergence and Collision of Indian Plate with Eurasian Plate—Eocene

The Indian Plate migrated rapidly northwards during the Late Cretaceous and the Paleocene (Acton 1999). According to Acton's (1999) model, northwards advance of the Indian Plate in between 120 and 73 Ma was ~6.6 cm/year, which increased to ~21.1 cm/year between 73 and 57 Ma. At ~57 Ma, the plate motion abruptly decreased to 9.5 cm/year, which continued until 20–30 Ma when there was a further major slowdown of the rate of movement. According to Ali and Aitchison (2008) and Baxter et al. (2016), the '57 Ma event' mark the collision of the Indian Plate with a Neotethyan intra-oceanic arc, but not with the Eurasian Plate. Ali and Aitchison (2004, 2006) and Aitchison et al. (2007) calculate the position of the new pole that sits 5°–9° further from the present-day North Pole. This new pole places the southern Lhasa Block at ~28°N at 55 Ma. Since then, this part of the Eurasian Plate has rotated ~21° clockwise relative to the spin axis and migrated 1100 km east. Ali and Aitchison (2008) suggest that evidence related to the collision of the oceanic part of the Indian Plate with the Dazhuqu Arc in the earliest Eocene (~55 Ma) has been found along the 2500-km-long Indus–Yarlung Tsangpo suture zone (trending ~E-W in the eastern part and ~NW-SE in the western part; Aitchison et al. 2007). The Dazhuqu Arc, also known as the Kohistan–Ladakh Arc (Metcalfe 2013) may have been formed as early as ~135 Ma, i.e. in the Early Cretaceous (Bosch et al. 2011). A number of island arc fragments that had obducted onto India after it collided with the Dazhuqu Arc at ~55 Ma are present in this suture zone. At the Middle Eocene (~45 Ma) the northern edge of the Indian Plate was still at some distance south of the Lhasa Block, i.e. southern front of the Eurasian Plate. Isotopic studies of igneous rocks and the biostratigraphic investigations of sedimentary rocks from the suture zone, led to the conclusion that the hard collision between continental part of the Indian Plate and the Eurasian Plate (southern Tibet, i.e. Lhasa Block), i.e. continent-continent collision happened at the very end of the Eocene, ~35 Ma (Ali and Aitchison 2008; Baxter et al. 2016). Therefore, Aitchison et al. (2007), Ali and Aitchison (2008) and Baxter et al. (2016)

reasonably argued that oceanic crust of the Indian Plate collided into a sub-equatorially located intra-oceanic arc (Neotethyan arc) at ~55 Ma, and later the continental crust of the Indian Plate impacted with the Tibet part of the Eurasian Plate starting around 35 Ma (Table 1). Approximately at the Eocene-Oligocene boundary (~35 Ma), northern Indian Plate entered the north-dipping subduction zone beneath southern Tibet, thus marking the initiation of continent–continent collision (Baxter et al. 2016). To the east, the northeast corner of the Indian Plate making a glancing contact with Sumatra Block, followed by the Burmese Plate from ~45 Ma, i.e. Early Eocene onwards (Ali and Aitchison 2008).

3.2.2 Tectonic Evolution of the Bengal Basin

The basin evolved through two major tectonic episodes. First, it initiated as an intra-cratonic rift basin within Gondwana landmass during Late Paleozoic–Mid Mesozoic and received the continental Gondwana sediments. This episode of basin development ended with widespread volcanism as continental flood basalts known as the Rajmahal Trap covered the Gondwana sediments. The second episode of basin development began in the Late Mesozoic with the break-up of Gondwana and is still going on (Alam 1989). At this stage, the tectonic evolution of the greater Bengal Basin is fundamentally related to the collision pattern of the Indian Plate with the Eurasian Plate to the north and the Burmese Plate to the east. The first uplift in the Himalayan and Indo-Burman region commenced in the Oligocene and the Early Miocene, respectively (Bender 1983). In its northern journey, the Indian Plate collided with the Eurasian Plate and caused folding and thrusting forming the Himalayan orogenic belt. Further movement of the Indian Plate was in the north-easterly direction, resulted in collision of northeastern end of the Indian Plate with the Burmese Plate, the former subducting below the latter, and the sedimentary accretionary prism gave rise to the formation of the CTFB and the IBR (Alam et al. 2003; Steckler et al. 2008; Wang et al. 2014).

The peri-cratonic part on the eastern margin subsided continuously and received voluminous sediments from the Late Mesozoic through Tertiary to the Recent times. This foreland basin containing a succession of dominantly deltaic sediments derived primarily from the erosion of the Himalaya and the IBR. Sediments accumulate in the basin through the Ganges, Brahmaputra and Meghna (GBM) River systems, and are dispersed into the Bay of Bengal, forming the largest submarine fan in the world (Curray et al. 2003). According to Curiale et al. (2002) and Curray et al. (2003), sediment contributions to the basin began to arrive primarily from the Himalaya and the Indo-Burman Ranges around the Early Oligocene (~35 Ma), and the Early Miocene (~20), respectively, and have been prograding southward presently. To keep the isostatic equilibrium (Mukherjee 2017), the arriving mass of these huge sediments loaded and depressed the underlying lithosphere further, producing additional accommodation space for deltaic sediments. The additional lithospheric depression and accommodation space have resulted from the southward thrusting of the Shillong Plateau over the basin through the past

5 Ma in the north, and from the westward thrusting of the Indo-Burman Ranges toward the basin in the east (Johnson and Alam 1991).

Geodynamic development of the previously mentioned three geotectonic provinces and their successive tectonic evolution are now discussed under the light of the India and Asia collision to the north, and the India and Burma collision to the east. Among the three geotectonic provinces (Alam et al. 2003), the development of the Geotectonic Province 1—Stable Shelf and Geotectonic Province 2—central Foredeep Basin are related to the collision between the Indian Plate to the south and the Eurasian Plate to the north, whereas development of Geotectonic Province 3—CTFB is related to the collision of the Indian Plate to the west and the Burmese Plate to the east.

The Early Cretaceous rifting of the Indian Plate from Gondwana and concomitant volcanic eruptions along the northeastern margin of the Indian Plate initiated the tectonic evolution of the Geotectonic Province 1 and 2 (Alam et al. 2003). In Geotectonic Province 1, pre-rift Permo-Carboniferous sediments are encountered within the subsurface graben basins on top of the Precambrian Basement rock (Khan 1991a, b; Reimann 1993). SW–NE trending lineaments are observed in both the aeromagnetic anomaly map (Rahman et al. 1990a) and in the Bouguer gravity anomaly map (Rahman et al. 1990b; Khan and Rahman 1992). This reflects grabens on the continental crust part of the Geoteconic Province 2 (Surma basin) beneath the thick Late Cenozoic sediment cover, which may also hold the Permo-Carboniferous sediments at depth. The basement rock beneath the Geotectonic Province 1 is characterized by tensional tectonics. Tension created a complex graben system with tilted and/or downthrown blocks, which were intersected by many normal faults (Khan and Agarwal 1993).

Along the Eocene Hinge Zone and its immediate east, alternating high and low magnetic values could represent ($\sim$750–950 nT; Rahman et al. 1990a) a probable transition from continental to oceanic crust (Curray 2014). The Barisal–Chandpur Gravity High probably indicates the rift valley formed during the initial break-up of Gondwana and formation of the Indian Plate along which transitional zone between continent-ocean crust of the Indian Plate has been marked by pulses of basalt flow (Desikachar 1974; Lohmann 1995; Alam et al. 2003; Curray 2014). Further to the east, seismic signature on top of the basement is characterized by numerous cross-cutting and down-going reflectors associated with basalt flows. They indicate oceanic crust below. Alam et al. (2003) also suggest that the continent–ocean crust boundary bends eastward beneath the Surma basin, and probably continues towards the juncture of the Dauki-Haflong Thrust. Hence, north of this Hinge Zone, at least northern part of the Surma basin is floored by the continental crust, whereas southeast of this line, basement rocks of the Bengal Basin are oceanic rather than of continental origin (BOGMC 1997). In addition, it is believed that position of the Eocene Hinge Zone is within Geotectonic Province 1, and it truncates against the Dauki Fault in the northeast boundary of the deeper foredeep centre.

Up to the Oligocene, the Geotectonic Province 1 and 2 are believed to be undergone similar tectonic and sedimentary evolution (Alam et al. 2003). From the Late Oligocene onward, when the remnant ocean basin (Ingersoll et al. 1995) took shape due to the collisional orogeny of the Barail–Cachar Hills at the northeastern corner of India (Nandy 1986), Geotectonic Province 2 has undergone its own tectonic evolution. A major change in sedimentation pattern here probably occurred in the Mid Pliocene by the major thrust related uplift (upthrust) of the Shillong Plateau (Najman et al. 2016) along the Dauki Fault (Bilham and England 2001; Biswas and Grasemann 2005a) in the south and Oldham Fault (Yin et al. 2010) in the north. In the Shillong Plateau, the Archaean Basement rocks are found ~2 km high, whereas equivalent rocks occur ~16–18 and 4–5 km below sea level, in the south and north of the plateau, respectively (Alam et al. 2003; Najman et al. 2016). The Geotectonic Province 3—CTFB started to develop during the Late Pliocene (Maurin and Rangin 2009a) within the upper parts of the thick deltaic sequence in the eastern margin of the Bengal Basin. At the same time, the Geotectonic Province 2 started evolving as a foreland basin at the centre; whereas in the Geotectonic Province 1, the authors of this chapter believes foredeep sediments will thrust northwestward in near future. Oblique subduction of the Indian Plate beneath the Burmese Plate in an arc-trench setting developed accretionary prism as well as major east-dipping thrusts (Kaladan Fault to the east, and Chittagong Coastal Fault to the west), which largely controlled the structural evolution of the Geotectonic Province 3 (Gani and Alam 1999; Acharyya 2007; Maurin and Rangin 2009a; Wang et al. 2014). In this province, the sediments in the upper part (~up to 5 km) within the major individual thrust deformed above a de'collement level by combined thin-skinned and thick-skinned tectonic processes during the last 2 Ma (Uddin and Lundberg 2004; Maurin and Rangin 2009a), giving rise to a series of elongate, N–S trending curvilinear folds (Sikder and Alam 2003). Through a balanced cross section across this geotectonic province, Maurin and Rangin (2009a) estimated a total E-W shortening of about 11 km in the past 2 Ma, which suggests a shortening rate of about 0.5 cm/year. In this regard, recent investigations suggest that the fold belt has grown progressively westward and many of the folds in the westernmost part of this province have been active only from the Late Pliocene or even later (Table 2) (Johnson and Alam 1991; Khan et al. 2005; Steckler et al. 2008; Maurin and Rangin 2009a; Wang et al. 2014; Khan et al. 2015, 2018).

4 Stratigraphy

The foremost depositional phase in the Bengal Basin began following the separation of the Indian Plate from Antarctica at about the beginning of the Late Cretaceous. This is in exception of minor Carboniferous coal that preserved at least locally on Precambrian continental crust (Sclater and Fisher 1974; Molnar and Tapponnier 1975; Uddin and Lundberg 2004). In the beginning of the Late Eocene, thick Tertiary clastic sediments accumulated in the basin with deposition accelerating with the arrival of clearly orogenic sediments in the earliest Miocene (Bender 1983; Uddin and Lundberg 1998, 2004). According to Alam et al. (2003), the present day active sediment depocentre is the Hatia Trough in the Geotectonic Province 2 as well as its extension to the south into the Bengal Fan.

The sediments preserved in the Stable Shelf part of the Bengal Basin, i.e. Geotectonic Province 1 (Tables 3 and 4) show a progressive thickening towards north and south from the shallowest part of the basement at the Rangpur Saddle from which it continues to slope progressively in both directions. Sediments attain a thickness of 3–4.5 km, and ~3.5 km to the northern and southern edge of the Stable Shelf, respectively (Uddin and Lundberg 2004; Guha et al. 2010). The stratigraphic unit on top of the Stable Shelf as well as in both southern and northern slopes varies in thickness, spatial extent and lithology over geologic time (Reimann 1993). The basement grabens are filled up by the Gondwana sediments comprising sandstone, shale and coal. The Rajmahal Trap lies above the Gondwana sediments and are overlain by the Tura Sandstone, followed by the fossliferous Eocene Sylhet Limestone and then the post Eocene sediments comprising sands, gravels and clay. The facies evolution ranges from continental and volcanics through brackish and lagoonal to shallow open marine to estuarine and fresh water environments (Reimann 1993; Alam et al. 2003).

The sediments preserved within the Foredeep part, i.e., Geotectonic Province 2 (Table 5) and 3 (Table 6) are ca.16–22 km thick sequence of Cenozoic sediments (Curray and Moore 1971; Curray 1994; Gani and Alam 1999; Uddin and Lundberg 2004), where facies evolution ranges from marine through deltaic to fluvial environments. From oldest to youngest, the stratigraphic units are the Tura Sandstone, Sylhet Limestone, Kopili Shale, Barail, Bhuban, Bokabil, Tipam Sandstone, Girujan Clay, Dupi Tila Sandstone, and Dihing Formations. Although Rahman et al. (2017) have tried to correlate sediment provenance with different Himalayan tectonic units, the relative contributions of the sediments to the basin from the Himalaya, CTFB-IBR, and Indian Shield and Shillong Plateau are much debated (Johnson and Alam 1991; Gani and Alam 1999; Uddin and Lundberg 2004; Najman et al. 2008, 2016).

5 Geotectonic Provinces, Tectonic Elements, and Their Related Structures

The India-Eurasia and India-Burma (Myanmar) collisions have affected the basin since the Tertiary (Bender 1983). Conventionally, the basin can be divided into two large divisions based on overall geotectonic settings: Stable Continental Shelf to the west and Bengal Foredeep to the east separated by the Continental Slope also known as the Eocene Hinge Zone (Bakhtine 1966; Guha 1978; Bender 1983; Reimann 1993; Uddin and Lundberg 2004). The Bengal Foredeep occupies the enormous area between the Eocene Hinge Zone to the west and the Indo-Burman Ranges to the east and plays important role in the tectonic evolution of the Bengal Basin. In the foredeep part of the Bengal Basin, a thick pile of sedimentary strata overlies a deeply subsided basement.

Table 3 Stratigraphic succession of the Geotectonic Province-1A: Southern Slope of the Rangpur Saddle, Stable Shelf of the Bengal Basin in Bangladesh

Age (approx)	Group	Formation	Lithology	Thickness Max (m)	Depositional Environment	Tectonic Events
Holocene		Alluvium	Silt, clay, sand and gravel			Folding in eastern Bengal Basin
Recent - Pleistocene	Barind	Barind Clay	Silty clay, clay and silty sand with minor sand	50	Fluvial-alluvial and rapidly prograding delta	
		Dihing	Oxidized sand with clay and silicified wood fragments	150	Fluvial and prograding delta-shelf	
Pleistocene - Early Pliocene	Dupi Tila	Dupi Tila	Sandstone, siltstone, claystone and gravel	280		
Early to Middle Miocene	Jamalganj	Jamalganj	Alternating sandstone, siltstone and shale	415	Delta front to shelf and slope	Folding in Indo-Burman Ranges
Oligocene	Bogra	Bogra	Siltstone, carbonaceous shale and sandstone	165	Increasing sedimentation rates	
Late Eocene	Jaintia	Kopili Shale	Shale, sandstone, locally glauconitic and fossiliferous; with calcareous bands	240	Deltaic slope	
Middle Eocene		Sylhet Limestone	Nummulitic limestone with sandstone interbeds	250	Carbonate platform	Hard Collision
Early Eocene		Tura Sandstone	Sandstone, shale, and coal	245	Deltaic to outer shelf	Soft Collision Close Sutures
Palaeocene						
Middle to Late Cretaceous	Rajmahal	Shibganj Trapwash	Sandstone, volcanic materials with clay	230	Coastal to fluvio-alluvial	
Early Cretaceous - Jurassic		Rajmahal Traps	Amygdaloidal basalt, andesite, serpentinized shale and agglomerate	610	Sub-aerial lava flows Fluvio-deltaic to shallow marine	Main Break-up
Late Permian	Gondwana	Paharpur	Feldspathic sandstone with thick coal seams	465	Fluvial to delta plain, coal swamps	Rapid Extension
Permian - Carboniferous		Kuchma	Sandstone with coal beds	490		Graben Formation
Precambrian		Basement	Precambrian metamorphic and igneous basement rocks			Stable Gondwana continent

Modified after Zaher and Rahman (1980), Khan (1991a, b), Lindsay et al. (1991), Reimann (1993), Alam (1997), BOGMC (1997) and Alam et al. (2003)

Table 4 Stratigraphic succession of the Geotectonic Province-1B: Southern Slope of the Rangpur Saddle, Stable Shelf of the Bengal Basin in West Bengal, India

Age		Lithological Description	Environment	
			Shelf Facies	Basin Facies
Recent - Pleistocene			Bengal Alluvium	Bengal Alluvium
Pliocene		Silty sandstone with clay	Debagram Formation Brakish to marshy Lagoonal to littoral	Ranaghat Formation Brakish to marshy Lagoonal to littoral
Miocene	Late Middle	Siltstone with sandstone and claystone	Pandua Formation Brakish to marshy Lagoonal to littoral Shallow marine	Matla Formation Brakish to marshy Lagoonal to littoral Shallow marine
Miocene / Oligocene	Early (Miocene) Late (Oligocene)	Siltstone sandstone and carbonaceous shale	Memari/Burdwan Formation Brakish to marshy Lagoonal to littoral	
Oligocene	Early	Sandstone with coal and claystone	Shallow marine Contineta Brackish deltaic	
Eocene	Late	Calcareous shale	Kopili Shale Brakish to marshy Lagoonal to littoral Shallow marine	
Eocene	Middle Early	Foraminiferal and algal limestone with sandstone	Sylhet Limestone Brakish to marshy Lagoonal to littoral Shallow marine	
Paleocene / Cretaceous	Late, Early (Paleocene) Late (Cretaceous)	Coarse to medium grained sandstone with lignite and shale	Jalangi Formation Continental Brakish to marshy Lagoonal to littoral	
Cretaceous	Early	Kaolinitic sandstone with shale	Bholpur/Ghatal Formation Estuarine Contineta	
Cretaceous	Early	Basalt	Rajmahal Trap	
Permo-Carboniferous		Sandstone, shale, carbonaceous shale, coal and sandstone	Pre-Trappean/Gondwana	
Precambrian		Granite and gneiss	Basement	

Modified after Biswas (1963), Sengupta (1966) and National Data Repository (2015)

Based on the gravity studies, the foredeep part is divided into a northeast-southwest deeper Foredeep Basin or Western Platform Flank of the foredeep, just east of the Continental Slope, and an Eastern Folded Flank of the foredeep that comprises the CTFB (Guha 1978; Khandoker 1989; Khan 1991a, b; Reimann 1993; Uddin and

Table 5 Stratigraphic succession of the Surma basin (northern part of the Geotectonic Province-2), central deeper Foredeep Basin, Bengal Basin

Age (approx)	Group	Formation	Lithology	Thickness Max (m)	Depositional Environment	Tectonic Events
Holocene		Alluvium			Fluvial	
Pleistocene	Dihing	Dihing	Poorly consolidated sandstone and clayey sandstone.	3350		
Late Pliocene	Dupi Tila	Upper Dupi Tila Lower Dupi Tila	Medium to coarse ferruginous sandstone with layers of quartz pebbles and siltstone with lignitic fragments and petrified wood.		Fluvial	Folding in the eastern Bengal Basin
Mid Pliocene	Tipam	Girujan Clay	Clay and siltstone.	3500	Alluvial	Dauki Fault
		Tipam Sandstone	Coarse-grained, pebbly, cross bedded sandstone.			
Early Pliocene Miocene	Surma	Upper Marine Shale	Shale.	3900	Fluvial tidal-deltaic estuarine	Uplfit of the Himalayas.
		Boka Bil	Dark grey pyrite-bearing shale, sandy shale, siltstone, and sandstones.			Orogeny
		Bhuban	Sandstone and sandy shale, sitstone.		Shallow marine	Subduction and folding Indo-Burman Ranges
Oligocene	Barail	Undifferentiated Barail	Pale grey, multi-coloured, fine grained sandstone and siltstone, grey carbonaceous siltstone and silty sandstone and shale.	7200	Shallow marine to basinal shale	Hard collision, closing of the Eurasian Tethys
Eocene Paleocene	Jaintia	Kopili Shale	Sandstone, glauconitic shale, carbonaceous and highly fossiliferous limestone; calcareous sandstone, carbonaceous shale and coal seams.		Shelf edge limestone	Soft collision Northward movement of India and sea-floor spreading
		Sylhet Limestone				
		Tura Sandstone	Medium to coarse grained grey sandstone.		Marginal part of the shelf	
Pre-Paleocene	Undifferentiated sedimentary rocks (with some volcanics?)					

Modified after Evans (1964), Holtrop and Keizer (1970), Khan et al. (1988), Hiller and Elahi (1988), Khan (1991a, b), Reimann (1993), Shamsuddin et al. (2001) and Alam et al. (2003)

Table 6 Stratigraphic succession of the Chittagong-Tripura Fold Belt (Geotectonic Province-3), Folded Flank, Bengal Basin

Age (approx)		Group	Formation	Lithology	Thickness Max (m)	Depositional Environment	Tectonic Events
Holocene			Alluvium			Fluvial	Folding in the eastern Bengal Basin
Plio-Pleistocene			Dupi Tila	Coarse ferruginous sandstone with layers of quartz pebbles and siltstone with lignitic fragments and petrified wood.	1,600	Fluvial	
		Tipam	Girujan Clay	Clay and siltstone.		Alluvial	Dauki Fault
			Tipam Sandstone	Coarse-grained, pebbly, cross-bedded sandstone.			
Miocene	Late Miocene	Surma	Boka Bil	Dark grey pyrite-bearing shale, sandy shale and sandstone.	1600	Fluvial tidal-deltaic estuarine	Uplift of the Himalayas
	Middle Miocene		Bhuban	Sandstone and pebbly sandstone at the top and sandy shale at the bottom	1500	Shallow marine	

Modified after Evans (1964), Khan (1991a, b), Reimann (1993), Gani and Alam (1999, 2003), Alam et al. (2003)

Lundberg 2004). The Western Platform Flank or Foredeep Basin shows small-amplitude, isometric or geographically equant Bouguer anomalies, whereas the folded flank or the CTFB exhibits large-amplitude, linear or elongate Bouguer anomalies (Bakhtine 1966; Uddin and Lundberg 2004). As earlier mentioned in Sect. 3.1, based on the result of geophysical investigations (gravity, aeromagnetic, and seismic), field geological mappings and on well data, Bengal Basin can be divided into three major geotectonic provinces (Fig. 1): (i) the Stable Shelf to the northwest or the Geotectonic Province 1—Passive to extensional cratonic margin, (ii) the Foredeep Basin to the center or the Geotectonic Province 2—remnant ocean basin, and (iii) Folded Flank to the east or the Geotectonic Province 3—the CTFB (Bakhtine 1966; Alam 1972; Khan and Rahman 1992; Khan and Agarwal 1993; Reimann 1993; Shamsuddin and Abdullah 1997; Alam et al. 2003). These three geotectonic provinces are synthesized in the following paragraphs.

5.1 *The Stable Shelf—Geotectonic Province 1*

The Stable Shelf (Fig. 4) to the northwest, also known as the Stable Pre-Cambrian Platform or Western Flank of the Bengal Basin, comprises two major tectonic elements: (i) the Foreland Shelf to the west, and (ii) the Eocene Hinge Zone or shelf break to the east (Bakhtine 1966; Guha 1978; Matin et al. 1983). The Bouguer anomaly map of the Stable Foreland Shelf is characterized by large negative anomalies (0 to −140 mGal), mostly related to the basement structural features (Fig. 5) (Rahman et al. 1990b; Khan and Rahman 1992). Khan and Rahman (1992) identified four tectonic zones on the basis of the trend, shape and magnitude of the Bouguer gravity anomaly contours and the basement faults from the aeromagnetic data. Except the Eocene Hinge Zone, alternating gravity highs and lows with closed contours indicate the presence of numerous graben and half-graben structures in most of the Foreland Shelf, where Gondwana coal/sediments have been deposited (Khan 1978; Khan and Rahman 1992; Khan and Chouhan 1996).

5.1.1 The Foreland Shelf

The Foreland Shelf is characterized by the presence of complex graben system, transected by many synthetic and antithetic faults (Khan and Rahman 1992; Khan and Agarwal 1993). Based on geophysical and bore- hole data, the Foreland Shelf is divided into four tectonic elements which are oriented approximately ENE. Towards the south, the four tectonic elements are the Himalyan Foredeep, the Dinajpur Shelf, the Rangpur Saddle or the Platform Flank, and the Bogra Shelf (Fig. 4) (Bakhtine 1966; Sengupta 1966; Guha 1978; Matin et al. 1983; Khan and Rahman 1992; Khan and Agarwal 1993; Reimann 1993; Guha et al. 2010). Recent GPS derived geodetic data suggest an overall very low subsidence rate (<1 mm/year) of the Foreland Shelf compared to the other geotectonic provinces of

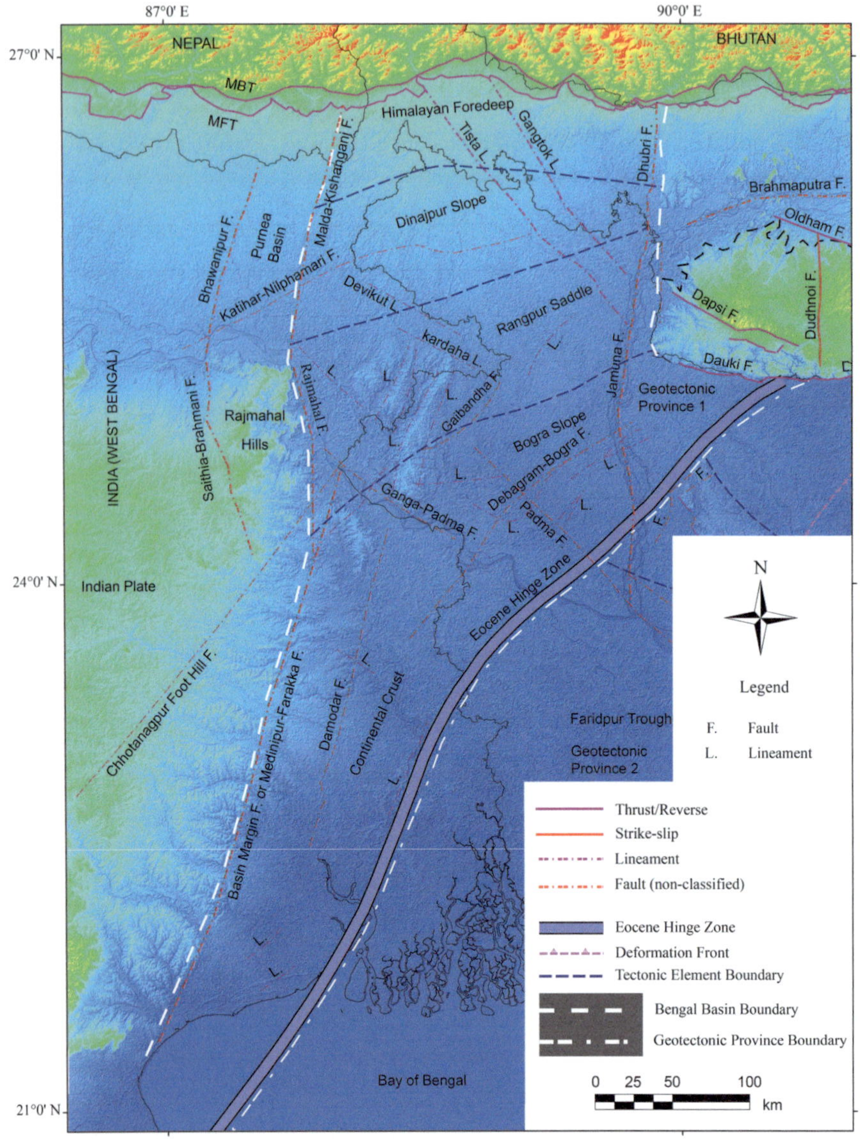

Fig. 4 Simplified tectonic map of the Geotectonic Province 1 of the Bengal Basin and its surroundings. The tectonic element boundaries are based on Bakhtine (1966), Guha (1978) and Matin et al. (1983). Faults and lineaments are mostly taken from Kayal (2008). MBT: Main Boundary Thrust; MFT: Main Frontal Thrust

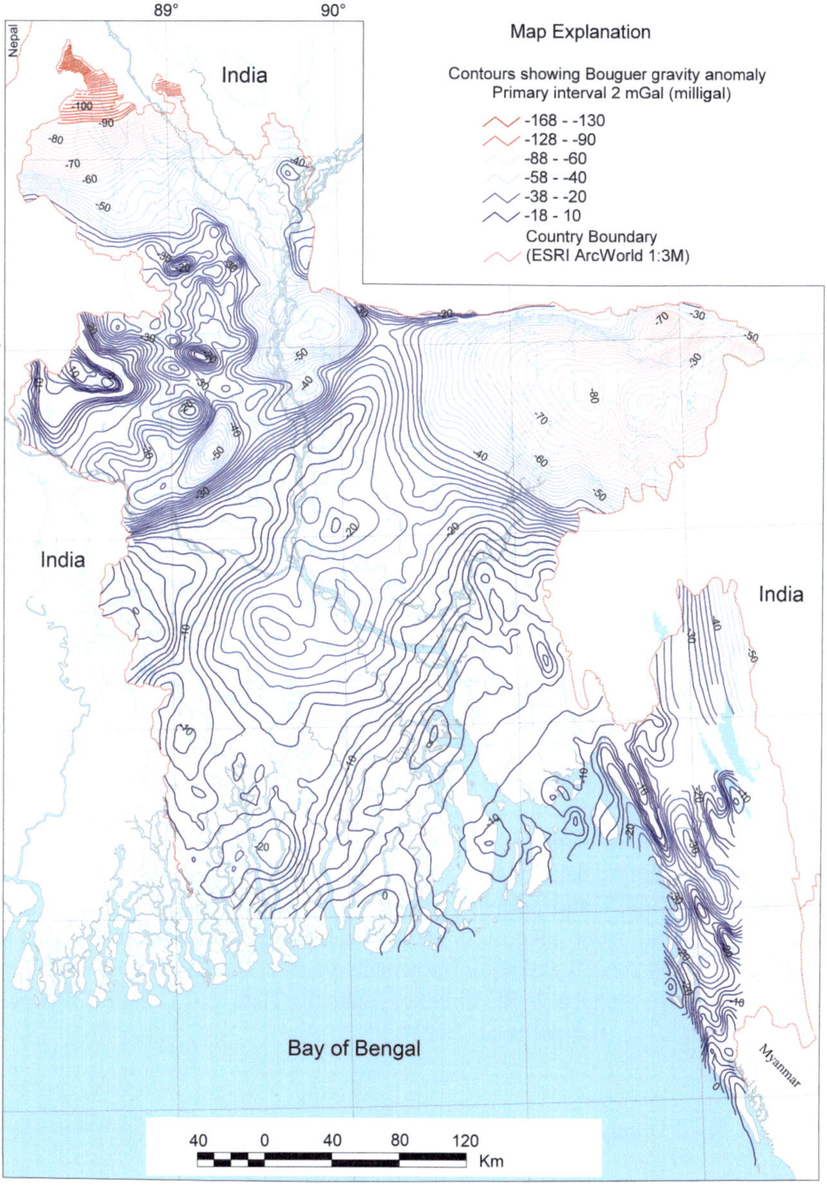

Fig. 5 Bouguer gravity anomaly map of Bangladesh, which covered major portion of the Bengal Basin. Projection used in this map is the Lambert Conformal Conic with Everest 1969 spheroid (modified after Rahman et al. 1990b)

the Bengal Basin (Reitz et al. 2015). Starting from the northernmost tectonic element, the sequential description of these tectonic elements of the foreland shelf is given with their tectonic evolution history. Note as the N-S oriented Rajmahal Hills is situated outside the Basin Margin Fault Zone (Raman et al. 1986; Roy and Chatterjee 2015) and at the western edge of the Foreland Shelf (Raman et al. 1986), it is not a part of the Bengal Basin.

Himalayan Foredeep

This tectonic element forms the northwestern most part of the Bengal Basin, lies south of the Main Frontal Thrust (MFT) and is covered by the Recent to sub-Recent piedmont plain deposits. This element is marked by the high negative Bouguer gravity anomaly from −110 to −150 mGal within a short distance in the extreme northwest of Bangladesh, and suggests noticeable thickening of basinal strata northward into the Siwalik foreland basin of the northeastern Himalaya (Khan and Rahman 1992; Rabbani et al. 2000; Uddin and Lundberg 2004). The −110 mGal contour near Panchagarh represents the approximate southern boundary of the Himalayan Foredeep with the Dinajpur Slope. The width of this tectonic element at the centre is <50 km, but gradually increases towards both east and west.

The Neogene sediments consisting of sandstones, subordinate shales and clays, and gravel beds are well developed in this tectonic element and attain a thickness of 3–4.5 km (Rabbani et al. 2000; Guha et al. 2010). The only well drilled in this tectonic element is located on the north-western most tip of Bangladesh at Salbanhat in Tetulia by Shell Oil Co. in 1988. The well touched the basement at 2518 m depth penetrating the Mio-Pliocene sequence but did not encounter the Eocene Limestone (Rabbani et al. 2000). The northern edge of this tectonic element along the MFT shows clear evidence of an active deformation zone as indicated by active faulting, uplift and surface-rupture, and earthquakes (Thakur 2004). The main deformation front related to the India-Eurasia collision slowly migrated southward in this zone. Two major lineaments transect the Himalayan Foredeep of the Bengal Basin along NW-SE direction, namely Tista and Gangtok lineaments (Chopra et al. 2013; Baruah et al. 2016).

Dinajpur Shelf/Dinajpur Slope

The Dinajpur Slope is characterized by almost E-W trending linear Bouguer gravity contours, with values ranging between −50 and −110 mGal and sloping NNW between Panchagarh to the north and Nilphamari to the south (Khan and Rahman 1992; Rabbani et al. 2000). The Neogene sediments are deposited in this tectonically down-warped part of the Foreland Shelf, which is known as the northern slope of the Rangpur Saddle, or Dinajpur Slope or Northern Foreland Shelf (Reimann 1993) of the Bengal Basin. Although Reimann (1993) included Dinajpur Slope into the Himalayan Foredeep, Guha et al. (2010) placed Dinajpur Slope as a separate tectonic element in between Himalayan Foredeep to the north and Rangpur Saddle to the south. The width of this zone ranges between 100 and 70 km from WSW to ENE, and gently plunges northward to the Himalyan Foredeep approximately at 1°–3°. The Dinajpur Slope turns ~E-W north of the Garo Hills of the Shillong Plateau and then connects to the Assam Basin. The Dhubri Fault with approximately N-S orientation separates the Dinajpur Slope to the west and the Assam Basin to the east. One major fault and two major lineaments transect this tectonic element in which the Katihar-Nilphamari Fault along ~E-W, and the Tista and the Gangtok lineaments along NW-SE (Chopra et al. 2013; Baruah et al. 2016). Moreover, N-S graben structures with Gondwana fill are also present (Khan and Rahman 1992; Reimann 1993).

Rangpur Saddle

The Rangpur Saddle is characterized by irregular shaped Bouguer gravity contours with several closed gravity highs and lows typical of a shallow basement, extending from south of Nilphamari and Dinajpur up to Gaibandha and Panchbibi (i.e. in and around Rangpur region). This tectonic element is a part of ENE prolongation of the Indian Shield to the Shillong Plateau and Mikir Hills (Bakhtine 1966; Guha 1978; Khan 1978; Matin et al. 1983; Hossain et al. 2017) within the so-called Garo-Rajmahal gap (Desikachar 1974; Khan and Chouhan 1996), and separates the Bogra Shelf to the south and Dinajpur Shelf to the north. The approximate boundary of the Rangpur Saddle with the northern and the southern tectonic elements has been marked at ~−700 m contour line on the basement. The eastern margin of the concealed block along the Garo Hills is a fault zone connected to the east-west-oriented Dauki Fault. This dome-shaped tectonic element is the shallowest part within the Stable Shelf bounded by N-S trending faults in the east and the west (Khan and Rahman 1992). In its shallowest portion, the basement is uplifted to a depth of 128 m from the surface (Rahman 1987) and is overlain by thin sedimentary deposits of the Plio-Pleistocene Dupi Tila Sandstone and Madhupur Clay. Geophysical methods traced approximately N-S oriented Dhubri-Jamuna Fault to the western end of the Garo Hills of the Shillong Plateau, and an approximately 160 km long N-S oriented Rajmahal Fault to the eastern edge of the Rajmahal Hills (Vaccari et al. 2011; Mohanty et al. 2013, 2014; Roy and Chatterjee 2015). These two faults form the eastern and western margins of the Garo-Rajmahal Gap, therefore, enclose the Rangpur Saddle.

In the eastern part of this tectonic element, the SE end of the Tista and Gangtok lineaments is terminated by the N-S oriented Jamuna Fault. In SRTM 30 m resolution image, surficial expression of three graben structures with NNE-SSW orientation has been observed in the western part of this tectonic element. Geophysical and well data indicate that this tectonic element (Desikachar 1974; Khan and Chouhan 1996) is underlain by the Permian Gondwana sediments followed upward by the Late Cretaceous Rajmahal tholeiitic basalts trap flows (Baksi 1995) and the

Cretaceous-Tertiary sediments. Near N-S oriented graben structures with Gondwana fill are also indicated by the above data set (Khan and Rahman 1992).

Although the concealed basement rocks of the Rangpur Saddle and northern part of the Bogra Slope are generally considered to be an eastward subsurface continuation of the Indian Shield, Ameen et al. (2007, 2016) and Tapu et al. (2016) considered them as separate microcontinental blocks. Evidently diorite, tonalite and granodiorite constitute the basement of the Rangpur Saddle, which is dissected by pegmatites and mafic/ultramafic dykes (Ameen et al. 1998; Kabir et al. 2001; Hossain et al. 2007). Based on SHRIMP Zircon U-Pb age dating, both Ameen et al. (2007) and Hossain et al. (2007) reported identical Paleoproterozoic ages 1.72, and 1.73 Ga of the tonalite, and diorite from Maddhapara, respectively. A pegmatite dyke from Maddhapara has also given age similar to the host rock (1.72 Ga—based on SHRIMP Zircon U-Pb age dating; Hossain et al. 2017). It may be mentioned that the undeformed tonalite/diorite pluton with age similar to 1.72 Ga has been identified neither in the Chhotanagpur Granite-Gneiss Complex (CGGC) nor in the Shillong Plateau. Thus, based on the 1.72 Ga age from an undeformed calc-alkaline I-type granitoid (tonalite) in Maddhapara and tentative gravity enclosure surrounding the region Ameen et al. (2007) and Tapu et al. (2016) suggest that Maddhapara and surrounding diorite-tonalite basement may have constituted a Paleoproterozoic micro-continental block, which they referred to as Dinajpur Block. Few latest investigations (Chatterjee 2017; Hossain et al. 2017) however, argued that the diorite-tonalite basement of the Rangpur Saddle is a continuation of the Central Indian Tectonic Zone (CITZ). Nevertheless, further isotopic age dating of basement samples from different locations, detailed geochemical and structural studies, robust geodynamic and tectonic model related to its emplacement history are required to check the validity of either hypothesis.

Bogra Shelf/Bogra Slope

The northern edge of the Bogra slope is characterized by the elongate close-spaced gravity contours of 0 to −30 mGa1, extending from Nawabganj to Jamalganj up to Gaibandha. From south of the Nawabganj-Gaibandha up to Singra area, several prominent closed anomaly contours are observed reflecting gravity highs and lows within −20 to −50 mGal range representing intra-cratonic basins, i.e. graben structures where Gondwana sediments deposited (Khan and Rahman 1992). This tectonic element is also known as Southern slope of the Rangpur Saddle or the Western Foreland Shelf of the Bengal Basin, which contains various Permian to Recent rocks, laid down on the Precambrian Basement rocks (Khan and Rahman 1992; Reimann 1993). The width of this zone ranges 60–125 km from NE to SW, and gently plunges southeast to the shelf edge at 1°–3° is clearly marked by the Eocene Sylhet Limestone on the seismic section (Salt et al. 1986; Reimann 1993). Seismic contours on top of the Eocene Limestone reveal a number of NE-SW trending faults of which the Debagram-Bogra Fault is the most prominent (Fig. 4). Another two major faults are also present in this tectonic element and these are the NW-SE oriented Ganges-Padma Fault to the west and the N-S oriented Jamuna Fault to the east. Recent GPS derived geodetic data suggest an overall very low subsidence rate of less than 1 mm/year in the area (Reitz et al. 2015). The Bogra Slope (Fig. 4) turns approximately E-W near the Garo Hills of the Shillong Plateau and then traverses along the southern slope of the plateau towards the Khasi and Jaintia Hills and reaches up to the southern fringe of the Mikir Hills. This tectonic element started developing at the time of rifting of the Gondwana during the Carboniferous-Permian when graben and half-graben structures were formed, followed by the drifting of the East Gondwana from the remainder during the Jurassic-Cretaceous. The later event results the Rajmahal lava flows and formed the Rajmahal Trap, which crops out in the Rajmahal Hills. From the Late Cretaceous to Eocene, carbonate sediments deposited at the lower edge of the shelf, whereas arenaceous sediment were deposited at the upper edge (Reimann 1993).

5.1.2 Eocene Hinge Zone

This zone is also known as shelf break or paleo-continental slope or trace of the Eocene shelf edge (Bakhtine 1966; Sengupta 1966; Bender 1983; Matin et al. 1983; Rahman et al. 1990a; Reimann 1993; Uddin and Lundberg 2004; Singh et al. 2016) and lies between the Stable Shelf/Foreland Shelf to the west and the Foredeep Basin to the east. This tectonic element is characterized by almost ENE-WSW trending linear Bouguer gravity contours, with values ranging approximately between −30 to −15 mGal and sloping towards southeast (Khan and Rahman 1992). Precambrian basement rock dips southeast abruptly from 2°–3° to 6°–12° at the contact of the stable platform and the Eocene Hinge Zone to the west, and then dips more gently 1°–2° again in the southeast at the contact of the Hinge Zone and deeper foredeep basin to the east (Uddin and Lundberg 2004). At its upper northwestern edge, the recorded seismic depth on top of the Eocene Sylhet Limestone is 3500 m, whereas at its lower southeastern edge, the recorded seismic depth on top of the same limestone is 5000 m. Although Sengupta (1966) and Khandoker (1989) named this NNE-SSW running narrow 25–100 km Hinge zone as the ‘Calcutta–Mymensingh gravity high’, more recent data (Khan and Agarwal 1993) suggest that this term is to some extent confusing and named more appropriately as Shelf-Break (Uddin and Lundberg 2004; Roy and Chatterjee 2015). Although it is usually shown to truncate against the Dauki Fault in the northeast, the other school of thoughts suggest that the Hinge Zone progressively convexes basin-ward to the northeast and then passes somewhere through the northern end of the Surma basin and possibly continues towards the Haflong Thrust at the northeastern corner of the Bengal Basin (Alam et al. 2003).

The Eocene Hinge Zone marks the structural as well as depositional transition between central foredeep basin to the southeast and stable shelf to the northwest. This tectonic element possibly marks the transition from the thick continental crust (west) to extended thinned crust of the continental margin (east) (Singh et al. 2016). Eocene Sylhet Limestone, the most prominent seismic reflector of the Bengal Basin has been interpreted to imply this shelf break (Reimann 1993). However, results of the recent seismic investigations suggest NE alignment of this Hinge Zone up to 90° 40′E and 25° 15′N, where the E-W oriented Dauki Fault presumably truncates this structural element. More towards east, this paleo-continental slope is not recognizable even in the deeper seismic section (6 s TWT). This probably indicates that the southern portion of the shelf as well as slope are downthrown by the Dauki Fault and now hidden under the thick pile of the Neogene sediments deposited in the northern part of the Surma basin (Salt et al. 1986). Although significant increase of sedimentary thickness from $\sim$3 to $\sim$17 km is noticed across the Hinge Zone (west to east), the crust thins from 38 km at the Indian Craton to 34 km at the Hinge Zone to 16–19 km at the central deeper foredeep basin (Singh et al. 2016). Such thickness of crust at the central deeper foredeep basin may result from igneous activity related to the Kerguelen Plume at the time of rifting (Ray et al. 2005; Singh et al. 2016).

5.2 *The Foredeep Basin—Geotectonic Province 2*

The central deeper Foredeep Basin or the Western Platform Flank occupies the huge area between the Eocene Hinge Zone to the west, the Shillong Plateau to the north, and the CTFB to the east. This deeper foredeep basin is approximately 200 km wide to the north, narrowed at the middle, and then gradually widens to about 500 km to the south, shows an overall NE trend (Fig. 6) (Reimann 1993). The basin shows small-amplitude, isometric or geographically equant Bouguer gravity anomalies (Bakhtine 1966; Uddin and Lundberg 2004). The deeper foredeep basin is a mosaic of few sub-basins (lows) and buried highs/ridges. This deeper foredeep basin narrowed north-eastward by the Madhupur High in west and the Tripura Uplift in the east, and broadly divides the basin into two parts. The oval shaped Surma basin occupies the northern part of the deeper foredeep basin whereas southern part is known as southern sub-basin. A NE-SW trending gravity and magnetic anomaly known as the Barisal-Chandpur High separates the southern part of the deeper foredeep basin into another two sub-basins, the Faridpur Trough in the west and the Hatia Trough in the east (Alam 1972; Das Gupta 1977; Guha 1978; Reimann 1993). The deeper foredeep basin, therefore, can be divided into six tectonic elements, and from the north to the south these are the Surma basin/Sylhet Trough, the Madhupur-Tripura High/Tangail-Tripura High, the Faridpur Trough, the Barisal-Chandpur High, the Hatia Trough, and the Bay of Bengal (Fig. 6) (Bakhtine 1966; Guha 1978; Mirkhamidov and Mannan 1981; Matin et al. 1983; Uddin and Lundberg 1999; Mukherjee et al. 2009).

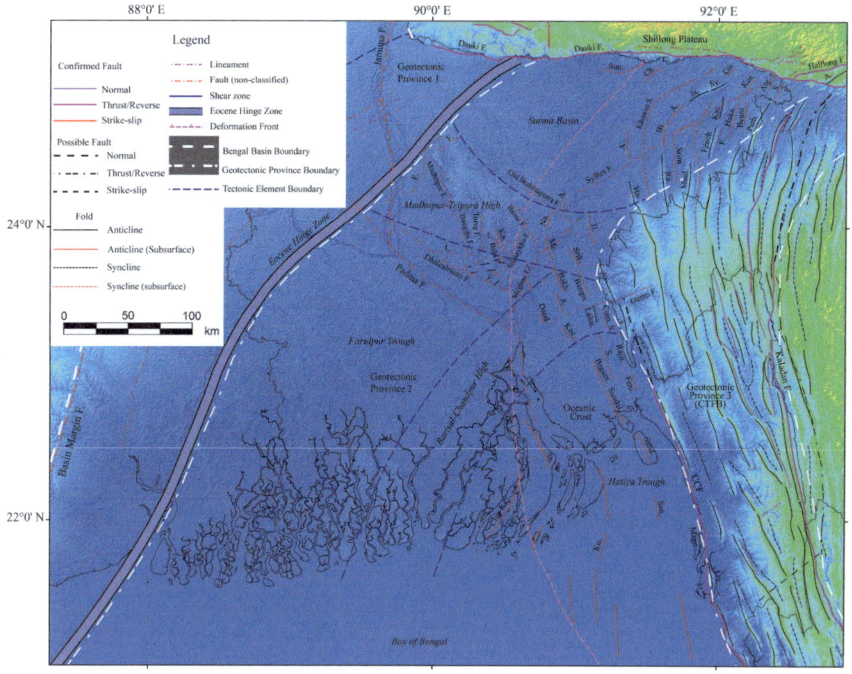

Fig. 6 Simplified tectonic map of the Geotectonic Province 2 of the Bengal Basin and its surroundings. The tectonic elements boundaries are based on Bakhtine (1966), Guha (1978) and Matin et al. (1983). Faults and lineaments are mostly taken from Biswas and Grasemann (2005a), Kayal (2008), Khan et al. (2011), and Wang et al. (2014). A: Anticline; Atg: Atgram; Bakh: Bakhrabad; Bang: Bangora; Beani: Beanibazar; Begum: Begumganj; Bi: Bibiyana; CCF: Chittagong Coastal Fault; Ch: Chhatak; Com: Comilla; Daud: Daudkandi; F: Fault; Fench: Fenchuganj; Go: Gowainghat; Ha: Habiganj; Haka: Hakaluki; Ja: Jalalabad; Ka: Kamta; Kac: Kachua; Kai: Kailas Tila; Ku: Kutubdia; Kus: Kushiara; Lalm: Lalmai; Magn: Magnama; Me: Meghna; Moul: Moulvibazar; Na: Narsingdi; Path: Patharia; Ra: Rashidpur; Rajn: Rajnagar; S: Syncline; San: Sangu; Sand: Sandwip; Sh: Shahbazpur; Sri: Srikail; Srim: Srimangal; Sun: Sunamganj; Sundal: Sundalpur; Sy: Sylhet and Ti: Titas

5.2.1 Surma Basin

The Surma basin/Sylhet Trough is a depositional basin of thick sedimentary strata along the northeastern part of the Bengal Basin, just south of the Shillong Plateau with a structural relief of about 20 km between the trough and the neighboring plateau separated by the E-W running Dauki Fault (Fig. 7) (Mirkhamidov and Mannan 1981; Murphy and Staff BOGMC 1988; Johnson and Alam 1991). It is bounded to the north by the striking Barail Range of the Shillong Plateau, to the east by the Barail-Imphal Ridge (southwestern margin) of Assam, to the south by the Madhupur-Tripura High, and to the west by the Eocene Hinge Zone (Fig. 6). Although, much accepted sediment thickness in the Surma basin exceeds 20 km (Hiller and Elahi 1984; Ali and Raghava 1985; Hiller and Elahi 1988; Murphy and

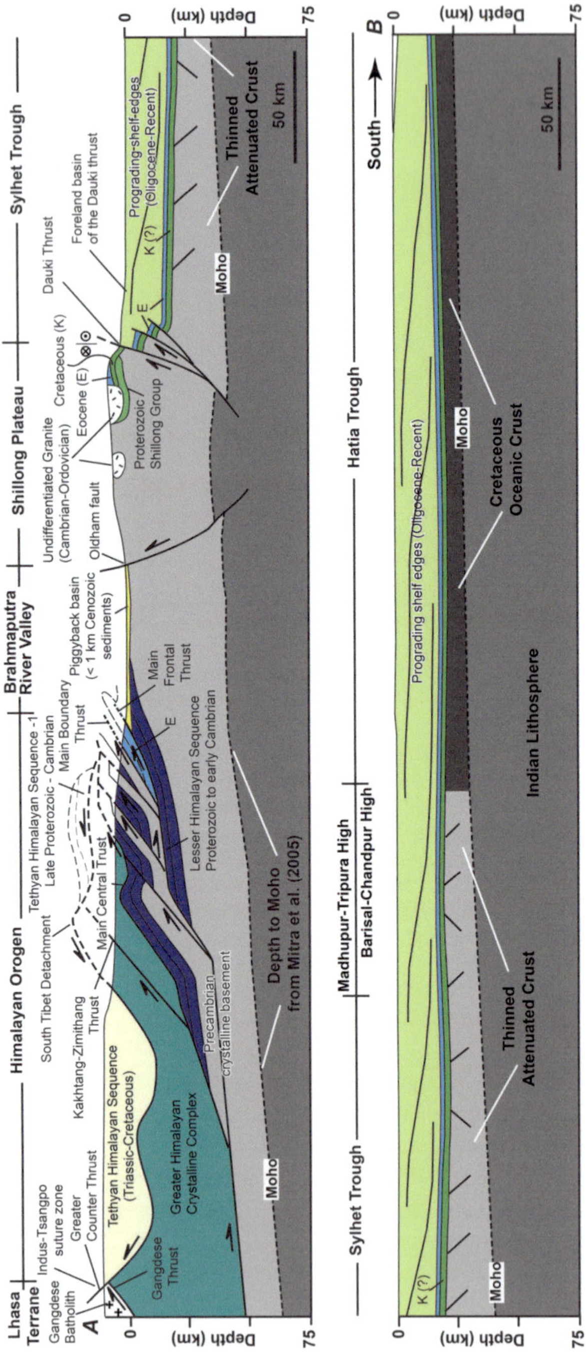

Fig. 7 Schematic cross-section across the Bengal Basin and its surroundings (drawn approximately N-S along the line AB as shown in Fig. 1) (modified after Murphy and Staff of BOGMC 1988; BOGMC 1997 and Webb et al. 2013)

Staff BOGMC 1988; Curray 1994; Uddin and Lundberg 2004), thickness <18 km is also reported (Paul and Lian 1975; Banerji 1979; Guha 1978; Matin et al. 1986; Singh et al. 2016). Beneath this thick sedimentary pile, type and thickness of the crust of the Indian Plate is still debated (Singh et al. 2016). However, it is more or less accepted that beneath the northwestern half of the Surma basin, the crust is continental to transitional type; whereas beneath southeastern half of the basin, the crust is oceanic. The Bouguer gravity map of Bangladesh suggests a minimum Bouguer anomaly (−80 mgl) for the Surma basin (Ali and Raghava 1985; Uddin and Lundberg 2004). Along N-S section, more or less flat and tectonically less disturbed nature of the Moho has been observed at similar depth of ~ 37 km beneath this tectonic element as well as the Shillong Plateau (Singh et al. 2016). Except for the western border made by the Eocene Hinge Zone, all other structural features framing the Surma basin in north, east and south are the results of mostly the Pliocene-Recent compressional tectonics (Reimann 1993). It is assumed that the major subsidence of the Surma basin did not take place during the Miocene, rather the Pliocene uplift of the Shillong Plateau and gradual westward encroachment of the IBR might have subsided the Surma basin during the Pliocene (Johnson and Alam 1991; Uddin and Lundberg 2004). Along the southern edge of the basin, a major NE-SW oriented fault with a length of >150 km has been reported as the Sylhet Fault (Kayal 1998; Bhattacharya et al. 2008; Kayal 2008; Angelier and Baruah 2009; Vaccari et al. 2011; Mohanty et al. 2013, 2014), which probably controls the course of the Surma River (Ovi et al. 2014).

Several investigations have noticed that the Surma basin, a complex conspicuous sub- basin of the Bengal Basin with thick sedimentary fill (Hiller and Elahi 1988; Murphy and Staff BOGMC 1988; Uddin and Lundberg 2004; Singh et al. 2016), has evolved from a passive continental margin with sedimentary packages thickening south, to a flexural basin with sedimentary packages thickening north (Johnson and Alam 1991; Uddin and Lundberg 2004; Bracciali et al. 2015). These changes of thickness of sedimentary packages are interpreted to be related to loading from the adjacent uplifting Shillong Plateau. According to Najman et al. (2016), passive margin configuration of the Surma basin with southward thickening of the sedimentary packages ceased at the end of the Bokabil Formation deposition (3.5 Ma). Transition of the Surma basin from passive margin to flexural basin occurred during 3.5 to ~ 2 Ma. At this time, the Tipam Formation deposited, which does not show any lateral variation in thickness, and has sub-horizontal dips in the basin centre. This also marks the onset of flexural loading by the Shillong Plateau. The IBR also propagated westward in this area at this time, as evidenced by thinning of Tipam Sandstone over the IBR anticlines. The combined influence of the uplifting Shillong Plateau and westward encroachment of the IBR together resulted the palaeo-Brahmaputra diversion away from the Surma basin, i.e. from the east to west of the Shillong Plateau by the end of the Tipam Formation deposition ~ 2 Ma (Najman et al. 2012, 2016). After the drainage diversion, sedimentation in the flexural Surma basin continued with the meandering facies of the Dupi Tila Formation, sourced primarily by recycling of Himalayan-derived materials from the sedimentary cover of the rising Shillong Plateau to the north (Evans 1932; Biswas

1961; Hiller and Elahi 1984; Najman et al. 2016). From the Dupi Tila Formation onward (~2 Ma–Recent), sedimentary packages in the flexural Surma basin has been thickening as well as dipping to the north. On the other hand, the pre-Oligocene rock within the Surma basin is the Jaintia Group consisting of Paleocene Tura Sandstone, Middle Eocene Sylhet Limestone and Upper Eocene Kopili Shale, and all of which crop out in the northern margin of the Surma basin (Reimann 1993; Alam et al. 2003). The Tura Sandstone, Sylhet Limestone and Kopili Shale are interpreted to be deposited in shallow-marine to marine, shallow-marine carbonate, and deltaic to slope depositional environments, respectively (Alam et al. 2003). Moreover, Oligocene Barial Group (also exposed along the northern fringe) is believed to be deposited in a predominantly tide-dominated shelf environment (Alam 1991). In addition, the Surma Group consisting of the Lower Bhuban and the Upper Boka Bil Formations (Holtrop and Keizer 1970; Hiller and Elahi 1988; Khan et al. 1988) crop out along the northeastern margin of the basin. Those deposited in a large, mud-rich delta system that might have drained a significant portion of the eastern Himalaya through the palaeo-Brahmaputra River (Johnson and Alam 1991). Sultana and Alam (2001) and Alam et al. (2003) mentioned that the Surma Group was deposited within a cyclic transgressive–regressive regime ranging from shallow marine to tide-dominated coastal settings. The Tipam Sandstone also crops out in the same area and is interpreted to have deposited in bed-load dominated braided-fluvial systems (Johnson and Alam 1991). According to Uddin and Lundberg (2004), paleo-Brahmaputra, paleo-Meghna, paleo-Karnafuli are most likely responsible for the consistent increase in the thickness of the Miocene sediments in this part of the Bengal Basin as these rivers delivered detritus from the eastern Himalaya and IBR straight to the Bengal Fan through the Surma basin and forming a major delta complex at the northeastern part of the Bengal Basin.

According to Holtrop and Keizer (1970) and Hiller and Elahi (1984), the Surma basin was structurally engraved in almost sub-Recent geological time, i.e. since 3–6 Ma and the tectonic movements are presumably still going on, as indicated by shallow lakes between the anticlinal ridges, especially during the monsoon period. Recent geodetic measurements (e.g., Nielsen et al. 2004; Vernant et al. 2014; Steckler et al. 2016) and geochronology studies (e.g., Biswas and Grasemann 2005a; Khan et al. 2006) also suggest that oblique subduction of the Indian Plate to the NE direction is going on and the basin is still tectonically active. This tectonic convergence results in complex anticlinal and synclinal systems as well as overlapping thrust systems in and around the Surma basin due to overall E-W and N-S shortening of the basin at the rate of 7 mm/year and 18 mm/year to the northern and eastern margins, respectively (Akhter et al. 2010; Steckler et al. 2012; Bulbul 2015). Moreover, the northern margin of the basin is subjected to flexural loading due to the Shillong Plateau. The above tectonic activities also result overall subsidence of the Surma basin currently at the rate of approximately 7–12 mm/year (Reitz et al. 2015). Primary structures of the basin is dominantly anticlines whose orientation changes from N-S (e.g., Rashidpur, Maulvi Bazar, Fenchuganj, Habiganj, Kailas Tila) at the southern part to the ENE-WSW (e.g., Patharia, Sylhet,

Jalalabad, Atgram) to the middle-eastern part (Fig. 6). Most of these anticlines are plunging to the north. To the northern part of the basin, the anticlines show approximately E-W oriented Type 1 folding (e.g., Chhatak, Gobamura, Dupi Tila) probably due to two sets of compressional forces, i.e. shortening direction one is N-S and the other is E-W, acting simultaneously (Biswas and Grasemann 2005b; Ovi et al. 2014). Some of these anticlines are fault controlled (Sikder and Alam 2003; Steckler et al. 2008; Najman et al. 2012), and few of them are dissected by the approximately N-S and E-W oriented faults (Hiller and Elahi 1984). The Surma basin has been studied extensively as a result of successful hydrocarbon explorations (Holtrop and Keizer 1970; Woodside 1983; Shamsuddin 1989; Chowdhury et al. 1996; Khanam et al. 2017). Anticlinal structures are the principal type of hydrocarbon trap in the Surma basin (Lietz and Kabir 1982). The Miocene to early-Pliocene shales up to a depth of 4.5 km are the source of gas/condensate accumulated in the sandstone of the Bhuban and Bokabil Formations (Shamsuddin 1989).

5.2.2 Madhupur-Tripura High

The Madhupur High to west and the Tripura Uplift to the east are connected by a threshold known as the Madhupur-Tripura High/Tangail-Trupura Uplift. This separates the Surma basin to the north and the Faridpur Trough to the southwest (Mirkhamidov and Mannan 1981; Reimann 1993; Guha et al. 2010). The Madhupur High, aligned approximately NW-SE, forms the south-western margin of the Surma basin which is the northernmost sub-basin of the Bengal Foredeep. The Madhupur High is a Pleistocene Terrace (Morgan and McIntire 1959), which represents slightly undulated and elevated topography from the adjacent flood plains, and is an exposed Quaternary interfluve between two pathways for the Brahmaputra River. Gravity and aero-magnetic data suggest that the basement is relatively uplifted beneath this area (Reimann 1993; Guha et al. 2010). Dissection of the initial Madhupur High is resulted from the Late Pleistocene monsoon climatic episodes that caused a tremendous current flow over the plain (Monsur 1995). According to Guha et al. (2010), the Nagarpur dome and the Nandina high are considered as protrusions from the basement without deforming the overlying sedimentary packages above this tectonic element. Although its upper part is heavily eroded, the crests of the hill keep flat surfaces which are gently tilted to the east (CDMP II 2013). A series of en-echelon faults, known as the Madhupur Fault flanks the western side of the uplifted Madhupur High/Madhupur Tract. These en-echelon faulting resulted either from torsion of the region, possibly as rotational faulting (Mukherjee and Khonsari 2017) or from the effect of shear (e.g., Mukherjee 2014a) along an assumed blind fault or possibly a combination of both (Morgan and McIntire 1959). The upliftment of the Madhupur High may have exerted a significant control over the avulsion history of the Brahmaputra/Jamuna River. This avulsion history is cyclic, with a periodicity of $\sim$1800 years, and the position of the Brahmaputra has fluctuated between east and west of the Madhupur High (Pickering et al. 2014). It is assumed that the Holocene Brahmaputra avulsion history, to some extent, is related to competition between tectonic deformation and transverse sediment input (Grimaud et al. 2017). Therefore, any future seismicity can cause river avulsions and related flooding (Reitz et al. 2012; Hossain et al. 2015). However, GPS derived geodetic data suggest an overall very low subsidence rate (<1 mm/year) for the Madhupur High (Reitz et al. 2015) compared to the other parts of the centre deeper foredeep basin. Moreover, Pickering et al. (2017) claimed that the Madhupur High as a terrace is resulted from the highstand floodplain deposition of eustasy-driven cycles, and subsequent lowstand exposure and weathering as well as active tectonics.

Though Madhupur High/Tract is popularly known and accepted to be a Pleistocene Terrace, but based on the morphology of the tract and other geomorphic signatures (e.g., relief features, river morphology, and river incision), the authors of this chapter assume that this Tract is the expression of a buried anticline. However, as the uplift did not expose the older bedrocks (open fold), the younger sediments are very soft to survive, and the structure is very young in age, this anticline is not so pronounced. Moreover, the Madhupur High is located just immediate west of the eastern deformation front. Such close proximity to the active deformation front justifies the possibility of a tectonically developing young subsurface anticline whose surficial expression is known as Madhupur High.

On the other hand, the Tripura Uplift/High forms a topographically high zone from where the terrain slopes northward as well as southward, as characterized by the nature of drainage network of the region. However, overall decreasing trend of the elevation is towards west. Beneath the Agartala region, a dome-shaped structure has been recognized based on seismic survey (Ganguly 1993), suggesting that the area up-arched (Das et al. 2011b). The western part of the Tripura High occupies the frontal fold belt of the CTFB, comprises of sedimentary rocks with linear ridges and basins resulting due to E-W convergence between the Indian Plate and the Burmese Plate (Desikachar 1974; Ganguly 1984; Nandy 2001). As this part of the tectonic element is considerably young both in terms of geology and tectonics (Das et al. 2011b), hence it is likely to exhibit more prominent morphological expressions due to continued tectonics. The region is characterized by a series of approximately N-S oriented sub-parallel, elongated, doubly plunging folds exhibiting an arcuate shape with mild westward convexity (Ganguly 1993). The Tripura High mostly comprising the Akhaura Terrace, is primarily deeply dissected by valleys running out of the adjoining hills (Brammer 2012, 2014). It also exhibits complex mixture of uplifted terrace and piedmont, anticlines, synclines, and river floodplains. Overall, the Madhupur High to the west and the Tripura High to the east together form a threshold region known as the Madhupur-Tripura High tectonic element to the south of the Surma basin and the northeast of the Faridpur Trough (Figs. 5 and 6).

5.2.3 Faridpur Trough

The Faridpur Trough/Patuakhali depression (Mirkhamidov and Mannan 1981), located southeast of the Hinge Zone, is characterised by a general gravity low of < −20 (Ali and Raghava 1985; GSB, 1980) with development of the Neogene sequence. To the immediate southeast of the Eocene Hinge Zone near the confluence of the Padma and the Jamuna rivers, the depth of the Sylhet Limestone is 6500 m (Guha et al. 2010). This tectonic element is bounded to the north by the Madhupur-Tripura High and to the southeast by the Barisal-Chandpur High (Fig. 8). Within this tectonic element, Chalna and Bagerhat are the two notable structural highs of very low amplitude. While Khan and Agarwal (1993) and Khan and Chouhan (1996) suggest that this tectonic element is underlain by oceanic crust. The recent data reveal that oceanic crust is situated more to the east below the Barisal–Chandpur Gravity High (Alam et al. 2003; Curray 2014; Singh et al. 2016), and consequently, this tectonic element is underlain most likely by transitional crust. Sediment thickness in the Faridpur Trough is more than 16 km, and beneath this sediment, thickness of the transitional extended crust of the Indian Plate is $\sim$16–19 km (Singh et al. 2016). The more or less flat and tectonically less disturbed nature of the Moho has been observed at 31–32 km depth beneath this tectonic element (Fig. 8) (Khan and Agarwal 1993; Singh et al. 2016). However, GPS derived geodetic data from Khulna area indicate that the overall subsidence rate is high ($\sim$8 mm/year) here (Reitz et al. 2015). Higgins et al. (2014) assumed that this high rate appears to be related to the sediment properties as the Holocene organic-rich muds: notably compaction (Mukherjee and Kumar in press; Dasgupta and Mukherjee submitted).

5.2.4 Barisal-Chandpur High

It is a tectonically uplifted zone of the Bengal Foredeep Basin that separates the Faridpur Trough to the northwest and the Hatia Trough to the southeast (Fig. 8) (Bakhtine 1966; Guha 1978; Uddin and Lundberg 1999). Reimann (1993) mentioned that the Barisal-Chandpur High can be interpreted as a gravity and magnetic anomaly caused by a magnetic body at a great depth. The gravity maxima of the Barisal-Chandpur High shows NE-SW trend. The average width of this tectonic element is $\sim$70 km, and corresponds to a slightly uplift of the overlying sedimentary cover. Muladi, Kamta and Daudkandi are the three main gentle anticlinal structures located in this tectonic element. The arc shape deformation front (Maurin and Rangin 2009a; Steckler et al. 2016) of the Bengal Basin passes approximately N-S through this tectonic element with slight concavity to the E. It is assumed that the continent-oceanic crust boundary of the Indian Plate in the Bengal Basin is located beneath the Barisal–Chandpur High (Curray 2014; Singh et al. 2016). Recent GPS derived geodetic data suggest variable subsidence rate of this High (Reitz et al. 2015). The northern part of this tectonic element (in Dhaka) shows high subsidence rate (>10 mm/year) compared to the southern part near Patuakhali ($\sim$3 mm/year). The very high subsidence around Dhaka city, which is situated at the southern edge of the Madhupur High may be related to the extensive groundwater extraction in and around the city and consequent rapid fall of the groundwater table (Bhuiyan and Hossain 2006; Hoque et al. 2007; Akhter et al. 2010), but cannot represent the natural subsidence rates over the Holocene timescale (Reitz et al. 2015). Nevertheless, if

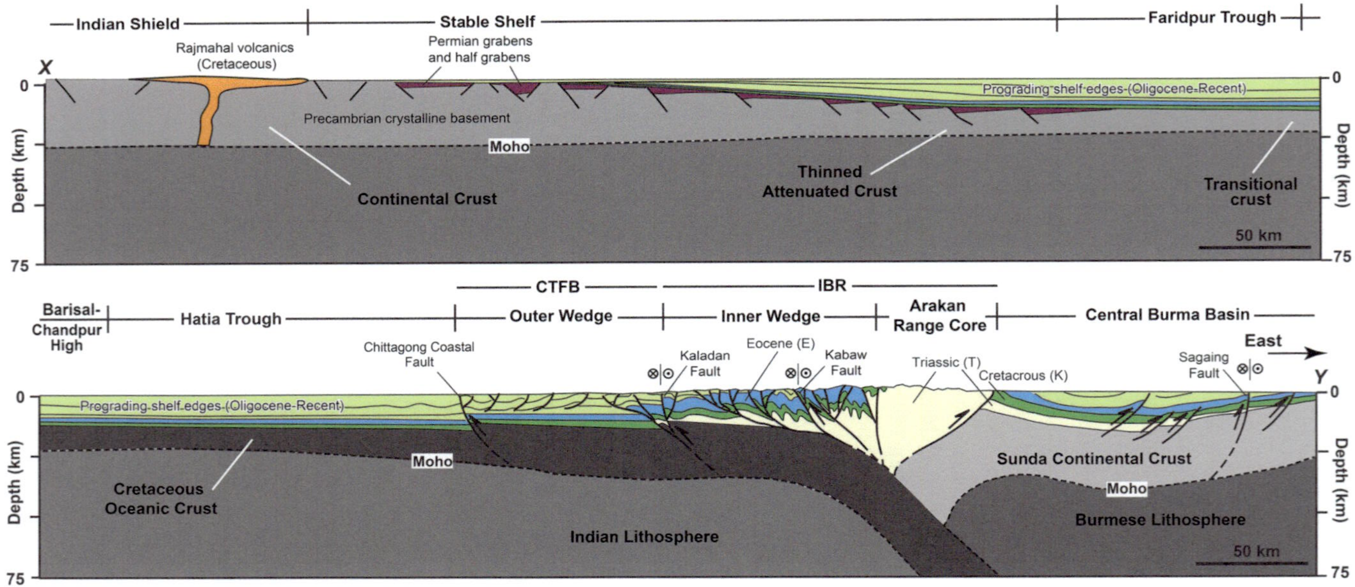

Fig. 8 Schematic cross-section across the Bengal Basin and its surroundings (drawn approximately W-E along the line XY as shown in Fig. 1) (modified after Murphy and Staff of BOGMC 1988; BOGMC 1997; Maurin and Rangin 2009a and Wang et al. 2014)

such high rate of subsidence continues, channel steering behavior (Reitz et al. 2015) is most likely going to occur in near future around the Dhaka city. However, based on the study of the Synthetic Aperture Radar (SAR) images, Higgins et al. (2014) points out that the ground subsidence in and around Dhaka varies significantly, and appears to be related to the sediment properties with the lowest rates observed in the Pleistocene Madhupur Clay and the highest rates in Holocene organic-rich muds. Within the city, the subsidence rate ranges 0 to >10 mm/year, and outside the city the subsidence rate ranges 0 to >18 mm/year.

5.2.5 Hatia Trough

The Hatia Trough of the Bengal Basin is located in the southeastern part of the basin and forms the outermost part of the west-propagating CTFB, situated in between CTFB to the east and the Barisal-Chandpur High to the west (Fig. 6). The base of this tectonic element is formed by oceanic crust of the Indian Plate (Khan and Chouhan 1996; Curray 2014; Singh et al. 2016). This tectonic element represents the deepest trough of the Bengal Foredeep Basin and has received the highest accretion of clastic sediments. Sediment thickness in the Hatia Trough exceeds 18 km, and beneath this sediment, the Indian Plate is not more than 16 km thick (Singh et al. 2016). As pointed by Najman et al. (2016), the Himalaya forms the dominant source for the Late Neogene-Recent sediments in the Hatia Trough. Small amount of arc-derived sediments from the Trans-Himalaya or Paleogene Indo-Burman Ranges are also reported. Data from the Shahbazpur 1 well in the Hatia Trough indicate the presence of more than 2000 m of Plio-Pliestocene sediments overlain by 480 m thick Holocene sediments (Alam et al. 2003). As in Faridpur Trough, the more or less flat and tectonically less disturbed Moho has also been observed at 31–32 km depth beneath this tectonic element (Khan and Agarwal 1993; Singh et al. 2016). Guha et al. (2010) claimed that the axis of the Bengal Foredeep runs through the apex of this tectonic element. Char Kajal, Shahbazpur, Kutubdia/Magnama, Sangu and a number offshore anticlinal structures are located in this tectonic element, of which Sangu and Shahbazpur yield hydrocarbons. Recently obtained GPS derived geodetic data suggest variable subsidence rate for the Hatia Trough (Reitz et al. 2015). The northern onshore part of this tectonic element (in Raipur, Noakhali) shows lower subsidence rate ($\sim$3 mm/year) than the southern offshore part far from the Magnama structure ($\sim$5 mm/year).

5.2.6 Bay of Bengal

The Bay of Bengal is the northeastern lobe of the Indian Ocean (Curray 2014) bordered to the west and the north by the passive continental margin of the Indian east coast and southern edge of the Bengal Basin, respectively (Fig. 2). It is bordered by the oblique subduction zone between the Indian Plate to the west and the Burmese/Sunda Plate to the east, and extends into the northernmost part of the

Indian Ocean to the south. The entire Bay of Bengal is floored by the Bengal Fan (Curray and Moore 1971; Curray 2014). Along 16°N, the Bay of Bengal is about 1400 km in width. According to Desa et al. (2013), the Bay of Bengal is the drifted half part of the Early Cretaceous Enderby Basin formed as a result of the Kerguelen hotspot (Banerjee et al. 1995) as Australia drifted away from India during Early Cretaceous (~ 132 Ma) with a fan-shaped expanse of ocean floor, separating the two plates by gradually widening to the east (Ali and Aitchison 2008). Therefore, the crust of the Bay of Bengal can be considered as a drifted part (Rangin and Sibuet 2017). Gravity data revealed the presence of NE-SW trending highs and lows in the western part of the Bay of Bengal along the coast of the Mahanadi-Krishna-Godavari (KG)-Cauvery basins. NNW-SSE transverse oceanic fracture zones truncated and offset the NE-SW trending gravity lows, which probably correspond to the graben infilling of the detrital sediments (Rangin and Sibuet 2017). Along the western part of the Bay of Bengal oldest sediments deposited are of Early Cretaceous in age (Desa et al. 2013).

Two aseismic ridges namely 90°E Ridge and 85°E Ridge are observed on the floor of the Bay of Bengal. The 90°E Ridge is largely present in the eastern part of the Bay of Bengal. Quaternary to Eocene sedimentary sequence was locally deposited on top of the volcanic rocks that form 90°E Ridge. The volcanic rocks in the 90°E Ridge are attributed to the drifted Kerguelen Hotspot Igneous Province (Coffin et al. 2002), which show a general decreasing age trend (90–38 Ma) from N to S (Royer et al. 1988). Presence of exhumed gabbro along the 90°E Ridge suggests that the Ridge was formed during the Early Cenozoic along the previously existing transform boundary due to injection of underplated spreading centre volcanism (Rangin and Sibuet 2017). The northern end of this ridge is traceable up to 19°N (Maurin and Rangin 2009b), but further northward of the Bay of Bengal, the gravity as well as topographic signature of the ridge disappear completely (Rangin and Sibuet 2017).

The 85°E Ridge (Sar et al. 2009) is an approximately linear and aseismic ridge west of the 90°E Ridge, and named for its near-parallel strike along the 85°E meridian (Fig. 2) (Bastia et al. 2010). The ridge extends from the Mahanadi Basin in the north, shifts westwards by ~ 250 km around 5°N, southeast of Sri Lanka and then continues to the south in the Indian Ocean. In the northern part, it is blind beneath the thick sediments of the Bay of Bengal and shows a negative gravity anomaly and in the southern part, it occasionally rises above the sea floor and is associated with a positive gravity anomaly (Michael and Krishna 2011). The volcanic rocks in the 85°E Ridge may be attributed to the drifted short lived hotspot, which show a general decreasing age trend ranging from ~ 80 Ma near the Mahanadi Basin to ~ 55 Ma near the Afanasy Nikitin Seamount (Michael and Krishna 2011). Curray and Munasinghe (1991) proposed a model to explain the origin of the 85°E Ridge. They proposed that the origin of Afanasy Nikitin Seamount, 85°E Ridge, and Rajmahal and Sylhet Traps as the trace of the Kerguelen hotspot that now lies beneath the Crozet Islands in the southern Indian Ocean. Moreover, they placed the Kerguelen hotspot, which was also responsible for the origin of the 90°E Ridge, at the triple junction between Greater India,

Australia, and Antarctica before the breakup of eastern Gondwana at ~120 Ma. Based on marine magnetic data combining with petrology, seismic reflection, deep seismic sounding, and magnetic information of the Bengal Basin, Talwani et al. (2016) also suggest that Afanasy Nikitin Seamount, 85°E Ridge, and Rajmahal and Sylhet Traps are related to the Kerguelen hotspot activity. Two sets of oceanic fracture zones have been found in the Bay of Bengal. The oldest fractures are transform faults with orientation S50–S60°E, are considered to be of Early Cretaceous, and occured during an early spreading episode. The youngest fractures are submeridian and resulted from the northward drift of the Indian Plate since the end of the Mesozoic (Maurin and Rangin 2009a). Moreover, between 20° and 18°N, the tilted clastic sequences dip southwest and are truncated by the dominant NE-SW trending listric normal faults (Rangin and Sibuet 2017). It is more or less well established that the basement in the northern part of the Bay of Bengal is of oceanic crust. However, Sibuet et al. (2016) and Rangin and Sibuet (2017) argued that the crust in the northern Bay of Bengal is thinned continental in nature intruded by volcanic products linked to the passage of the Kerguelen hotspot after the continental rifting of the northern Bay of Bengal. The Moho is ~15 km deep at the northernmost part of the Bay of Bengal, and it gradually goes deeper to ~37 km below the northern edge of the Bengal Basin (Singh et al. 2016).

In the Bay of Bengal, another major geomorphic unit is observed known as the Swatch of No Ground (SoNG) (Fig. 2), an active submarine canyon represents the present submarine drainage system of the Bay of Bengal (Curray et al. 2003). It serves as a point source for one of the major active fan valley deposits into the Bay of Bengal/Bengal Fan. In the northern end, the head of the canyon lies in shallow water depth (~38 m) at 21° 27′N, 89° 41′E, and then continues south for 160 km as a long, straight trough with an average gradient of 8.2 m/km. At ~20°12′N and 89° 12′E, it reaches ~1400 m depth. At midshelf, the broad U-shaped trough has a width of ~20 km from rim to rim, and below 862 m from the rims, the width is 8 km (Curray et al. 2003). The canyon continues to the south as an active channel and reaches as far as the 0° latitude (Curray 2014). The location of the mouth of this canyon is presumably related to the confluence of the Ganges and the Brahmaputra rivers during the last glacial low stand sea level, because an active submarine canyon is related to river mouth migrations, sedimentation, and the Quaternary sea level fluctuations. Due to sea level rise, the confluence of the Ganges and the Brahmaputra rivers shifted ~90 km further east, therefore, the canyon receives very little sediments at present (Curray et al. 2003). Currently, the turbidity currents generated within the SoNG are relatively small, and reach only to the middle Bengal Fan, due to present highstand sea level (Curray et al. 2003; Curray 2014). Besides, large influx of sediments mainly from Himalaya and to some extent from the Indian peninsula is transported by the Ganges and the Brahmaputra rivers, especially during the northeast and the southwest monsoons (Chauhan and Vogelsang 2006). The origin of these sediments is dominantly detrital rather than biogenic (Achyuthan et al. 2014).

5.3 The Folded Flank (CTFB)—Geotectonic Province 3

The Folded Flank of the foredeep (Fig. 9) generally known as CTFB developed within the upper parts of the thick deltaic sequence, and exhibits large-amplitude, linear or elongated gravity anomalies (Bakhtine 1966). Structural complexity of the Folded Flank increases towards east (Khan 1991a, b) and to the farther east, it merges into the Indo-Burman Ranges (IBR)/Paleogene Inner Burman Ranges. This folded belt is bounded by the central Foredeep Basin to the west and the IBR to the east. Although the CTFB gradually merges into the IBR/Mizo Fold Belt, they are structurally separated by approximately NNW-SSE running Kaladan Fault (Ram and Venkataraman 1984; Reimann 1993). The strata exposed in this belt are broadly lithologically similar to those exposed along the southern edge of the Sub-Himalayan belt of the Himalayas. Although the timing of initiation of the IBR, in the east, is widely debated, the estimates range from the Late Eocene–Early Oligocene (Mitchell 1993) to the Late Miocene (Ni et al. 1989), thus costraining the timing of development of CTFB (Najman et al. 2016). According to Maurin and Rangin (2009a), a very rapid development and westward propagation of the CTFB since 2 Ma is facilitated from uplift of the Shillong Plateau, and consequent deepening and filling of the Surma basin with a large amount of unconsolidated sediments. The progressive westward overprinting of thin-skinned tectonics by thick-skinned tectonics within the IBR was due to rapid propagation of the accretionary wedge westward since Pliocene (Maurin and Rangin 2009a; Najman et al. 2016). Thus, the age of the structures gradually decreases from east (IBR) to west (CTFB) and the youngest structures are observed at the western margin of the CTFB, i.e. near the deformation front. This looks, therefore, like an 'in-sequence' deformation. According Wang et al. (2014), many of the folds in the western part of the CTFB activated only from the Late Pliocene onward, but some of the folds in the eastern CTFB appear inactive. In this regard, GPS derived geodetic data suggest variable subsidence and uplift rates for the CTFB (Reitz et al. 2015). In general, the eastern margin of the CTFB shows very small uplift signature, whereas the western margin demonstrates small subsidence signature. Although Reitz et al. (2015) suggest that these variable rates depend on whether the GPS sites are located in synclines or anticlines, this is unlikely to get subsidence in the CTFB as the area subjected to continuous convergence collision. It is most likely that the GPS stations used in their study are positioned in the underthrust flanks (culminate to synclines) of the anticlines as seems to be the case at least for the Sitakund Anticline. Based on balanced cross-section across the CTFB, Maurin and Rangin (2009a) estimated about 11 km total east-west shortening in the past 2 Ma, which in turn indicates $\sim$5 mm/year long-term shortening rate.

The CTFB consists of sediments that were deposited in the Bengal Basin, and were subsequently uplifted and incorporated into the Neogene accretionary prism due to the oblique subduction of the oceanic crust of the Indian Plate beneath the Burmese Plate to the east during Pliocene (Gani and Alam 1999; Najman et al. 2012). It is one of the active orogenic belts in the world susceptible to produce moderate to large earthquakes and holds potential for hydrocarbon exploration. Several investigations have been carried out in this area mainly focusing on structural modeling (Sikder and Alam 2003; Mandal et al. 2004; Kabir and Hossain 2009; Maurin and Rangin 2009a; Najman et al. 2012; Abdullah et al. 2015; Khan et al. 2018), active tectonics (Goodbred and Kuehl 1999; Wang et al. 2014; Khan et al. 2005, 2015, 2018), earthquake and landslide hazards studies (Ansary et al. 2000; Cummins 2007; Steckler et al. 2008; Wang et al. 2014; Steckler et al. 2016), geochemistry and provenance studies (Najman et al. 2008; Dina et al. 2016; Rahman et al. 2017).

In general, gradual increase of negative gravity values to east reflects the deepening of the Moho, apparently due to lithospheric flexure caused by loading of the IBR (Uddin and Lundberg 2004). However, this is comparatively much less distinct than the sharp increase of negative gravity values of the Himalayan Foredeep in northwestern part of the Bengal Basin (Khan and Rahman 1992). The Neogene CTFB reveals large-amplitude, linear or elongated gravity anomalies, and structural complexity increases towards east and merges into the IBR farther east along approximately N-S running Kaladan Fault (Bakhtine 1966; Khan 1991a, b; Najman et al. 2012). To the west, deformation of the CTFB ceases along the deformation front. From 24°N up to Shillong Plateau, the alignment of the fold belt shows an overall arc shaped broad curvature, starting as NNW-SSE trend from the south near Rakhine-Teknaf-Chittagong to approximately N-S trend near Tripura in the middle and then finally NNE-SSW in the Surma basin to the north (Khan and Chouhan 1996; Reimann 1993). Rapid westward propagation of the deformation front is responsible for this broad curvature of the CTFB (Wang et al. 2014).

The CTFB has developed exemplary linear NNW-SSE to N–S to NNE-SSW oriented doubly plunging folds, which are arranged in a set of alternating valley-forming synclines and ridge-forming anticlines (Das et al. 2010). Some of the folds in the CTFB region are elliptical and cusp shaped synclines bounded by fault-affected ridges. According to Das et al. (2010) multi-curved geometry of some of the CTFB fold ridges are attributed to the free westward migration of fold, multi-phase deformation of soft sediments, and longitudinal thrust as well as transverse faulting. The surficial expressions of the CTFB anticlines to the west are limited in the offshore along the line drawn western edge of the St. Martin's Island to the shore of the Sandwip Island. To the onshore the line continues NNW towards Lalmai anticline and then bends to follow the NE-strike and finally ends at the southeastern edge of the Shillong Plateau near the northeast corner of the Surma basin (Reimann 1993). Based on the geometric shape and degree of intensity of folding, Bakhtine (1966) divided the CTFB tectonic element into three subdivisions. Towards west, these are eastern highly compressed disturbed zone, middle asymmetric thrust faulted zone, and western quiet zone.

Eastern Highly Compressed Zone: The eastern intensely folded narrow ridge-shaped zone is highly compressed, and comprises of sharp anticlines with steeply dipping limbs. The overall structural trend is NNW-SSE, and axial planes are steeply inclined either to the east or to the west. The asymmetric anticlines plunge very gently. Adjacent to the Kaladan Fault, the folds are locally faulted box

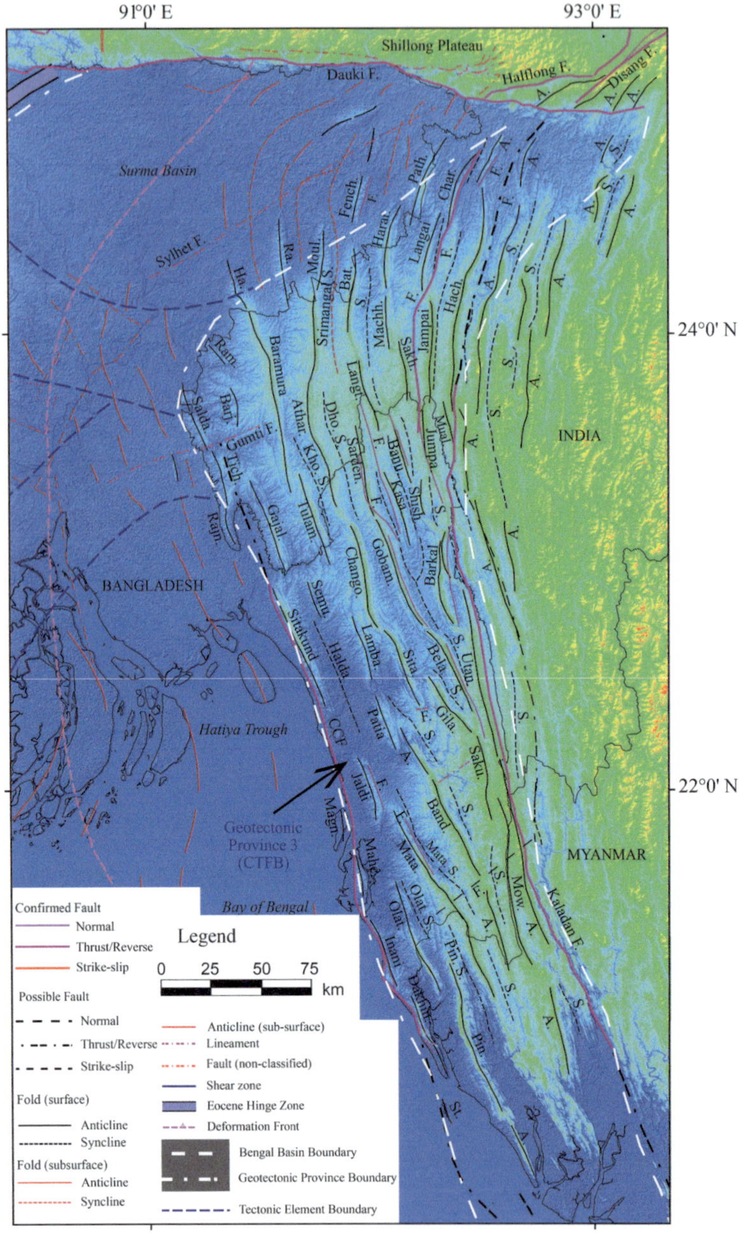

anticlines with pop-up and en echelon character (Maurin and Rangin 2009a). The widths of the synclines are generally broad in the north compared to those of the south. The geometrical analysis of the bedding planes of the anticlinal structures shows that the axial plane of the anticlines strikes 330°–345°, dips 75°–85° either towards NE or SW, and having interlimb angle of 75°–95°. Plunge varies 5°–7°,

◂**Fig. 9** Simplified tectonic map of the Geotectonic Province-3 of the Bengal Basin and its surroundings (modified after Bakhtine 1966). Faults and lineaments are mostly taken from Biswas and Grasemann (2005a), Kayal (2008), and Wang et al. (2014). A: Anticline; Athar: Atharamura; Band: Bandarban; Banu: Banuachari; Barj: Barjala; Bat: Batchia; Bela: Belasari; CCF: Chittagong Coastal Fault; Chango: Changotaung; Char: Chargola; Dakhin: Dakhin Nhila; Dho: Dholai; F: Fault; Fench: Fenchuganj; Gajal: Gajalia; Gila: Gilasari; Gobam: Gobamura; Ha: Habiganj; Hach: Hach Har; Harar: Harargaj; Jumpa: Jumpai/Botiang; Kasa: Kasalong/Kasarang; Kho: Khowai; Lamba: Lambaghona/Sarta; Langt: Langtari; Machh: Machhlithum; Magn: Magnama; Mahe: Maheskhali; Mata: Matamuhuri; Moul: Moulvibazar; Mow: Mowdak; Mual: Mualuawm; Olat: Olathang/Waylataung; Path: Patharia; Pin: Pincha; Ra: Rashidpur; Rajn: Rajnagar; Ram: Ramjtia; S: Syncline; Sakh: Sakhan; Saku: Sakudaung; Salda: Saldanadi; Sarden: Sardeng/Sabtaung; Semu: Semutang; Shish: Shishuk/Sisbak; Sita: Sitapahar; St: St. Martin's; Tich: Tichna; Tulam: Tulamura and Utan: Uttan Chatra

either to the north or to the south. Some of the prominent anticlines in this zone are Mowdak, Sakudaung, Uttan Chatra, Barkal, Shishuk/Sisbak, Banuachari, Jumpai/Botiang.

Middle Asymmetric Thrust Faulted Zone: This zone is situated west of the highly compressed zone, where overall structural trend is ~NNW-SSE. Generally, in this zone axial plane is steeply inclined to the west, and structures plunge very gentle towards N. As in the eastern highly compressed zone, the widths of the synclines are commonly broad in the north than those to the south. Steep limbs of anticlines are generally faulted. The geometrical analysis of the bedding planes of the anticlinal structures shows that the axial plane of the anticlines strikes 338°–350°, dips 70°–78° towards NE, and have interlimb angle of 75°–95°. Plunge varies 5°–10°, either to the north or to the south (Chowdhury 1970; Hossain et al. 2018). Some of the prominent anticlines in this zone are Matamuhuri, Bandarban, Gilasari, Belasari, Sitapahar (Fig. 10a), Gobamura, Changotaung, Kasalong/Kasarang, Sardeng/Sabtaung.

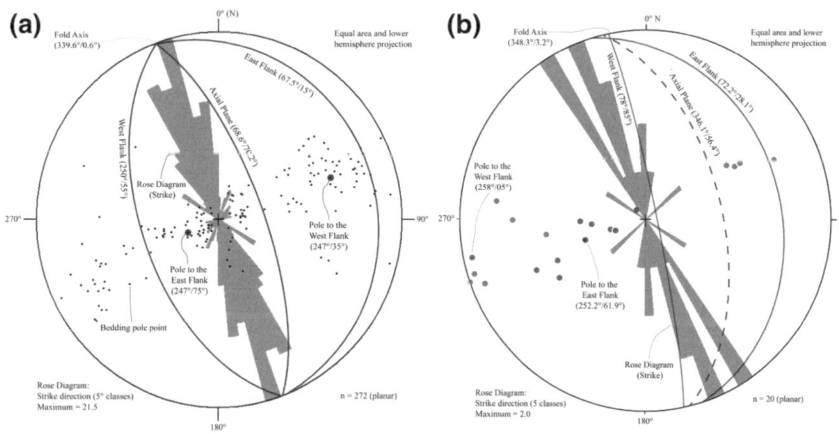

Fig. 10 Lower hemisphere equal area projection of the bedding attitude data shows the orientation of the fold geometric elements of the **a** Sitapahar and **b** Sitakund Anticlines of the Chittagong Tripura Fold Belt (CTFB), Bengal Basin. Both the anticlines are west verging and asymmetrical, and show similar structural trend, but the Sitakund anticline is overturned in nature (Hossain et al. 2018)

Western Quiet Zone: This is the westernmost zone of the CTFB with similar structural trends as in the other two zones. Most of the structures are doubly plunging with axial planes steeply inclined to the east, i.e. those are asymmetric overturned structures. The structures have very gentle dips in the crest and steep dips in the flank. In general, the western flanks of the structures are underthrust, e.g., Sitakund Structure (Abdullah et al. 2015; Hossain and Akhter 1983). The geometrical analysis of the bedding planes of the anticlinal structures shows that the axial plane of the anticlines strikes 335°–350°, dips 55°–65° in the NE direction, having interlimb angle of 60°–70°. Plunge amount varies 3°–10°, either to the north or to the south (Hossain et al. 2018). Some of the prominent anticlines in this zone are St Martin's Island, Dakhin Nhila, Inani, Olathang/Waylataung, Maheskhali, Jaldi, Patia, Sarta/Lambaghona, Sitakund (Fig. 10b), and Semutaung. A major fault is present in this zone running NNW-SSE along the western edge of the Dakhin Nhila-Inani-Maheskhali-Sitakund structures. Among the three zones of the CTFB, the western quiet zone is the most important and prospective tectonic sub-element for oil and gas in Bangladesh, where more than 15 gas fields have been discovered. Towards north some of these are Feni, Bakhrabad, Titas, Habiganj, Bibiyana, Rashidpur, Moulvibazar, Fenchuganj, and Beanibazar.

However, recent studies (Wang et al. 2014; Steckler et al. 2016) suggest that another fold belt exist in between western quiet zone and deformation front, and stretches from offshore to the south and onshore to the north. This very low intensity fold belt is characterized by long-wavelength de'collement folds, locally west verging, and subparallel to the western quiet zone. The amplitude of these folds decreases westward in accordance to the westward propagation of the accretionary wedge and die out near deformation front to the west (Maurin and Rangin 2009a).

6 Neotectonics and Active Tectonics of the Bengal Basin

Neotectonics and active tectonics of the Bengal Basin are directly related to the oblique collision of the Indian Plate with the Eurasian Plate to the north and the Burmese Plate to the east. The collision developed the Himalayan arc to the north, and the Burmese–Andaman arc to the east. Based on recent seismic studies, it is presumed that subduction has clogged below the Himalayan arc and is only actively continuing below the Burmese–Andaman arc (Biswas and Das Gupta 1989; Biswas et al. 1992; Biswas and Majumdar 1997). Due to the resistance to subduction below the Eurasian Plate and plate convergence, tectonic deformation and seismic activity occur in the Indian intra-plate region, including the stable shelf in the western part of the Bengal Basin, and result in folding and brittle deformation (Mukherjee 2013c, 2014b) in conjunction with high-angle basement faults of variable trends (NE-SW, E-W, NW-SE) (Khan and Rahman 1992; Khan and Agarwal 1993; Biswas and Majumdar 1997; Mukherjee et al. 2009). To the north, tectonic loading is accommodated along the Dauki Fault zone (Bilham and England 2001; Biswas et al. 2007), which consists of a set of high angle, deepseated reverse faults. To the east, the Cenozoic tectonic evolution of the eastern parts of the Bengal Basin is related mostly to the NE oblique subduction of the Indian Plate beneath the Burmese Plate. The collision also results westward migration of accretionary prism complexes as well as the deformation front (Gani and Alam 1999; Alam et al. 2003; Wang et al. 2014; Steckler et al. 2016). The CDMP II (2013) reported that the rate of convergence gradually decreases towards west. These multifaceted neotectonic evolutions of the Bengal Basin result in several active faults in and around the Bengal Basin, especially along its northern and eastern margins. Some of these faults are regional, and capable of generating moderate to great earthquakes. The rate of motion and the time since the last rupture indicate that the probability of earthquakes from the existing active faults is high (CDMP II 2013). Overall, the geotectonic setting of the Bengal Basin at the juncture of three converging lithospheric plates with the presence of regional scale seismogenic faults is conducive for the frequent and recurrent earthquakes.

A general trend of increasing seismic/earthquake activity has been observed in the Bengal Basin (Khan and Chouhan 1996; CDMP II 2013). Although the apparent increase of tectonic activity may be attributed to modern and increased recording facilities, the historical records and the present-day census suggest a relative increase in tectonic activity (Khan and Chouhan 1996). This seeming increase of seismicity in the Bengal Basin indicates new fractures/faults propagation from the preceding seismically undisturbed zones or the reactivation of some earlier fractures/faults in pre-existing seismically active zones. Catalogues of seismicity of high intensity earthquakes (Table 7) in and around the Bengal Basin (Bilham 2004; Akhter 2010; Martin and Szeliga 2010; Szeliga et al. 2010; Kundu and Gahalaut 2012; Olympa and Abhishek 2015; Berthet et al. 2014; Kayal 2014; Wang et al. 2014) show a number of earthquakes that caused not only catastrophic damages of the property and life but also shifting of the large river courses as well as dynamic change in landscape. Among such incidents, the Chittagong earthquake of 1762 (R 8.5+), Sirajganj earthquake of 1787 (MM X), Cachar earthquake of 1869 (R 7.5), Bengal earthquake of 1885 (Mw 6.8), Great Indian/Assam earthquake of 1897 (Mw 8.1), Srimangal earthquake of 1918 (Mw 7.6), Meghalaya earthquake of 1923 (Ms 7.1), Dubri earthquake of 1930 (Mw 7.1), Bihar-Nepal earthquake of 1934 (Mw 8.1), Assam earthquake of 1950 (R 8.5), and Nepal earthquake of 2015 (Mw 7.8) are well known. The maximum felt intensity of these earthquakes is localised in tectonically uplifted region, especially in the Bengal Basin (Martin and Szeliga 2010). Majority of the earthquakes in the Bengal Basin generally occur at shallow depth (<30 km). To the east, below the IBR, focal depths gradually increase in the depth range of 30–70 km and define an eastward gently dipping subducting plate (Kundu and Gahalaut 2012). Focal depths to the north also show an increasing depth trend. In general, most of the seismic activities in the Bengal Basin commonly occur along the regional active faults. The following discussion on neotectonics and active tectonic of the Bengal Basin are in terms of the neotectonic activities along these faults.

Table 7 Seismicity catalogues of high intensity (Mw > 5) earthquakes in the Bengal Basin and its surroundings

Name of the Earthquake	Year of occurrence	Magnitude (Mw)
Seismicity to the East (CTFB)		
Compiled from Fergusson (1863, Ansary et al. (2000), Cummins (2007), Akhter (2010), Martin and Szeliga (2010), Szeliga et al. (2010), Kundu and Gahalaut (2012) and Wang et al. (2014)		
Chauk Earthquake (Myanmar)	2016	~6.8
Thabeikkyin Earthquake (Myanmar)	2012	~6.8
Thabeikkyin Earthquake (Myanmar)	2012	~6.8
Bandarban Earthquake (Bangladesh)	2003	~6.1
Moheskhali Earthquake (Bangladesh)	1999	~5.2
Kaladan Earthquake (Bangladesh-Myanmar border)	1955	~6.2
Chittagong/Arakan Earthquake (Bangladesh-Myanmar border)	1762	R 8.5+
Seismicity to the North (Surma basin and adjacent region)		
Compiled from Wesnousky et al. (1999), Bilham and England (2001), Morino et al. (2011), Chopra et al. (2013), Olympa and Abhishek (2015), Berthet et al. (2014), Morino et al. (2014), Baruah et al. (2016), Bilham et al. (2017) and Rajendran et al. (2017)		
Sikkim Earthquake (India)	2011	~6.9
Assam Earthquake (India)	1950	~8.4
Assam Earthquake (India)	1943	~7.2
Bihar–Nepal Earthquake (India-Nepal border)	1934	~8.1
Dhubri Earthquake (India)	1930	~7.1
Kangra Earthquake (India)	1905	~7.8
Assam Earthquake (India)	1897	~8.1
Cachar Earthquake (India)	1869	~7.5
Dauki Earthquake (Bangladesh-India border)	1548	?
Nepal Earthquake (Nepal)	1505	~8.7
Seismicity to the West (Foreland Shelf)		
Compiled from Bilham (2004), Akhter (2010), Martin and Szeliga (2010), Nath et al. (2014), Olympa and Abhishek (2015)		
Pabna Earthquake (Bangladesh)	1935	~6.2
Hinge Zone Earthquake (Bangladesh)	1935	~6.2
Rajshahi Earthquake (Bangladesh)	1842	~7.3
Hinge Zone Earthquake (Bangladesh)	1842	~7.3
Sirajganj Earthquake (Bangladesh)	1787	MM X
Kolkata Earthquake (India)	1737	?
Seismicity at the Centre of the Bengal Basin		
Compiled from Akhter (2010), Martin and Szeliga (2010) and Nath et al. (2014)		
Mymensingh Earthquake (Bangladesh)	2008	~5.1
Bengal (Manikganj) Earthquake (Bangladesh)	1885	~7
Mymensingh Earthquake (Bangladesh)	1846	~6.2
Sirajganj Earthquake (Bangladesh)	1845	~7.1
Dhaka Earthquake (Bangladesh)	1812	MM VIII

6.1 Active Tectonics to the East (CTFB)

The CTFB and the IBR fold belt regions are poorly explored due to the complex geology, dense vegetation and mountainous terrains thus making it challenging for field observation as well as geological and geophysical data acquisition. The CTFB or Geotectonic Province 3 is an active tectonic element in the eastern part of the Bengal Basin (Cummins 2007; Steckler et al. 2016). In the CTFB and the IBR fold belt regions, five regional faults with ~N-S to NNE-SSW orientation are recorded (Nandy 2001; Acharyya 2007; Maurin and Rangin 2009a; Kundu and Gahalaut 2012; Hossain et al. 2014; Wang et al. 2014), which mostly accommodate this convergence. Towards west, these are Sagaing Fault—west of Shan Plateau and east of Central Burma Basin (CBM) as well as Chindwin Basin, Kabaw Fault—west of CBM and east of IBR, Churachandpur Mao Fault (CMF)—at the core/centre of the IBR, Kaladan Fault—west of the IBR and east of the CTFB, and the Chittagong Coastal Fault (CCF)—at the western margin of the CTFB and east of the Hatiya Trough (Figs. 1 and 8). CMF and CCF are also known as the Lelon Fault (Maurin and Rangin 2009a), and the Chittagong-Cox's Bazar Fault, respectively. Geotectonic evolution and neotectonic activities in the eastern Bengal Basin, Central Burmese Basin, Chindwin Basin and Rakhine Basin of Myanmar are directly related to these four major faults which are parallel to the CTFB-IBR fold belts. A shallow dipping megathrust structurally connects with the CCF, Kaladan and CMF faults (Wang et al. 2014; Steckler et al. 2016). Although variable convergence rate is reported (CDMP II 2013; Barman et al. 2016; Steckler et al. 2016), the rate gradually decreases from east (IBR) to west (CTFB). Among the total convergence of ~46 mm/year, 13–17 mm/year is taken by the CTFB-IBR fold belt and the shallow dipping megathrust, and rest being accommodated by the dextral-slip motion of the Sagaing fault (~20 mm/year) and CMF (~10 mm/year). The rate of convergence to the east of the Kaladan Fault is ~13 mm/year, in between the Kaladan Fault and CCF is ~5 mm/year, and between CCF and deformation front (Figs. 1 and 9) is ~2 mm/year (CDMP II 2013). This rate of convergence results ~200 km advancement of the outer part of the accretionary prism (represented by the CTFB) over the last 2 Ma (Maurin and Rangin 2009a). The orientation of maximum compressive stress rotates from NE–SW across the inner and northern IBR to E–W near the western CTFB of the Bengal Basin (Angelier and Baruah 2009). The rotation of maximum stress orientation is consistent with the deformation partitioning related to oblique collision and is reflected in the rotation of relative displacement vectors across the region (CTFB and IBR). This deformation partitioning results a variety of movements across the fold belt. Towards west, right-lateral slip is observed in the east of the IBR (Sagaing Fault) to oblique reverse-dextral slip in the middle and western edge of the IBR (Kabaw Fault) and pure thrusting in the CTFB (CCF) (Fig. 8). Regardless of the highly oblique plate motion and thick Tertiary sediments, the GPS measurements suggest that subduction in this region to be active (Steckler et al. 2016). GPS studies also suggest a locked megathrust signifies an underappreciated large seismic hazard in the CTFB and IBR regions.

The ~N-S oriented and ~1400 km long Sagaing Fault is the major locus of dextral-slip motion related to the northward translation of the Indian Plate against the Burmese Plate (Wang et al. 2014). The northern end of this transform fault develops horse-tail appearance and terminates into the eastern Himalayan syntaxis (Fig. 1). This fault accommodates about 20 mm/year of the north component of the relative motion of the oblique convergence between the Indian and the Burmese Plates (Barman et al. 2016). The fault is well expressed geomorphically along most of its length due to such high slip rate. Earthquake catalogues (Table 7) show that about half of the Sagaing Fault has ruptured during many large earthquakes over the decades. Based on bends, splays, terminations of historical ruptures, and distinct secondary features, Wang et al. (2014) divide the Sagaing Fault into six distinct segments. From south to north, these segments are Bago, Pyu, Nay Pyi Taw, Meiktila, Sagaing, and Northern Sagaing. The Northern Sagaing segment is again divided into three sub-segments, and these are Tawma-Ban Mauk, In Daw-Mawlu and Shaduzup-Kamaing-Mogang. Most of these segments of the Sagaing Fault are seismically active and produce major earthquakes except the Meiktila segments (Wang et al. 2014). About 13 major earthquakes (Mw 6.8–7.7) have been recorded since 1929 along this mega fault and most of them occurred at shallow depth (<25 km). The most recent Thabeikkyin earthquake with 6.8 Mw occurred in November 2012 in the Sagaing segment, which caused damage and panic as far as in the Chittagong City of Bangladesh, approximately 400 km west of the epicentre. A notable seismic gap along the Sagaing Fault is observed in the Meiktila segment, between the new capital of Myanmar (Nay Pyi Taw) and the large city of Mandalay. If this 220 km long Meiktila segment ruptures at once, it is capable of producing an earthquake as large as Mw 7.8–7.9 (Blaser et al. 2010).

The oblique reverse-dextral slip and the east-dipping Kabaw Fault (Fig. 1) traverses the eastern side of the Kabaw valley for ~800 km. The shortening rate to the locked Kabaw Fault is ~13 mm/year (Steckler et al. 2016). Based on the youthful appearance (fresh/unweathered exposure) on SRTM and ASTER images, a segment of this fault between 22 and 24.8°N is considered as (seismically) active (Wang et al. 2014). If this ~280 km segment of the Kabaw Fault ruptures at once, it could generate an Mw 8.4 earthquake. The most recent earthquake with 6.8 Mw occurred in August 2016 along this fault near Chauk City of Myanmar. The tremors were felt as far as in Bangladesh, Bihar (India), and Thailand and damaged infrastructures and life.

The N-S to NNE-SSW oriented and ~300 km long right-lateral Churachandpur-Mao Fault (Fig. 1) is the most prominent at the core/centre of the IBR (Fig. 1). This steeply east dipping fault accommodates dextral velocities of ~10–18 mm/year between the Indian and Burmese Plates through dextral strike-slip motion, predominantly through aseismic slip (Gahalaut et al. 2013; Steckler et al. 2016). However, both aseismic and seismic slips have been observed along this fault. The aseismic slip part of this fault could be related to minor to moderate earthquakes. Wang et al. (2014) suggest that this fault creeps only at sub-surface, and that its northern part is seismic. If this fault segment ruptures, it could produce an Mw 7.6 earthquake. Although earthquake catalogue does not show any historical seismic events that could be attributed to rupture of the Churachandpur-Mao Fault (Szeliga et al. 2010), the Imphal earthquake with 6.8 Mw occurred in January 2016 in the northern segment of this fault is a proof of seismic slip. This earthquake was strongly felt in Bangladesh and caused significant damage.

The ~450 km long west verging Kaladan Fault (Figs. 1 and 9), which mostly follows the course of the Kaladan River is basically the contact between the Barail and Surma range and is named after the Kaladan River (Nandy 2001; Maurin and Rangin 2009a; Barman et al. 2016). Kaladan Fault marks the eastern boundary of the CTFB and extends from the Arakan coast in the south to the northern-most part of the CTFB and the IBR contact (Sikder and Alam 2003; Maurin and Rangin 2009a; Kundu and Gahalaut 2012; Wang et al. 2014). Along 20° 30′N, the dip of the fault plane is 70°E with an 8°–10°S pitch, and shows a clear right-lateral offset. In this area, the orientation of the maximum compression is ~N80–N60°E (Maurin and Rangin 2009a). Although, the Kaladan Fault shows both dextral strike-slip and west verging thrust components, it is generally considered as thrust fault, and seismically the fault is sparsely active. The right-lateral slip dominates its northern termination whereas in the middle portion pure E-W or WSW-ENE thrusting focal mechanisms are observed, suggesting that active motion is partitioned within a very short distances along this fault. To the south, Ramree thrust is most likely the southward continuation of the Kaladan Fault (Maurin and Rangin 2009a). While Mw 6.2 earthquake in December 1955 along the Kaladan Fault near Bangladesh-Myanmar border region at a depth of 35 km is a proof of seismic slip, mainly shallow focus and low intensity earthquakes are observed along this fault (Kundu and Gahalaut 2012).

Based on the geomorphic signatures on 30 m resolution SRTM image and geological field reconnaissance, a ~300 km long major fault can be inferred along the western boundary of the CTFB in between ~19° 40′N and 23° 30′N is known as the Chittagong Coastal Fault (CCF). Multichannel seismic survey in the offshore area also suggests the existence of the CCF as a crustal dextral strike-slip fault with an appreciable thrust component, and is resultant product of thin-skinned and thick-skinned tectonics (Maurin and Rangin 2009a). The fault is seismically less active. To the north, the fault splays before disappearing near the Tichna anticline in Tripura, just south of the Surma basin. The fault is probably younger than the Kaladan Fault as the accretionary wedge is progressing westward during the last 2 Ma, and it is not affected by sufficient erosion yet. Maurin and Rangin (2009a) suggest that both the CCF and the Kaladan Faults can be interpreted as deep-seated basement reverse fault with dextral strike-slip component. The restoration of the compressive fold belt and megathrust yields ~11 km of shortening of the CTFB including the CCF and the Kaladan Faults. Although published earthquake catalogues do not show any historical seismic events that can be attributed to rupture of the CCF, Moheskhali earthquake with 5.2 Mw occurred along the CCF in Moheskhali Island in July 1999 is a proof of seismic slip. This earthquake was strongly felt around Bangladesh and caused significant damage locally (Ansary et al. 2000). Moreover, occurrence of any large earthquake in the offshore area along the east coast of the northern Bay of Bengal is likely to produce great tsunami

(Cummins 2007). One of such large earthquakes (R 8.5+) occurred in 1762 along the Arakan coast (i.e. Arakan subduction zone) known as Chittagong/Arakan earthquake (Fergusson 1863; Cummins 2007; Martin and Szeliga 2010; Kundu and Gahalaut 2012; Wang et al. 2014). It is believed that rupture of the megathrust of the Ramree domain through the southern portion of the Dhaka domain caused this earthquake. The fault model to reproduce the observed subsidence and uplift associated with this earthquake along the eastern coast of the northern Bay of Bengal suggests that the fault's upper edge coincides with the deformation front. Cummins (2007) concluded that the evidence of oblique convergence and associated active subduction in conjunction with thrust earthquake activity along the coastal region suggest the risk of a major tsunami in the northern Bay of Bengal.

The CTFB and IBR are also segmented by two major conjugate oblique/transverse faults. The NW–SE oblique Mat Fault is sinistral, whereas the NE–SW oblique Gumti Fault is dextral (Fig. 9) (Nandy 2001; Barman et al. 2016). However, seismicity along these faults is not reported. In addition, medium to small scale dip-slip, strike-slip and thrusts are also present in the CTFB area that have played important role not only in shaping the present topography of the folded flank of the Bengal Basin, but also controlling the stream courses. The dip-slip faults are mainly oblique to the strike of the anticlines, whereas strike-slip and thrust faults are mostly parallel to the strike. The latter faults mostly affected the folds of the CTFB regions (Sikder and Alam 2003; Mandal et al. 2004; Kabir and Hossain 2009; Maurin and Rangin 2009a; Najman et al. 2012; Abdullah et al. 2015; Hossain et al. 2018). A single large and few moderate magnitude earthquakes are recorded in the CTFB region, which may be connected with these faults. Among such events, the Srimangal earthquake in 1918 (Mw 7.5), the Bandarban earthquake in 1997 (Mw 6.1), and Barkal earthquake in 2003 (Mw 5.7) are well known (Akhter 2010).

On the other hand, based on the presence of fault scarp on the young fluvial/alluvial surface at the flanks of the anticlines, some of the anticlinal structures in the CTFB region are considered as tectonically active (CDMP II 2013). Among such structures NNW trending Harargaj, Sitakund, Jaldi, Sandwip, Maheshkhali, and Dakhin Nhila anticlines are well recognized. The seismic reflection profiles across few of these anticlinal structures reveal growth strata that constrain the initiation of folding (Mandal et al. 2004; Steckler et al. 2008; Najman et al. 2016). The age of the growth strata of the Habiganj, Jaldi, and Kutubdia structures constrains the initiation of anticlinal growth during Pliocene or younger (Johnson and Alam 1991; Steckler et al. 2008; Maurin and Rangin 2009a; Wang et al. 2014). According to Khan et al. (2015), the Dakhin Nhila structure is tectonically active and show different uplift rates spatially and temporally due to faulting for the last 55,000 years. More recent activity recorded from young- aged soils from the crestal portion of the Maheshkhali and Jaldi anticlines suggest uplifts of 18,000 and 35,000 years respectively (Khan et al. 2005). Farther north, the upper surface of the Sandwip structure is estimated to be about 7000 years old as indicated by radiocarbon date (Goodbred and Kuehl 1999). Similar young neotectonic activities are also reported by Ansary et al. (2000), Steckler et al. (2008), Khan et al. (2017, 2018) in this region.

6.2 Active Tectonics to the North (Surma Basin and Adjacent Region)

The geodynamics and the neotectonics of the northern part of the Bengal Basin (northern part of the Geotectonic Province 1 and 2) are directly related to the tectonic evolution of the 2500 km long Himalayan frontal belt as well as to the Main Frontal Thrust. The Himalayan frontal belt is the youngest portion of the Himalayan Orogeny (Thakur 2004). The northward convergence of the Indian Plate resulted in crustal shortening of the northern margin of the Indian continent, accommodated by south-verging Main Central Thrust (MCT), Main Boundary Thrust (MBT) and Main Frontal Thrust (MFT) (Fig. 7) (Thakur 2004). From MCT to MFT, these thrusts are progressively younger in age and shallowing at depth in a southward direction, suggesting southward migration of the Himalayan deformation front. Exceptions to this are the Himalayan active/out-of-sequence structures, most notably faults (Mukherjee 2015), some of which are even backthrust-faults (Bose and Mukherjee, submitted-a, b). The Himalayan frontal belt is limited to the south by the Main Frontal Thrust (MFT). Most of the collisional force in the Himalayan frontal belt is accommodated by the uplift and convergence along the MFT (Wesnousky et al. 1999). Although variable convergence rates (11–21 mm/year) are observed across the Himalayan frontal belt, a general increasing trend is observed towards the centre which is known as Nepal Himalaya (Wesnousky et al. 1999; Berthet et al. 2014; Bilham et al. 2017). Towards east, the Himalaya, except in Pakistan, can be divided into Dehra Dun Himalaya (western Himalaya), Nepal Himalaya (central Himalaya) and Bhutan-Arunachal Himalaya (Eastern Himalaya) (Thakur 2004). Here after Sikkim, Bhutan and western Arunachal are collectively referred as the Bhutan Himalaya. The rate of convergence along the Nepal Himalaya is $\sim$20 mm/year and along the Bhutan Himalaya (directly north of the Bengal Basin) is $\sim$16–18 mm/year. The Himalayan frontal belt is one of the few collisional orogenic belts on-land, where earthquakes with great magnitudes analogous to those of oceanic subduction zones generally occur (Berthet et al. 2014).

The northern part of the Bengal Basin mainly comprises of the Surma basin to the east, and northern edge of the Stable Shelf to the northwest. Assam Basin and Shillong Plateau are situated in between Himalayan frontal belt to the north and Surma basin to the south. Broadly, E-W oriented Brahmaputra fault is situated at the southern margin of the Assam Basin. Oldham, Brahmaputra and few other faults separate the Assam Basin from the Shillong Plateau, whereas the Dauki Fault separates the Shillong Plateau from the Surma basin. Southern portion of the Surma basin is transacted by two major faults, the Sylhet Fault to the southeast and the Old Brahmaputra Fault to the southwest (Fig. 6). In the west, the Shillong Plateau is separated from the Bengal Basin by the Jamuna Fault (Dhubri Fault). In the east, the Shillong Plateau and the Assam Basin separate from the Naga domain/Naga Hills by the Naga and Disang thrusts. In addition, a few other faults and lineaments also play important role in neotectonic activities as well as geomorphic evolution of this part of the Bengal Basin, and these are Dapsi (NW-SE), Haflong ($\sim$E-W), Dhudnoi

(N-S) and Kopili Faults (NW-SE), and Tista and Gangtok Lineaments (NW-SE) (Fig. 1).

Approximately E-W oriented MFT separates the Assam Basin from the Himalayan frontal belt. The MFT is associated with active low-angle thrusting and associated uplift (Berthet et al. 2014). It represents a zone of active deformation between the Sub-Himalaya and the Indo-Gangetic Alluvial Plain (IGAP) of the Peninsular India, thereby thrusting sediments of the Siwalik Group southward over the IGAP (Wesnousky et al. 1999). The MFT comprises of mainly homoclinal strata with northerly dips generally ranging 20°–30°. In the Bhutan Himalaya, the dip of the MFT is ~25°–30° (Wesnousky et al. 1999; Long et al. 2011; Berthet et al. 2014) and the Holocene slip rate along the fault is ~20–28 mm/year (Burgess et al. 2012; Berthet et al. 2014). The observed uplift rate across the MFT is ~10–13 mm/year (Wesnousky et al. 1999; Berthet et al. 2014). The MFT is tectonically active in the Bhutan-Arunachal, Nepal, and Dehra Dun Himalaya (Wesnousky et al. 1999; Berthet et al. 2014). Although some of the N-S plate convergence motion is accommodated by the upliftment of the Shillong Plateau, however they do not have any significant effect on the high activity of the MFT in the Bhutan Himalaya. The rupture zones of 1905 Kangra earthquake (Mw ~7.8), 1505 western Nepal earthquake (Mw ~8.7), 1934 Bihar–Nepal earthquake (Mw ~8.1), and 1950 Assam earthquake (Mw ~8.4) extend N–S right across the MFT zone to the MBT zone (Fig. 11) (Wesnousky et al. 1999; Berthet et al. 2014; Bilham et al. 2017; Rajendran et al. 2017). In this context, the ~350 km long Bhutan Himalaya zone

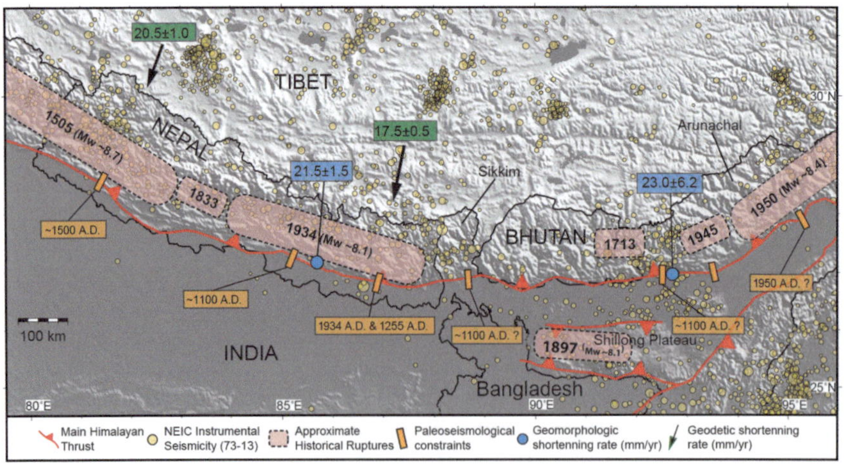

Fig. 11 Active tectonic map of the Central and Eastern Himalaya (to the north of the Bengal Basin). Pink and orange rectangles show the historical events and the ages of major surface-rupturing events, respectively. Blue and green rectangles give the Holocene shortening rates and the geodetic slip rates, respectively. Distributions of the earthquake epicentres from 1973 to 2013 recorded by the National Earthquake Information Center (NEIC) are shown in yellow circles (modified after Berthet et al. 2014)

of the MFT, which is located between the epicentres of the 1934 Bihar–Nepal earthquake and 1950 Assam earthquake, appears to be an area where no similar great earthquakes has ever been documented. These observations and the lack of primary surface rupture suggest a potential for great earthquakes along the Himalayan Frontal Thrust in the Bhutan Himalaya zone (Wesnousky et al. 1999; Berthet et al. 2014). An active zone of deformation, upliftment and slip observed along the MFT, especially along the Bhutan Himalaya suggests increased seismic hazard to the adjoining northern part of the Bengal Basin.

The WNW and ESE oriented Oldham Fault separates the Assam Basin from the Shillong Plateau. The Oldham Fault is a ~120 km long north-verging and south dipping reverse fault, and generates co-seismic displacement. Although the 1897 devastating Assam earthquake (Mw ~8.1) occurred due to co-seismic displacement along this fault (Bilham and England 2001), the source is also attributed to multiple faults around the Shillong Plateau including the Dauki, Dapsi and Oldham Faults (Bilham and England 2001; Morino et al. 2014; Barman et al. 2016; Singh et al. 2016). The Oldham Fault together with the Dauki Fault to the south of the Shillong Plateau constitutes a high-angle conjugate crustal-scale shear zone, uplifting the plateau as a pop-up structure (Biswas and Grasemann 2005a). Both the hanging wall (i.e. Shillong Plateau) and the foot wall (i.e. southern margin of the Assam Basin/Himalayan Foreland Basin) of the Oldham Fault are associated with uplift of the Shillong Plateau. However the rate of uplift of the footwall block is found to be less than that of the hanging wall block. The rate of vertical displacement along the Oldham Fault is 0–0.68 mm/year (Biswas et al. 2007). By considering ~45°–50° dipping for the fault plane, average fault slip rate is estimated to be ~2–4 mm/year and the horizontal shortening rate accommodated by the plateau is ~2–5 mm/year (Bilham and England 2001). However, Biswas et al. (2007) suggest much lower shortening rate (0.65–2.3 mm/year) that accounts for 10–15% contraction rate of the Himalaya. The GPS measurements indicate the southward movement of the central Shillong Plateau at a rate of ~6.5–7 mm/year with respect to the reference points in southern and central India (Barman et al. 2016). From these rates, Bilham and England (2001) conclude that the recurrence interval for earthquakes resembling the 1897 Assam earthquake to be 3000–8000 years. However, examination of the past four high-intensity seismic events along the northern edge of the plateau indicates a 500 years recurrence interval but these may record shaking from moderate local earthquakes and large seismic events in the adjoining Bhutan Himalaya. Geological and geophysical data suggest the presence of another large fault at the north of the Oldham Fault with E-W orientation known as Brahmaputra Thrust (Ghosh et al. 2015) that marks the northern most edge of the Shillong Plateau, and may have played an important role in the plateau uplifting. This fault is an active thrust and important for the crustal resettlement in this part of the Indian Plate.

The ~320 km-long E-W trending south-verging and north dipping Dauki Fault, a major plate boundary-influenced fault marks the boundary between the Surma basin to the south and the Shillong Plateau to the north (Hossain et al. 2016). The strike of this fault is parallel to the MFT, which forms a plate boundary fault whereas the Dauki Fault is an intra-plate active reverse fault. The continuous subsidence of the southern footwall of the Dauki Fault has resulted in the Surma basin, where structural relief between highest point of the plateau and lowest point of the basin is estimated to be around 20 km (Mirkhamidov and Mannan 1981; Hiller and Elahi 1984; Shamsuddin and Abdullah 1997; Biswas et al. 2007). Across the Dauki Fault, a striking negative Bouguer gravity anomaly exceeding −70 mGal has been recorded for the Surma basin, whereas the Shillong Plateau shows a positive Bouguer gravity anomaly >+20 mGal (Rahman et al. 1990b; Rajendran et al. 2004). Hence, the Dauki Fault might have been formed at the onset of the exhumation of the Shillong Plateau, and the total vertical displacement of the basement top along this fault is ~10 km (Biswas et al. 2007). During the Late Tertiary, the rate of upliftment of the hanging wall rock (i.e. the Shillong Plateau) ranged 0.32–0.80 mm/year, whereas the rate of subsidence of the footwall rock (i.e. the Surma basin) is ~0.45 mm/year. Consequently, the resultant mean vertical displacement rate along the Dauki Fault ranges 0.77–1.25 mm/year (Biswas et al. 2007). Higher vertical displacement rate along the Dauki Fault than the Oldham fault is partially responsible for the ~2°–3° northward tilt of the Shillong Plateau surface. The Dauki Fault is segmented to at least 4 parts along its entire length (Biswas and Grasemann 2005a). Geomorphic indices like mountain front sinuosity, stream length gradient index, valley floor width to height, drainage basin asymmetry, and transverse topographic symmetry factor suggest that the tectonic activity of the fault is much higher in the eastern part than the western portion (Biswas and Grasemann 2005a; Sharma and Sarma 2017).

Rapid movement along the Dauki Fault started in the Pliocene, resulting in significant topographic rise and basin loading in the north and south, respectively (Najman et al. 2016). The flexure related to seismic event has been identified along the Dauki Fault near Jaflong (Sylhet) and is inferred to be AD 840–920 (Morino et al. 2014). Another rupturing in the Dauki Fault estimated to be AD 1548 has been identified in a trench at Gabrakhari village, Haluaghat, Mymensingh (Morino et al. 2011, 2014). It is also believed that the 1897 Assam earthquake that occurred along the Oldham Fault also caused slip along the Dauki Fault. Although it was previously thought that the interval between these giant plateau-building earthquakes exceeds 3000 years (Bilham and England 2001), the signature of the above seismic events indicates recurrence interval of 350–700 years (Morino et al. 2014). Moreover, the active fan development along the southern foothills of the Shillong Plateau is controlled by the neotectonic activities along the Dauki Fault (Alam and Islam 2017).

In the southeastern margin of the Shillong Plateau, a branched extension of the Dauki Fault with approximately E-W strike is known as the Haflong thrust fault. The Haflong Fault marks the edge of the Shillong Plateau before it links up with the Naga thrust to the east (Biswas and Grasemann 2005a). The WNW–ESE trending ~90–100 km long Dapsi thrust fault (Fig. 1) is located within the southwestern edge of the Shillong Plateau. This fault is the north-western extension of the Dauki Fault that separates the Cretaceous–Tertiary sediments to the south and the Precambrian Gneissic complex to the north (Barman et al. 2016). The N-S trending

Dudhnoi Fault lying towards the northwest of the Shillong Plateau was activated during the 1897 Assam earthquake (Barman et al. 2016). A significant difference in the rate of motions has been observed on the opposite side of the Dudhnoi Fault. The central Shillong Plateau to the east shows 7 mm/year southward motion, whereas the western Shillong Plateau shows 2.1–2.7 mm/year southward motion, suggesting different tectonic settings of the plateau and dextral motion along the Dudhnoi Fault (Barman et al. 2016). In addition, the development of the Shillong Plateau as a pop-up structure is also facilitated to the west by the N-S trending Dhubri and Jamuna Faults. The epicenter of the 1930 Dhubri earthquake (Mw ~7.1) was along the Dhubri Fault near Dhubri, Assam. To the south, the earthquake also caused slip in the Jamuna Fault, and west of the Garo Hills, fissures and sand vents were reported (Olympa and Abhishek 2015).

The NW-SE trending strike-slip Kopili Fault located between the Shillong Plateau and the Mikir Hills records a dextral slip of 4.7 ± 1.3 mm/year indicating the fragmentation of Assam Basin/Brahmaputra valley across this fault (Barman et al. 2016). The eastern edge of the Mikir Hills marks by the Bomdila Lineament (Baruah and Hazarika 2008; Angelier and Baruah 2009). The transverse Kopili fault may be a surface expression of the deep-seated seismogenic structure (Bhattacharya et al. 2008). The western edge of the Mikir Hills appears to be locked currently at 10.2 ± 1.4 km depth with the Assam Basin indicating strain accumulation. This strain accumulation may cause seismic event in near future along this seismically highly active 300 km long fault that reaches into the MFT system to the north and Naga thrust to the south (Bhattacharya et al. 2008; Barman et al. 2016). The rupture area of the 1869 Cachar earthquake (Mw ~7.5) and 1943 Assam earthquake (Mw ~7.2) are located within the Kopili fault. The earthquakes damaged over hundreds of kilometres across the region (Olympa and Abhishek 2015).

At eastern Arunachal Himalaya over the Assam Basin and the Naga Hills, three major plates interact along two convergent boundaries: the Himalayan frontal belt and the IBR, which meet at the 430 km-long Assam Syntaxis (Angelier and Baruah 2009; Wang et al. 2014). The northern end of the IBR (i.e. Naga Hills) and its underlying lithosphere experience nearly arc perpendicular extension with ESE–WNW trends, whereas easternmost Himalayan frontal belt (i.e. Arunachal Himalaya) experiences WNW-ESE compression related to convergence between the outer IBR and the Indian Plate. Together, the convergence and the extension structures forms a complicated geotectonic system (Nandy 2001) that causes the northeastern portion of the Indian Plate to be deformed and also results in the development of the Naga and Disang Thrusts (Evans 1964; Bilham and England 2001; Angelier and Baruah 2009). The NE-SW oriented Naga Thrust with several arcuate lobes separates the Assam Basin from the northern outermost edge of the Naga Hills (Fig. 1). The fault with clear linear scrap extends from the Haflong Thrust to the southwest and terminates at ~96°E where it appears to cut by a ~NW-striking Mishmi Thrust and associated anticlines (Vaccari et al. 2011; Mohanty et al. 2013, 2014). The fault shows variable dips (moderate to steep) and breaks the surface at some positions (Wang et al. 2014). The lower Miocene rocks constituting the Naga Hills are thrusted and folded over undeformed upper Pliocene to Quaternary beds and alluvium of the Assam Basin at the mountain front along the Naga Thrust (Aier et al. 2011). The clear geomorphological expression of some anticlines and younger terrace with 20–50 m upliftment imply that it is still active and the dip of the thrust fault decreases at depth. Additionally the slip on the Naga Thrust dies out over ~100 km from the crest of the Naga Hills (Wang et al. 2014). In the immediate south, another linear thrust fault running parallel to the Naga Thrust is known as the Disang Thrust. The narrow linear belt of imbricate thrust slices of 8–25 km wide in between the Naga and Disang Thrusts is known as the Schuppen Belt. Morphotectonic analysis reveals that the overall tectonic activity of the Schuppen Belt (i.e. along the Naga and Disang Thrusts) is higher than the Dauki Fault area (Sharma and Sarma 2017).

The NNW–SSE trending Tista and Gangtok Lineaments are the two prominent tectonic features, extending from the Sikkim Himalaya to the north and Rangpur Saddle to the south. The southern ends of these lineaments terminate against the Jamuna Fault. The focal mechanisms in this area indicate right-lateral strike-slip along the Tista and Gangtok Lineaments. The 2011 Sikkim earthquake (Mw ~6.9) may have been caused by the seismic slip along the Tista Lineament (Chopra et al. 2013; Baruah et al. 2016). Several smaller earthquakes are recorded along these lineaments, especially along the Tista Lineament suggesting that the both lineaments are tectonically active (Baruah et al. 2016).

The Surma basin is an active tectonic element forming in response to the Shillong Plateau upliftment to the north, and subduction of the Indian Plate beneath the Burmese Plate to the east. Tectonic evolution and neotectonic activities of this basin are controlled by the Dauki Fault to the north, gradual westward encroachment of the CTFB and IBR through Haflong, Naga-Disang thrust to the east, the Sylhet Fault and the CTFB to the south, and the Old Brahmaputra and Jamuna Faults to the west. Except to the west, the rest of the structural features framing the Surma basin are the results of mostly compressional tectonic movements in Pliocene to Recent times (Reimann 1993). The 150-km long Sylhet Fault with NE-SW orientation, passes through the southern edge of the basin, is a major fault (Kayal 1998; Bhattacharya et al. 2008; Kayal 2008; Angelier and Baruah 2009; Vaccari et al. 2011; Mohanty et al. 2013, 2014). This fault also known as Hail Hayalua Lineament is probably the southwest extension of the Naga-Disang Thrust (Ghosh et al. 2015) which is offset by the Dauki Fault through its dextral slip component. In November 2017, a shallow focus (30.3 km depth) earthquake with Mw 4.9 occurred at distance of 27 km SSW of Habiganj, Bangladesh is probably related to the Sylhet Fault. Moreover, based on the presence of fault scarp on the young fluvial/alluvial surface at the flank of the anticline, the Fenchunganj, Sylhet, Habiganj and Rashidpur anticlines are considered as active structures within the Surma basin region. In addition, neotectonic activities are also observed in the south of the Dauki Fault in the Jaintiapur area. In the eastern part, to the immediate south of the Dauki Fault, some small hillocks present in the Jaintiapur area (Chowdhury et al. 1996), Sylhet which are capped by Dihing Formation. The Optically Stimulated Luminescence (OSL) dating of this formation indicates young depositional age (73–24 ka), and variable uplift rate (Khan et al. 2006). This suggests that the area is tectonically active and gone through a differential upliftments in the recent past (Khan et al. 2018).

6.3 Active Tectonics to the West (Foreland Shelf)

The western part of the Bengal Basin (Geotectonic Province 1) mainly comprises the Stable Shelf, and contains two major tectonic elements: the Foreland Shelf to the west, and the Eocene Hinge Zone or shelf break or shelf slope to the east (Reimann 1993). Geographically, the area comprises the western part of Bangladesh and the West Bengal state of India. Along its longitudinal direction, the Stable Shelf can be divided structurally into four tectonic elements from northwest to southeast: (i) the North Bengal Foreland, (ii) the Basin Margin Fault Zone/ western scarp zone, (iii) the Central Stable Shelf, and (iv) the Eocene Hinge Zone (Nath et al. 2014). Except the Eocene Hinge Zone, alternating gravity highs and lows indicate the presence of numerous graben and half-graben structures in most of the Stable Shelf (Khan and Rahman 1992; Khan and Chouhan 1996). This part of the Bengal Basin is intersected by several faults and lineaments. These are Katihar-Nilphamari Fault, Jangipur–Gaibandha Fault, Debagram–Bogra Fault, Rajmahal Fault, Ganga-Padma Fault, Malda–Kishanganj Fault, Medinipur-Farakka Fault, Chhotanagpur Foot Hill Fault, Damodar Fault, Saithia–Bahmani Fault, Purulia Shear Zone, Kardaha lineament, and Devikut lineament (Fig. 4) (Nandy 2001; Vaccari et al. 2011; Mohanty et al. 2013; Nath et al. 2014; Roy and Chatterjee 2015; Prasad and Pundir 2017). The Medinipur-Farakka Fault is also known as the Basin Margin Fault Zone (Raman et al. 1986; Roy and Chatterjee 2015). This fault zone broadly includes the Medinipur-Farakka Fault to the east, and Chhotanagpur Foothill Fault to the west. The Rajmahal and Saithia-Brahmani Faults are situated to the east and west of the Rajmahal Hills, respectively (Ghose et al. 2017). The approximately N-S running Malda-Kishanganj Fault (Nandy 2001; Vaccari et al. 2011; Mohanty et al. 2013, 2014; Prasad and Pundir 2017) reaches up to the east and northeastern margin of the Rajmahal Hills. Rajmahal Fault to the east of the Rajmahal Hills (Fig. 4) is considered as the northwestern margin of the Bengal Basin (Mohanty et al. 2013; Roy and Chatterjee 2015).

Nath et al. (2014) have identified several tectonic features including faults and lineaments that have the potential to generate earthquakes of Mw 3.5 and above in this part of the Bengal Basin. Major active structures are the Jangipur–Gaibanda Fault, the Katihar-Nilphamari Fault, and the Devikut Lineament. The Eocene Hinge Zone is also identified as a seismically active tectonic element. This hinge is reportedly triggered two earthquakes of magnitude Mw 7.3 and Mw 6.2 in 1842, and 1935, respectively (Akhter 2010; Martin and Szeliga 2010; Nath et al. 2014). Although this part of the Bengal Basin is not subjected to any direct seismic event with high magnitude, the younger unconsolidated fluvial sediments are prone to liquefaction under favourable ground shaking from distal seismic sources. The 1737

Kolkata earthquake, the 1787 Sirajganj earthquake (MM X), the 1842 Rajshahi earthquake (Mw 7.3), the 1897 Assam earthquake (Mw 8.1), the 1934 Bihar-Nepal earthquake (Mw 8.1), and the 1935 Pabna earthquake (Mw 6.2) had widely affected the western part of the Bengal Basin (Bilham 2004; Akhter 2010; Martin and Szeliga 2010; Olympa and Abhishek 2015; Nath et al. 2014). However, none of these destructive earthquakes are reported to have caused any co-seismic surface ruptures in this part of the Bengal Basin (Nath et al. 2014). Comparatively, this part of the Bengal Basin is seismically less active, except in the southernmost part where few low to moderate intensity seismic events are recorded.

6.4 Active Tectonics at the Centre of the Bengal Basin

The centre of the Bengal Basin comprises the deeper Foredeep Basin or Geotectonic Province 2 excluding the northern Surma basin. This central part of the basin is narrow at the northern end near the Madhupur-Tripura High, and then gradually widens to $\sim$500 km to the south. A NE-SW Barisal-Chandpur High separates the southern part of the basin into two sub-basins, the Faridpur Trough in the west and the Hatia Trough in the east (Alam 1972; Das Gupta 1977; Guha 1978; Reimann 1993). Topographically, this part of the basin is mainly composed of low-lying flat land with the exception of slightly elevated Pleistocene uplands in the northern and southeastern parts. The northern Pleistocene upland is known as the Madhupur High/Madhupur Tract, whereas the southeastern Pleistocene upland is known as the Lalmai Terrace (Bakr 1977; Khan et al. 2018). The Madhupur High is forming the surface expression of a fault controlled buried anticline surrounded by the Jamuna Fault to the west and the Old Brahmaputra Fault to the east. The High is tilted toward the east and is faulted in the west (Morgan and McIntire 1959; Steckler et al. 2008). The capital city Dhaka is situated at the southern edge of the Madhupur High. GPS derived geodetic data suggest an overall very low subsidence rate (<1 mm/year) for the Madhupur High (Reitz et al. 2015) compared to the other part of the deeper Foredeep Basin. In and around the Dhaka City, variable but rapid rates of subsidence are observed which probably related to usual sedimentary compaction and extensive groundwater extraction. The coastal belt of this Geotectonic Province 2 also shows moderate but variable rates of subsidence within 3–8 mm/year.

Except a few faults, tectonic deformation in this part of the basin is mainly accommodated by the young and gently folded buried anticlines. Some of the buried anticlines are Begumganj, Shahbazpur, Kachua, Daudkandi, Kamta, and Narshingdi. All these folds are actively developing within the area bordered by the CTFB to the east and the deformation front to the west (Fig. 9). The major active faults in this part of the Bengal Basin are Madhupur, Dhaleswari, Padma, and Meghna. Some other active faults and lineaments are also present in and around the Dhaka City (Khan et al. 2011). These tectonic structures include the Bangshi and Turag Faults to the west, Tongi Khal Fault to the north, Balu, Sitalaykha, Banar, and Arial Khan Faults to the east, and Buriganga Fault to the south. The 1812 Dhaka earthquake (MM VIII), the 1845 Sirajganj earthquake (Mw 7.1), the 1846 Mymensingh earthquake (Mw 6.2), the 1885 Bengal (Manikganj) earthquake (Mw 7), the 2001 Dhaka earthquake (Mw 4.5), the 2006 Narail earthquake (Mw 4.2), the 2008 Manikganj earthquake (Mw 3.8), the 2008 Mymensingh earthquake (Mw 5.1), and the 2008 Chandpur earthquake (Mw 4.5) caused damage, especially in Dhaka City (Akhter 2010; Martin and Szeliga 2010; Nath et al. 2014). Among these, the 1846 and 2008 Mymensingh earthquakes probably occurred along the Old Brahmaputra Fault, the 1885 and 2008 Manikganj earthquakes probably occurred along the Madhupur Fault, and the 2001 Dhaka earthquake probably occurred along the Buriganga Fault.

The blind Madhupur Fault is constituted of a series of en-echelon faults, and flanks the western side of the uplifted Madhupur High. It is believed that this fault resulted either from torsion of the region or from the effect of shear along an assumed buried fault or possibly a combination of both (Morgan and McIntire 1959). This thrust fault is east- dipping and is considered as an important structure for the seismic hazard assessment of the central part of the Bengal Basin (CDMP II 2013). The uplift of the Madhupur High may have exerted a significant control on the avulsion history of the Brahmaputra/Jamuna River. The cyclic avulsion history of the Brahmaputra River with a periodicity of about 1800 years (Pickering et al. 2014) is most likely related to the uplift of the Madhupur High due to slip along the Madhupur Fault.

In order to have a better understanding of seismicity and seismotectonics of the Bengal Basin, especially in Bangladesh region it is necessary to focus on multi-disciplinary and integrated studies. *For an improved probabilistic and deterministic seismic hazard mapping, it is indispensable to install and run short-period seismic stations under a single seismological network covering the whole of Bengal Basin and to elucidate sub-surface structures across the region.*

7 Present Day Deformation Front

The deformation front may exhibit distinctive oceanic subduction features such as accretionary prism, forearc wedge, forearc basins and volcanic arc. In general, deformation front line indicates a hypothetical approximate vertical surficial projection of the subsurface de'collement level or basal detachment where slip diminishes to zero. In accordance, the Indo-Burmese deformation front also represents a near-vertical surficial projection of the western front of the subsurface de'collement level where slip gradually diminishes to zero towards west. The overall curvature of the Indo-Burmese deformation front might be related to the differences in sediment thickness as well as the fabric of the subducting oceanic crust of the eastern margin of the Indian Plate (Fig. 1) (Maurin and Rangin 2009a). Although, Steckler et al. (2016) consider the deformation front (updip limit of the large downdip extent locked detachment zone) potential for a very large earthquake, it is more likely that the rupture deformation would be terminated updip more to the east as an imbricate fault in the wedge rather than propagating up to the Indo-Burmese deformation front. Although, Reitz et al. (2015) suggest that the Indo-Burmese deformation front is blind and buried by active delta sedimentation, however, deformation front is exposed on land as huge thickness of sediments entering the Indo-Burmese subduction zone (Steckler et al. 2008).

Wang et al. (2014) suggest the southern limit of the deformation front from ~14°N. However, the present authors would like to propose it's southern limit at ~18°N in the offshore where the Sunda Arc trench appears to die out. From offshore, deformation front coming ashore near the Mehgna estuary at 23°N (Cummins 2007). In the north, the deformation front is terminated by the Dauki Fault at the northern edge (middle) of the Surma basin (Steckler et al. 2016). Nevertheless, some authors also suggest the northern end of the active deformation front reaches as far as to the western end of the Dauki Fault (Steckler et al. 2008; Maurin and Rangin 2009a). In the context of the east-west placement of the deformation front, Acharyya (1998) placed the present subduction zone on the western side of the CTFB, in the middle of the deeper foredeep basin (Geotectonic Province 2). However, this is certainly the deformation front of the subduction zone (Mukhopadhyay and Dasgupta 1988; Khan 1991a, b), but not the actual subduction zone which lies further east beneath the CTFB-IBR. Wang et al. (2014) suggest that the active deformation front has propagated as far west as the active GBM delta. Deep seismic reflection survey is indicative of the blind deformation front more west around the mouth of Padma River near Dhaka (Maurin and Rangin 2009a; CDMP II 2013; Steckler et al. 2016). It is necessary to clarify here that the deformation front does not represent the plate boundary, rather active faults on the eastern margin of the CTFB represent the approximate plate boundary. The whole length of the deformation front is characterized by the presence of Wadati-Benioff zone extending about 300–450 km down dip to the east beneath the Burmese Plate. High resolution SRTM images and seismic data display clear evidence of tectonic shortening at and adjacent to the deformation front. Faults and folds along these deformation front show right-stepping en-echelon character, i.e. dextral-reverse slip (Wang et al. 2014). Accretion of sediment is particularly clear at the southern end in the offshore part of the deformation front. To the east of this region, along the coast, abundant evidence for youthful folding, and uplifted coastal terraces indicate the tectonic activity along this front. In the Dhaka domain area (Wang et al. 2014), i.e. in the middle part, most of the active faults and anticlines lie within 120 km east of the deformation front. In this domain, structural and stratigraphic studies show progressive westward growth of the fold belt, toward the deformation front (Johnson and Alam 1991; Steckler et al. 2008; Maurin and Rangin 2009a; Wang et al. 2014). It is worthwhile to mention here that the deformation front line both in the offshore and onshore, marks the boundary where evidence of only youthful and buried very gentle folds are present, but no surficial expression of the folding is visible. This is an indication that the slip required for the structural deformation above the de'collemen surface diminished to zero at the deformation front (Maurin and Rangin 2009a).

As Maurin and Rangin (2009a) claim, deformation front propagated at ~100 mm/year through the past 2 Ma, which is significantly greater than the current convergence rate 9 mm/year across the CTFB (Nielsen et al. 2004). In general, the deformation front of the eastern Himalayas advanced more rapidly southward, whereas the deformation front of the CTFB-IBR probably advances more gradually westward to the Bengal Basin (Uddin and Lundberg 2004). This fast propagation of the deformation front is the key to explain the broad curvature of the northern end of the CTFB in the Surma basin immediate south of the Shillong Plateau. Over the past 2 Ma, the fold belt has propagated only into the basin, but not over the Shillong Plateau. The plateau has been rising as continental basement pop-up structure since the Pliocene (Biswas and Grasemann 2005a, b; Najman et al. 2016). The curvature of the fold belt thus reflects the impediment to west and northward propagation imposed by the Shillong Plateau (Wang et al. 2014).

In the Ramree domain, gradual increase of seismic activity has been observed along the subducting Indian Plate towards north. The deepest hypocenters in the Wadati-Benioff zone are ~70 km deep and 140 km east of the deformation front in the south, whereas, hypocenters reach depths of ~120 km and extend to ~300 km east from the deformation front in the north. The Indian Plate and the Burmese Plate are subjected to dextral-oblique convergence across the subduction zone with rates between 14 and 26 mm/year, and the ratio of convergence to dextral slip is ~3:2 (Wang et al. 2014). In the eastern part of the Dhaka domain (i.e. beneath the IBR), a well-expressed Wadati-Benioff zone illuminates the subducting Indian oceanic lithosphere beneath the Burmese Plate (Ni et al. 1989; Guzmán-Speziale and Ni 2000). In this domain, hypocentres occur as deep as 200 km and up to 450 km east from the deformation front. Recent relative motions between the Indian Plate and the Burmese Plate suggest oblique convergence across the subduction zone beneath the CTFB and IBR. Moreover, to the northern end of Dhaka domain, convergence rate ranges 11–20 mm/year, which implies significant active right-lateral strike-slip faulting or clockwise rotations. Although the northern end of the deformation front is truncated by the Dauki Fault at the northeastern corner of the Surma basin, geomorphology shows stark contrast between Dhaka and Naga domains, whereas the nature of the higher IBR does not change markedly across the domain boundary. The Wadati-Benioff zone beneath the Naga Hills reaches ~160 km depth, and extends ~150 km southeast from the Naga mountain front. The fault trace appears from the geomorphological expression of the Naga Thrust as several arcuate lobes along the range front (Mathur and Evans 1964).

8 Summary and Conclusions

The Bengal Basin is located at the eastern part of the Indogangetic Plain and extends up to the Bengal Deep Sea Fan in the Bay of Bengal. The development of the Bengal Basin is controlled and limited by the interplay between the Eurasian, the Indian and the Burmese Plates. The E-W oriented Himalayan Mobile Belt and the N-S oriented Indo-Burman Ranges developed as the Indian Plate subducted below the Eurasian Plate in the north and below the Burmese Plate in the east, respectively with a syntexial bend conjugating the two belts in between. The geometry, structural features and sediments of the Bengal Basin furnished an archive of the past history of the area.

At the initial stage (Precambrian) only the Geotectonic Province 1 (Stable Shelf of the Bengal Basin) was part of Indian Plate when the Indian Plate was a part of the Gondwana Supercontinent. Subsequently, the Indian Plate occupied a central position in the Gondwana Supercontinent until the beginning of Middle Jurassic ($\sim$170–175 Ma) when the break-up of the Gondwana Supercontinent started. From the Middle Jurassic onward up to the end of the Palaeocene (55.9 Ma) the Indian Plate have drifted and isolated from its neighbouring cratonic blocks. During the Late Paleozoic–Mid Mesozoic time the basin developed as an intra-cratonic rift basin as evidenced by the continental Gondwana sediments deposition in the graben, half-graben structures in the Geotectonic Province 1. This episode of basin development was followed by the Kerguelen igneous activity spreaded the SE Indian Ocean. The Geotectonic Province 1 experienced widespread volcanism as continental flood basalts known as the Rajmahal Trap covered the Gondwana sediments. At that time, the peri-cratonic part on the eastern margin (the present position of the Geotectonic Province 2) might have been influenced by marine environment and also affected by this volcanic activity. Within the Geotectonic Province 2 the Barisal–Chandpur Gravity High may have been formed as a rift valley during the initial break-up of the Gondwana and formation of the Indian Plate along which transitional zone between continent-ocean crust of the Indian Plate has been marked by several phases of basalt flow. Subsequently, the Geotectonic Province 2 continuously subsided and received a massive volume of sediments from the Late Mesozoic through Tertiary to Recent times.

During the Late Cretaceous and Paleocene (between 120 and 157 Ma) the Indian Plate migrated rapidly northwards until the oceanic part of the Indian Plate collided with a Neotethyan intra-oceanic arc. With further migration, the continental crust of the Indian Plate collided with the Tibetan part of the Eurasian Plate around the Eocene-Oligocene boundary ($\sim$35 Ma) causing the northern part of Indian plate to subduct beneath the southern Tibet (initiation of continent–continent collision). This continental convergence resulted in folding and thrusting in the Himalayan region.

The Bengal Basin was an integral part of the major basin that extended from the Burmese Plate in the east to the Indian Shield in the west during the Paleocene. During this Period, the Indian Plate moved northwards and the Tethys and the young Indian Ocean were interconnected across the Upper Assam Shelf. Subsequently, northern migration of the Indian plate was replaced by north-eastward movement and resulted in the collision of north-eastern end of the Indian Plate with the Burmese Plate uplifting the Indo–Burman Ranges (IBR) by the Late Oligocene. The uplift of the IBR as an accretionary wedge produced a twin gulf that separated the Burmese Tertiary Basins on the eastern and the Bengal–Assam Basins on the western and northern flanks, respectively. These basins were fully separated with the rise of the Himalayas and IBR in the Miocene and the Geotectonic Province 2 or the central Foredeep Basin came into existence by ~23 Ma considered as the 'remnant ocean basin'.

The structural development of the Geotectonic Province 3 (CTFB) is related with the development of IBR whereas the stratigraphy exposed in the region are broadly similar to those exposed along the southern sub-Himalayan Belt. As collision and further subduction of the Indian Plate beneath the Burmese Plate in an arc-trench setting continued, it resulted in the development of accretionary prism and major east dipping thrusts (Kaladan Fault to the east, and Chittagong Coastal Fault to the west) that largely control the structural evolution of the Geotectonic Province 3. The age and intensity of folding and elevation of the structures increase in the easterly direction towards the IBR and the thrust component increases westward.

The CTFB comprises NNW-SSE trending low amplitude anticlinal and synclinal folds. The western boundary of this belt is the Chittagong Coastal Fault (CCF) and eastern boundary is Kaladan Thrust Fault. The Kaladan Thrust zone, sub-parallel to anticlines mark the tectonic boundary between CTFB of the Bengal Basin and the Indo-Burman Range. The CTFB comprises of Miocene-Recent sedimentary rock successions. During the Miocene, the depositional environments changed from marine to transitional marine. Deltaic and fluviatile conditions prevailed in the Plio-Pleistocene. Structural (folding, faulting) development in the CTFB may have started during Pliocene and is still going on.

A detail tectonic update of the Indian western part (Dasgupta and Mukherjee 2017), Himalaya (Mukherjee and Koyi 2010a, b; Mukherjee et al. 2012; Mukherjee and Mulchrone 2012; Banerjee et al. 2019; Mahato et al. 2019) coupled with such works/reviews from the eastern sector can provide a more detail tectonic evolution of India.

Acknowledgements Thanks are due to Prof. Dr. S. M. Mahbubul Ameen for discussion and suggestion during the writing of this manuscript. Dr. Soumyajit Mukherjee and an anonymous reviewer are greatly appreciated for their insightful comments and thoughtful suggestions on the initial version of this manuscript. The Springer proof-reading team and Annett Buettner are thanked for their cooperation. Mukherjee (2019) summarizes this work.

References

Abdullah R, Yeasmin R, Ameen SMM, Khanam F, Bari Z (2015) 2D structural modelling and hydrocarbon potentiality of the Sitakund structure, Chittagong Tripura Fold Belt, Bengal Basin, Bangladesh. Journal Geological Society of India 85, 697–705

Abrajevitch AV, Ali JR, Aitchison JC, Badengzhu, Davis AM, Liu J, Ziabrev SV (2005) Neotethys and the India-Eurasia collision: insights from a palaeomagnetic study of the Dazhuqu ophiolite, southern Tibet. Earth and Planetary Science Letters 233, 87–102

Acharyya SK (1994) The Cenozoic foreland basin and tectonics of the Eastern Himalaya: problems and perspectives. Himalayan Geology 15, 3–21

Acharyya SK (1998) Break-up of the greater Indo-Australian continent and accretion of blocks framing South and East Asia. Journal of Geodynamics 26, 149–170

Acharyya SK (2003) A plate tectonic model for Proterozoic crustal evolution of central Indian tectonic zone. Gondwana Geological Magazine 7, 9–31
Acharyya SK (2007) Collisional emplacement history of the Naga-Andaman ophiolites and the position of the eastern Indian suture. Journal of Asian Earth Sciences 29, 229–242
Acharyya SK, Mitra ND, Nandy DR (1986) Regional geology and tectonic setting of Northeast India and adjoining region. Memoir of the Geological Survey of India 119, 6–12
Achyuthan H, Nagasundaram M, Gourlan AT, Eastoe C, Ahmad SM, Padmakumari VM (2014) Mid-Holocene Indian summer monsoon variability off the Andaman Islands, Bay of Bengal. Quaternary International 349, 232–244
Acton GD (1999) Apparent polar wander of India since the Cretaceous with implications for regional tectonics and true polar wander. In: Radhakrishna T, Piper JDA (eds) The Indian subcontinent and Gondwana: a palaeomagnetic and rock magnetic perspective. Memoir of Geological Society of India, vol 44, pp 129–175
Ahmed ST (1968) Cenozoic Fauna of the Cox's Bazar coastal cliff. Unpublished M.Sc. thesis, University of Dhaka, Dhaka, 68 p
Aier I, Luirei K, Bhakuni S, Thong GT, Kothyari GC (2011) Geomorphic evolution of Medziphema intermontane basin and quaternary deformation in the Schuppen belt, Nagaland, NE India. Geomorphology 55, 247–265
Aitchison JC, Ali JR, Davis AM (2007) When and where did India and Asia collide? Journal of Geophysical Research 112, B05423
Akhter SH (2010) Earthquakes of Dhaka. In: Islam MA, Ahmed SU, Rabbani AKMG (eds) Environment of capital Dhaka: Plants wildlife gardens park open spaces air water earthquake. Asiatic Society of Bangladesh, pp 401–426
Akhter SH, Steckler MS, Seeber L, Agostinetti NP, Kogan MG (2010) GPS velocities and structure across the Burma accretionary prism and Shillong Plateau in Bangladesh. AGU Fall Meeting Supplement. Abstract T22A-08
Alam M (1972) Tectonic classification of the Bengal Basin. Geological Society of America Bulletin 83, 519–522
Alam M (1989) Geology and depositional history of Cenozoic sediments of the Bengal Basin of Bangladesh. Palaeogeography, Palaeoclimetology, Palaeoecology 69, 125–139
Alam MM (1991) Paleoenvironmental study of the Barail succession exposed in northeastern Sylhet, Bangladesh. Bangladesh Journal of Scientific Research 9, 25–32
Alam M (1997) Bangladesh. In: Moores EM, Fairbridge RW (eds) Encyclopedia of European and Asian regional geology. Chapman & Hall, London, pp 64–72
Alam AKMK, Islam MB (2017) Recent changes in Jadukata fan (Bangladesh) in response to Holocene tectonics. Quaternary International 462, 226–235
Alam M, Alam MM, Curray JR, Chowdhury MLR, Gani MR (2003) An overview of the sedimentary geology of the Bengal Basin in relation to the regional tectonic framework and basin-fill history. Sedimentary Geology 155, 179–208
Ali JR, Aitchison JC (2004) Problem of positioning Palaeogene Eurasia: a review, efforts to resolve the issue, implications for the India–Asia collision. In: Clift PD, Wang P, Khunt W, Hayes DE (eds) Continent–ocean interactions within the East Asia Marginal Seas. AGU Monograph 149, 23–35
Ali JR, Aitchison JC (2005) Greater India. Earth-Science Reviews 72, 169–188
Ali JR, Aitchison JC (2006) Positioning Paleogene Eurasia problem: solution for 60–50 Ma and broader tectonic implications. Earth Planetary Science and Letters 251, 148–155
Ali JR, Aitchison JC (2008) Gondwana to Asia: plate tectonics, paleogeography and the biological connectivity of the Indian sub-continent from the middle jurassic through latest Eocene (166–35 Ma). Earth Science Reviews 88, 145–166
Ali SMM, Raghava MSV (1985) The Bouguer gravity map of Bangladesh and its tectono-geologic implications. Bangladesh Journal of Geology 4, 43–56

Allégre CJ, Courtillot V, Tapponnier P, Hirn A, Mattauer M, Coulon C, Jaeger JJ, Achache J, Schärer U, Marcoux J, Burg JP, Girardeau J, Armijo R, Garie´ py C, Göpel C, Li TD, Xiao XC, Chang CF, Li GQ, Lin BY, Teng JW, Wang NW, Chen GM, Han TL, Wang XB, Den WM, Sheng HB, Cao YG, Zhou J, Qiu HR, Bao PS, Wang SC, Wang BX, Zhou YX, and Xu RH (1984) Structure and evolution of the Himalayan-Tibet orogenic belt. Nature, 307, 17–22

Ameen SMM, Khan MSH, Akon E, Kazi AI (1998) Petrography and major oxide chemistry of some Precambrian crystalline rocks from Maddhapara, Dinajpur. Bangladesh Geoscience Journal 4, 1–19

Ameen SMM, Wilde SA, Kabir MZ, Akon E, Chowdhury KR, Khan MSH (2007) Paleoproterozoic granitoids in the basement of Bangladesh: a piece of the Indian Shield or an exotic fragment of the Gondwana jigsaw? Gondwana Research 12, 280–387

Ameen SMM, Hossain MS, Hossain MS, Abdullah R, Bari Z, Uddin MN, Das SC, Tapu AT, Jahan H, Shahriar MS (2016) Deciphering the Precambrian crust between Shillong Massif and northeast India. In: International Association of Gondwana Research (IAGR) 2016 Convention, Trivandum, India. Abstract volume, pp 75–79

Angelier J, Baruah S (2009) Seismotectonics in Northeast India: a stress analysis of focal mechanism solutions of earthquakes and its kinematic implications. Geophysical Journal International 178, 303–326

Ansary MA, Al-Hussaini TM, Sharfuddin M (2000) Damage assessment of July 22, 1999 Moheshkhali earthquake, Bangladesh. Paper presented at 8th ASCE specialty conference on probabilistic mechanics and structural reliability. Indiana, USA, 6 p

Bage AK, Maurya VP, Shalivahan VP, Singh S, Tripathi A (2014) Preliminary magnetotellurics results over Rajmahal traps. Extended Abstract, 22nd EM induction workshop, Weimar, Germany, pp 1–4

Bakhtine MI (1966) Major tectonic features of Pakistan: part II. The Eastern Province. Science & Industry 4, 89–100

Bakr MA (1977) Quaternary geomorphic evolution of the Brahmanbaria–Noakhali area, Comilla and Noakhali districts, vol 1. Records of the Geological Survey of Bangladesh, Bangladesh, p 44

Baksi AK (1995) Petrogenesis and timing of volcanism in the Rajmahal flood basalt province, northeastern India. Chemical Geology 121, 73–90

Baksi AK, Barman T, Paul D, Farrar E (1987) Widespread Early Cretaceous flood basalt volcanism in eastern India: geochemical data from the Rajmahal–Bengal–Sylhet Traps. Chemical Geology 63, 133–141

Ball V (1877) Geology of the Rajmahal Hills. Memoir of the Geological Survey of India 13, 155–248

Banerjee B, Sengupta BJ, Banerjee PK (1995) Signals of Barremian (116 Ma) or younger oceanic crust beneath the Bay of Bengal along 14°N latitude between 81°E and 93°E. Marine Geology 128, 17–23

Banerjee P, Burgmann R, Nagarajan B, Apel E (2008) Intraplate deformation of the Indian subcontinent. Geophysical Research Letters 35, L18301

Banerjee S, Bose N, Mukherjee S (2019) Field structural geological studies around Kurseong, Darjeeling-Sikkim Himalaya, India. In: Mukherjee S (ed) Tectonics and structural geology: Indian context. Springer International Publishing AG, Cham, pp 425–440. ISBN: 978-3-319-99340-9

Banerji RK (1979) Disang Shale, its stratigraphy, sedimentation history and basin configuration in northeastern India and Burma. Eighth Colloquium Indian Micropalaeontology and Stratigraphy. Baroda, India, 12 p

Banerji RK (1981) Cretaceous-Eocene sedimentation, tectonism and biofacies in the Bengal Basin, India. Palaeogeography, Palaeoclimetology, Palaeoecology 34, 57–85

Barman P, Jade S, Shrungeshwara TS, Kumar A, Bhattacharyya S, Ray JD, Jagannathan S, Jamir WM (2016) Crustal deformation rates in Assam Valley, Shillong Plateau, Eastern Himalaya, and Indo-Burmese region from 11 years (2002–2013) of GPS measurements. International Journal of Earth Sciences 106, 2025–2038

Baruah S, Hazarika D (2008) A GIS based tectonic map of north eastern India. Current Science 95, 176–177

Baruah S, Saikia S, Baruah S, Bora PK, Tatevossian R, Kayal JR (2016) The september 2011 Sikkim Himalaya earthquake Mw 6.9: is it a plane of detachment earthquake? Geomatics, Natural Hazards and Risk 7, 248–263
Bastia R, Radhakrishna M, Das S, Kale AS, Catuneanu C (2010) Delineation of the 85°E ridge and its structure in the Mahanadi Offshore Basin, Eastern Continental Margin of India (ECMI), from seismic reflection imaging. Marine and Petroleum Geology 27, 1841–1848
Baxter AT, Aitchison JC, Ali JR, Chan JS-L, Chan GHN (2016) Detrital chrome spinel evidence for a Neotethyan intra-oceanic island arc collision with India in the Paleocene. Journal of Asian Earth Sciences 128, 90–104
Bender F (1983) Geology of Burma. Borntraeger, Berlin, 260 p
Berthet T, Ritz J-F, Ferry M, Pelgay P, Cattin R, Drukpa D, Braucher R, Hetényi G (2014) Active tectonics of the eastern Himalaya: new constraints from the first tectonic geomorphology study in southern Bhutan. Geology 42, 427–430
Bhattacharya PM, Mukhopadhyay S, Majumdar RK, Kayal JR (2008) 3-D seismic structure of the northeast India region and its implications for local and regional tectonics. Journal of Asian Earth Sciences 33, 25–41
Bhuiyan AH, Hossain MS (2006) Groundwater constrains and opportunity: a key for future urban planning of Dhaka City, Bangladesh. Journal of Environmental Science (Dhaka) 4, 63–75
Bilham R (2004) Earthquakes in India and the Himalaya: tectonics, geodesy and history. Annales Geophysicae 47, 839–858
Bilham R, England PC (2001) Plateau “pop-up” in the great 1897 Assam earth-quake. Nature 410, 806–809
Bilham R, Mencin D, Bendick R, Bürgmann R (2017) Implications for elastic energy storage in the Himalaya from the Gorkha 2015 earthquake and other incomplete ruptures of the Main Himalayan Thrust. Quaternary International 462, 3–21
Biswas B (1961) Geology of the Bengal Basin with special reference to stratigraphy and micropaleontology. Dissretation, University of Calcutta, 138 p
Biswas B (1963) Results of exploration of petroleum in western part of the Bengal basin, India. In: Proceedings of the 2nd symposium on development of petroleum resources ECAFE, mineral resource development series no. 18, 241–250. Flood basalt province, northeast India. Chemical Geology 121, 133–141
Biswas S, Das Gupta A (1989) Distribution of stresses in the Himalayan and the Burmese arcs. Gerlands Beiträge zur Geophysik 98, 223–239
Biswas S, Grasemann B (2005a) Quantitative morphotectonics of the southern Shillong Plateau (Bangladesh/India). Australian Journal of Earth Science 97, 82–93
Biswas S, Grasemann B (2005b) Structural Modelling of the Subsurface Geology of the Sylhet Trough, Bengal Basin. Bangladesh Geoscience Journal 11, 19–33
Biswas S, Majumdar RK (1997) Seismicity and tectonics of the Bay of Bengal: evidence for intraplate deformation of the northern Indian plate. Tectonophysics 269, 323–336
Biswas S, Coutand I, Grujic D, Hager C, Stockli D, Grasemann B (2007) Exhumation and uplift of the Shillong Plateau and its influence on the eastern Himalayas: new constraints from apatite and zircon (U–Th–[Sm])/He and apatite fission track analysis. Tectonics 26, TC6013
Biswas S, Majumdar RK, Das Gupta A (1992) Distribution of stress axes orientation in the Andaman-Nicobar island region: a possible stress model and its significance for extensional tectonics of the Andaman Sea. Physics of Earth and Planetary Interior 70, 57–63
Blaser L, Krüger F, Ohrnberger M, Scherbaum F (2010) Scaling relations of earthquake source parameter estimates with special focus on subduction environment. Bulletin of Seismological Society of America 100, 2914–2926
BOGMC (1997) Petroleum exploration opportunities in Bangladesh. Bangladesh Oil, Gas and Mineral Corporation (Petrobangla), Dhaka, March, 1997
Bosch D, Garrido CJ, Bruguier O, Dhuime B, Bodinier J-L, Padròn-Navarta JA, Galland B (2011) Building an island-arc crustal section: time constraints from a LA-ICP-MS zircon study. Earth and Planetary Science Letters 309, 268–279

Bose N, Mukherjee S (submitted-a) Field documentation and genesis of the back-structures from a part of the Garhwal Lesser Himalaya, Uttarakhand, India: Tectonic implications. In: Sharma, Villa IM, Kumar S (eds) Crustal architecture and evolution of the Himalaya-Karakoram-Tibet Orogen. Geological Society of London Special Publications

Bose N, Mukherjee S (submitted-b) Field documentation and genesis of back-structures from the foreland part of a collisional orogen: Examples from the Darjeeling-Sikkim Lesser Himalaya, Sikkim, India. Gondwana Research

Bracciali L, Najman Y, Parrish RR, Akhter SH, Millar I (2015) The Brahmaputra tale of tectonics and erosion: early Miocene river capture in the Eastern Himalaya. Earth and Planetary Science Letters 415, 25–37

Brammer H (2012) The physical geography of Bangladesh. University Press Ltd., Dhaka, p 547

Brammer H (2014) Bangladesh's dynamic coastal regions and sea-level rise. Climate Risk Management 1, 51–62

Bulbul MAU (2015) Neotectonics of the Surma Basin, Bangladesh from GPS analysis. AGU Fall Meeting Supplement. Abstract T41B-2884

Burgess PW, Yin A, Dubey CS, Shen Z-K, Kelty TK (2012) Holocene shortening across the main frontal thrust zone in the eastern Himalaya. Earth and Planetary Science Letters 357–358, 152–167

CDMP II (2013) Report of active fault mapping in Bangladesh: paleo-seismological study of the Dauki Fault and the Indian-Burman plate boundary fault. Comprehensive Disaster Management Programme (CDMP II), Ministry of Disaster Management and Relief, Bangladesh, 67 p

Chatterjee N (2017) Constraints from monazite and xenotime growth modelling in the MnCKFMASH-PYCe system on the *P–T* path of a metapelite from Shillong-Meghalaya Plateau: implications for the Indian shield assembly. Journal of Metamorphic Geology 35, 393–412

Chauhan OS, Vogelsang E (2006) Climate induced changes in the circulation and dispersal patterns of the fluvial sources during late quaternary in the middle Bengal Fan. Journal Earth System and Science 115, 379–386

Chopra S, Sharma J, Sutar A, Bansal BK (2013) Estimation of source parameters of M_w 6.9 Sikkim Earthquake and modeling of ground motions to determine causative fault. Pure and Applied Geophysics 171, 1311–1328

Chowdhury KR (1970) Strcture and petrology of the Tertiary sedimentary sequence of the northern part of the Changotaung Hill Range, Chittagong Hill Tracts. M.Sc. thesis, University of Dhaka, 192 p

Chowdhury KR, Biswas S, Ahmed AMM (1996) The structural and tectonic set-up of Jaintiapur and adjacent areas, Sylhet District, Bangladesh. Bangladesh Geoscience Journal 2, 1–14

Chowdhury KR, Huq NE, Ibna-Hamid ML, Ameen SMM (2003) Evidence of Mud intrusion into the deposits of the Surma Group, Bengal Basin, Bangladesh. Bangladesh Geoscience Journal 9, 173–178

Chowdhury SQ (1982) Palynostratigraphy of the Neogene sediments of the Sitapahar anticline (western flank), Chittagong Hill Tracts, Bangladesh. Bangladesh Journal of Geology 1, 35–49

Coffin MF, Pringle MS, Duncan RA, Gladczenko TP, Storey M, Müller RD, Gahagan LA (2002) Kerguelen hotspot magma output since 130 Ma. Journal of Petrology 43, 1121–1137

Cummins PR (2007) The potential for giant tsunamigenic earthquakes in the northern Bay of Bengal. Nature 449, 75–78

Curiale JA, Covington GH, Shamsuddin AHM, Morelos JA, Shamsuddin AKM (2002) Origin of petroleum in Bangladesh. AAPG Bulletin 86, 625–652

Curray JR (1991) Possible green schist metamorphism at the base of a 22 km sedimentary section, Bay of Bengal. Geology 19, 1097–1100

Curray JR (1994) Sediment volume and mass beneath the Bay of Bengal. Earth and Planetary Science Letters 125, 371–383

Curray JR (2014) The Bengal depositional system: from rift to orogeny. Marine Geology 352, 59–69

Curray JR, Moore DG (1971) Growth of the Bengal deep-sea fan and denudation in the Himalayas. Geological Society of America Bulletin 82, 563–572
Curray JR, Moore DG (1974) Sedimentary and Tectonic processes in the Bengal deep-sea fan and geosyncline. In: Burk CA, Drake CL (eds) The geology of continental margins. Springer, Berlin, Heidelberg
Curray JR, Munasinghe T (1991) Origin of the Rajmahal Traps and the 85°E ridge: preliminary reconstructions of the trace of the Crozet hotspot. Geology 19, 1237–1240
Curray JR, Emmel FJ, Moore DG (2003) The Bengal Fan: morphology, geometry, stratigraphy, history and processes. Marine and Petroleum Geology 19, 1191–1223
Das Gupta AB (1977) Geology of Assam-Arakan region. Oil Commentary, India 15, 4–34
Das JD, Saraf AK, Shujat Yazdana (2010) A remote sensing technique for identifying geometry and geomorphological features of the Indo-Burman frontal fold belt. International Journal of Remote Sensing 31, 4481–4503
Das JD, Shujat Y, Saraf AK (2011a) Spatial technologies in deriving the morphotectonic characteristics of tectonically active western Tripura region, Northeast India. Journal of Indian Society of Remote Sensing 39, 249–258
Das JD, Shujat Y, Saraf AK, Rawat V, Sharma K (2011b) Morphotectonic features and fault propagation folding of Bhuban Hills, NE India using satellite image and DEM. Journal of Indian Society of Remote Sensing 39, 73–81
Dasgupta S, Mukherjee S (2017) Brittle shear tectonics in a narrow continental rift: asymmetric non-volcanic Barmer basin (Rajasthan, India). Journal of Geology 125, 561–591
Dasgupta T, Mukherjee S (submitted) Sediment compaction and applications in petroleum geoscience. In: Swenner R (ed) Springer series: advances in oil and gas exploration & production. ISSN: 2509-372X
DeCelles PG (2012) Foreland basin systems revisited: variations in response to tectonic settings. In: Busby C, Pérez A (eds) Tectonics of sedimentary basins: recent advances. Wiley, Chichester, UK, pp 405–426
DeCelles PG, Kapp P, Gehrels GE, Ding L (2014) Paleocene-Eocene foreland basin evolution in the Himalaya of southern Tibet and Nepal: implications for the age of initial India-Asia collision. Tectonics 33, 824–849
DeCelles PG, Robinson DM, Zandt G (2002) Implications of shortening in the Himalayan fold-thrust belt for uplift of the Tibetan Plateau. Tectonics 21, 1062
Desa MA, Ramana MV, Ramprasad T, Anuradha M, Lall MV, Kumar BJP (2013) Geophysical signatures over and around the northern segment of the 85°E Ridge, Mahanadi offshore, Eastern Continental Margin of India: tectonic implications. Journal of Asian Earth Sciences 73, 460–472
Desikachar SV (1974) A review of the tectonic and geologic history of eastern India in terms of plate tectonics theory. Geological Society of India Journal 15, 137–149
Dietz RS (1953) Possible deep-sea turbidity-current channels in the Indian Ocean. Geological Society of America Bulletin 64, 375–378
Dina NT, Rahman MJJ, Hossain MS, Sayem ASM (2016) Provenance of the Neogene succession in the Bandarban structure, South-East Bengal Basin, Bangladesh: insights from petrography and petrofacies. Himalayan Geology 37, 141–152
Emmel FJ, Curray JR (1985) Bengal Fan, Indian Ocean. In: Bouma AH, Barnes NE, Normark WR (eds) Submarine fans and related turbidite sequences. Springer, New York, pp 107–112
Evans P (1932) Tertiary succession in Assam. Transactions of the Mining and Geological Institute of India 27, 155–260
Evans P (1964) The tectonic framework of Assam. Journal Geological Society of India 5, 80–96
Farhaduzzaman M, Abdullah WH, Islam MA (2014) Hydrocarbon source potential and depositional environment of the Surma Group shales of Bengal basin, Bangladesh. Journal of Geological Society of India 83, 433–446
Fergusson J (1863) On recent changes in the delta of the Ganges. Quarterly Journal of the Geological Society 19, 321–354

Fournier L, Fauquembergue K, Zaragosi S, Zorzi C, Malaizé B, Bassinot F, Joussain R, Colin C, Moreno E, Leparmentier F (2016) The Bengal fan: external controls on the Holocene active channel turbidite activity. The Holocene 27, 900–913

Gahalaut VK, Kundu B, Laishram SS, Catherine J, Kumar A, Singh MD, Tiwari R, Chadha R, Samanta S, Ambikapathy A (2013) Aseismic plate boundary in the Indo-Burmese wedge, northwest Sunda Arc. Geology 41, 235–238

Ganguly S (1984) Tectonic evolution of the orogenic belt of Tripura. Quarternary Journal of Geological Mining and Metal Society India 56, 128–137

Ganguly S (1993) Stratigraphy, sedimentation and hydrocarbon prospects of Tertiary succession of Tripura and Cachar. Indian Journal of Geology 65, 145–180

Ganguly S (1997) Petroleum geology and exploration history of the Bengal basin in India and Bangladesh. Indian Journal of Geology 69, 1–25

Gani MR, Alam MM (1999) Trench-slope controlled deep-sea clastics in the exposed lower Surma group in the southeastern fold belt of the Bengal Basin, Bangladesh. Sedimentary Geology 127, 221–236

Gani MR, Alam MM (2003) Sedimentation and basin-fill history of the Neogene clastic succession exposed in the southeastern fold belt of the Bengal Basin, Bangladesh: a high-resolution sequence stratigraphic approach. Sedimentary Geology 155, 227–270

Garzanti E, Baud A, Mascle G (1987) Sedimentary record of the northward flight of India and its collision with Eurasia (Ladakh Himalaya, India). Geodinamica Acta 1, 297–312

Ghatak A, Basu AR (2011) Vestiges of the Kereguelen plume in the Sylhet Traps, northeastern India. Earth and Planetary Science Letters 308, 52–64

Ghose NC, Chatterjee N, Windley BF (2017) Subaqueous early eruptive phase of the Late Aptian Rajmahal volcanism, India: evidence from volcaniclastic rocks, bentonite, black shales, and oolite. Geoscience Frontiers 8, 809–822

Ghosh GK, Dasgupta R, Reddy BJ, Singh SN (2015) Gravity data interpretation across the Brahmaputra Thrust and Dauki Fault in the north-eastern India. Journal of Geophysics XXXVI, 31–38

Gilbert OE Jr (2001) Structural geology and regional tectonics of the Chittagong Hills Fold Belt, Eastern Bangladesh, GSA Annual Meeting, Houston, Texas, Abstract 21770

Goodbred SL Jr, Kuehl SA (1998) Floodplain processes in the Bengal Basin and the storage of Ganges-Brahmaputra river sediment: an accretion study using ^{137}Cs and ^{210}Pb geochronology. Sedimentary Geology 121, 239–258

Goodbred SL Jr, Kuehl SA (1999) Holocene and modern sediment budgets for the Ganges Brahmaputra river system: evidence for highstand dispersal to floodplain, shelf and deep-sea depocenters. Geology 27, 559–562

Goodbred SL Jr, Kuehl SA, Steckler MS, Sarkar MH (2003) Controls on facies distribution and stratigraphic preservation in the Ganges-Brahmaputra delta sequence. Sedimentary Geology 155, 301–316

Grimaud J-L, Paola C, Ellis C (2017) Competition between uplift and transverse sedimentation in an experimental delta. Journal of Geophysical Research: Earth Surface 122, 1339–1354

Guha DK (1978) Tectonic framework and oil & gas prospects of Bangladesh. In: Proceedings of the 4th annual conference Bangladesh Geological Society, pp 65–75

Guha DK, Henkel H, Imam B (2010) Geothermal potential in Bangladesh—results from investigations of abandoned deep wells. In: Proceedings world geothermal congress 2010, Bali, Indonesia, pp 1–8

Guzmán-Speziale M, Ni JF (2000) Comment on "Subduction in the Indo-Burma Region: is it still active?" by SP Satyabala. Geophysical Research Letters 27, 1065–1066

Hall R (2002) Cenozoic geological and plate tectonic evolution of SE Asia and the SW Pacific: computer-based reconstructions, model and animations. Journal of Asian Earth Scinces 20, 353–431

Higgins SA, Overeem I, Steckler MS, Syvitski JPM, Seeber L, Akhter SH (2014) InSAR measurements of compaction and subsidence in the Ganges-Brahmaputra Delta, Bangladesh. Journal of Geophysical Research Earth Surface 119, 1768–1781

Hiller K, Elahi M (1984) Structural development and hydrocarbon entrapment in the Surma Basin. In: Bangladesh (northwest Indo-Burman fold belt): Fifth Offshore Southwest Conference, Singapore, pp 656–663

Hiller K, Elahi M (1988) Structural growth and hydrocarbon entrapment in the Surma basin, Bangladesh. In: Wagner HC, Wagner LC, Wang FFH, Wong FL (eds) Petroleum resources of China and related subjects, Houston, Texas. Circum-Pacific council for energy and mineral resources earth science series, vol 10, pp 657–669

Hodges KV (2000) Tectonics of the Himalaya and southern Tibet from two perspectives. GSA Bulletin 112, 324–350

Holtrop JF, Keizer J (1970) Some aspects of the stratigraphy and correlation of the Surma Basin wells, East Pakistan. In: ECAFE mineral resources development series, vol 36. United Nations, New York, pp 143–154

Hoque MA, Hoque MM, Ahmed KM (2007) Declining groundwater level and aquifer dewatering in Dhaka metropolitan area, Bangladesh: causes and quantification. Hydrogeology Journal 15, 1523–1534

Hossain A, Steckler MS, Akhter SH (2015) Resistivity imaging of strata and faults in Bangladesh. AGU Fall Meeting Supplement. Abstract T41B-2886

Hossain I, Tsunogae T, Rajesh HM, Chen B, Arakawa Y (2007) Palaeoproterozoic U-Pb SHRIMP zircon age from basement rocks in Bangladesh: a possible remnant of the Columbia supercontinent. Comptes Rendus Geosciences 339, 979–986

Hossain I, Tsunogae T, Tsutsumi Y, Takahashi K (2017) Petrology, geochemistry and LA-ICP-MS U-Pb geochronology of Paleoproterozoic basement rocks in Bangladesh: an evaluation of calc-alkaline magmatism and implication for Columbia supercontinent amalgamation. Journal of Asian Earth Sciences 157, 22–39

Hossain KM (1985) Plate tectonic theory and the development of the folds of Chittagong and Chittagong Hill Tracts. Bangladesh Journal of Geology 4, 57–67

Hossain KM, Akhter SH (1983) Structural behaviour of the Sitakund Hill range. Bangladesh Journal of Geology 2, 17–27

Hossain MS, Chowdhury KR, Khan MSH, Abdullah R (2016) Geotectonic settings of the Dauki fault—a highly potential source for a significant seismic threat. In: Kruhl JH (ed) International conference Humboldt Kolleg on living under threat of earthquake—Kathmandu, Nepal. Abstract volume, p 25

Hossain MS, Islam MM, Islam O, Chowdhury KR, Khan MSH (2018) Deformation characteristics of the Chittagong Tripura fold belt—an insight from the geometrical analysis of folded structures (in preparation)

Hossain MS, Khan MSH, Chowdhury KR, Afrooz M (2014) Morpho-structural classification of the Indo-Burman ranges and the adjacent regions. In: National conference on rock deformation & structures (RDS-III), Assam, India. Abstract volume, p 31

Ingersoll RN, Graham SA, Dickinson WR (1995) Remnant ocean basins. In: Busby CJ, Ingersoll RV (eds) Tectonics of sedimentary basins. Blackwell, Oxford, pp 363–391

Ismail M (1978) Stratigraphical position of Bogra limestone of the platform area of Bangladesh. In: Proceedings of the 4th annual conference of the Bangladesh geological society, pp 19–26

Johnson SY, Alam AMN (1991) Sedimentation and tectonics of the Sylhet trough, Bangladesh. Geological Society of America Bulletin 103, 1513–1527

Kabir ASS, Hossain D (2009) Geophysical interpretation of the Rashidpur structure Surma Basin, Bangladesh. Journal of Geological Society of India 74, 39–48

Kabir MZ, Chowdhury KR, Akon E, Kazi AI, Ameen SMM (2001) Petrogenetic study of Precambrian basement rocks from Maddhapara, Dinajpur, Bangladesh. Bangladesh Geoscience Journal 7, 1–18

Kayal JR (1998) Seismicity of northeast India and surroundings—development over the past 100 years. Journal of Geophysics 19, 9–34

Kayal JR (2008) Microearthquake seismology and seismotectonics of South Asia. Springer, Dordrecht, p 449

Kayal JR (2014) Seismotectonics of the great and large earthquakes in Himalaya. Current Science 106, 188–197

Kayal JR, Arefiev SS, Barua S, Hazarika D, Gogoi N, Kumar A, Chowdhury SN, Kalita S (2006) Shillong plateau earthquakes in northeast India region: complex tectonic model. Current Science 91, 109–114

Kent RW, Saunders AD, Kempton PD, Ghose NC (1997) Rajmahal Basalts, Eastern India: Mantle sources and melt distribution at a volcanic rifted margin. In: Mahoney JJ, Coffin MF (eds) Large igneous provinces: continental, oceanic, and planetary flood volcanism. American Geophysical Union, Washington, DC

Kent RW, Pringle MS, Muller RD, Saunders AD, Ghose NC (2002) $^{40}Ar/^{39}Ar$ geochronology of the Rajmahal Basalts, India, and their relationship to the Kerguelen Plateau. Journal of Petrology 43, 1141–1153

Khan AA (1991a) Tectonics of the Bengal Basin. Journal of Himalayan Geology 2, 91–101

Khan AA, Agarwal BNP (1993) The crustal structure of western Bangladesh from gravity data. Tectonophysics 219, 341–353

Khan AA, Chouhan RKS (1996) The crustal dynamics and the tectonic trends in the Bengal Basin. Journal of Geodynamics 22, 267–286

Khan AA, Rahman T (1992) An analysis of gravity field and tectonic evaluation of the northwestern part of Bangladesh. Tectonophysics 206, 351–364

Khan AA, Sattar GS, Rahman T (1994) Tectogenesis of the Gondwana rifted basins of Bangladesh in the so-called Garo-Rajmahal gap and their pre-drift regional tectonic correlation. In: Proceedings 9th international Gondwana symposium. Geological Survey of India, Calcutta, India

Khan FH (1991b) Geology of Bangladesh. The University Press Limited, Dhaka, 203 p

Khan MAM (1978) Geology of the eastern and northeastern parts of Sadar subdivision, Sylhet district. Records of the Geological Survey of Bangladesh, II(IV), Bangladesh

Khan, MAM, Ismail M, Ahmad M (1988) Geology and hydrocarbon prospects of the Surma Basin. In: Seventh offshore southeast Asia conference, Singapore, Bangladesh, pp 364–387

Khan MSH, Biswas S, Singh S, Pati P (2006) OSL choronology of Dihing formation and recent upliftment rate along the Dauki Fault, NE Bangladesh. Bangladesh Geoscience Journal 12, 1–11

Khan MSH, Haque MM, Pati P, Chowdhury KR, Biswas S (2015) OSL derived uplift rate of Dakhin Nhila anticline along the southeastern coast of the Bay of Bengal, Bangladesh. Himalayan Geology 36, 143–152

Khan MSH, Hossain MS, Chowdhury KR (2017) Geomorphic Implications and active tectonics of the Sitapahar Anticline – CTFB, Bangladesh. Bangladesh Geoscience Journal 23, 1–20

Khan MSH, Hossain MS, Uddin MA (2018) Geology and active tectonics of the Lalmai Hill area—overview from Chittagong Tripura fold belt perspective. Journal of Geological Society India (accepted)

Khan MSH, Parkash B, Kumar S (2005) Soil–landform development of a part of the fold belt along the eastern coast of Bangladesh. Geomorphology 71, 310–327

Khan MSH, Saha RK, Akhter S (2011) Soil geomorphic evolution of Shitalakhya-Dhaleswari interfluve. Bangladesh Geoscience Journal 17, 53–72

Khanam F, Rahman MJJ, Alam MM, Abdullah R (2017) Facies characterization of the Surma Group (Miocene) sediments from Jalalabad gas field, Sylhet Trough, Bangladesh: study from cores and wireline log. Journal of Geological Society of India 89, 155–164

Khandoker RA (1989) Development of major tectonic elements of the Bengal Basin: a plate tectonic appraisal. Bangladesh Journal of Scientific Research 7, 221–232

Klootwijk CT, Gee JS, Peirce JW, Smith GM, Mcfadden PL (1992) A early India-Asia contact: palaeomagnetic constraints from Ninetyeast Ridge, ODP Leg 121. Geology 20, 395–398

Kuehl SA, Hairu TM, Moore WS (1989) Shelf sedimentation off the Ganges–Brahmaputra river system—evidence for sediment bypassing to the Bengal Fan. Geology 17, 1132–1135

Kumar S, Rino V, Hayasaka Y, Kimura K, Raju S, Terada K, Pathak M (2017) Contribution of Columbia and Gondwana supercontinent assembly- and growth-related magmatism in the

evolution of the Meghalaya Plateau and the Mikir Hills, Northeast India: constraints from U-Pb SHRIMP zircon geochronology and geochemistry. Lithos 277, 356–375
Kundu B, Gahalaut VK (2012) Earthquake occurrence processes in the Indo-Burmese wedge and Sagaing fault region. Tectonophysics 524–525, 135–146
Lietz JK, Kabir J (1982) Prospects and constraints of oil exploration in Bangladesh. In: Proceedings of the 4th offshore Southeast Asia conference, Singapore, pp 1–6
Lindsay JF, Holliday DW, Hulbert AG (1991) Sequence stratigraphy and the evolution of the Ganges–Brahmaputra Delta complex. AAPG Bulletin 75, 1233–1254
Lohmann HH (1995) On the tectonics of Bangladesh. Swiss Association of Petroleum Geologists and Engineers Bulletin 62, 29–48
Long S, McQuarrie N, Tobgay T, Grujic D, Hollister L (2011) Geological map of Bhutan. Journal of Maps 7, 184–192
Mahato S, Mukherjee S, Bose N (2019) Documentation of brittle structures (back shear and arc-parallel shear) from Sategal and Dhanaulti regions of the Garhwal Lesser Himalaya (Uttarakhand, India). In: Mukherjee S (ed) Tectonics and structural geology: Indian context. Springer International Publishing AG, Cham, pp 411–423. ISBN: 978-3-319-99340-9
Mandal BC, Woobidullah ASM, Guha DK (2004) Structural style analysis of the Semutang anticline, Chittagong Hill tracts, Eastern Fold Belt of the Bengal Basin, Bangladesh. Journal of Geological Society India 64, 211–222
Mannan A (2002) Stratigraphic evolution and geochemistry of the Neogene Surma Group, Surma Basin, Sylhet, Bangladesh. Published doctoral dissertation. Department of Geology, University of Oulu, p 190. ISBN: 951-42-6711-7
Martin S, Szeliga W (2010) A catalog of felt intensity data for 589 earthquakes in India, 1636–2008. Bulletin of the Seismological Society of America 100, 562–569
Mathur LP, Evans P (1964) Oil in India. 22nd session international geological congress proceedings. New Delhi, India
Matin A, Misra S (2009) Repeated cataclasis in a reactivated fault zone—an example from Bengal Basin Margin Fault, Jharkhand, India. Journal of Virtual Explorer 32
Matin MA, Fariduddin M, Hussain MMT, Khan MAM, Boul MA, Kononov AI (1986) New concepts on the tectonic zonation of the Bengal Foredeep. In: 6th offshore South East Asia conference and exhibition, 28–31 Jan 1986, pp 51–54, Singapore
Matin MA, Khan MAM, Fariduddin M, Boul MA, Hossain MMT, Kononov AI (1983) The tectonic map of Bangladesh—past & present. Bangladesh Journal of Geology 2, 29–36
Maurin T, Rangin C (2009a) Structure and kinematics of the Indo-Burmese Wedge: recent and fast growth of the outer wedge. Tectonics 28, TC2010
Maurin T, Rangin C (2009b) Impact of the 90°E ridge at the Indo-Burmese subduction zone imaged from deep seismic reflection data. Marine Geology 266, 143–155
McDougall I, McElhinny MW (1970) The Rajmahal traps of India—KAr ages and palaeomagnetism. Earth and Planetary Science Letters 9, 371–378
Metcalfe I (2013) Gondwana dispersion and Asian accretion: tectonic and palaeogeographic evolution of eastern Tethys. Journal of Asian Earth Sciences 66, 1–33
Michael L, Krishna KS (2011) Dating of the 85°E ridge (northeastern Indian Ocean) using marine magnetic anomalies. Current Science 100, 1314–1322
Milliman JD, Rutkowski C, Meybeck M (1995) River discharge to sea: a global river index (GLORI). LOICZ reports and studies. LOICZ Core Project Office, Institute for Sea Research (NIOZ), 125 p
Mirkhamidov FM, Mannan MM (1981) The nature of gravitional field and it's relation with geotectonics of Bangladesh. Unpublished report. Petrobangla, Dhaka
Misra AA, Mukherjee S (2015) Tectonic inheritance in continental rifts and passive margins. Springer Briefs in Earth Sciences. ISBN 978–3-319-20576-2
Misra AA, Bhattacharya G, Mukherjee S, Bose N (2014) Near N-S paleo-extension in the western Deccan region in India: does it link strike-slip tectonics with India-Seychelles rifting? International Journal of Earth Sciences 103, 1645–1680

Misra AA, Sinha N, Mukherjee S (2015) Repeat ridge jumps and microcontinent separation: insights from NE Arabian Sea. Marine and Petroleum Geology 59, 406–428

Misra S (2006) Precambrian chronostratigraphic growth of Singhbhum-Orissa Craton, Eastern Indian Shield: an alternative model. Journal of Geological Society of India 67, 356–378

Mitchell AHG (1993) Cretaceous-Cenozoic tectonic events in the western Myanmar (Burma)–Assam region. Journal of Geological Society 150, 1089–1102

Mitra S, Priestley K, Bhattacharyya AK, Gaur VK (2005) Crustal structure and earthquake focal depths beneath northeastern India and southern Tibet. Geophysical Journal International 160, 227–248

Mohanty WK, Mohapatra AK, Verma AK, Tiampo KF, Kislay K (2014) Earthquake forecasting and its verification in northeast India. Geomatics, Natural Hazards and Risk 7, 194–214

Mohanty WK, Verma AK, Vaccari F, Panza GF (2013) Influence of epicentral distance on local seismic response in Kolkata city, India. Journal of Earth System Science 122, 321–338

Molnar P, Tapponnier P (1975) Cenozoic tectonics of Asia: effects of a continental collision. Science 189, 419–426

Monsur MH (1995) An introduction to the Quaternary geology of Bangladesh. City Press and Publications, Dhaka, Bangladesh, 70 p

Morgan JP, McIntire WG (1959) Quaternary geology of the Bengal Basin, East Pakistan and India. Bulletin of the Geological Society of America 70, 319–341

Morino M, Kamal ASMM, Akhter SH, Rahman MZ, Ali RME, Talukder A, Khan MMH, Matsuo J, Kaneko F (2014) A paleo-seismological study of the Dauki fault at Jaflong, Sylhet, Bangladesh: historical seismic events and an attempted rupture segmentation model. Journal of Asian Earth Sciences 91, 218–226

Morino M, Kamal ASMM, Muslim D, Ali RME, Kamal MA, Rahman MZ, Kaneko F (2011) Seismic event of the Dauki Fault in 16th century confirmed by Trench investigation at Gabrakhari Village, Haluaghat, Mymensingh, Bangladesh. Journal of Asian Earth Sciences 42, 492–498

Mukherjee S (2013a) Channel flow extrusion model to constrain dynamic viscosity and Prandtl number of the higher Himalayan Shear Zone. International Journal of Earth Sciences 102, 1811–1835

Mukherjee S (2013b) Higher Himalaya in the Bhagirathi section (NW Himalaya, India): its structures, backthrusts and extrusion mechanism by both channel flow and critical taper mechanisms. International Journal of Earth Sciences 102, 1851–1870

Mukherjee S (2013c) Deformation microstructures in rocks. Springer Geochemistry/Mineralogy, Berlin. 111 p

Mukherjee S (2014a) Review of flanking structures in meso- and micro-scales. Geological Magazine 151, 957–974

Mukherjee S (2014b) Atlas of Shear Zone structures in meso-scale. Springer Geology, Cham. 124 p

Mukherjee S (2015) A review on out-of-sequence deformation in the Himalaya. In: Mukherjee S, Carosi R, van der Beek P, Mukherjee BK, Robinson D (eds) Tectonics of the Himalaya. Geological Society Special publications, London, vol 412, pp 67–109

Mukherjee S (2017) Airy's isostatic model: a proposal for a realistic case. Arabian Journal of Geosciences 10, 268

Mukherjee S (2019) Introduction to "Tectonics and Structural Geology: Indian Context". In: Mukherjee S (ed) Tectonics and structural geology: Indian context. Springer International Publishing AG, Cham, pp 1–5. ISBN: 978-3-319-99340-9

Mukherjee S, Khonsari MM (2017) Brittle rotational faults and the associated shear heating. Marine and Petroleum Geology 88, 551–554

Mukherjee S, Koyi HA (2010a) Higher Himalayan Shear Zone, Zanskar section-microstructural studies & extrusion mechanism by a combination of simple shear & channel flow. International Journal of Earth Sciences 99, 1083–1110

Mukherjee S, Koyi HA (2010b) Higher Himalayan Shear Zone, Sutlej section-structural geology & extrusion mechanism by various combinations of simple shear, pure shear & channel flow in shifting modes. International Journal of Earth Sciences 99, 1267–1303
Mukherjee S, Kumar N (in press) A first-order model for temperature rise for uniform and differential compression of sediments in basins. Int J Earth Sci https://doi.org/10.1007/s00531-018-1634-6
Mukherjee A, Fryar AE, Thomas WA (2009) Geologic, geomorphic and hydrologic framework and evolution of the Bengal basin, India and Bangladesh. Journal of Asian Earth Sciences 34, 227–244
Mukherjee S, Mulchrone K (2012) Estimating the viscosity and Prandtl number of the Tso Morari Gneiss Dome, western Indian Himalaya. International Journal of Earth Sciences 101, 1929–1947
Mukherjee S, Carosi R, van der Beek PA, Mukherjee BK, Robinson DM (2015) In: Mukherjee S, Carosi R, van der Beek P, Mukherjee BK, Robinson D (eds) Tectonics of the Himalaya: an introduction. Geological Society Special publications, London, vol 412, pp 1–3
Mukherjee S, Koyi HA, Talbot CJ (2012) Implications of channel flow analogue models in extrusion of the higher Himalayan Shear Zone with special reference to the out-of-sequence thrusting. International Journal of Earth Sciences 101, 253–272
Mukherjee S, Misra AA, Calvès G, Nemčok M (2017) Tectonics of the Deccan Large Igneous Province: an introduction. In: Mukherjee S, Misra AA, Calvès G, Nemčok M (eds) Tectonics of the Deccan Large Igneous Province. Geological Society Special publications, London, vol 445, pp 1–9
Mukherjee S, Mukherjee B, Thiede R (2013) Geosciences of the Himalaya-Karakoram-Tibet orogen. International Journal of Earth Sciences 102, 1757–1758
Mukhopadhyay M, Dasgupta S (1988) Deep structure and tectonics of the Burmese arc: constraints from earthquake and gravity data. Tectonophysics 149, 299–322
Mukhopadhyay M, Verma RK, Ashraf MH (1986) Gravity field and structures of Rajmahal hills: examples of the Palaeo-Mesozoic continental margin in eastern India. Tectonophysics 131, 353–367
Murphy RW, Staff of BOGMC (1988) Bangladesh enters the oil era. Tulsa. Oil & Gas Journal, 76–82
Nag S, Gaur RK, Paul T (2001) Late Cretaceous-Tertiary sediments and associated faults in southern Meghalaya Plateau of India vis-a-vis South Tibet: their interrelationships and regional implications. Journal of Geological Society of India 57, 327–338
Najman Y (2006) The detrital record of orogenesis: a review of approaches and techniques used in the Himalayan sedimentary basins. Earth-Science Reviews 74, 1–72
Najman Y, Allen R, Willett EAF, Carter A, Barford D, Garzanti E, Wijbrans J, Bickle M, Vezzoli G, Ando S, Oliver G, Uddin M (2012) The record of Himalayan erosion preserved in the sedimentary rocks of the Hatia Trough of the Bengal Basin and the Chittagong Hill tracts, Bangladesh. Basin Research 24, 499–519
Najman Y, Bickle M, BouDagher-Fadel M, Carter A, Garzanti E, Paul M, Wijbrans J, Willett E, Oliver G, Parrish R, Akhter SH, Allen R, Ando S, Chisty E, Reisberg L, Vezzoli G (2008) The Paleogene record of Himalayan erosion: Bengal Basin, Bangladesh. Earth and Planetary Science Letters 273, 1–14
Najman Y, Bracciali L, Parrish RR, Chisty E, Copley A (2016) Evolving strain partitioning in the Eastern Himalaya: the growth of the Shillong Plateau. Earth and Planetary Science Letters 433, 1–9
Najman Y, Jenks D, Godin L, Boudagher-Fadel M, Millar I, Garzanti E, Horstwood M, Bracciali L (2017) The Tethyan Himalayan detrital record shows that India-Asia terminal collision occurred by 54 Ma in the Western Himalaya. Earth and Planetary Science Letters 459, 301–310
Nandy DR (1986) Tectonics, seismicity and gravity of Northeastern India and adjoining region. Memoir Geological Society of India 119, 13–16

Nandy DR (2001) Geodynamics of Northeast India and the adjoining region. ABC Publications, Calcutta, pp 1–209

Naqvi SM (2005) Geology and evolution of the Indian plate. Capital Publishing Company, New Delhi, p 450

Nath SK, Adhikari MD, Maiti SK, Devaraj N, Srivastava N, Mohapatra LD (2014) Earthquake scenario in West Bengal with emphasis on seismic hazard microzonation of the city of Kolkata, India. Natural Hazards and Earth System Sciences 14, 2549–2575

National Data Repository (2015) Directorate general of hydrocarbons. Under Ministry of Petroleum & Natural Gas, India

Ni JF, Bevis M, Holt WE, Wallace TC, Seager WR (1989) Accretionary tectonics of Burma and the three-dimensional geometry of the Burma subduction zone. Geology 17, 68–71

Nielsen C, Chamot-Rooke N, Rangin C, Cruise Team ANDAMAN (2004) From partial to full strain partitioning along the Indo-Burmese hyper-oblique subduction. Marine Geology 209, 303–327

Norton IO, Sclater JG (1979) A model for the evolution of the Indian Ocean and the break-up of Gondwanaland. Journal of Geophysical Research 84, 6803–6830

Oldham RD (1893) A Mannual of the geology of India and Burma, 2nd edn, vol 1, 483 p

Olympa B, Abhishek K (2015) A review on the tectonic setting and seismic activity of the Shillong Plateau in the light of past studies. Disaster Advances 8, 34–45

Ovi MMH, Khan MSH, Haque MM (2014) Geomorphic signature of active tectonics from Sylhet City and adjoining areas, Surma Basin, Bangladesh. Bangladesh Geoscience Journal 20, 19–34

Pascoe EHA (1975) A manual of geology of India and Burma, vol 2, 3rd edn. Controller of Publications, New Delhi

Paul DD, Lian HM (1975) Offshore basins of southwest Asia—Bay of Bengal to South Sea. In: Proceedings of the 9th world petrol congress, vol 3. Tokyo, pp 1107–1121

Pickering JL, Goodbred SL, Beam JC, Ayers JC, Covey AK, Rajapara HM, Singhvi AK (2017) Terrace formation in the upper Bengal basin since the Middle Pleistocene: Brahmaputra fan delta construction during multiple highstands. Basin Research 30, 550–567

Pickering JL, Goodbred SL, Reitz MD, Hartzog TR, Mondal DR, Hossain MS (2014) Late Quaternary sedimentary record and Holocene channel avulsions of the Jamuna and Old Brahmaputra River valleys in the upper Bengal delta plain. Geomorphology 227, 123–136

Prasad B, Pundir BS (2017) Gondwana biostratigraphy of the Purnea Basin (Eastern Bihar, India), and its correlation with Rajmahal and Bengal Gondwana Basins. Journal Geological Society of India 90, 405–427

Rabbani MG, Chowdhury KR, Huq MM (2000) Stratigraphic analysis by interpretation of seismic and drill hole data of Rangpur Dinajpur area, Bangladesh. Bangladesh Geoscience Journal 6, 1–16

Rabinowitz PD, Woods S (2006) The Africa-Madagascar connection and mammalian migrations. Journal of African Earth Sciences 40, 270–276

Racey A, Ridd MF (2015) Petroleum geology of Myanmar. Geological Society, London, Memoirs 45, 93–108

Rahman A (1987) Geology of Maddhapara area, Dinajpur district, Bangladesh. Records of the Geological Survey Bangladesh 5, 1–61

Rahman MA, Blank HR, Kleinkopf MD, Kucks RP (1990a) Aeromagmetic anomaly map of Bangladesh, scale 1:1000000. Geological Survey, Bangladesh, Dhaka

Rahman MA, Mannan MA, Blank HR, Kleinkopf MD, Kucks RP (1990b) Bouguer gravity anomaly map of Bangladesh, scale 1:1000000. Geol. Surv, Bangladesh, Dhaka

Rahman MJJ, Suzuki S (2007) Geochemistry of sandstones from the Miocene Surma Group, Bengal Basin, Bangladesh: implications for provenance, tectonic setting and weathering. Geochemical Journal 41, 415–428

Rahman MJJ, Worden RH (2016) Diagenesis and its impact on the reservoir quality of Miocene sandstones (Surma Group) from the Bengal Basin, Bangladesh. Marine and Petroleum Geology 77, 898–915

Rahman MJJ, Xiao W, McCann T, Songjian A (2017) Provenance of the Neogene Surma Group from the Chittagong Tripura Fold Belt, southeast Bengal Basin, Bangladesh: constraints from whole-rock geochemistry and detrital zircon U-Pb ages. Journal of Asian Earth Sciences 148, 277–293

Rajendran CP, Rajendran K, Duarah BP, Baruah S, Earnest A (2004) Interpreting the style of faulting and paleoseismicity associated with the 1897 Shillong, northeast India, earthquake: implications for regional tectonism. Tectonics, 23, TC4009

Rajendran K, Parameswaran RM, Rajendran CP (2017) Seismotectonic perspectives on the Himalayan arc and contiguous areas: inferences from past and recent earthquakes. Earth-Science Reviews 173, 1–30

Ram J, Venkataraman B (1984) Tectonic framework and hydrocarbon prospects of Mizoram. Petroliferous basins of India—II. Petroleum Asia Journal VII, 60–65

Raman, KS, Kumar S, Neogi BB (1986) Exploration in Bengal Basin India—an overview. In: Proceedings of the South East Asia petroleum exploration society, vol VII, pp 180–191

Ramana MV, Nair RR, Sarma KVLNS, Ramprassad T, Krishna KS, Subramanyam V, D'Cruz M, Subramanyam C, Paul J, Subramanyam AS, Chandra Sekhar DV (1994) Mesozoic anomalies in the Bay of Bengal. Earth and Planetary Science Letters 121, 469–475

Rangin C, Sibuet J-C (2017) Structure of the northern Bay of Bengal offshore Bangladesh: evidences from new multi-channel seismic data. Marine and Petroleum Geology 84, 64–75

Ray JS, Pattanayak SK, Pande K (2005) Rapid emplacement of the Kerguelen plume related Sylhet traps, eastern India: evidence from 40Ar–39Ar geochronology. Geophysical Research Letters 32, L10303

Reimann K-U (1993) Geology of Bangladesh. Gebrueder Borntraeger, Berlin, 160 p

Reitz MD, Pickering JL, Goodbred SL, Paola C, Steckler MS, Seeber L, Akhter SH (2015) Effects of tectonic deformation and sea level on river path selection: theory and application to the Ganges-Brahmaputra-Meghna River Delta. Journal of Geophysical Research Earth Surface 120, 671–689

Reitz MD, Steckler MS, Paola C, Seeber L (2012) Modeling coupled avulsion and earthquake timescale dynamics. AGU Fall Meeting Supplement. Abstract EP41A-3506

Richards S, Lister G, Kennett B (2007) A slab in depth: three-dimensional geometry and evolution of the Indo-Australian plate. Geochemistry, Geophysics, Geosystems 8, Q12003

Rowley DB (1996) Age of initiation of collision between India and Asia; a review of stratigraphic data. Earth and Planetary Science Letters 145, 1–13

Roy AB (2014) Indian subcontinent, reference module in earth systems and environmental sciences. Elsevier

Roy AB, Chatterjee A (2015) Tectonic framework and evolutionary history of the Bengal Basin in the Indian subcontinent. Current Science 109, 271–279

Royer JY, Patriat P, Bergh HW, Scotese CR (1988) Evolution of the southwest Indian Ridge from the Late Cretaceous (anomaly 34) to the Middle Eocene (anomaly 20). Tectonophysics 155(1), 235–260

Sager WW, Zhang J, Korenaga J, Sano T, Koppers AAP, Mahony JJ (2013) An immense shield volcano within the Shatsky rise oceanic plateau, northwest Pacific Ocean. Nature Geoscience 6, 976–981

Salt CA, Alam MM, Hossain MM (1986) Bengal Basin: current exploration of the hinge zone area of south-western Bangladesh. In: 6th offshore SE Asia (SEAPEX) conference, Singapore, pp 55–67

Sar D, Maheshwari MK, Rangarajan S, Bahuguna CS (2009) Eighty five degrees east ridge & its hydrocarbon potential. Geohorizons Society of Petroleum Geophysicists, India 15, 15–23

Schettino A, Scotese CR (2005) Apparent polar wander paths for the major continents (200 Ma to the present day): a palaeomagnetic reference frame for global plate tectonic reconstructions. Geophysical Journal International 163, 727–759

Sclater JG, Fisher RL (1974) Evolution of the east central Indian Ocean, with emphasis on the tectonic setting of the Ninetyeast Ridge. GSA Bulletin 85, 683–702

Searle MP, Corfield RI, Stephenson B, Mccarron J (1997) Structure of the north Indian continental margin in the Ladakh-Zanskar Himalayas: implications for the timing and obduction of the Spontang Ophiolite, India-Asia collision and deformation events in the Himalaya. Geological Magazine 134, 297–316

Sengupta S (1966) Geological and geophysical studies in the western part of the Bengal Basin, India. AAPG Bulletin 50, 1001–1017

Shamsuddin AHM (1989) Organic geochemistry of oligocene-miocene deposits of the Bengal Foredeep, Bangladesh. Journal of Geological Society of India 33, 332–352

Shamsuddin AHM, Abdullah SKM (1997) Geological evolution of the Bengal Basin and its implication in hydrocarbon exploration in Bangladesh. Indian Journal of Geology 69, 93–121

Shamsuddin AHM, Brown TA, Lee S, Curiale J (2001) Petroleum systems of Bangladesh. Proc 13th Southeast Asia Petroleum Exploration Conference, 4 – 6th April, Singapore

Shanmugam G (2016) Submarine fans: a critical retrospective (1950–2015). Journal of Palaeogeography 5, 110–184

Sharma RS (2010) Cratons and fold belts of India. In: Lecture notes in earth sciences, Springer, Berlin, Heidelberg

Sharma S, Sarma JN (2017) Application of Drainage basin morphotectonic analysis for assessment of tectonic activities over two regional structures of the northeast India. Journal of the Geological Society of India 89, 271–280

Sibuet J-C, Klingelhoefer F, Huang Y-P, Yeh Y-C, Rangin C, Lee C-S, Hsu S-K (2016) Thinned continental crust intruded by volcanics beneath the northern Bay of Bengal. Marine and Petroleum Geology 77, 471–486

Sikder AM (1998) Tectonic evolution of eastern folded belt of Bengal Basin. Ph.D. thesis. Dhaka University, Dhaka, 175 p

Sikder AM, Alam MM (2003) 2-D modelling of the anticlinal structures and structural development of the eastern fold belt of the Bengal Basin, Bangladesh. Sedimentary Geology 155, 209–226

Singh A, Bhushan K, Singh C, Steckler MS, Akhter SH, Seeber L, Kim W-Y, Tiwari AK, Biswas R (2016) Crustal structure and tectonics of Bangladesh: new constraints from inversion of receiver functions. Tectonophysics 680, 99–112

Singh A, Singh C, Kennett BLN (2015) A review of crust and upper mantle structure beneath the Indian subcontinent. Tectonophysics 644–645, 1–21

Singh AP, Kumar N, Singh B (2004) Magmatic underplating beneath the Rajmahal Traps: gravity signature and derived 3-D configuration. Journal of Earth System Science 113, 759–769

Smith AG, Hallam A (1970) The fit of the southern continents. Nature 225, 139–144

Socquet A, Vigny C, Chamot-Rooke N, Simons W, Rangin C, Ambrosius B (2006) India and Sunda plates motion and deformation along their boundary in Myanmar determined by GPS. Journal of Geophysical Research 111, B05406

Steckler MS, Akhter SH, Seeber L (2008) Collision of the Ganges-Brahmaputra Delta with the Burma Arc. Earth and Planetary Science Letters 273, 367–378

Steckler MS, Akhter SH, Seeber L, Bilham RG, Kogan MG, Masson F, Maurin T, Mondal D, Piana Agostinetti N, Rangin C, Saha P (2012) GPS velocities and structure across the Burma accretionary prism and Shillong anticline in Bangladesh. AGU Fall Meeting Supplement. Abstract T51F-2667

Steckler MS, Mondal D, Akhter SH, Seeber L, Feng L, Gale J (2016) Locked and loading megathrust linked to active subduction beneath the Indo-Burman ranges. Nature Geoscience 9, 615–618

Storey M, Mahoney JJ, Saunders AD, Duncan RA, Kelley SO, Coffin MF (1995) Timing of hot-spot related volcanism and the breakup of Madagascar and India. Science 267, 852–855

Sultana DN, Alam MM (2001) Facies analysis of the Neogene Surma group succession in the subsurface of the Sylhet Trough, Bengal Basin, Bangladesh. In: 10th geological confernce. Bangladesh Geological Society, Dhaka, Abstract, p 70

Syvitski JP, Vörösmarty CJ, Kettner AJ, Green P (2005) Impact of humans on the flux of terrestrial sediment to the global coastal ocean. Science 308, 376–380

Szeliga W, Hough S, Martin S, Bilham R (2010) Intensity, magnitude, location and attenuation in India for felt earthquakes since 1762. Bulletin of Seismological Society of America 100, 570–584

Talwani M, Desa MA, Ismaiel M, Krishna KS (2016) The Tectonic origin of the Bay of Bengal and Bangladesh. Journal of Geophysical Research Solid Earth 121, 4836–4851

Tapu AT, Ameen SMM, Abdullah R, Zaman MN (2016) Geochemical evaluation of the diorite basement in Barapaharpur, Rangpur, northwest Bangladesh. Bangladesh Geoscience Journal 22, 17–36

Thakur VC (2004) Active tectonics of Himalayan Frontal Thrust and Seismic Hazard to Ganga plain. Current Science 86, 1554–1560

Torsvik TH, Tucker RD, Ashwal LD, Carter LM, Jamtveit B, Vidyadharan KT, Venkataramana P (2000) Late cretaceous India-Madagascar fit and timing of break-up related magmatism. Terra Nova 12, 220–224

Uddin A, Lundberg N (1998) Cenozoic history of the Himalayan-Bengal system: sand composition in the Bengal Basin, Bangladesh. Geological Society of America Bulletin 11, 497–511

Uddin A, Lundberg N (1999) A paleo-Brahmaputra? Subsurface lithofacies analysis of Miocene deltaic sediments in the Himalayan-Bengal system, Bangladesh. Sedimentary Geology 123, 239–254

Uddin A, Lundberg N (2004) Miocene sedimentation and subsidence during continent–continent collision, Bengal basin, Bangladesh. Sedimentary Geology 164, 131–146

Uddin MN, Ahmed Z (1989) Palynology of the Kopili formation at GDH-31, Gaibandha District, Bangladesh. Bangladesh Journal of Geology 8, 31–42

Vaccari F, Walling MY, Mohanty WK, Nath SK, Verma AK, Sengupta A, Panza GF (2011) Site-specific modeling of SH and P-SV waves for microzonation study of Kolkata metropolitan city, India. Pure and Applied Geophysics 168, 479–493

Valdiya KS (2010) The making of India: geodynamic evolution. Macmillan Publishers India. 816 p

Valdiya KS (2016) The making of India: geodynamic evolution. In: Tripathi SC (ed) Society of earth scientists series, 2nd edn. Springer, Cham, 945 p

Vernant P, Bilham R, Szeliga W, Drupka D, Kalita S, Bhattacharyya AK, Gaur VK, Pelgay P, Cattin R, Berthet T (2014) Clockwise rotation of the Brahmaputra Valley relative to India: tectonic convergence in the eastern Himalaya, Naga Hills, and Shillong Plateau. Journal of Geophysical Research Solid Earth 119, 6558–6571

Wadia DN (1953) Geology of India. Rev Ed, Macmillan, 552 pp

Wang Y, Sieh K, Tun ST, Lai K-Y, Myint T (2014) Active tectonics and earthquake potential of the Myanmar region. Journal of Geophysical Research 119, 3767–3822

Webb AAG, Yin A, Dubey CS (2013) U-Pb zircon geochronology of major lithologic units in the eastern Himalaya: implications for the origin and assembly of Himalayan rocks. Geological Society of America Bulletin 125, 499–522

Wesnousky S, Kumar GS, Mohindra R, Thakur VC (1999) Uplift and convergence along the Himalayan frontal thrust of India. Tectonics 18, 967–976

Woodside PR (1983) Petroleum geology of Bangladesh. Oil & Gas Journal 81, 149–155

Yin A, Dubey CS, Webb AAG, Kelty TK, Grove M, Gehrels GE, Burgess WP (2010) Geologic correlation of the Himalayan orogen and Indian craton: part 1. Structural geology, U-Pb zircon geochronology, and tectonic evolution of the Shillong Plateau and its neighboring regions in NE India. Geological Society of America Bulletin 122, 336–359

Zaher MA, Rahman A (1980) Prospects and investigations for minerals in the northwestern part of Bangladesh. Petroleum and Mineral Resources of Bangladesh. Seminar and Exhibition, Dhaka, pp 9–18

Zhu BD, Kidd WSF, Rowley DB, Currie BS, Shaffique N (2005) Age of initiation of the India-Asia collision in the east-central Himalaya. Journal of Geology 113. 265–285

Fold-Thrust Belt Architecture and Structural Evolution of the Northern Part of the Nallamalai Fold Belt, Cuddapah Basin, Andhra Pradesh, India

Vikash Tripathy, Satyapal, S. K. Mitra and V. V. Sesha Sai

1 Introduction

Fold and thrust belts play an important role in the understanding of fragmentation and amalgamation of continental fragments such as those of Columbia and Rodinia supercontinents. Temporal changes in structural patterns of fold belts may indicate major tectonics, which is relatable to the global tectonic episodes. The 400 km long poly-deformed Mesoproterozoic Nallamalai Fold Belt (NFB) is an arcuate orogeny, also called as orocline, whose eastern and western boundaries are bounded by thrust zones. NFB is thrust to the western part of the Cuddapah basin along Nallamalai thrust/Rudravaram line and in the east by the Nellore Schist Belt (NSB) along Vellikonda thrust (Fig. 1). The two thrusts restrict the NFB in the eastern part of the Cuddapah basin displaying structural features, which are incomparable to the structures recorded from rest of the basin. The structural history of NFB in conjugation with their relation to various regional tectonic events has widely been discussed earlier (Nagaraja Rao et al. 1987; Meijerink et al. 1984; Venkatakrishna and Dotiwala 1987; Matin and Guha 1996; Mukherjee 2001; Saha 1994, 2002; Saha et al. 2010; Tripathy and Saha 2013, 2015; Matin 2014). However, the importance of the thrust zones in the northern part of the NFB has not yet been fully realized because of limited data. We present new detailed field data to highlight the changes in structural fabric and their significance in the evolution of the northern part of the NFB.

V. Tripathy (✉)
Geological Survey of India, Training Institute, Hyderabad, India
e-mail: vikashtripathy@gmail.com

Satyapal
Geological Survey of India, Western Region, Jaipur, India

S. K. Mitra · V. V. Sesha Sai
Geological Survey of India, Southern Region, Hyderabad, India

S. Mukherjee (ed.), *Tectonics and Structural Geology: Indian Context*,
Springer Geology, https://doi.org/10.1007/978-3-319-99341-6_7

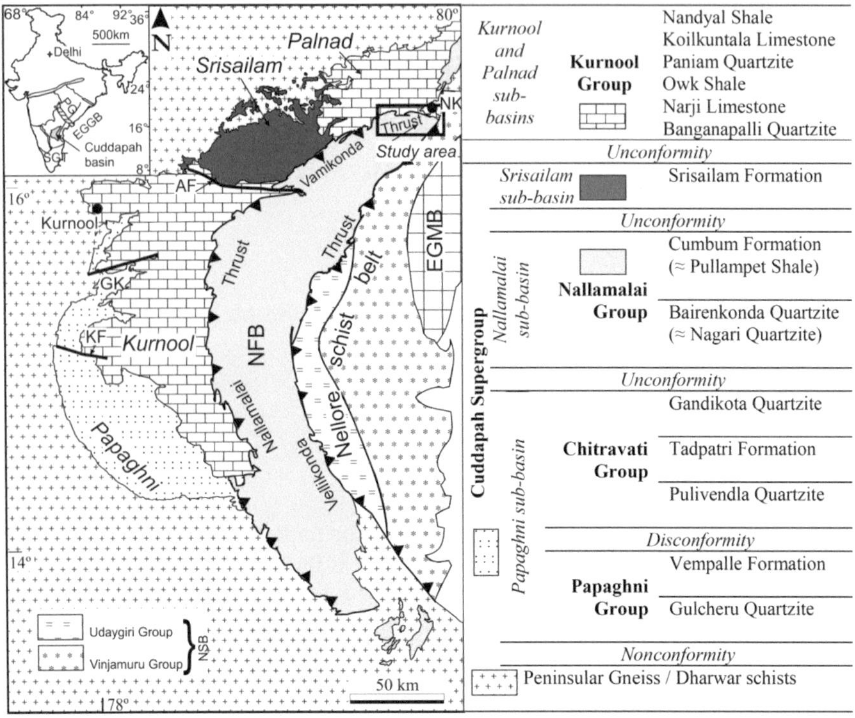

Fig. 1 Regional geology and abridged stratigraphic successions of the Cuddapah basin, modified after Nagaraja Rao et al. (1987) and Geological Survey of India (1990). AF: Atmakuru fault; EGMB: Eastern Ghats Mobile Belt; GK: Gani-Kalva fault; KF: Kona fault; NFB: Nallamalai Fold Belt; NSB: Nellore Schist Belt; NK: Nekarikallu; PG: Pranhita-Godavari Basin; SGT: Southern Granulite Terrain

Attention is also given towards the nature and style of deformation, partitioned in space and time as observed in the rock units of the Nallamalai and the Kurnool Groups in the northern extent of the NFB, and hitherto unreported klippe and window structures. Overall, the area displays structures which are of pre-Kurnool and post-Kurnool affinity. The pre-Kurnool deformations are restricted within the Nallamalai Group while the post-Kurnool deformations are either pre-thrust or post thrust, later of which are imprinted both on the Nallamalai and the Kurnool Groups. This paper puts together the structural and stratigraphic data recorded from the northern part of NFB i.e., around Vamikonda thrust (Venkatakrishnan and Dotiwalla 1987) to establish thrusting of NFB over the Palnad sub-basin that display thin-skinned tectonics. These observations are important in the context of constraining the temporal evolution of the NFB and in regional correlation with adjoining basins and mobile belt. In this contribution, we discuss structural fabrics in the northern part of the NFB and correlate them with tectonics related to possible amalgamation of SE India and East Antarctica fragments.

2 Geological Setting

The eastern Dharwar craton is a collage of a number of tectonic blocks juxtaposed during various Archaean-Proterozoic tectonic events. It is bounded in the east by the Eastern Ghat Mobile Belt (EGMB) which is an exhumed remnant of a Proterozoic collisional zone resulted due to East Antarctica collision (Bhattacharya 1996; Dasgupta and Sengupta 2003; Dobmeier and Raith 2003; Vijaya Kumar and Leelanandam 2008; Mukhopadhyay and Basak 2009). West of this mobile belt, Nellore Schist Belt (NSB), Nallamalai Fold Belt (NFB) and less deformed sedimentary deposits of Papaghni, Srisailam, Palnad and Kurnool sub-basins are present overlying the basement granitic complex and Archaean schist belts such as Gadwal, Tsundupalle, Veligallu and Kadiri. The lower Cuddapah sequence is usually undeformed with sub-horizontal dips towards E except around fault zones such as the Gani-Kalva and the Kona faults where beds show steep dips (Tripathy 2011; Tripathy and Saha 2013, 2015). In the western part of the Cuddapah basin the Paleoproterozoic lithounits (e.g., ~1900 Ma, French et al. 2008) are overlain unconformably by the younger Neoproterozoic Kurnool Group (check Tripathy and Saha 2015 and references therein), both of which deformed subsequently (Tripathy and Saha 2013, 2015). The structural variation between the NFB and the western part of Cuddapah basin is marked by a thrust zone variously been known as the Rudravaram Line (Meijerink et al. 1984) or the Nallamalai Thrust Front (Venkatakrishnan and Dotiwalla 1987) or the Maidukuru thrust (Chakraborti and Saha 2009). The eastern boundary of NFB is considered as a thrust zone named as the Vellikonda Thrust Front (Venkatakrishnan and Dotiwalla 1987) where the Nellore Schist Belt (NSB) is thrust over the NFB (Fig. 1). Similarly, Landsat image interpretation in the northern part of the NFB reveals a thrust zone along the NFB-Palnad sub-basin contact named as the Vamikonda Thrust Front (Venkatakrishnan and Dotiwalla 1987).

Stratigraphically, the oldest lithounits i.e., the Papaghni and the Chitravati Groups of the Cuddapah Supergroup are restricted in the southwestern and western part of the basin (Fig. 1). Radiometric ages derived from mafic sills of the Tadpatri Formation suggest the lower Cuddapah sedimentation to be older than ~1900 Ma (Bhaskar Rao et al. 1995; Anand et al. 2003; French et al. 2008; reviewed by Sesha Sai et al. 2017). Based on the occurrences of red bed sequence, Sesha Sai et al. (2016) considered Gulcheru to be deposited at around ~2.1 Ga. The Paleoproterozoic to Mesoproterozoic sedimentation of the Nallamalai Group is restricted within the NFB and consists of lower arenaceous unit known as the Bairenkonda Quartzite (Nagari Quartzite of southern NFB) and the upper argillaceous dominant Cumbum Formation (Pullampet Shale of southern NFB). The fold belt records two phases of folding which has been cut through by Chelima lamproite (1354 Ma, Chalapathi Rao et al. 1999), Zangamarajupalle lamproite and syenite plugs (e.g. Giddalur and 1326 Ma Racherla syenite; Chalapathi Rao et al. 2012) in its western part. Existence of granitic intrusions (e.g. Vellaturu, Ipuru, Vinukonda) along the eastern margin of the NFB with radiometric ages of ~1575 Ma for Vellaturu Granite (Crawford and Compston 1973); ~1615 Ma WR isochron age and ~1589 Ma U-Pb zircon age for Vinukonda Granite (Gupta et al. 1984; Dobmeier et al. 2006) provide an upper age limit for the deposition and deformation for the Nallamalai Group. U-Pb detrital zircon age from the sediments of the Nallamalai Group constrain the sediments to have deposited between 1659 and 1590 Ma (Collins et al. 2015).

In the northern part of the NFB, along the Vamikonda range, the Cumbum Formation is exposed consisting chiefly of phyllite, slate, quartz phyllite, siltstone, argillaceous carbonate i.e., intercalated phyllite and carbonate, metamarl, quartzite and carbonate bands along with occasional chert (Rajurkar 1972; Ramlinga Swamy 1972). The phyllite is steel grey, pink to purple and variegated colours. The quartzite is thin to thickly bedded, ripple marked, trough to planar cross bedded with unidirectional to bidirectional flow pattern and rare occurrence of desiccation cracks suggesting sediment deposition in subtidal to peritidal set up (Saha and Tripathy 2012; Dasgupta and Biswas 2006).

The Kurnool Group represents the youngest lithounits of the Cuddapah basin with a maximum stratigraphic thickness of about 450 m and is exposed in the Kurnool and Palnad sub-basins (Fig. 1; Nagaraja Rao et al. 1987). The Group is represented by two cycles of sandstone, limestone and shale sequence. In the southern part of the Palnad sub-basin in the Durgi-Karampudi tract, the Narji Limestone forms the lower most exposed unit of the group which are steel grey to buff grey coloured pyritiferrous limestone. The limestone unit is conformably overlain by non-calcareous plane laminated variegated white, purple to yellow-ochre Owk Shale. The shale further grades to Paniam Quartzite, which is coarse to medium-grained with occasional gritty layers.

3 Structural Heterogeneity in the Northern Part of NFB

Intracontinental compression deformation within the NFB broadly occurred twice. The first phase is related to ~N-S trending doubly plunging F_1 fold and thrust as recorded throughout the fold belt (Nagaraja Rao et al. 1987; Matin and Guha 1996; Mukherjee 2001; Saha 1994, 2002; Chakraborti and Saha 2009; Tripathy and Saha 2010; Chetty 2011; Matin 2014). The F_1 folds vary geometrically due to thrust-related modifications. Progressive development of F_1 fold and fault along the Vellikonda thrust evolved in a general shear deformation suggesting an inclined transpressional tectonics (Tripathy and Saha 2010; Saha et al. 2010). The second phase in the central part of the NFB is marked by moderately to steeply dipping ~N-S trending crenulation cleavage (S_{n2}) that overprints the S_{n1} cleavage (Tripathy and Saha 2010; Tripathy 2010).

In the western part of the NFB along the Nallamalai thrust front (the Rudravaram line), the fold belt acts as hanging wall and are thrust over the Papaghni, Kurnool, Srisailam and Palnad sub-basins. In the northern part, the Palnad nappe is exposed that represents part of the thrust NFB over the Kurnool Group of Palnad sub-basin

(Natarajan and Rajagopalan Nair 1977; Saha and Chakraborty 2003; Chakraborti 2006; Chakraborti and Saha 2009; Saha et al. 2010). The overall arcuate pattern of the fold belt is an outcome of all the major deformation events, which are evaluated in the present work.

An aerial view of the area shows NE trending lineaments (Fig. 2), which has marred the folding pattern within the NFB. The folds at such locations shows juxtaposition of simultaneous antiforms along thrust plane, so that the downthrown synform lies below the antiform. Here the fold belt shows changes in the orientation of the fabric from NE-SW to E-W and juxtaposition of the two thrust fronts i.e., the Nallamalai and the Vellikonda thrust near Nekarikallu (Fig. 3). Matin (2014) suggests the NFB to be a syncline with eastern limb being overturned and an eroded anticline. In such a model the synclinal closure is envisaged in the northern end of the NFB near Nekarikallu. Matin (2014) viewed the deformations within the NFB related to contemporaneous faulting and folding in front of Vellikonda thrust and modeled the Nallamalai thrust as its imbricate. The deformation is understood to have resulted due to fault propagation folding. In the present discourse we bring forth the structural geometry and architecture of NFB to provide a plausible structural model which explains the oroclinal nature of NFB. The details of the structural fabric in northern part of NFB is discussed in detail (cf. Matin 2014) where it changes from NE-SW to ~E-W trending near to the thrust boundary with Palnad sub-basin. The data presented here from northern part of NFB is in detail as compared to Matin 2014 and gives better understanding of tectonic set up of the area.

The northern part of the NFB falling in the Survey of India toposheet no. 56P/11 and 56P/15 south of Durgi and Nekarikallu is mapped (Fig. 3). The paper provides

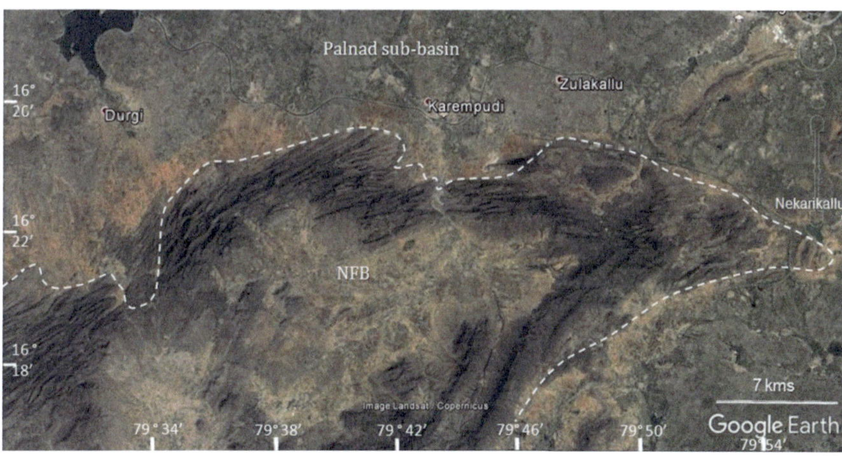

Fig. 2 Google Earth image of the northern part of the Nallamalai Fold Belt showing prominent NE to E trending lineaments. Dotted line is the boundary of the NFB with Palnad sub-basin and granitic gneiss on east. Imagery Date: 31/12/2005

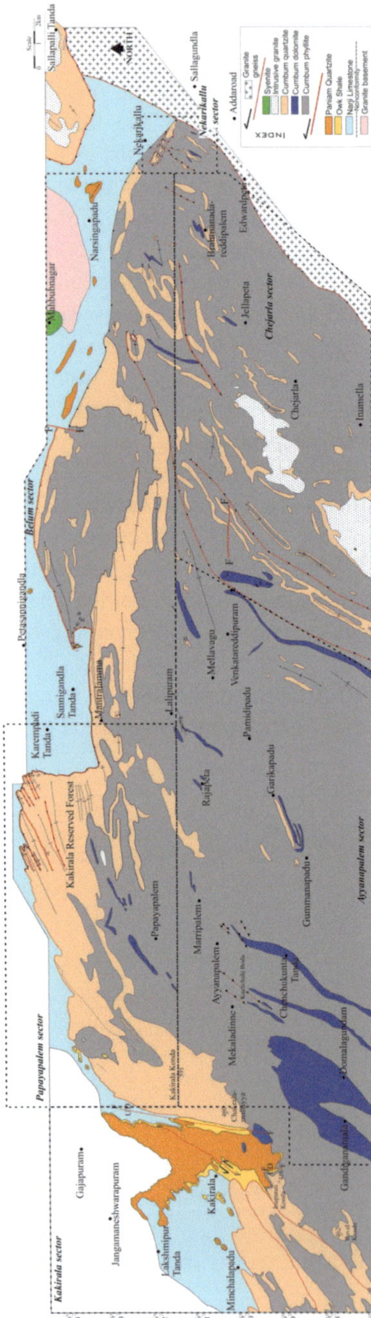

Fig. 3 Geological map of the northern part of the Nallamalai Fold Belt, Cuddapah basin showing boundaries of various sectors

an account of various structural features from the area which is divided in six sectors towards east (Fig. 3) as described below.

The studied area can be divided into two major domains i.e., the footwall and hanging wall of the major thrust plane (Satyapal and Tripathy 2014). The Nallamalai Group of NFB acts as hanging wall while the Kurnool Group of Palnad sub-basin is footwall. Of these sectors, the Ayyanapalem and Chejarla sectors are represented by NE-SW trending Cumbum phyllite and minor Cumbum quartzite and dolomite bands of Nallamalai Group. This sector is part of the hanging wall of the major thrust. viz., Vamikonda thrust. The Chejarla sector is intruded by younger granitoids (Sesha Sai 2004, 2013).

3.1 Ayyanapalem Sector

The ***Ayyanapalem sector*** in Gandiganamala-Ayyanapalem-Jayantiramapuram south of Papayapalem sector (Fig. 4a). The sector is dominated by phyllite, carbonate and phyllite-carbonate intercalations with NE trending major foliations. The bedding and S_{n1} cleavage transpose with tight to isoclinal to overturned F_{n1} folding as also evident from stereoplots (Fig. 4a, b). The D_{n1} deformations are overprinted by gently southeast dipping ductile shear zones with thrust movement (Fig. 4b, c). Southeast and southwest of Ayyanapalem, a number of such shear zones (S_c) overprint the earlier S_{n0}/S_{n1} foliation (Fig. 4c). Later thrust-related D_{n2} deformation tightened and reoriented the folds within the phyllite e.g., north of Domalagudem and around Mekaladinne. The folds are dragged with a top-to-NW movement to produce overturned tight to isoclinals folds (Mukherjee et al. 2015 for mechanism) where the S_{n0} and S_{n1} transpose (Fig. 4a, b, e). Juxtaposition of F_{n1} antiforms due to the development of break thrust characterizes the structural makeup of the Cumbum Quartzite. The quartzite is locally intercalated with phyllite showing close to tight folds.

3.2 Chejarla Sector

In the ***Chejarla sector***, Cumbum phyllite, quartzite and dolomite are exposed (Fig. 5a). Here, Chejarla granite expose with intrusive contact with the Cumbum quartzite and phyllite. The sedimentary rock units in this sector are overturned to recumbent (F_{n1} fold) with northwest vergence (Fig. 5b–f). The S_{n0} and S_{n1} transpose locally (Fig. 5e, f) with fold axis plunging gently in variable directions (Fig. 5c, d, and f). These F_{n1} folds developed in earlier deformation phase and reoriented to recumbent by later deformation. The recumbent F_{n1} folds are overprinted by inclined asymmetric F_{n2} folds which are the last folding event recorded from the area (Fig. 5g). The superposition of F_1 and F_2 folds form hook shaped Type 3 interference pattern (Ramsay 1967).

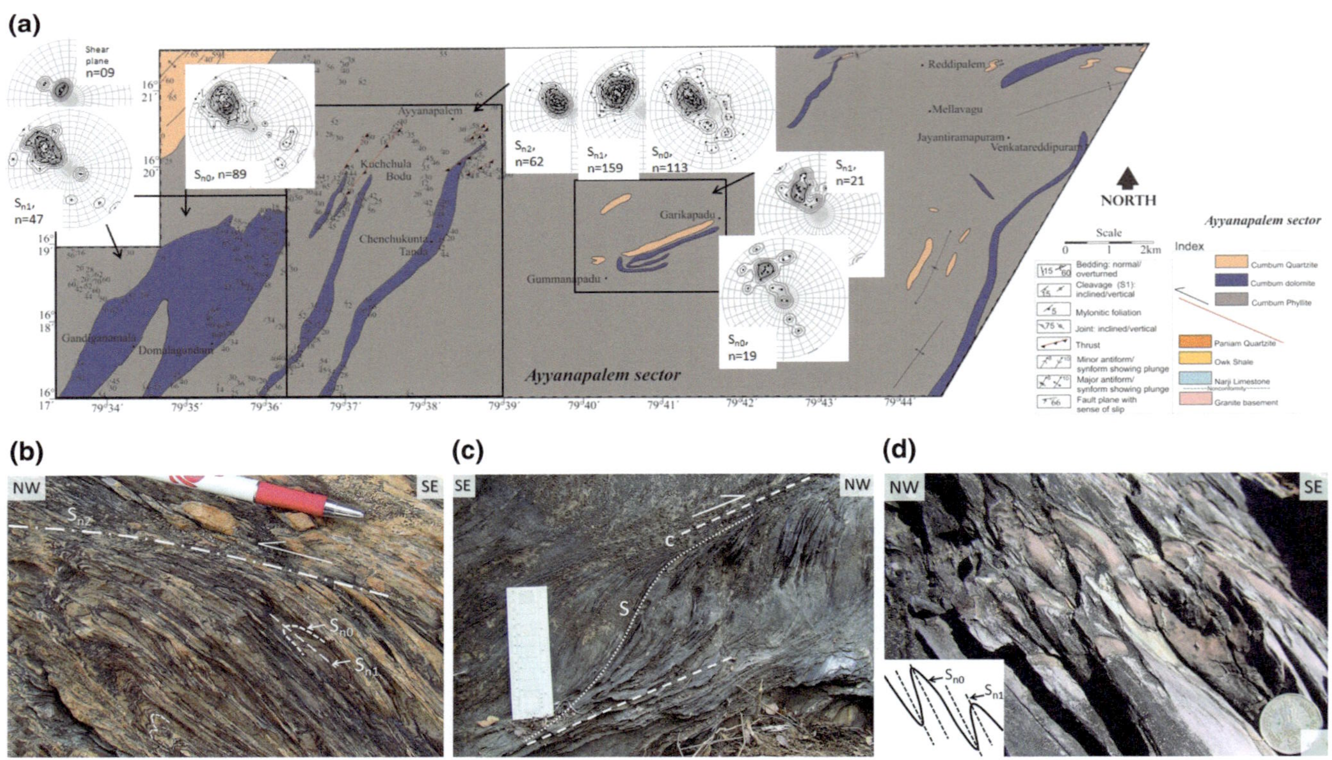

Fig. 4 **a** Geological map of Ayyanapalem sector. **b** Field photograph of S_{n2} fabric cutting through the transposed S_{n0}–S_{n1} in Cumbum Phyllite, western flank of Kuchuchula Bodu. **c** Field photograph of sheared dolomite intercalated with phyllite, Kuchuchula Bodu. S- and C-shear planes prominent (see Mukherjee 2011, 2012, 2013, 2014b, 2015 for kinematics). **d** Field photograph of SE dipping overturned isoclinal folds in phyllite intercalated with dolomite, N of Mekaladinne. Outcrop facing SW

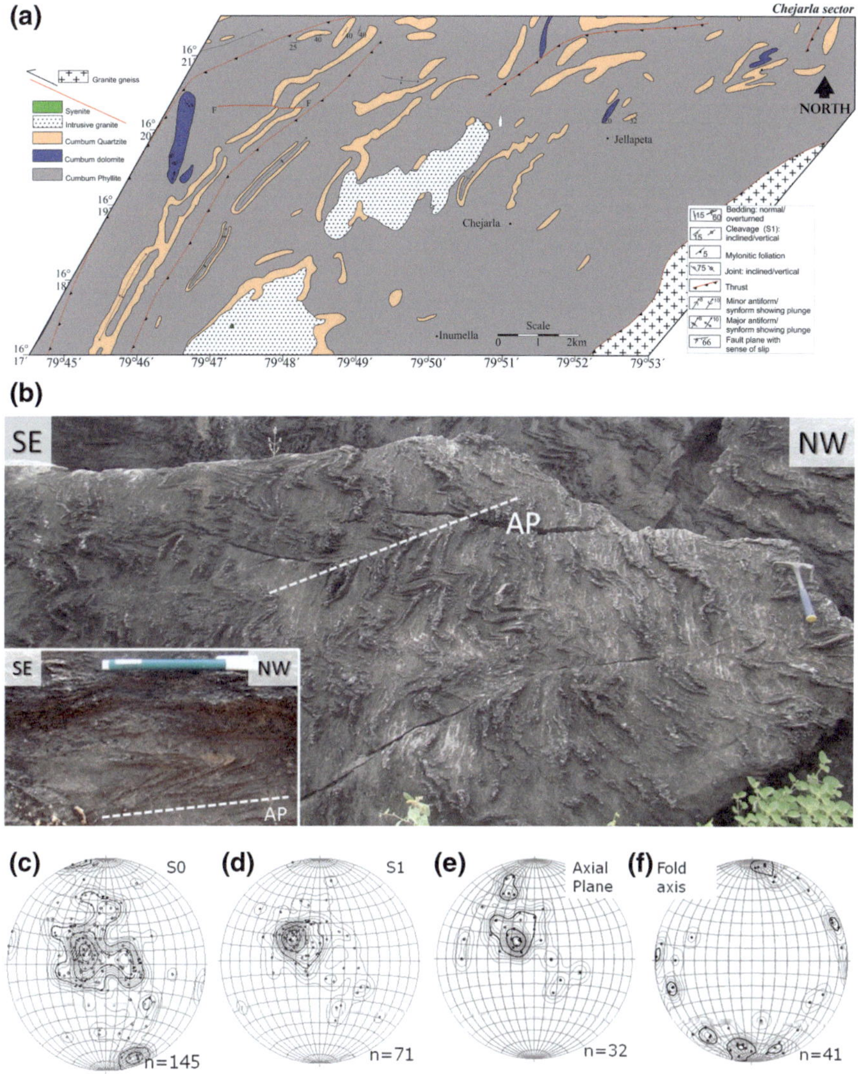

Fig. 5 **a** Geological map of Chejerla sector. **b** Field photograph of inclined folds in Cumbum Dolomite SE of Venkatareddipuram. Inset: Field photograph of recumbent folds in Cumbum Dolomite SE of Venkatareddipuram. **c–f** Stereographic projection of S_{n0}, S_{n1}, axial plane and fold axis of F_{n1} respectively. **g** Field photograph of hook shaped pattern formed by superposition of F_{n1} and F_{n2} folds giving rise to Type-3 interference pattern (Ramsay 1967). Inset: sketch of the field photograph

Southeast of Venkatareddipuram, NNE trending dolomite band show recumbent to inclined folds along with flat and ramp structures suggesting thrust related movement. Thrust related structures on minor scale are shear band, tightening and rotation of folds.

Fig. 5 (continued)

The Kakirala, Papayapalem, Belum and Nekarikallu sectors hosts part of ~E-W trending south dipping Vamikonda thrust, south of which of the hanging wall is exposed and northward footwall is outcropped. The changes in structural disposition is well marked and these variations are discussed.

3.3 *Kakirala Sector*

This sector in the western most part of the studied area (Fig. 6a) and hosts the Narji Limestone, Owk Shale and Paniam Quartzite in the footwall part while the hanging wall is represented by phyllite and quartzite of Cumbum Formation of Nallamalai Group. The multiple deformed Nallamalai Group has thrust over the Kurnool Group along the ~E-W trending south dipping Vamikonda thrust. The thrusting of NFB over the Palnad sub-basin is marked by deformations both in footwall and hanging wall. Evidence of thrust related deformation is manifested by its overprinting on earlier events of deformations of NFB and deformation of generally undeformed Kurnool Group.

3.3.1 Hanging Wall Deformation

In the area around Kakirala, the major outcrop trends NE-SW due to the deformation of the Cumbum Formation which differentiates itself from the structural disposition recorded from the rock units of the Kurnool Group. The D_{n1}

deformation is represented by large-scale NE trending, doubly plunging, inclined F_{n1} folds deforming the phyllite and quartzite of the Cumbum Formation (Fig. 6a, c). These folds define the primary outcrop pattern of the area, which is associated with SE dipping slaty to disjunctive axial planar cleavage (Fig. 6f), well pronounced in the Cumbum phyllite as opposed to the Cumbum quartzite which displays large amplitude folds.

Southwest of Kakirala, the Cumbum phyllite is thrust over the sub-horizontal Kurnool Group whose deformational impressions are recorded on both side of the thrust sheet. The hanging wall deformation is characterized by NNW vergent tight to isoclinal overturned folds overlying the sub-horizontal Owk Shale. Slickenside, oblique stylolites and tension gashes are other evidences which overprint earlier D_{n1} deformation (Fig. 6b, c). Near the thrust, the Owk shale shows overturned folds (S_{k2}; Fig. 6b) similar to those of the Cumbum Quartzite (Fig. 6c). These are called shear-induced drag folds (Mukherjee 2014a). Thrusting has also caused inversion (Fig. 6d) of the antiformal limb and its juxtaposition to another antiform. Such map-scale deformational feature as exposed SW of Kakirala and NW of Inuparti Konda are due to the drag within the limb caused by thrusting (Fig. 6e). Imbricate thrusting of the Cumbum Formation associated with sole thrust over the Kurnool Group is also evident from this part of the fold belt.

3.3.2 Footwall Deformation

The footwall deformation in the Kurnool Group is characterized by NE-SW trending gentle F_{k1} folds which are close to open along with associated moderately to steeply southeast dipping S_{k1} slaty to disjunctive cleavage (Fig. 6a). The S_{k1} is also present, at places, as crenulation cleavage within the Narji Limestone (Fig. 6b). The first generation F_{k1} folding within the Kurnool Group is also evident from aerial photographs, satellite images (Figs. 2 and 3, triangular outcrop of the Paniam Quartzite south of Durgi) and contour pattern, around the southern part of the Gundala Vagu. The Paniam Quartzite overlying the Owk Shale shows NE-SW trending doubly plunging folds which changes its orientation along later faults.

Later N-S trending steeply dipping normal faults has exposed the Owk Shale above the Paniam Quartzite, 2 km south of Kakirala (Fig. 6a). Here, the brecciated Paniam Quartzite defines a faulted contact with the Owk Shale. The horizontal Owk Shale is overlain by overturned and cleaved Cumbum phyllite thus, defining a thrust contact between the two lithounits coinciding with Vamikonda thrust.

3.3.3 Late Structures

Faults such as the NNE trending Chakralamallayya fault (Ramalinga Swamy 1972) in particular southeast of Gajapuram, deformed both the Nallamalai and Kurnool Groups. This steep NNE trending fault dragged the NE trending doubly plunging folds and axial planar cleavage with a left-lateral sense, within the NFB and

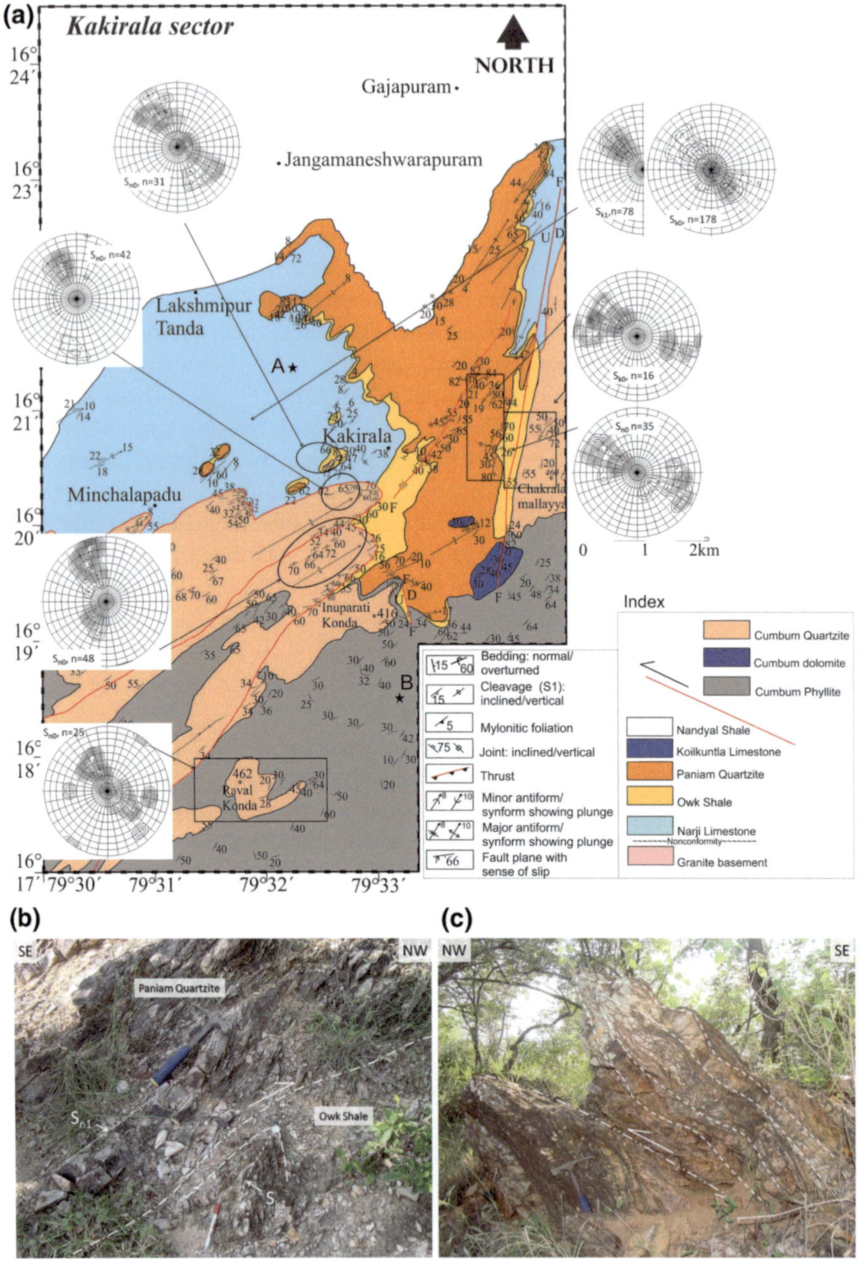
(a)
Kakirala sector
NORTH
Gajapuram
Jangamaneshwarapuram
Lakshmipur Tanda
Kakirala
Minchalapadu
Chakrala mallayya
Inuparati Konda
Raval Konda
A
B
16° 24′
16° 23′
16° 21′
16° 20′
16° 19′
16° 18′
16° 17′
79°30′
79°31′
79°32′
79°33′
0 1 2km
Index
Cumbum Quartzite
Cumbum dolomite
Cumbum Phyllite
Nandyal Shale
Koilkuntla Limestone
Paniam Quartzite
Owk Shale
Narji Limestone
Nonconformity
Granite basement
Bedding: normal/ overturned
Cleavage (S1): inclined/vertical
Mylonitic foliation
Joint: inclined/vertical
Thrust
Minor antiform/ synform showing plunge
Major antiform/ synform showing plunge
Fault plane with sense of slip
(b)
SE
NW
Paniam Quartzite
Owk Shale
(c)
NW
SE

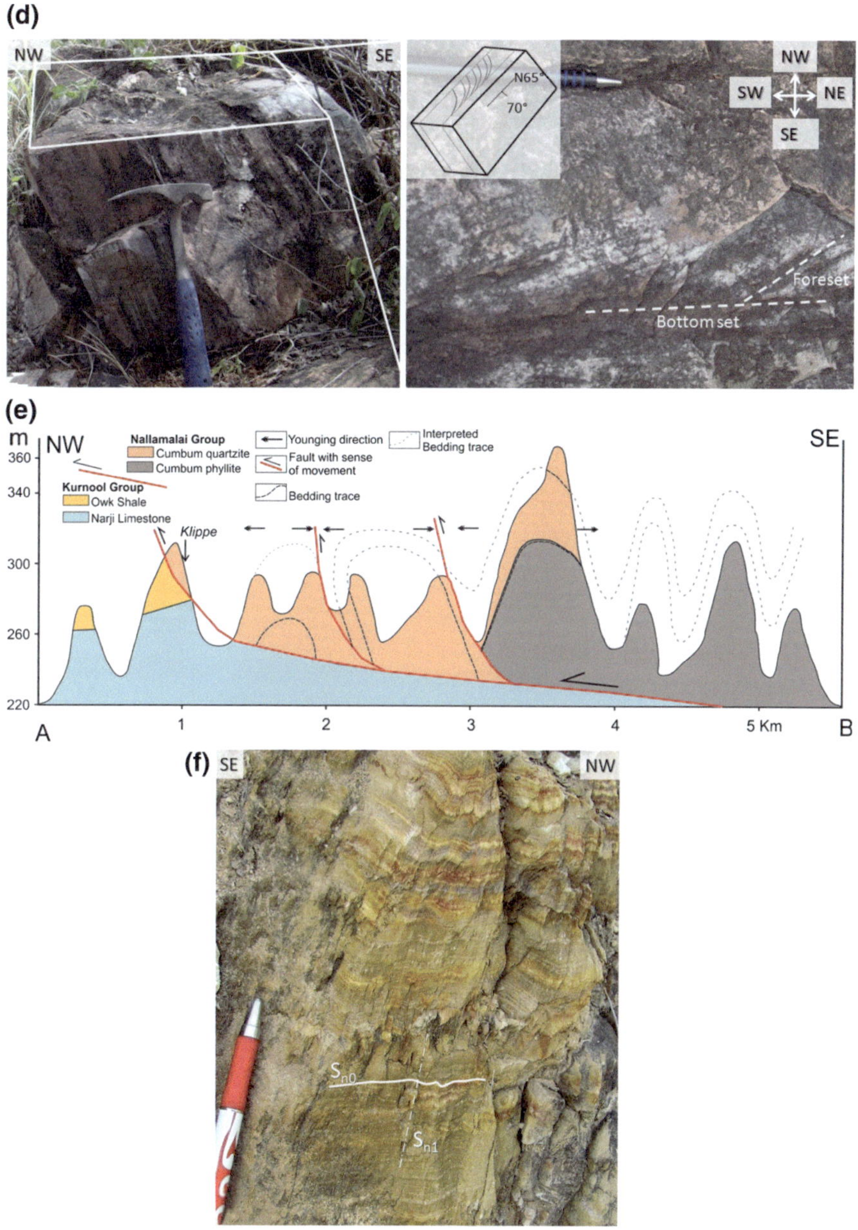

Fig. 6 (continued)

◂**Fig. 6** **a** Geological map of Kakirala sector along with bedding plane poles of the Cumbum Formation (S_{n0}) and Kurnool Group (S_{k0}) plotted in equal area stereo-net. **b** Overturned folds in Owk Shale with axial plane S_{k1}, underlying the folded Cumbum Quartzite (S_{n1}), 500 m SW of Kakirala. **c** Overturned and thrust sheet with NW vergence in Cumbum Quartzite near the thrust zone, part of *klippe*, SW of Kakirala. **d** Overturned limb in Cumbum Quartzite as evident by the younging direction of the cross beds, S of Kakirala. **e** Geological cross section along A-B, west of Kakirala. A-B marked in Fig. 6a. **f** Varigated laminations and near orthogonal SE dipping slaty to disjunctive axial planar cleavage in Cumbum phyllite

Kurnool Group, in NNE trend (Fig. 6a). As the fault is sub-vertical with beds dipping both in eastern and western side its hanging wall and footwall is difficult to establish. However, the stratigraphic positioning suggests that the upthrown western block has brought the Kurnool Group in juxtaposition to the Cumbum Formation. Along this zone the Owk Shale shows vertical to sub-vertical dips on both east and west directions along which ferruginous enrichment has taken place. Another later fault with NE trend cutting through the NE trending folded Paniam Quartzite east of Kakirala overprint earlier deformations. This fault is recognized on the basis of slickensides, brecciation, steepening of beds, tension gashes etc.

3.4 Papayapalem Sector

This sector is bounded between east of Chakralamallayya fault and west of Mantralamma has exposures of folded and thrust Cumbum phyllite, quartzite and carbonate bands (Fig. 7a). Some of the dolomite bands has lead-zinc mineralisation in form of galena and sphalerite. The NE to ENE trending Cumbum Formation is thrust part of the hanging wall of the major south dipping thrust over the Palnad sub-basin. The Narji Limestone is outcropped in the northern part of this sector and is part of the footwall.

3.4.1 Hanging Wall Deformation

The Cumbum Formation rocks show doubly plunging folds that are NE to ENE trend with beds dipping either towards NW or SE (Fig. 7a). Outcrop pattern at many places indicate thrust between alternate antiforms (e.g. in Kakirala Reserved forest Fig. 7a). Such structures are well documented in the west and northwest of Karempudi Tanda where alternate antiforms are stacked due to the formation of break thrust along the forelimb of the folds forming an imbricate thrust system. However, the areas south of Karempudi Tanda and around Mantralamma has open folds with one limb dipping vertical to sub-vertical towards NNW/N while other limb dip gently to moderately towards south and southeast. Variations in pattern and style of folds in strike extension i.e., west of Karempudi Tanda and around Mantralamma is possibly because of basement configuration and topography at the

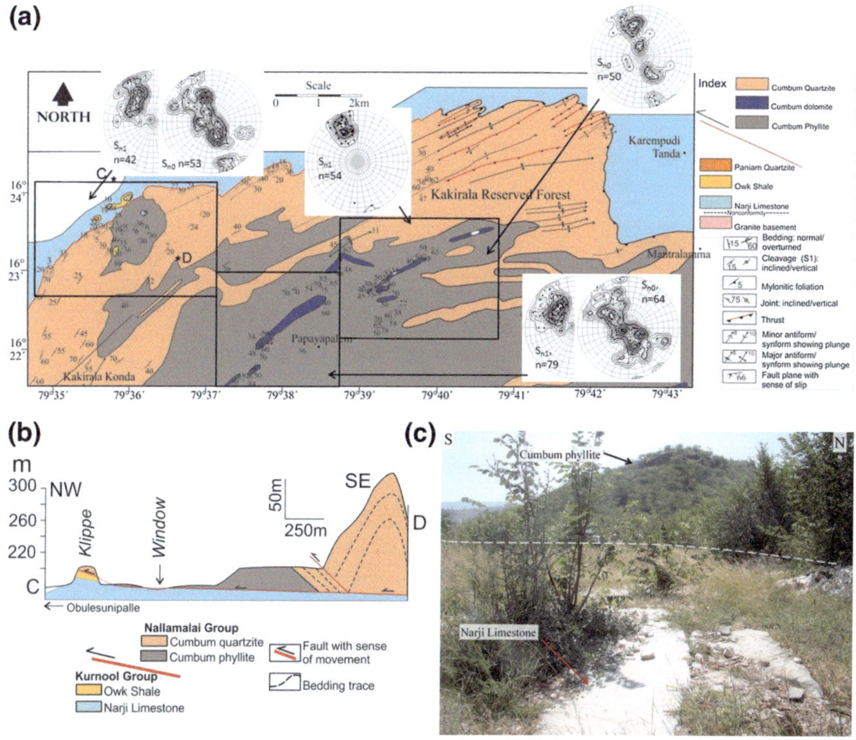

Fig. 7 **a** Geological map of Papayapalem sector showing S_{n0} and S_{n1} foliations. **b** Geological cross section along C-D, east of Gajapuram (Fig. 7a). **c** Sub-horizontal Narji Limestone as window, surrounded by deformed Cumbum phyllite, east of Gajapuram

time of thrusting. The quartzite exposed west of Karempudi Tanda represents a possible décollement associated with imbricate thrusts. However, due to possible topographic height of competent basement (granite and sedimentary cover of Kurnool Group) the NFB were restricted around Mantralamma with development of inclined to overturned folds.

3.4.2 Window and Klippe

Satellite image study (Fig. 2) revealed NE-SW trending doubly plunging folds (Fig. 2) in the southeast of Durgi. The SE limb of the fold has a thrust contact with southern lithounits. Extensive erosion in this part exposes of older Cumbum phyllite as an *klippe* and Narji Limestone and Owk Shale as *window* lying below the Nallamalai Group (Fig. 7a–c). The thrust stacks represent imbricate thrust system while the presence of *klippe* and *window* structures are understood to be sole thrust along which the NFB is thrust over the Palnad sub-basin (Fig. 7b).

3.5 Belum Sector

This sector stretches between Mantralamma to Narsingpadu in northeastern part of the mapped area (Fig. 8a) where the quartzite and phyllite of Cumbum Formation is thrust over the Narji Limestone of Kurnool Group.

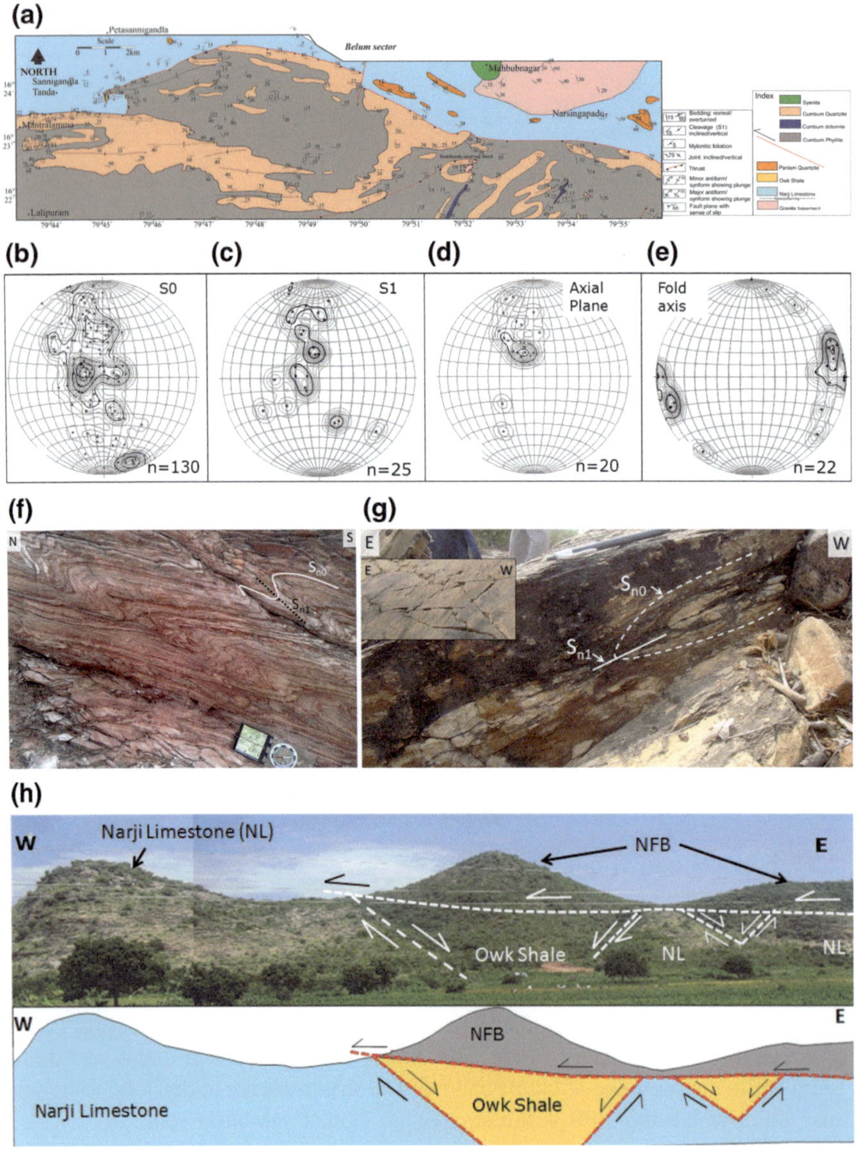

◂**Fig. 8** **a** Geological map of the Belum sector. **b–e** Stereographic projection of S_{n0}, S_{n1}, axial plane and fold axis from Cumbum Formation exposed in the Belum sector. **f** Field photograph of the overturned north vergent folds in Cumbum phyllite east of Sanigandla Tanda. **g** Field photograph of the overturned isoclinal folds in transverse section of Cumbum Quartzite at the contact with granite (not seen in this photograph). Inset shows ridge and groove on XY plane. **h** Panaromic view of the E hill range east of Sanigandla Tanda. Notice the variations in the vegetation density and the lithological changes as viewed from the cross section. Lower part of figure shows sectional view of the hill range with earlier formed horst and graben structures in the Kurnool Group overthrust by the Cumbum Formation of NFB. (i) Stereographic projection showing relation between S_{n0} and fold axis from the thrust Cumbum Formation. Data collected from east of Sanigandla Tanda (**h**). **j** Stereographic projection of bedding plane, S_{k0} from Narji Limestone showing gentle dip and gently plunging stretching lineation. **k** Sectional view of the sub-horizontal Narji Limestone showing bedding parallel shear, SE of Sanigandla Tanda. Note the plan view in inset showing bedding parallel stretching lineation in Narji Limestone

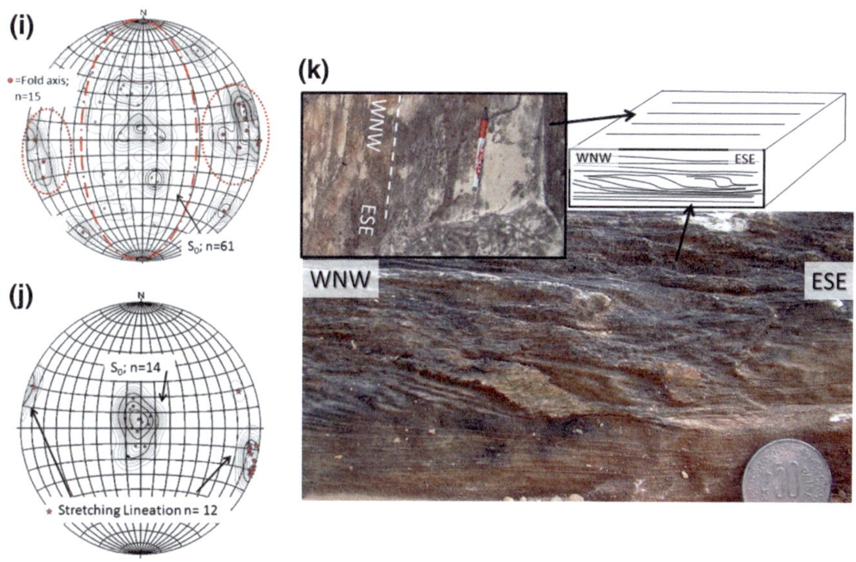

Fig. 8 (continued)

3.5.1 Hanging Wall Deformation

The doubly plunging asymmetrical E trending folded lithounits of the Cumbum Formation has north vergent axial planar cleavage (Fig. 8b–f). The S_{n1} cleavage is slaty to disjunctive with south to southeast dip near to the thrust zone (Fig. 8c). However, further southward the bedding and cleavage show near to orthogonal relationship. The change in axial plane of folds from S to N in this sector indicates drag of hanging wall near to the thrust zone as a result of top-to-north shear component.

In the northern part of this sector the granitic basement for the Kurnool Group in the Palnad sub-basin is exposed west of Gopalapuram. The granite is phaneritic with porphyry of K-feldspar, quartz and schlieren of biotite and minor hornblende. About 3.5 km south of Mahbubnagar, within the Guttikonda reserved forest phaneritic granite is exposed which has thrust contact with the Cumbum Formation (Fig. 8g). The quartzite is isoclinally folded and overturned with west vergent gently dipping axial plane (Fig. 8g). Slickenside on fold limb is represented with the development of ridge and groove (similar to the ridge and groove of Power and Tullis 1989; Means 1987 and Steps—Sheltering trails of Doblas 1998) suggesting faulted contact between the Cumbum quartzite and the granite (Fig. 8g inset).

3.5.2 Footwall Deformation

Close to the northernmost exposed Cumbum quartzite and $\sim$1.5 km east of Sannigandla Tanda, NW of the Belum sector, the Kurnool Group shows pre-thrust and syn-thrust deformations. The Narji Limestone, Owk Shale and Paniam Quartzite of the Kurnool Group are sub-horizontally exposed here with development of horst and graben structures (D_{k1}), which is later thrust by northwest vergent overturned Cumbum phyllite plunging both towards east and west (Fig. 8h). Deformations related to this thrusting has generated stretching lineation parallel to bedding plane of the Narji Limestone (Fig. 8i–k) along with development of sigmoids as evident from transverse sections (Fig. 8k).

3.6 *Nekarikallu Sector*

This sector is the easternmost part of the study area and is important because of the presence of the junction of the Vamikonda (a.k.a Nallamalai) and Vellikonda thrusts (Fig. 9a). The later imprints of deformations recorded in this sector is dependent up on the orientation of the structural fabric.

The ESE trending ridge of Cumbum Formation S of Nekarikallu is represented by NW trending folds (F_{n1}, Fig. 9b–d) and south to southeast dipping axial planar cleavage (S_{n1}, Fig. 9e) with SE plunging fold axis (Fig. 9f). The ridge is cut through by a number of imbricate thrust (D_2, Fig. 9b, c) whose movement direction parallels stretching lineation and fold axis trend (Fig. 9g). Due to this thrusting, the isoclinal recumbent folds (F_{n1}) is overprinted by later NW vergent crenulations (F_{n2}) and mylonite related to the D_{n3} deformation. Near the thrust zones the ESE trending S_{n1} ($\|S_{n0}$) cleavage is folded/crenulated to generate reclined folds (Fig. 9d, f–h). These imbricate thrusts are linked to a sole thrust as envisaged from gently southeast dipping thrust zone (Fig. 9a, i) and cross section (Fig. 9b). The sole thrust is evident by the presence of thrust contact between the overturned folded phyllite over the asymmetrical folds (Fig. 9i). The formation of recumbent folds is possibly due to the drag of F_{n1} folds along the thrust zone. These late thrusts overprint earlier structural fabrics and generate F_{n2} folds and thrust slices.

However, the NE trending ridge NW of Adda road show transposition of S_{n0} and S_{n1} (Fig. 9j, k). The folds are west vergent overturned to recumbent (Fig. 9m, n) and vary in style as observed from the ESE-WNW ridge. These structures are overprinted by shear band with top-to-west thrust movement (Fig. 9o). The late thrust related structures include development of gently east plunging stretching lineation (Fig. 9l).

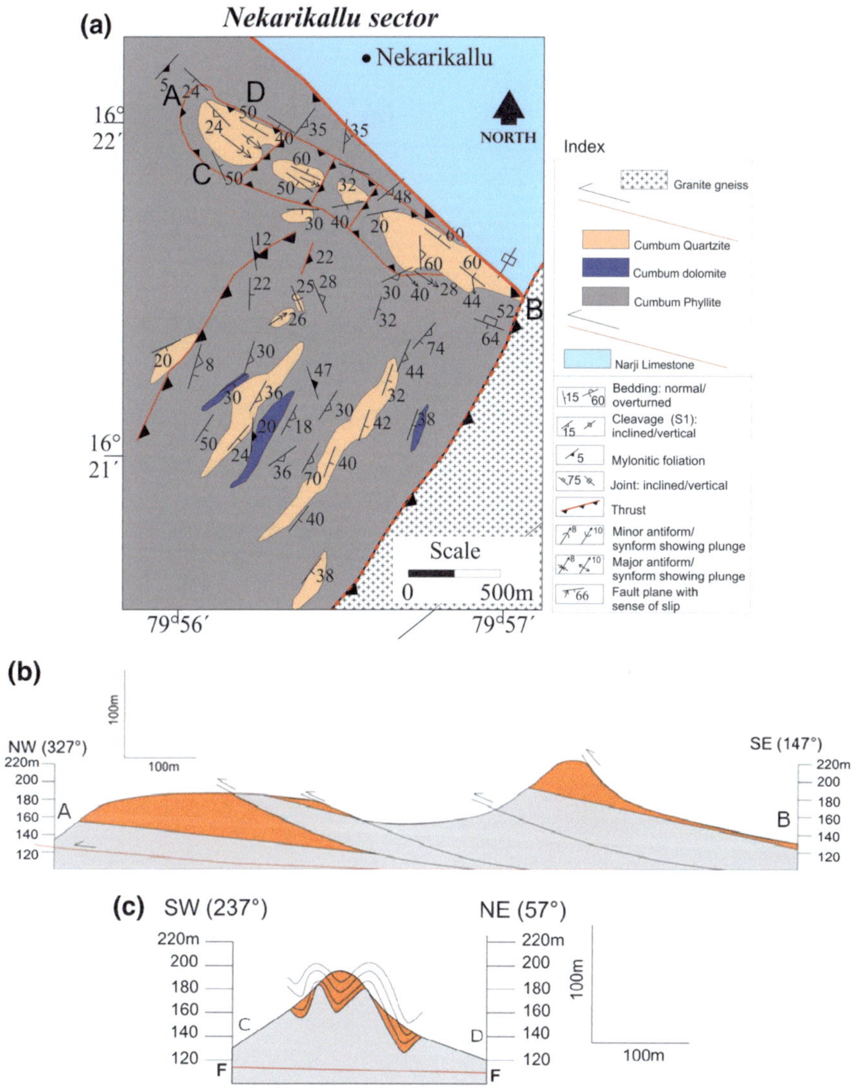

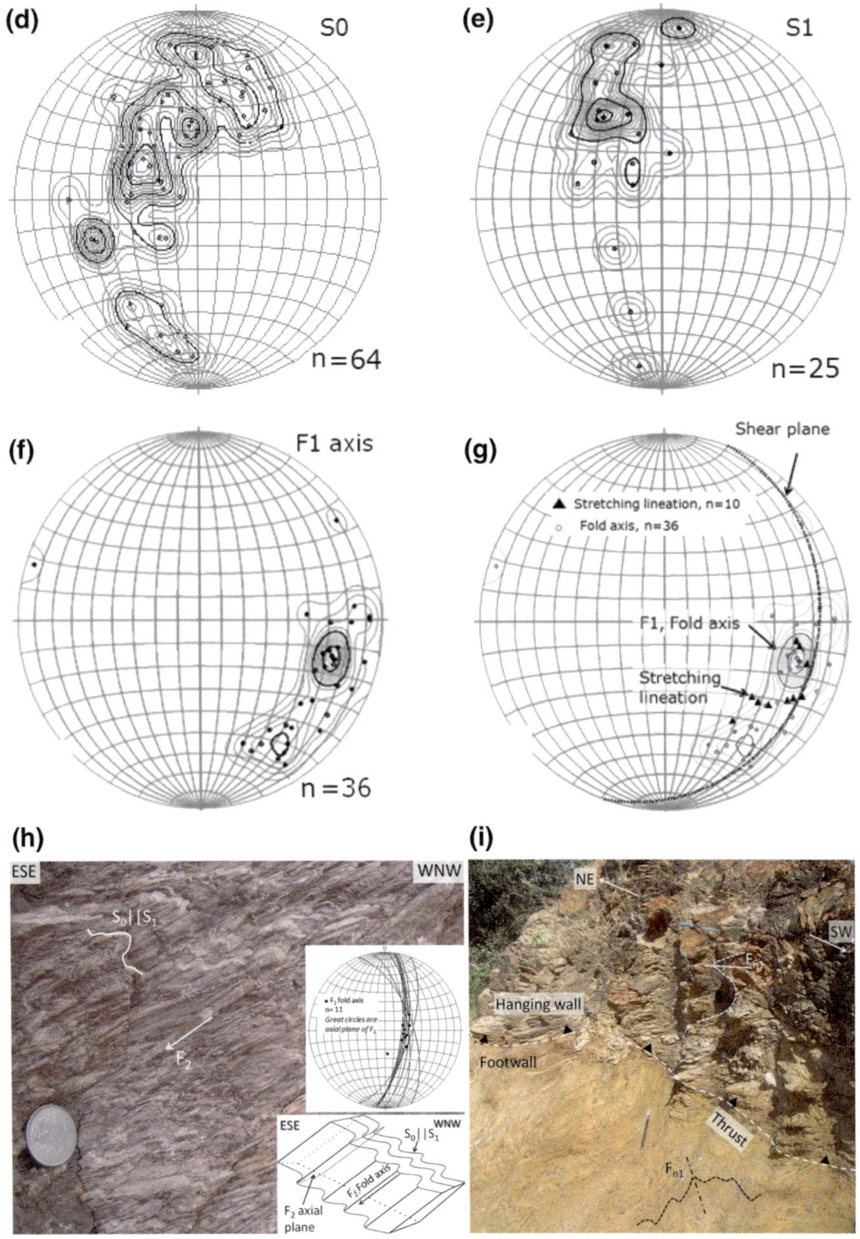

Fig. 9 (continued)

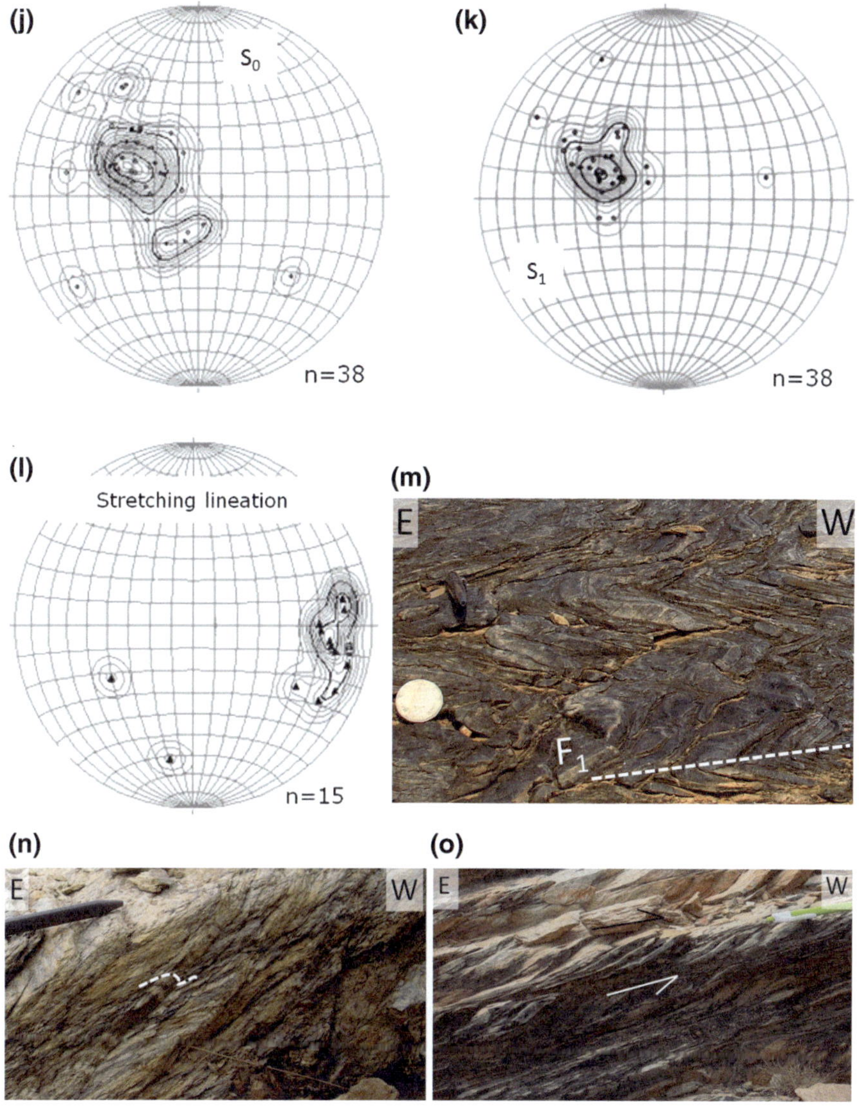

Fig. 9 (continued)

◂**Fig. 9** **a** Geological Map of the Nekarikallu sector. **b** Cross section along A-B, along the ridge axis of ESE ridge, south of Nekarikallu. **c** Cross section along C-D, across the ridge axis of ESE ridge, south of Nekarikallu. **d–f** Stereographic projection of S_{n0}, S_{n1} and F_{n1} fold axis; **g** Stereographic projection showing relation between stretching lineation, F_{n1} fold axis and shear plane. **h** Field photograph of Cumbum phyllite along the NE-SW imbricate thrust zone south of Nekarikallu with development of reclined folds (F_{n2}) due to the folding of transposed S_{n0}/S_{n1}. **i** Field photograph of Sole thrust SW of Nekarikallu marked by the thrust recumbent folds over the asymmetrical folds of Cumbum phyllite. Note variations in orientation of F_{n1} folds in hanging wall and footwall. **j–k** Stereographic projection of S_{n0} and S_{n1} showing transposition of the both the fabrics, along the NE trending ridge, NW of Adda Road. **l** Stereographic projection of the stretching lineation along the shear zone (**n**) showing consistent ESE plunge. **m** Field photograph of recumbent folds in Cumbum phyllite close to the NE-SW trending thrust zone SW of Nekarikallu. **n** Field photograph of west vergent overturned tight folds (F_{n1}) in heterolithic Cumbum quartz-phyllite, along the NE trending ridge NW of Adda Road. **o** Field photograph of shear zone cutting through the overturned F_{n1} folds (**m**) with a top-to-west tectonic movement in Cumbum dolomite, developed along the NE trending ridge NW of Adda Road

4 Discussions

4.1 Geometry, Kinematic and Relative Chronology of Deformation in NFB

The Nallamalai Group is folded and cut through by a number of thrust sheets, which are convex towards the Palnad sub-basin resembling the direction in which the thrust verge. The events of folding and faulting are categorized in time and space that generated changes in the fold pattern and geometry. The fold belt (NFB) acted as footwall with respect to the Nellore schist belt (NSB) along the Vellikonda thrust during D_{n2} stage of deformation (Mesoproterozoic, Saha 2002; Table 1) which is unrelated to the Nallamalai/Vamikonda thrust, against which the fold belt is acting as hanging wall ($D_{n3} \approx D_{k2}$). As the Vamikonda thrust is juxtaposing Mesoproterozoic NFB and Neoproterozoic Kurnool Group, it is envisaged that the thrust must at least be of Late Neoproterozoic age. The geometry of fold thus formed and modified during different deformations indicates multiple compressions in the Eastern part of the East Dharwar craton. The prime cause of this variation can be because of the vorticity within the shear zone related to thrusting, rheological differences and geometry of early formed folds (Carreras et al. 2005). To understand the structural evolution, the geometry of this belt is to be considered which is characterized by a number of W and NW vergent thrusts, prominent detached folds and oroclinal nature of the fold belt. An orthogonal change in the amount of curvature of the belt from N to E (Figs. 1 and 2) represent an oroclinal bend as theoretically suggested by Johnston et al. (2013). The presence of *window*, *klippe* and imbricate thrust system along the Vamikonda thrust are evidences for allochonous nature of NFB. Further, the presence of *Palnad Nappe* (Natarajan and Rajagopalan Nair 1977; Saha and Chakraborty 2003; Chakraborti 2006; Chakraborti and Saha 2009; Saha et al. 2010) suggests an exotic nature of NFB

Table 1 Summarised chronological sequence of deformations as recorded from northern Nallamalai Fold Belt (NFB)

Cambrian	(iii) Late faults such as in Kakirala sector displaces both the Kurnool and Nallamalai Group of rocks	Post-thrust
Neoproterozoic	(ii) Thrusting (D_{n3}) of NFB over Palnad sub-basin (Kurnool Group) • **Hanging wall (NFB)**: (a) Development of imbricate thrust in Cumbum Formation linked to sole and imbricate thrust (b) Reorientation of and overprinting over F_{n1} folds. Tight to isoclinal, overturned to recumbant/reclined fold (c) Presence of Klippe and window structures within NFB, suggesting thin-skinned tectonics • **Footwall (Palnad sub-basin)**: Localised folding (F_{k1}) and development of cleavage (S_{k1}) due to compression ($D_{k2} \approx D_{n3}$) related to thrusting of NFB	Syn-thrust
	(i) Localised extensional events • e.g., presence of horst and graben structures (D_{k1}) within Kurnool Group in Belum sector	Pre-thrust
	Deposition of Kurnool Group in Kurnool and Palnad sub-basins	
Mesoproterozoic and Paleoproterozoic	(ii) Thrusting of Nellore Schist Belt (NSB) along Velikonda Thrust • Development of asymmetrical folds and associated thrusts within NFB • Reorientation of F_{n1} folds and refolding at places ($\approx D_{n2}$ deformation) • Development of break thrust within Nallamalai quartzite (i) Folding of Nallamalai Group (F_{n1} folds D_{n1} deformation) • Doubly plunging open to close folds	

The term "thrust" used in last column viz., "pre-thrust; syn-thrust and post-thrust" are in relation to the formation of the Vamikonda thrust due to the thrusting of NFB over Palnad sub-basin

with respect to rest of Cuddapah basin (Saha 1994, 2002; Saha and Chakraborty 2003; Tripathy and Saha 2010).

Variations in fold styles in different lithounits are well documented in the northern part of NFB. The NE-SW trending open folds in Cumbum Formation, south of Minchalapadu in Kakirala sector, define earlier F_{n1} folds related to D_{n1} deformation (Fig. 6a). The forelimb of these folds are cut through by later thrusts resulting in the juxtaposition of two antiforms with synform being downthrown below the former during the D_{n2} deformation (Fig. 10a). These thrusts are connected to a sole thrust along which the NFB is thrust over the Kurnool Group of Palnad sub-basin as evidenced by the *klippe* SW of Kakirala (Fig. 6e). The NW vergent F_{n1} tight to isoclinal and overturned folds in Cumbum phyllite suggests overall top-to-NW shear component. The competency contrast between quartzite and phyllite is the principle cause for such modifications in fold style, which deformed predominantly by simple/general shear (Mukherjee 2012, 2013) during

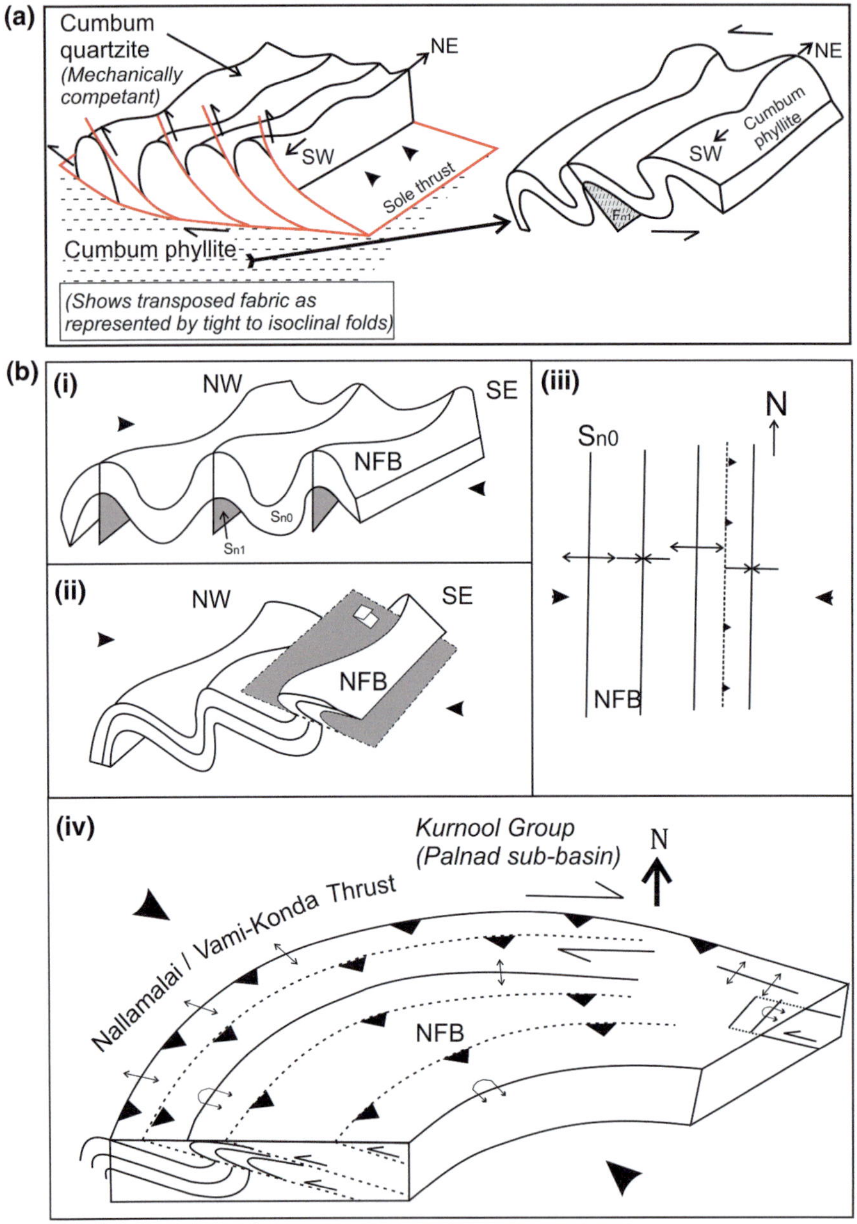

Fig. 10 **a** Cartoon showing variations in fold style related to rheological variations as observed from the field. Note that the phyllite being mechanically incompetent are tighly folded to transposed fabrics, glide along a detachment plane while quartzite has development of imbricate thrust. **b** (i) Model showing first phase of folding in NFB (ii) which been modified and reoriented by later phase of folding and thrusting. (iii) The first phase of folding and later reorientation is show in plan map. (iv) During syn-thrust phase, the fold belt developed oroclinal bending with development of right-lateral movement in northern part of NFB

thrusting. The existence of such mechanical instability in a simple to general shear environment modified the pre-existing folds (F_{n1}) and developed discrete shear zones (D_{n2}) within phyllite and carbonate e.g., NE trending gentle shear zones with top-to-NW thrusting SE of Ayyanapalem and Kuchchula Bodu in the Ayyanapalem sector. Similar thrust stacking are present in the Papayapalem sector where imbricate thrust system is envisaged based on window and klippe, east of Gajapuram along with juxtaposed antiforms. The phyllite and intercalated carbonates in the Ayyanapalem sector are isoclinal overturned to recumbent folds trending in NE as opposed to the Papayapalem sector where the structures trend E. The bulk geometry suggests maximum compression along NW-SE. Under such scenario, NE plunging asymmetrical folds, which are perpendicular to the shear and compressional components, development of Type- 3 (Ramsay 1967) interference pattern are evident (Fig. 5g). However, such structural interferences are not the same at each location as for example in Belum sector such overprinting caused northward vergence of fold and cleavage.

However, the structural geometry present in the eastern part of the area i.e., Chejarla sector, the fold geometry and architecture differ significantly. The folds bounded within the thrust sheet SE of Venkatareddipuram are recumbent to inclined (Fig. 5b; Ramsay et al. 1983). Changes in shape and orientation of folded layer and their axial surfaces are due to the superposition of simple shear deformation on earlier existing folds (Ramsay et al. 1983). Under conditions of intense shear, the folds rotated towards the plane of primary shear (or the C-plane) to develop overturned, asymmetrical to recumbent folds overriding the thrust, between south of Mahbubnagar and south of Venkatareddipuram.

Due to the oroclinal bent, the two thrust fronts i.e., the Nallamalai thrust (Vamikonda thrust in northern part of NFB) and the Vellikonda thrust, which activated at different time, juxtapose in the northeastern end of the NFB near Nekarikallu. The latest deformational imprints over the ESE and NW trending ridges in Nekarikallu sector are important in understanding the overall structure of the northern part of the NFB. The D_{n1} deformation in the ESE trending ridge (S of Nekarikallu, Fig. 1) is characterized by S to SE dipping axial planar cleavage in phyllite and associate with doubly plunging open folds in quartzite. However, the NE ridge (NW of Addaroad, Fig. 1) has SE dipping S_{n0} overturned isoclinal to tight folds, which are parallel to the axial planar cleavage (S_{n1}). Such structures are akin to those reported from Porumamilla situated in the east-central part of the NFB (Tripathy and Saha 2010; Saha et al. 2010). Imprints of later compression are documented from both the ridges. The presence of NE trending sole thrust and associated imbricate thrusts in the ESE ridge (Fig. 9a–c, i) and N-S trending shear zones (Fig. 9o), modified the NE-SW trending F_{n1} folds. The asymmetry of these NW vergence folds, supports a top-to-NW thrust component related to late horizontal compression. The evidences of structural discordance suggest a thrust contact between the NE and ESE ridges.

The sedimentary lithounits have rheological differences, and the mechanically weak layers are surfaces along which thrusts are localized. Many thrust dominated belts have similarities in them as major décollements are located in shale or salt unit overlain by competent units (Mitra 2003; Nemčock et al. 2005; Mukherjee et al. 2010). The deformations within the NFB imbricated thrust stacks and sole thrust along which rock units such as shale, phyllite and carbonate frequently associate with such weak horizons. Where thrust planes passes through competent lithounits e.g., quartzite, they steepens and form flat-ramp geometry. The F_{n1} folds are modified to overturned to recumbent in phyllite rich layers (e.g., Fig. 5b, 9 m) while in resistant quartzite layers the strain is accommodated by development of thrust surfaces.

The D_{n1} deformation is predominantly associated with axial planar slaty cleavage in phyllite and disjunctive cleavage in quartzite showing variations in orientations in different sectors due to later deformations such as thrusting and related folding. The thrust has reoriented the fold and cleavage and provided them the asymmetric geometry that defines the overall NW/N vergence interpreted in terms of a non-coaxial D_{n2} deformation. There are marked variation in the modifications and overprinting of D_{n2} deformation over the first generation structures. Depending upon the position of the fold relative to the thrust zone the axial plane of the F_{n1} fold are close to the plane of simple shear (e.g. in the dolomite bands east of Venkatareddipuram and heterolithic quartz-phyllite south west of Nekarikallu, Fig. 5b). Vergence of overturned to inclined nature of fold is common throughout the area with similar slip sense. The overthrusting of NFB over the Palnad sub-basin must had happened after the development of thrust restricted within the fold belt as thrust surfaces has also changed their trend from NE to E similar to the variations in the fold trends. Oroclines (Carey 1955; Johnston et al. 2013) are result of the action of later stress over existing orogen either in a transverse (progressive) or in a longitudinal (secondary) manner. The progressive orocline involves basal décollement with compressive stress perpendicular to the strike of the orogen which are responsible for bending of fold and thrust belt undergoing thin-skinned tectonics in response to ongoing crustal compression. The thin-skinned tectonic nature of NFB is envisaged by Chakraborti and Saha (2009) from western part of the NFB. The presence of window and klippe structures (Figs. 6e and 7b, c) from present study area supports the thin-skinned nature of the NFB. Thus, the geometry and architecture of the NFB presented here corroborate the progressive orocline as (i) the belt shows thin-skinned nature and (ii) the late compressional structures are result of compressive stress perpendicular to the strike of the orogen. However, such late compressional structures in central and southern part of NFB where it trends ~N-S, are difficult to establish as it approximate Type 0 interference pattern (Ramsay 1967). The bending of the NFB fabric along Vamikonda thrust front is thus in response to the compression directed from SE to ESE. The presence of granitic basement and Kurnool Group in the Palnad sub-basin must had acted as a tectonic obstacle (i.e., a backstop) due to which the northern part of the fold-and-thrust belt dragged right laterally (Fig. 10b). The footwall uplifts as represented by horst and graben (Fig. 8h) south of Karempudi Tanda represented by hills of Narji Limestone must also had acted as obstacle due to which the Cumbum Quartzite in this part is stacked to develop overturned to inclined folds with one limb being vertical to sub-vertical as opposed to the open doubly plunging folds west of Karempudi Tanda of Papayapalem sector.

4.2 Deformations in Kurnool Group

The Kurnool Group in the Palnad sub-basin acted as footwall with respect to the Vamikonda thrust and bears various stages of deformation which are pre-, syn- and post-thrust (Table 1). The term thrust here is related to the Vamikonda thrust. The pre-thrust deformations are mostly related to extensional structures such as N-S trending normal faults. The horst and graben structures present in the Kunool Group are overthrust by the Cumbum Formation in Belum sector. The former structures represent such extensional events related to pre-thrust deformations (Fig. 8h).

The development of NE trending folds and SE dipping disjunctive to slaty cleavage in Narji Limestone, Owk Shale and Paniam Quartzite are well documented from the Kakirala sector as discussed earlier. Similar development of cleavage, gently ESE plunging bedding parallel stretching lineation and development of mylonite in Narji Limestone in Belum sector with north or northwest vergence are evidences of syn-thrust deformation (Fig. 8i–k).

The N-S trending Chakralamallayya fault cutting through the Cumbum Formation and Kurnool Group east of Kakirala reoriented the structural fabric of both the Groups in N-S direction. The steepening of Owk Shale and Paniam Quartzite along the fault and reorientation of folds and thrusts to N-S direction are evidences of post-thrust deformation. The post thrust deformation is also reported in the southern part of Cuddapah basin where NNE-SSW strike-slip faults are arranged in right stepover pattern (Goswami et al. 2016). The development of dome and basin structures of Type- 1 interference pattern (Ramsay 1967) in Paniam Quartzite along the fault zone in the study area are evidences of post-thrust deformation possibly related to compression that acted on early formed NE trending folds.

4.3 Tectonic Implications

Evidences of 1.8–1.5 Ga Columbia supercontinent are recorded from most of the continents (e.g. Rogers and Santosh 2002; Nance et al. 2014). Such evidences are also recorded from India such as the amalgamation of northern and southern India along the Central Indian Tectonic Zone (CITZ) at 1.8 Ga (Jain et al. 1991; Acharyya and Roy 2000; Bhowmik et al. 1999; Bhowmik and Roy 2003). Based on palaeomagnetic data for the North China Block during the time interval ca. 1780–1440 Ma, Zhang et al. (2012) placed the western India adjacent to the North China Block. Such Columbia reconstruction contains the southern part of Eastern Ghats Belt and Napier Complex to have juxtaposed at 1.6 Ga (Henderson et al. 2014).

In EGB, the Western Charnockite Zone (WCZ) (cf. Domain 1A of Dasgupta and Sengupta 2003; Ongole domain, Krishna Province of Dobmeier and Raith 2003) has suffered ~1.6 Ga granulite metamorphism without signs of Grenvillian overprint (Rickers et al. 2001; Dasgupta and Sengupta 2003; Dobmeier and Raith 2003; Bhattacharya et al. 2010) as opposed to its central and eastern parts. Such metamorphic imprints of Columbia affinity in Eastern Dharwar craton possibly affected the deformations further westward in NSB and NFB (Tripathy 2010; Tripathy and Saha 2013, 2015). The Nallamalai Group has undergone two major folding events as evident from different parts of the fold belt (Matin and Guha 1996; Mukherjee 2001; Saha 2002; Tripathy and Saha 2010; Saha et al. 2010). It is understood to have developed as a result of compressive stresses active during the Columbia supercontinent (Saha 2002, 2004; Tripathy and Saha 2013).

The thrust related emplacement of Kandra igneous complex and Kanigiri ophiolitic melange (Sesha Sai 2009; Saha 2011; Sesha Sai 2013) in southern Nellore schist belt (NSB) suggests a Paleoproterozoic (1847 + 14 Ma, Vijaya Kumar et al. 2010) plate convergence along the SE margin of India (Bhadra et al. 2004; Saha 2011; Saha and Patranabis-Deb 2014). The earlier phase of deformation in metapelites of younger Udaygiri Group of NSB attended presumably before 1933 ± 11 Ma based on retrograde metamorphic xenotime dating (Das et al. 2015). The 1.8–1.9 Ga deformation and metamorphism related to the amalgamation of these East Gondwana fragments deformed and metamorphosed the NSB and its thrusting to NFB along Vellikonda Thrust (Tripathy and Saha 2013). Various radiometric ages of igneous intrusions in NFB (reviewed in geological setting of this paper) cluster around Late Paleoproterozoic to Early Mesoproterozoic (1.5–1.6 Ga). The two major folding events in NFB are also understood to be prior to such igneous intrusions within them (Saha and Chakraborty 2003). It is thus plausible that the first deformation phase generating the F_{n1} folds and thrusts in NFB and later structures of NSB were related to compression of Late Paleoproterozoic of Columbia Supercontinental affinity. However, sediments deposited during ~1.9 Ga in Papaghni sub-basin in western part of Cuddapah basin lack regional deformation and thus suggest its variance from deformed NFB in terms of deformational chronology, tectonic set up and thus paleogeographic position.

Steep basement cutting faults such as the Gani-Kalva and the Kona faults in western part of the Cuddapah basin opened the Kurnool sub-basin (Tripathy and Saha 2013) possibly coeval to the emplacement of ~1090 Ma kimberlite pipes west of the basin margin (Kumar et al. 1993). Recent discoveries of 1.4–1.3 Ga lamproite dyke-sill complex in Tadpatri Shale are proposed as diamond provenance for Banganapalle conglomerate (Joy et al. 2012). The deposition of Kurnool Group thus has happened during Early to Mid Neoproterozoic. Such crustal-scale extension during the Late Mesoproterozoic to Early Neoproterozoic is evident from ~1300 Ma Prakasam alkaline province (PAkP), which has a number of alkaline intrusive bodies in the schist belt and considered to represent rift related magmatism (Leelanandam 1989; Upadhyay et al. 2006; Leelanandam et al. 2006; Vijaya Kumar and Leelanandam 2008; Dharma Rao et al. 2010).

The thrusting of NFB over the Palnad sub-basin and development of folds and cleavage in the Kurnool Group are related to compressional activities possibly related to Pan-African orogeny (Saha 2004; Tripathy and Saha 2013; Saha and Patranabis-Deb 2014; Tripathy and Saha 2015). In the eastern part of the belt, Nekarikallu sector, the N trending imbricate thrust are thus related to the compression of the Late Neoproterozoic, impressions of which are also recorded through syn-kinematically grown phengitic white mica in mylonite from Vinukonda granite during 501–474 Ma (Dobmeier et al. 2006). The first report of window and klippe along the Vamikonda Thrust from northern part of NFB represent such compression involving the sedimentary packages in a thin-skinned set up. In addition, an earlier extensional episode is recorded only within the Kurnool Group of rocks as envisaged by the presence of horst and graben structures in the Belum sector. These horst and graben affecting the Kurnool Group are overthrust by the folded Nallamalai Group of rocks (Fig. 8h), suggesting such extensional event to be of pre-thrust (related to Vamikonda thrust) in nature.

5 Conclusion

The present work suggests thrust stacking in eastern and northern part of NFB where interaction of the two temporally variant thrusts viz., Vellikonda and Nallamalai thrusts happened during Neoproterozoic-Paleozoic under the compressional stress regimes active during Gondwana assemblage. The chronological sequence of deformation is summarised in Table 1 and concluded below.

- Structures within the northern part of NFB are categorised here on their chronological occurrences in relation to the thrusting of NFB over Palnad sub-basin. Accordingly, pre-, syn- and post-thrust episodes are clubbed together.
- The pre- and syn-thrust deformation is responsible for the major regional structural trends presently hosted by the NFB.
- Following major structural changes are recorded in the northern NFB due to the syn-thrust deformation:

  i. Thrusting of NFB over Palnad sub-basin.
  ii. Bringing two spatially and temporally distinct thrusts i.e., the Vellikonda thrust and Vamikonda (Nallamalai) thrust in close vicinity at the NE part of the NFB near Nekarikallu.
  iii. Due to this syn-thrust compressional episode, during Late Neoproterozoic contemporaneous with Pan-African orogeny, the structural fabric (cleavage, fold axis, axial plane, thrust internal to NFB etc.) within NFB shows reorientation of its trend from N-S to NE-SW to E-W. It is understood to have formed due to the presence of granitic basement and Kurnool Group of rocks in Palnad sub-basin which must had acted as a tectonic barrier due to which the NFB in its northern part is dragged with a right lateral component.

iv. Change in the orientation of the major structural fabric on regional scale is evident all along the strike of the NFB. This change is explained with the model of progressive orocline which is a result of (i) thin-skinned nature of the fold belt and (ii) the late compressive stress perpendicular to the strike of the orogen bend produce the curvature in the fold belt.

Acknowledgements The authors are grateful to the Director General, Geological Survey of India and Additional Director General, GSI, SR, Hyderabad for kindly according permission to publish the work. This work is part of the GSI project for FS 2012-14 at state unit: Andhra Pradesh. The authors are grateful to Dr. Soumyajit Mukherjee (IIT, Bombay), Editor for his kind invitation for contribution to this book. This manuscript benefited greatly from the constructive reviews and criticisms by editor and an anonymous reviewer. The views expressed in this paper are of authors and not necessarily of GSI. Vide Mukherjee et al. (in press) for recent tectonic updates from the southern portion of the Cuddapah basin, and Goswami and Upadhyay (2019) in a similar context. Mukherjee (2019) summarizes this work.

References

Acharyya SK, Roy A (2000) Tectonothermal history of the Central Indian Tectonic Zone and reactivation of major faults/shear zones. Journal of the Geological Society of India 55:239–256

Anand M, Gibson SA, Subbarao KV, Kelley SP, Dickin AP (2003) Early proterozoic melt generation processes beneath the intra-cratonic Cuddapah basin, Southern India. Journal of Petrology 44, 2139–2171

Bhadra S, Gupta S, Banerjee M (2004) Structural evolution across the Eastern Ghats Mobile Belt Bastar craton boundary: hot over cold thrusting in an ancient collision zone. Journal of Structural Geology 26, 233–245

Bhaskar Rao YJ, Pantulu GVC, Damodar Reddy V, Gopalan K (1995) Time ofearly sedimentation and volcanism in the proterozoic Cuddapah basin, South India: evidence from Rb-Sr age of Pulivendla mafic sill. Geological Society of India Memoirs 33, 329–338

Bhattacharya S (1996) Eastern Ghats granulite terrain of India: an overview. Journal of Southeast Asian Earth Sciences 14, 165–174

Bhattacharya S, Das P, Chaudhary AK, Saw AK (2010) Mafic granulite xenoliths in the eastern ghats granulite belt: implications for lower crustal processes in the Southeastern Indian Peninsula. Indian Journal of Geology 80, 55–69

Bhowmik SK, Roy A (2003) Garnetiferous metabasites from the Sausar MobileBelt: petrology, P-T path and implications for the tectonothermal evolution of the Central Indian Tectonic Zone. Journal of Petrology 44, 387–420

Bhowmik SK, Pal T, Roy A, Pant NC (1999) Evidence for Pre-Grenvillian high-pressure granulite metamorphism from the northern margin of the Sausarmobile belt in Central India. Geological Society of India 53, 385–399

Carey SW (1955) The orocline concept in geotectonics. Proceedings of the Royal Society of Tasmania 89, 255–288

Carreras J, Druguet E, Griera A (2005) Shear zone-related folds. Journal of Structural Geology 27, 1229–1251

Chakraborti S (2006) Analysis of fold-and-thrust structures in parts of Neoproterozoic Palnad Sub-basin, Cuddapah basin, south India. Ph.D. thesis, University of Calcutta

Chakraborti S, Saha D (2009) Tectonic stresses and thin-skinned tectonics in a Proterozoic fold-and-thrust belt read from calcite mylonites in the Cuddapah basin, south India. Indian Journal of Geology 78, 37–54

Chalapathi Rao NV, Miller JA, Gibson SA, Pyle DM, Madhavan V (1999) Precise $^{40}Ar/^{39}Ar$ age determinations of the Kotakonda kimberlite and Chelima lamproite, India: implication for the mafic dyke swarm emplacement in the Eastern Dharwar craton. Journal of Geological Society of India 53, 425–432

Chalapathi Rao NV, Fu-Yuan Wu, Srinivas M (2012) Mesoproterozoic emplacement and enriched mantle derivation of the Racherla alkali syenite, Palaeo-Mesoproterozoic Cuddapah Basin, southern India: insights from in situ Sr–Nd isotopic analysis on apatite. Geological Society, London, Special Publication 365, 185–195

Chetty TRK (2011) Tectonics of proterozoic Cuddapah Basin, Southern India: a conceptual model. Journal Geological Society of India 78, 446–456

Collins A, Patranabis-Deb S, Alexander E, Bertram CN, Falster GM, Gore RJ, Mackintosh J, Dhang PC, Saha D, Payne JL, Jourdan F, Backé G, Halverson GP, Wade BP (2015) Detrital mineral age, radiogenic isotopic stratigraphy and tectonic significance of the Cuddapah Basin, India. Gondwana Research 28, 1294–1309

Crawford AR, Compston W (1973) The age of the Cuddapah and Kurnool systems, Southern India. Journal of Geological Society of Australia 19, 453–464

Das S, Shukla D, Bhattacharjee S, Mitra SK (2015) Age constraints of Udayagiri Domain of Nellore schist belt by xenotime dating around Pamuru, Prakasam district, Andhra Pradesh. Geological Society of India 85, 289–298

Dasgupta PK, Biswas A (2006) Rhythms in Proterozoic sedimentation: an example from Peninsular India. Satish Serial Publishing House, Delhi, p 340

Dasgupta S, Sengupta P (2003) Indo-Antarctic correlation: a perspective from the Eastern Ghats Belt. In: Yoshida M, Windley BF, Dasgupta S (eds) Proterozoic East Gondwana: supercontinent assembly and breakup. Geological Society London, Special Publication, 206, 131–143

Dharma Rao CV, Santosh M, Wu Yuan-Bao (2010) Mesoproterozoic ophiolitic mélange from the SE periphery of the Indian plate: U-Pb zircon ages and tectonic implications. Gondwana Research. https://doi.org/10.1016/j.gr.2010.06.007

Doblas M (1998) Slickenside kinematic indicators. Tectonophysics 295, 187–197

Dobmeier CJ, Raith MM (2003) Crustal architecture and evolution of the Eastern Ghats Belt and adjacent regions of India. In: Yoshida M, Windley BF, Dasgupta S (eds) Proterozoic East Gondwana: supercontinent assembly and breakup, 206. Geological Society, Special Publications, London, pp 145–168

Dobmeier C, Lütke S, Hammerschmidt H, Mezger K (2006) Emplacement and deformation of the Vinukonda granite—implications for the geological evolution of peninsular India and for Rodinia reconstructions. Precambrian Research 146, 165–178

French JE, Heaman LM, Chacko T, Srivastava RK (2008) 1891–1883 Ma SouthernBastar–Cuddapah mafic igneous events, India: a newly recognized large igneous province. Precambrian Research 160, 308–322

Geological Survey of India (1990) Geological quadrangle map of degree sheet No. 57I

Goswami S, Upadhyay PK (2019) Tectonic history of the granitoids and Kadiri schist belt in the SW of Cuddapah basin, Andhra Pradesh, India. In: Mukherjee S (ed) Tectonics and structural geology: Indian context. Springer International Publishing AG, Cham, pp 253–278. ISBN 978-3-319-99340-9

Goswami S, Mukherjee A, Zakaulla S, Rai AK (2016) Stress states, faulting and their effects on the Papaghni Group, Cuddapah Basin, India: a study along Giddankivaripalle-Madyalabodu tract. Indian Journal of Geosciences 70, 17–33

Gupta JN, Pandey BK, Chabria T, Banerjee DC, Jayaram KMV (1984) Rb–Sr geochronological studies on the granites of Vinukonda and Kanigiri, Prakasam district, Andhra Pradesh, India. Precambrian Research 26, 105–109

Henderson B, Collins AS, Payne J, Forbes C, Saha D (2014) Geologically constraining India in Columbia: the age, isotopic provenance and geochemistry of the protoliths of the Ongole Domain, Southern Eastern Ghats, India. Gondwana Research 26, 888–906

Jain SC, Yedekar DB, Nair KKK (1991) Central India shear zone: a major precambrian crustal boundary. Journal of Geological Society of India 37, 521–531

Johnston ST, Weil AB, Gutiérrez-Alonso G (2013) Oroclines: thick and thin. Geological Society of America Bulletin 125, 643–663

Joy S, Jelsma HA, Preston RF, Kota S (2012) Geology and diamond provenance of the Proterozoic Banganapalle conglomerates, Kurnool Group, India. From: In: Mazumder R, Saha D (eds) Palaeoproterozoic of India. Geological Society, London, Special Publications, 365, pp 197–218

Kumar Vijaya, Leelanandam C (2008) Evolution of the Eastern Ghats belt, India: a plate tectonic perspective. Journal of Geological Society of India 72, 720–749

Kumar A, Kumari VMP, Dayal AM, Murthy DSN, Gopalan K (1993) Rb–Sr ages of proterozoic kimberlites of india: evidence for contemporaneous emplacement. Precambrian Research 79, 363–369

Leelanandam C (1989) The Prakasam Alkaline Province in Andhra Pradesh, India. Journal Geological Society of India 34, 25–45

Leelanandam C, Burke K, Ashwal LD, Webb SJ (2006) Proterozoic mountain building in Peninsular India: an analysis based primarily on alkaline rock distribution. Geological Magazine 143, 195–212

Matin A (2014) Tectonics in the Cuddapah fold-thrust belt in the Indian shield, Andhra Pradesh, India and its implication on the crustal amalgamation of India and Rayner craton of Antarctica during Neoproterozoic orogenesis. International Journal of Earth Science 103, 7–22

Matin A, Guha J (1996) Structural geometry of the rocks of the southern part of the Nallamalai Fold Belt, Cuddapah Basin, Andhra Pradesh. Journal Geological Society of India 47, 535–545

Means WD (1987) A newly recognized type of slickenside striation. Journal of Sturctural Geology 9, 585–590

Meijerink AMJ, Rao DP, Rupke J (1984) Stratigraphic and Structural development of the Precambrian Cuddapah Basin, SE India. Precambrian Research 26, 57–104

Mitra S (2003) A unified kinematic model for the evolution of detachment folds. Journal of Structural Geology 25, 1659–1673

Mukherjee MK (2001) Structural pattern and kinematic framework of deformation in the southern Nallamalai fold-fault belt, Cuddapah district, Andhra Pradesh, Southern India. Journal Asian Earth Sciences 19, 1–15

Mukherjee S (2011) Mineral fish: their morphological classification, usefulness as shear sense indicators and genesis. International Journal of Earth Sciences 100, 1303–1314

Mukherjee S (2012) Simple shear is not so simple! Kinematics and shear senses in Newtonian viscous simple shear zones. Geological Magazine 149, 819–826

Mukherjee S (2013) Deformation microstructures in rocks. Springer Geochemistry/Mineralogy. Berlin. pp 1–111. ISBN 978-3-642-25608-0

Mukherjee S (2014a) Review of flanking structures in meso- and micro-scales. Geological Magazine 151, 957–974

Mukherjee S (2014b) Atlas of shear zone structures in meso-scale. Springer Geology, Cham, pp 1–124. ISBN 978-3-319-0088-6

Mukherjee S (2015) Atlas of structural geology. Elsevier, Amsterdam. ISBN 978-0-12-420152-1

Mukherjee S (2019) Introduction to "Tectonics and Structural Geology: Indian Context". In: Mukherjee S (ed) Tectonics and structural geology: Indian context. Springer International Publishing AG, Cham, pp 1–5. ISBN: 978-3-319-99340-9

Mukherjee S, Talbot CJ, Koyi HA (2010) Viscosity estimates of salt in the Hormuz and Namakdan salt diapirs, Persian Gulf. Geological Magazine 147, 497–507

Mukherjee S, Punekar J, Mahadani T, Mukherjee R (2015) A review on intrafolial folds and their morphologies from the detachments of the western Indian Higher Himalaya. In: Mukherjee S, Mulchrone KF (eds) Ductile shear zones: from micro- to macro-scales. Wiley Blackwell pp 182–205

Mukherjee S, Goswami S, Mukherjee A (in press) Structures and their tectonic implications form the southern part of the Cuddapah basin, Andhra Pradesh, India. Iranian Journal of Science and Technology, Transaction A: Science. http://doi.org/10.1007/s40995-018-0566-0

Mukhopadhyay D, Basak K (2009) The Eastern Ghats belt- a polycyclic granulite terrain. Journal Geological Society of India 73, 489–518

Nagaraja Rao BK, Rajurkar ST, Ramalingaswamy G, Ravindra Babu B (1987) Stratigraphy, structure and evolution of the Cuddapah Basin. In: Radhakrishna BP (ed) Purana basins of Peninsular India (Middle to Late Proterozoic). Geological Society of India, Memoir, 6, pp 33–86

Nance RD, Murphy JB, Santosh M (2014) The supercontinent cycle: a retrospective essay. Gondwana Research 25, 4–29

Natarajan V, Rajagopalan Nair S (1977) Post Kurnool thrust and structural features in the northeast part of the Palnad basin, Krishna district. A.P. Journal Geological Society of India 18, 111–116

Nemčock M, Schamel S, Gayer R (2005) Thrust belts structural architecture, thermal regimes and petroleum systems. Cambridge University press, 541 p

Power WL, Tullis TE (1989) The relationship between slickenside surfaces in fine-grained quartz and the seismic cycle. Journal of Structural Geology 11, 879–893

Rajurkar ST (1972) Geology of Vami Konda ranges and adjoining Palnad area, Cuddapah basin, Andhra Pradesh. Unpublished report, Geological Survey of India

Ramlinga Swamy G (1972) Geology of the Karempudi, Gummanampadu and adjoining areas in Cuddapah basin, Guntur district, Andhra Pradesh. Geological Survey of India, unpublished report

Ramsay JG (1967) Folding and fracturing of rocks. MacGraw Hill, New York, p 568p

Ramsay JG, Casey M, Kligfield R (1983) Role of shear in development of the helvetic fold-thrust belt of Switzerland. Geology 11, 439–442

Rickers K, Mezger K, Raith MM (2001) Evolution of the continental crust in the proterozoic Eastern Ghats Belt, India and new constraints for Rodinia reconstruction: implications from Sm–Nd, Rb–Sr and Pb–Pb isotopes. Precambrain Research 112, 183–210

Rogers JJW, Santosh M (2002) Configuration of Columbia, a mesoproterozoic supercontinent. Gondwana Research 5, 5–22

Saha D (1994) Fold-fault structures of the Nallamalai range, Diguvametta-Nandi Ka Nama Pass, Prakasam district, Andhra Pradesh, South India. Indian Journal of Geology 66, 203–213

Saha D (2002) Multi-stage deformation in the Nallamalai Fold Belt, Cuddapah Basin, South India —implications for mesoproterozoic tectonism along Southeastern margin of India. Gondwana Research 5, 701–719

Saha D (2004) Structural asymmetry and Plate tectonic set-up for a proterozoic fold and thrust belt: Nallamalai fold belt and adjoining terrane, South India. Geological Survey of India, Special Publication 84, 101–119

Saha D (2011) Dismembered ophiolites in Paleoproterozoic nappe complexes of Kandra and Gurramkonda, South India. Journal of Asian Earth Sciences 42, 158–175

Saha D, Chakraborty S (2003) Deformation pattern in the Kurnool and Nallamalai Groups in the Northeastern Part (Palnad Area) of the Cuddapah Basin, South India and its Implication on Rodinia/Gondwana. Tectonics, Gondwana Research 6, 573–583

Saha D, Patranabis-Deb S (2014) Proterozoic evolution of Eastern Dharwar and Bastar cratons, India—an overview of the intracratonic basins, craton margins and mobile belts. Journal of Asian Earth Sciences 91, 230–251

Saha D, Tripathy V (2012) Palaeoproterozoic sedimentation in the Cuddapah Basin south India and regional tectonics—a review. In: Mazumder R, Saha D (eds) Paleoproterozoic of India, vol 365. Geological Society of London Special Publication, pp 159–182

Saha D, Chakraborti S, Tripathy V (2010) Intracontinental thrusts and inclined transpression along Eastern margin of the East Dharwar craton, India. Journal Geological Society of India 75, 323–337

Satyapal, Tripathy V (2014) Specialized thematic mapping in northern part of Nallamalai Fold Belt, Cuddapah basin, Guntur and Prakasam districts, Andhra Pradesh. Unpublished Geological Report, Geological Survey of India, Hyderabad
Sesha Sai VV (2004) Petrographic and petrochemical charecterisation of Proterozoic granites in Nellore schist belt and northeastern fringes of Cuddapah basin. Records Geological Survey of India 137, 184–188
Sesha Sai VV (2009) Sheeted dykes in Kandra ophiolite complex, Nellore schist belt, Andhra Pradesh—vestiges of oceanic crust. Journal of Geological Society of India 74, 509–514
Sesha Sai VV (2013) Proterozoic granite magmatism along the terrane boundary tectonic zone to the East of Cuddapah Basin, Andhra Pradesh—petrotectonic implications for Precambrian crustal growth in Nellore Schist Belt of Eastern Dharwar Craton. Journal of Geological Society of India 81, 167–182
Sesha Sai VV, Khanna TC, Krishna Reddy NR (2016) Red beds in the Cuddapah Basin, eastern Dharwar craton, India: implications for the initiation of sedimentation during the proterozoic oxygenation event. Journal of Indian Geophysical Union 20, 342–350
Sesha Sai VV, Tripathy V, Bhattacharjee S, Khanna TC (2017) Paleoproterozoic magmatism in the Cuddapah basin, India. Journal of Indian Geophysical Union 21, 516–525
Tripathy V (2010) Brittle deformation in the Western Cuddapah Basin, South India and implications for intra-continental tectonics. Jadavpur University, pp 264 (Ph.D. thesis)
Tripathy (2011) Brittle deformation in the western Cuddapah basin, South India and implications for intra-continental tectonics. Unpublished Ph.D. thesis, Jadvapur University, 264 p
Tripathy V, Saha D (2010) Structure and low grade metamorphism of the east central part of the proterozoic Nallamalai fold belt, south India—thrust stacking and discontinuous metamorphic gradients along eastern margin of East Dharwar craton. Indian Journal of Geology 80(1–4), 173–188
Tripathy V, Saha D (2013) Plate margin paleostress variations and intracontinental deformations in the evolution of Cuddapah basin through the Proterozoic. Precambrian Research 235, 107–130
Tripathy V, Saha D (2015) Inversion of calcite twin data, paleostress reconstruction and multiphase weak deformation in cratonic interior—evidence from the Proterozoic Cuddapah basin, India. Journal of Structural Geology 77, 62–81
Upadhyay D, Raith MM, Mezger K, Hammerschmi K (2006) Mesoproterozoic rift-related alkaline magmatism at Elchuru, Prakasam Alkaline Province, SE India. Lithos 89, 447–477
Venkatakrishnan R, Dotiwala FE (1987) The Cuddapah salient: a tectonic model for the Cuddapah basin, India, based on Landsat image interpretation. Tectonophysics 136, 237–253
Vijaya Kumar K, Ernst WG, Leelanandam C, Wooden JL, Grove MJ (2010) First Paleoproterozoic ophiolite from Gondwana: geochronologic—geochemical documentation of ancient oceanic crust from Kandra, SE India. Tectonophysics 487, 22–32
Zhang S, Li Z-X, Evans DAD, Wu H, Li H, Dong J (2012) Pre-Rodinia supercontinent Nuna shaping up: a global synthesis with new paleomagnetic results from North China. Earth and Planetary Science Letters 353–354, 145–155

Tectonic History of the Granitoids and Kadiri Schist Belt in the SW of Cuddapah Basin, Andhra Pradesh, India

Sukanta Goswami and P. K. Upadhyay

1 Introduction

The Kadiri greenstone terrane in the eastern Dharwar craton has not yet been studied extensively. The petrotectonic associations in the Kadiri greenstone granite terrain reflect its tectonic setting and field observations give ideas on evolution of the belt. Several physical and chemical factors are to be considered to reconstruct the geologic history of Archean granite-greenstone terranes. The geological and geochemical features of this terrane provide clues to describe their genesis and evolution. Since, greenstone belt development cannot be considered in isolation from models for the origin of Archean highgrade granitic terranes because of close allience between Earth's thermal history and crustal evolution, all the rock types such as granitoids, greenstone belt rocks and dykes are to be considered. Based on geochemical studies of meta-volcanics of this schist belt, Satyanarayana et al. (2000) conclude island arc and active continental marginal environment. The calc-alkaline association of basalt, andesite, dacite and rhyolite (BADR) is the distinguishing volcanic rock suite of convergent margins. The low-grade green-schist facies metamorphism and associated structural deformations of rocks is noteworthy. The Kadiri schist locates below the unconformity plane defined by the overlying sediments of Cuddapah basin near the village Dorigallu (Fig. 1).

According to Jayananda et al. (2013b), the crystallization age of the felsic magmas of Kadiri schist belt rocks is 2556 ± 13 Ma, dated by U–Pb SIMS. In fact, the 2.7 Ga mafic dominated episode of greenstone volcanism is coeval to emplacement of TTG suites surrounding the greenstone belt. On the other hand, the 2.6–2.54 Ga volcanism was dominated by intermediate and particularly felsic lavas in the Dharwar craton (Jayananda et al. 2013a, b). The 2.6–2.54 Ga episode is

S. Goswami (✉) · P. K. Upadhyay
Department of Atomic Energy, Atomic Minerals Directorate for Exploration and Research, Bengaluru 560072, India
e-mail: sukantoyahoo@gmail.com

S. Mukherjee (ed.), *Tectonics and Structural Geology: Indian Context*, Springer Geology, https://doi.org/10.1007/978-3-319-99341-6_8

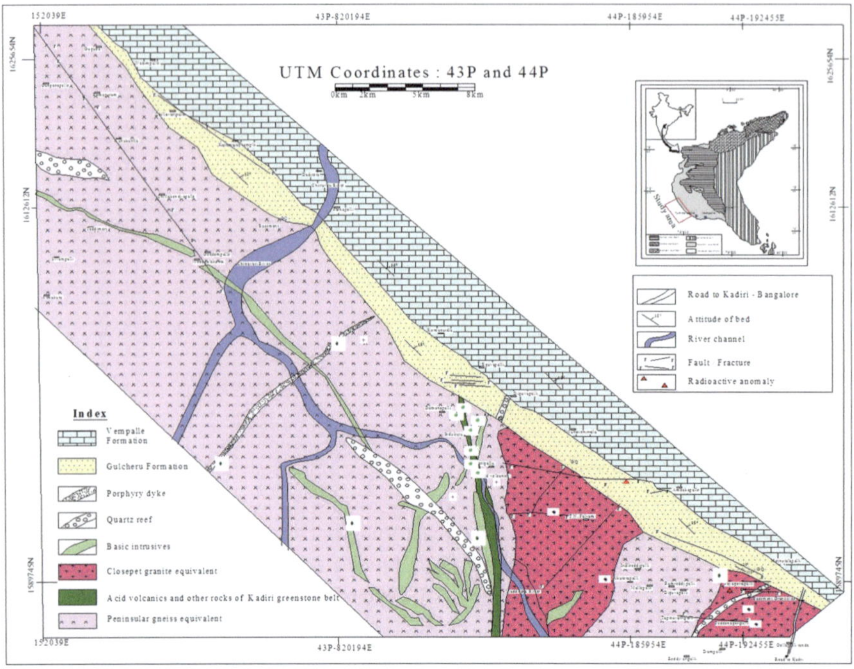

Fig. 1 Geological map of the northern extreme of Kadiri greenstone belt near the non-conformity contact with Cuddapah sediments

linked to juvenile calc-alkaline magmatism and crustal reworking throughout the EDC (Chadwick et al. 2007; Chardon and Jayananda 2008; Chardon et al. 2011). The NW ($\sim$2.21 Ga) and ENE ($\sim$2.36 Ga) trending mafic dyke swarms of the area were dated by U–Pb (Halls et al. 2007; French and Heaman 2010). The geological history of the Kadiri greenstone granite terrain is not yet well documented. The relationship between the schist belt and the surrounding gneisses and granitoids is also not well understood. Therefore, in this paper we approach all the rock types to describe the tectonic history of the terrain for a period from emplacement of TTG to rifting of Cuddapah basin (i.e., 2.7–1.9 Ga). For general ideas about rifting (from other terrains), consult Misra et al. (2014, 2015, 2018a, b), Misra and Mukherjee (2015, 2018) and Dasgupta and Mukherjee (2017).

2 Geology

The narrow linear Kadiri Greenstone Belt is located SW of Cuddapah Basin in the Eastern Dharwar Craton (EDC) covering parts of Anantapur and Chittoor districts located between 13° 45′ to 14° 7′N Latitude and 78° 2′ to 78° 15′E Longitude. Survey of India Toposheet Nos. 57J/3, 57J/4, 57K/1 and 57K/2 covers this region.

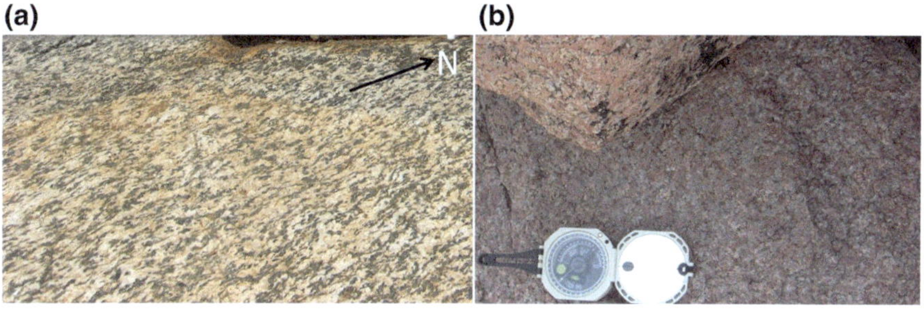

Fig. 2 Two different granitoids. **a** Peninsular gneiss equivalent granitoids with feeble N-S gneissic foliation. **b** Younger grey to pink closepet equivalent granitoids

The greenstone belt litho-units comprise mainly acid to intermediate meta-volcanics and associated minor meta-basics and BIFs along with granitic plutons. Large quartz reefs and porphyry dykes cut across the belt. The granitoids represent two catagories (Fig. 2): (1) dark grey coarse-grained hornblende-biotite granite with feeble gneissosity at places occupying relatively flat paneplaned areas; and (2) medium to coarse-grained granodiorite of pink to grey colour, that intrude the former, and found in hillocks. The Granitoids were defined as Tonalite-Trondhjemite-Granodiorite Gneiss, Tonalite-Granodiorite-Monzogranite and Monzogranite-Syenogranite (Nandy and Dey 2013; Sreenivasulu et al. 2014). In the schist belt sub-spherical accretionary lapilli made up of aggregated coarser ash, lapilli or aggregates of variable sized clasts (Fig. 3) occur commonly near the unconformity usually indicating magma–water interaction (Wohletz and Mcqueen 1984). Other than lapilli with 2–64 mm diameter, the pyroclasts with diameter

Fig. 3 Sub-elliptical accretionary lapilli made up of aggregates of variable sized clasts

Fig. 4 The solid fragments and rounded bomb indicates violent eruption of molten lava, which cooled into solid fragments before they reach the ground and acquired aerodynamic round shapes

>64 mm are also embedded on the lava (Fig. 4). The broken angular blocks indicate solid fragments and rounded clasts are bomb formed during violent eruption of molten lava, which cool into solid fragments before they reach the ground and acquired round shapes. Flow foliation is another characteristic feature of the felsic volcanics of the schist belt (Fig. 5) where flow beds occur with ~N strike, 5–10° dip towards E. Autoclastic fragmental rocks (Fig. 6) are identified to interpret in situ fragmentation of coherent magmatic bodies (Fisher and Schmincke 1984; Mc Phie et al. 1993). These autobreccias commonly form at the base and the top of lava flows (Fisher and Schmincke 1984; Cas and Wright 1988; McPhie et al. 1993). Feebly developed gneissosity is observed in the surrounding gneissic rocks. Granite-gneiss complexes comprise the dominant portion in this granite-greenstone terrane. The foliation in the gneissic complexes is N-S similar to the greenstone belt near greenstone-gneiss contacts and becomes exceedingly variable in intervening

Fig. 5 Flow foliation developed in acid volcanics due to high viscosity. The foliation planes show ~N-S strike, 2°–8°=>E dip

Fig. 6 Section of flow layers showing auto breccia with in layer suggest in situ fragmentation of coherent magmatic bodies

regions. Mechanical rotation and elongation of clast along preferred N-S trend and the internal deformation and crushing of grains at places by cataclasis during rotation due to E-W compression are noteworthy. The deformation signature corresponds to a shallow low temperature environment of regional metamorphism.

Isoclinal folds with steeply dipping axial planes and moderate to steep plunges are preserved. The cross-section through rhyolite lava exhibits flow banding of visually distinct layers of differing crystallinity or vesicularity. The youngest dykes and quartz reefs cut across both the granite and the greenstone belt. The linear andesite and dolerite dykes often run along top of granitic hillocks for km with loose angular blocks arranged apparently like unsorted dump. These younger intrusives are identified as blocky lava with very rough surface composed of often loose clinkers and rubble (Fig. 7). The Cuddapah sediments overlie the basement complex along the Eparchaean unconformity with basal conglomerates (Goswami et al. 2015). Above this nonconformity the polymictic conglomerate (Fig. 8) with abundant clasts of chert, jasper and BIF indicates supracrustal provenance with Archaean volcano-sedimentary rocks.

Fig. 7 Basic emplacement along NE-SW cut across granitoids exhibiting stack of pieces of andesitic blocks are characterised as blocky lava

Fig. 8 **a** Eparchaean unconformity between Archaean basement granitoids and Proterozoic Gulcheru Formation. **b** The lowermost Gulcheru Formation consists of polymictic conglomerate unit with abundant clasts of chert, jasper and BIF

3 Structures

E, NE, NW and N are the different fracture trends. The first two trends are more dominant (Fig. 9). Schistosity and gneissosity show a regional N and the younger granitic intrusions too are N-S. The granitoids are intruded by basic intrusions along NW, NE, WNW and ENE trends. Quartz reefs are also remarkable intrusive features along NW, NE and WNW. Hematitisation is noted locally near the contact zones of quartz reef and basement granitoids. The rocks are deformed with a specific texture and structure that records the deformation by developing the preferred clast orientation. The fabric appears to be a type of tectonites (Mukherjee 2015). Granitic apophyses in the deformed meta-volcanic matrix of the greenstone

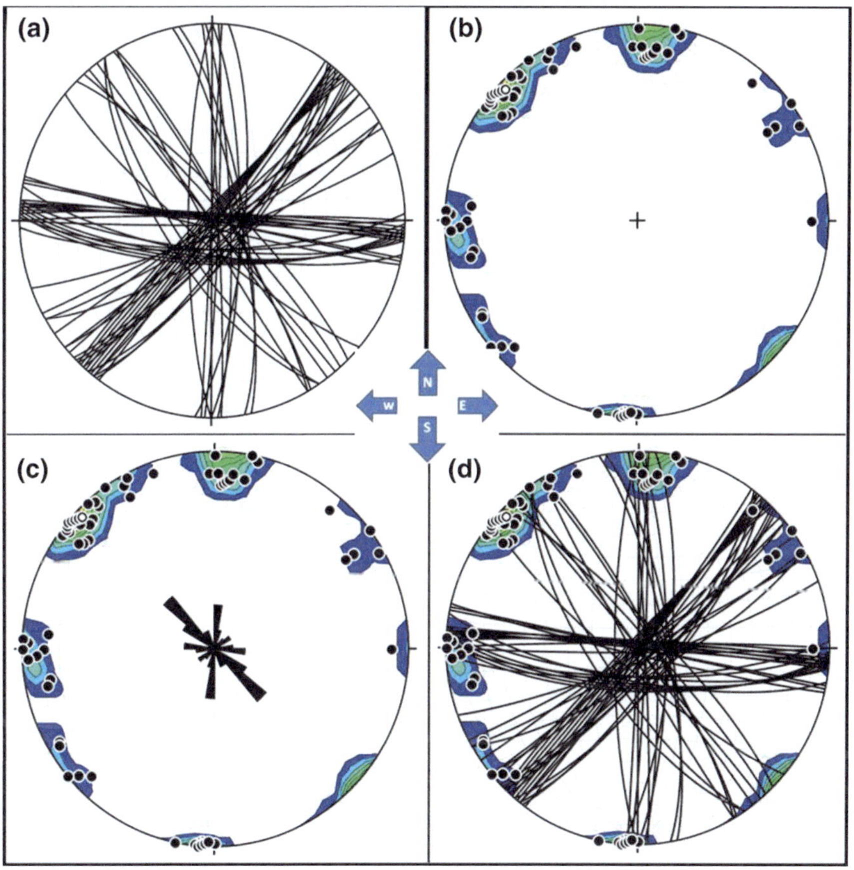

Fig. 9 **a** β diagram showing all fracture plane attitudes of the study area. **b** π diagram showing poles of all fracture planes. **c** π diagram superimposed with corresponding rose diagram to visualize the paleostress condition. **d** Superimposed pie and beta diagram of fracture shows NE trend to be most abundant followed by E

Fig. 10 **a** Outcrop view of L-S tectonites. **b** Zoomed in view of the three dimensional section of tectonite outcrop. The N-S stretching is compensated by E-W flattening and undisturbed intermediate stress direction. **c** Plan view of tectonite outcrop. **d** Later deformation affected tectonite

belt stretched to form tectonites at places related to dynamothermal metamorphism and deviatoric stress. Foliation development in these rock is due to $\sigma_1 > \sigma_2 \geq \sigma_3$ stress condition. Stretched pebbles in the deformed volcanic matrix gives L-S tectonite in which σ_1 is the maximum E-W compressive stress, σ_3 minimum compression or extension/stretching along N-S, and σ_2 vertical intermediate stress axis (Andersonian condition; Fig. 10a–c). Later deformation also manifest at places in some tectonite outcrops where the triaxial ellipsoid clasts are further cut by fault plane (Fig. 10d). The N-S fractures are oldest trend possibly related to surface manifestation of younger pluton emplacement that reactivated in nature. More specifically, younger ENE-WSW and WNW-ESE fracture sets are probably conjugate sets with 30–60° angle (Fig. 9). The E-W fracture is youngest and affects both basement as well as the Cuddapah sediments. The granitic apophyses in the greenstone schist belt is dominantly aligned along N-S fractures and after reactivation and brecciation they sufferred non-coaxial stress which is possibly related to shear during conjugate fracturing with ~E-W acute bisector directions.

4 Petrology

The Kadiri schist belt metavolcanics are very fine-grained and mostly are felsic. Minerals cannot be identified even under a hand lens. Under optical microscope quartz, plagioclase, orthoclase and chlorite are seen as the major minerals followed by sphene, epidote and calcite. Very fine-grained, foliated acid volcanics composed chiefly of quartz, sericite and chlorite and other flakey clay minerals and titanate minerals with flow foliation. Epidote, calcite, plagioclase feldspar and zoisite are the minor/accessory mineral phases. The systematic approach of studying tectonite fabric elements under microscope (Mukherjee and Chakraborty 2007; Mukherjee 2010a, b, Mukherjee 2011a, b, 2012, 2013, 2014a, b; Mukherjee and Koyi 2009, 2010a, b) is to study both clast and matrix components. The clasts of granitic apophyses and volcanic matrix are shown in Fig. 11.

The Peninsular gneiss equivalent and younger closepet granite equivalent rocks are recognised and differentiated as per these field guidelines:

A. Geomorphologically it is clear that the Peninsular gneiss equivalents forms the peneplane mostly soil covered area where as younger granitoids of Closepet forms ridges and mounds with circular to elongate masses of high and rugged relief.

Fig. 11 **a** Very fine grained foliated preferentially aligned quartz, clayey minerals and titanate. Minor quartz veins are visible (2X, TL, 2N). **b** Alignment of chlorite along foliation (10X, TL, 1N). **c** Phenocryst of plagioclase and sanidine in fine grained matrix (2X, TL, 2N). **d** Granitic texture with quartz, plagioclase, orthoclase and minor chlorite (2X, TL, 2N)

B. Presence of feebly developed N-S gneissic foliation in flat-lying older rocks compared to massive intrusives of younger closepet equivalence.
C. The younger granitoids are of two types, pink and grey coloured, with no distinct boundary between the two and consists of mostly quartz, plagioclase and potash feldspars with finer-grains than the older gneisses with coarse hornblende and biotite crystals.
D. Older rocks are more structurally disturbed affected by intense fracturing, venation, shear and faulting and associated mylonitization than the younger granites with relatively less deformation features.
E. Supracrustal enclaves are more numerous in older gneiss equivalent rocks.
F. Closepet equivalents are more uraniferous than the older Peninsular gneiss equivalents.

The dykes of porphyry andesites (Fig. 12) with euhedral plagioclase laths as phenocrysts indicate earlier stage slow cooling of magma, which was forced to the surface along linear fissures due to buoyancy and rapid cooling to form the remaining fine groundmass. The blocky nature of the lava is also seen due to the higher viscosity, which favors preservation of the ragged and spinose form of the clinker fragments (Fig. 7). Higher silica content of the dykes than the basalts provided higher viscosity that developed blocks schematically defined by Rowland and Walker (1990). Subsequent breakage of this lava along cooling joints then contributed planar-surfaced and non-vesicular debris at the flow top. The geomorphology of younger dykes and quartz reefs along top of the hills is because of younger age and relatively higher silica content, which could resist from weathering to provide linear ridge along top of granitic hills.

5 Geochemistry

The geochemical analyses were performed in the chemistry laboratory, Atomic Minerals Directorate for Exploration and Research, Bangalore (India). Major and selective minor and trace element data (Table 1 and 2) from collected samples are used to interpret the magma evolution and the Meso Archaean to Palaeo Proterozoic tectonic history. Harker diagrams (after 1909) of multiple major element oxide plots against SiO_2 display smooth, curvilinear trends for almost all rock data points indicate genetically related rock types (Figs. 13 and 14). The variations in major element indicate a common parent source from which diversification caused the major elements to either increase or decrease progressively with respect to SiO_2. Bowen's reaction series is reflected in a way in our data where MgO and FeO decrease with increasing SiO_2. Unlike compatible elements like Ni, Cr etc., incompatible elements also enrich in successively later differentiated part of magma. Hence, large ion lithiphile element like uranium is more enriched in later generated more differentiated younger granites. Most of the uranium anomalies are located in younger closepet equivalents. Upon fractionation, calc-alkaline magmas

Fig. 12 Andesite porphyry displaying **a** aphanitic porphyritic texture with subhedral plagioclase phenocrysts (white crystals) and the fine—grained aphanitic grey groundmass in outcrop scale. **b** Porphyritic texture under microscope (2X, TL, 2N). **c** Quartz, feldspar, biotite and minor chlorite rich groundmass (10X, TL, 1N). **d** Zone plagioclase phenocryst (10X, TL, 2N). **e** Deformed plagioclase phenocryst (10X, TL, 2N)

Table 1 Major element oxide (%) data of selective samples from Kadiri granite-greenstone terrain

S. No.	Rock	Symbol	SiO_2	Al_2O_3	TiO_2	Fe_2O_3	MnO	MgO	CaO	Na_2O	K_2O	P_2O_5
S-1.8/6	Altered Basic	•	33.03	16.19	3.86	25.85	0.2	16.29	2.01	<0.01	<0.01	0.31
S-1.8/9	Diorite	+	48	17.12	0.62	10.17	0.11	9.78	10.24	2.15	0.29	0.09
1.8SP1	Basalt porphyry	•	64.34	14.06	0.88	8.22	0.1	0.63	3.19	3.57	4.01	0.4
1.8SP2	Pink Granite	♦	73.9	13.27	0.23	1.92	0.03	0.34	0.93	3.65	5.45	0.11
1.8SP3	Diorite	+	57.72	14.8	0.72	7.19	0.1	5.01	5.36	4.01	2.15	0.29
1.8SP4	Basalt porphyry	•	65.05	15.83	0.54	4.87	0.06	0.99	3.78	5.51	1.56	0.22
1.8SP5	Basalt porphyry	•	54.87	13.46	0.71	8.78	0.13	7.94	5.97	4.24	1.67	0.32
1.8SP6	Granite	♦	73.14	13.25	0.14	1.41	0.03	0.55	0.5	4.74	5.15	0.07
1.8SP7	Gray granite	♦	73.59	13.45	0.13	1.12	0.01	0.15	0.11	4.59	5.44	0.07
1.8SP8	Diorite	+	56.36	15.69	0.7	8.55	0.1	5.38	6.15	4.37	1.82	0.27
1.8SP9	Basalt porphyry	•	62.67	13.76	0.98	7.36	0.1	3.46	4.59	3.46	2.26	0.22
1.8SP10	Granite	♦	72.96	13.58	0.11	13.58	0.03	0.21	0.92	4.6	4.88	0.08
KS1	Schist belt	•	76.69	13.23	0.22	1.25	0.04	0.89	0.69	3.15	3.29	0.18
KS2	Schist belt	•	62.56	21.02	0.31	4.41	0.07	0.37	0.62	1.79	7.39	0.14
KS3	Schist belt	•	73.36	16.03	0.25	1.60	0.05	0.72	0.49	2.40	4.57	0.17
KS4	Schist belt	•	72.05	13.85	0.31	2.36	0.05	1.13	1.96	4.36	1.20	0.19
KS5	Schist belt	•	77.54	12.57	0.14	1.10	0.03	0.92	3.41	0.81	1.53	0.18
KS6	Schist belt	•	58.93	16.26	0.81	6.48	0.07	3.65	7.16	4.00	0.77	0.21
KS7	Schist belt	•	67.90	15.50	0.31	3.35	0.05	0.68	1.26	6.05	2.92	0.15

record a progressive decrease in iron and magnesium with increasing SiO_2 and alkali concentrations. Available FeO and MgO are progressively removed from the melt through crystallization of olivine, pyroxenes, amphiboles, biotite and iron oxide minerals such as magnetite. The removal of ferromagnesian minerals enriches the melt in Na_2O, K_2O and SiO_2. The SiO_2 versus FeOt/MgO plot (Miyashiro 1974) and AFM diagram (after Irvine and Barager 1971) of data-points fall in the calc-alkaline field (Figs. 15 and 16). The progressive changes in chemical composition are plotted on the AFM (A, alkali; F, iron; M, magnesium wt%) diagram (Fig. 16) to study along with the Harker diagrams (Figs. 13 and 14). This is to demonstrate that the calc-alkaline rocks generated by diversification of andesitic parent magma. Thus, with increasing fractionation from magma, the calc—alkaline series exhibits increase in SiO_2, Na_2O and K_2O and decrease in CaO, MgO and the total iron (Bowen 1928; Miyashiro 1974; Grove and Kinzler 1986). However, due to successive metamorphism and alterations the addition or removal of certain oxides can show some discrepancy (Table 2).

As far as tectonic setting is concerned, the plutonic and hypabyssal rocks were separately studied and interpreted from volcanics. Felsic volcanic data suggest convergent plate margin for greenstone belt rocks (Fig. 17a) wherein younger batholiths and composite volcanoes develop above the subduction zone. It is likely that the parent melts for the acid volcanics produced by partial melting of a relatively undepleted mantle source. Older dioritic to granidioritic peninsular gneiss equivalent and younger closepet equivalent granitoids along with intermediate younger hypabyssal porphyries are plotted in the tectonic discrimination diagrams after Maniar and Piccoli (1989). The older gneiss equivalences plot in the post-orogenic granitoids (POG) field on the Al_2O_3 versus SiO_2 plot (Fig. 17b). Thus we can consider the Dharwar batholith to have formed during the crust stabilizing post-orogenic environment. Whereas the younger granitoids and dykes plot in the combined field (Fig. 17c), which must be clarified from other supportive discrimination diagrams. The geochemical data are also plotted into the Batchelor and Bowden's diagram (1985) that support the tectonic environment for older gniess equivalents as post-orogenic. Additionally it shows younger granitoids and porphyries to associate with post-plate collision uplift (Fig. 17d).

6 Discussions

The greenstone belt rocks near Dorigallu indicate submarine volcanism along fissures developed in convergent plate boundary involved the transfer of magma (together with phenocrysts, dissolved volatiles, and country-rock material entrained from the conduit walls or ground surface) from some depth. The breccias and agglomerates indicate fissure vent surrounding environment (Goswami et al. 2018a, b). Fragments in breccias are chiefly felsic volcanics and range up to 1 m across although generally averaging 10–30 cm. According to Goodwin et al. (1972) the ratio of pyroclastics to flows increases both with increasing stratigraphic height and with decreasing distance

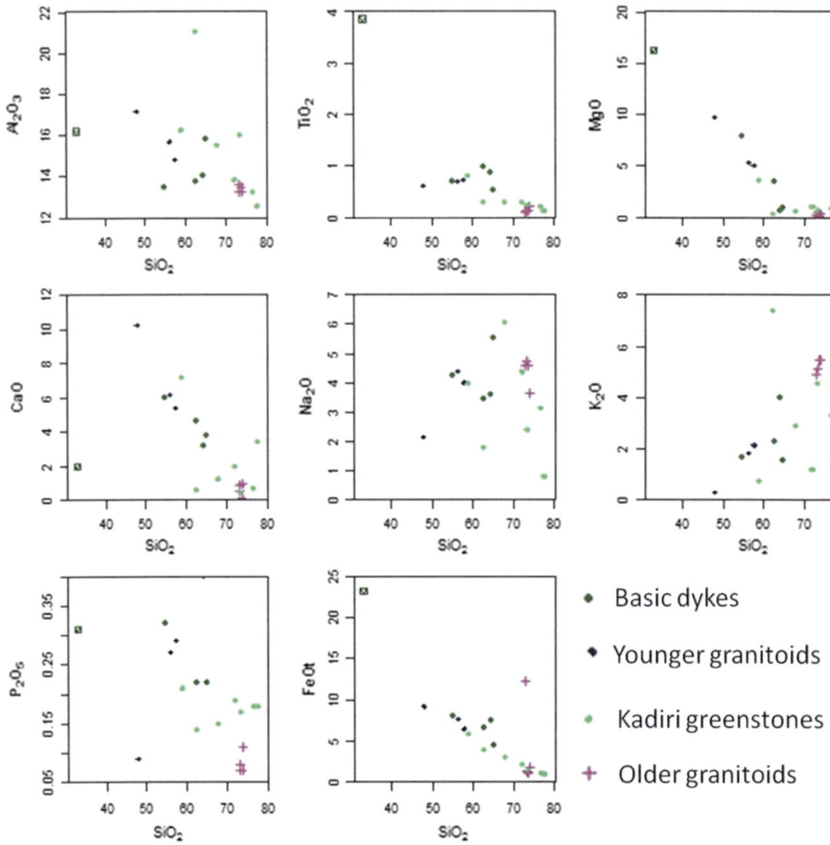

Fig. 13 Variation diagram (Harker Harker 1909) showing multiple oxide (%) plots against SiO_2 (%)

to eruptive centers. Thus the studied sections in Kadiri schist belt suggest middle portion of the greenstone succession. The intensive shearing and brecciation of granitic apophyses, intruded along schistosity plane in the northern part of Kadiri greenstone belt, causes pseudo appearance of agglomerate. The actual agglomerate units are not well-sorted sorted and fragments are generally angular although some units are composed of well-rounded bombs. The medium to fine ash-flow tuff units are welded and well bedded. Individual beds are 1 cm to few meters thick.

The basalt, andesite, dacite and rhyolite (BADR) is the signature suite of rocks indicates calc-alkaline volcanic association. This suite can act as a fingerprint of convergent plate margin. This setting constitutes possibly one of the most voluminous rock assemblages on the Earth after MORB (Perfit et al. 1980; Grove and Kinzler 1986). Harker diagram plots of major elements indicate a liquid line of descent from a common magma source. This BADR rock association are derived from a common parent magma of basaltic to andesitic composition. These calcalkaline volcanic rocks

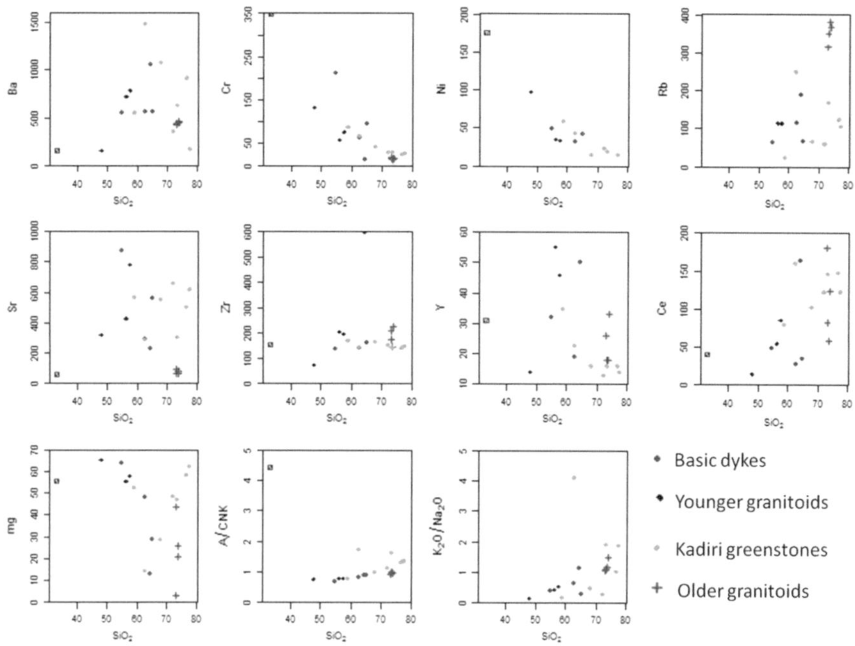

Fig. 14 Variation diagrams of selective trace elements (ppm) against SiO_2 (%)

Fig. 15 SiO_2 versus FeOt/MgO binary diagram (after Miyashiro 1974) distinguishing between tholeiitic and calc-alkaline rocks

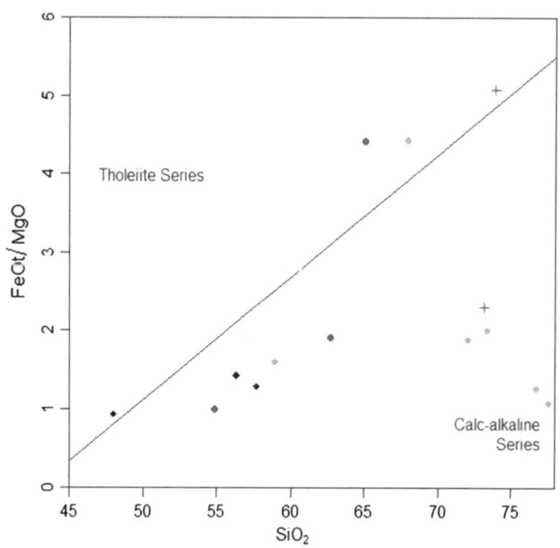

Fig. 16 AFM diagram (after Irvine and Barager 1971) serves to discriminate between calc-alkaline and tholeiitic subalkaline series. In this ternary diagram A = (Na_2O+K_2O), F = FeOt and M = MgO

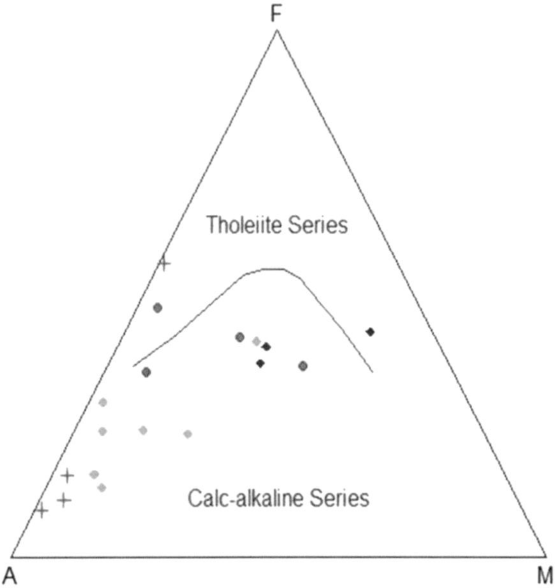

are formed at convergent margins where more silicic rocks represent enhanced fractionation. The calc-alkaline trend with ocean-continent convergence setting of the present data supports the observation of Chadwick et al. (2000). The calc-alkaline suite is also reflected from enrichment in SiO_2, alkalis (Na_2O and K_2O), LILE, volatiles and is relatively depleted in FeO, MgO, HFS concentrations (Miyashiro 1974; Hawkesworth et al. 1993; Pearce and Peate 1995) like the Phanerozoic convergent margins eg., Sierra Nevada (Fliedner et al. 2000; Ducea 2001) and ring of fire in present day. The peraluminous to metaluminous nature (Fig. 18a, b) with dominantly acidic composition indicates ocean-continent convergence that took place during Late Archaean time (~ 2.6 Ga). Oceanic or thinner continental lithosphere in the overlying plate generally produces metaluminous, mafic to intermediate rocks (Miyashiro 1974; Gill 1981). Thicker continental lithosphere overlying the subduction zones commonly yields peraluminous, potassic, intermediate to silicic rocks. The high K calc-alkaline and shoshonite i.e., high K andesite magma (Fig. 19a) indicate melting at the base of the lithospheric stack. In Th versus Co plot (Fig. 19b) as per Hastie et al. (2007), the high K calc-alkaline and shoshonite series indicate late stage ocean-continent subduction. The progression from tholeiite to calc-alkaline to shoshonite (trachyandesite) reflects increasing K_2O and K_2O/Na_2O and decreasing iron enrichment (Jakes and White 1972; Miyashiro 1974). The hot, buoyant Archean ocean lithosphere possibly favoured shallow low-angle subduction, which can be explained by Al_2O_3, K_2O and SiO_2 rich rhyolites, rhyodacites and shoshonites and plutonic rocks of increasingly granitic composition. This points to a relatively low-angle subduction (~ 25°) of thick continental lithosphere (>25 km) with higher degrees of partial melting (Davies 1992; Ernst 2007; Stern 2008).

Table 2 Selective trace element (ppm) data of the samples from Kadiri granite-greenstone terrain

S. No.	Rock	Symbol	Ba	Co	Cr	Cu	Nb	Ni	Pb	Rb	Sr	Th	Zn	Zr	Y	Ce
S-1.8/6	Altered Basic	•	153	72	348	<10	37	175	82	<10	62	<10	71	156	31	40
S-1.8/9	Diorite	+	155	37	133	59	22	97	<10	<10	323	<10	60	74	14	14
1.8SP1	Basalt porphyry	•	1058	19	13	<10	32	<10	18	188	226	<10	149	595	50	164
1.8SP2	Pink Granite	♦	449	24	16	<10	29	<10	45	367	65	82	37	225	33	124
1.8SP3	Diorite	+	784	22	77	48	21	33	16	114	781	<10	118	198	46	85
1.8SP4	Basalt porphyry	•	568	24	95	<10	<10	42	12	66	562	<10	93	163	<10	34
1.8SP5	Basalt porphyry	•	555	29	212	59	<10	49	10	65	873	<10	115	138	32	48
1.8SP6	Granite	♦	442	47	19	<10	27	<10	31	351	97	41	34	177	18	83
1.8SP7	Gray granite	♦	456	25	14	<10	22	<10	32	381	81	40	33	148	18	58
1.8SP8	Diorite	+	723	24	59	97	20	35	22	114	431	<10	125	205	55	55
1.8SP9	Basalt porphyry	•	566	37	63	26	<10	32	25	115	295	<10	45	141	19	27
1.8SP10	Granite	♦	430	26.8	16.3	<10	22	<10	48	315	64	71	43	209	26	180
KS1	Schist belt	•	919	15	26	11	21	15	21	125	508	35	31	142	16	148
KS2	Schist belt	•	1489	22	68	<10	28	43	<10	250	291	29	39	146	23	160
KS3	Schist belt	•	632	14	31	16	28	19	10	169	306	35	28	144	16	147
KS4	Schist belt	•	358	23	31	16	14	24	35	60	662	24	69	153	13	123
KS5	Schist belt	•	172	27	29	<10	17	<10	14	107	622	25	28	150	14	123
KS6	Schist belt	•	557	26	88	21	33	59	11	24	572	<10	97	170	35	80
KS7	Schist belt	•	1086	22	44	95	12	15	11	67	555	18	<10	166	16	102

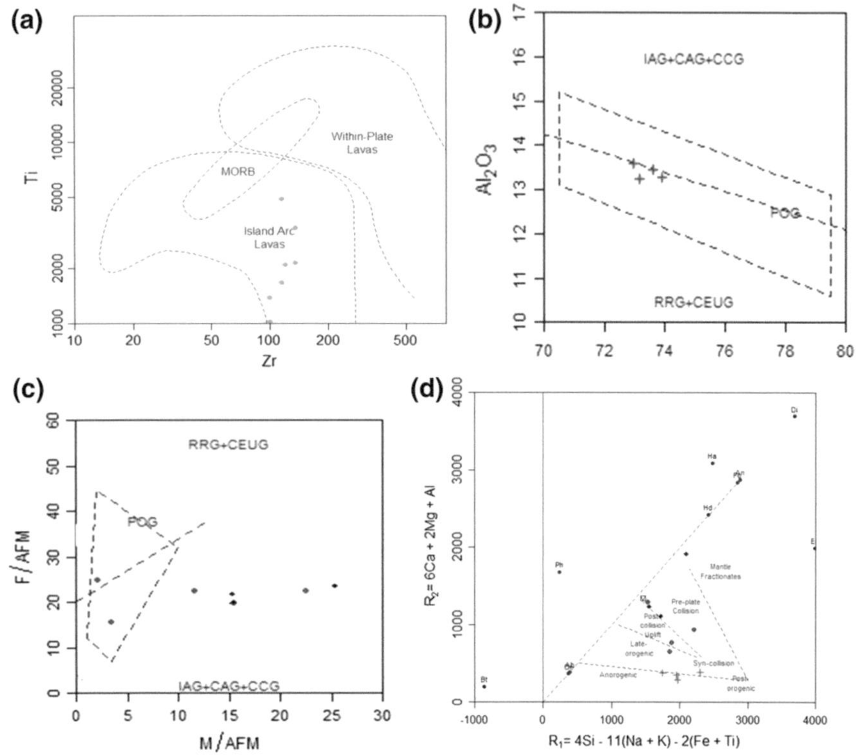

Fig. 17 **a** The logarithmic Zr versus Ti plot of data points fall in Island arc lava field which suggest convergent plate margin (after Pearce, Pearce 1982). **b** Al_2O_3 versus SiO_2 plot (after Maniar and Piccoli, Maniar and Piccoli 1989). **c** AFM ternary (after Maniar and Piccoli, Maniar and Piccoli 1989). Note: IAG = island arc granitoids, CAG = continental arc granitoids, CCG = continental collision granitoids, POG = post-orogenic granitoids, RRG = rift-related granitoids, CEUG = continental epirogenic uplift granitoids, OP = oceanic plagiogranites. **d** R1–R2 binary diagram (De La Roche et al. 1980) with geotectonic implications (after Batchelor and Bowden 1985)

To relate Archean high-grade gneiss with low grade (granite-greenstone) terrane we have to determine one of the following three situations (after Shackleton 1976). 1. Both age and tectonic setting differ. 2. Same age but different tectonic setting. 3. High-grade terranes are the deep crustal equivalents of granite greenstone terranes. Since, the age of greenstone/schist belt is different from that of granitoids, the second possibility eliminates. The term Dharwar Batholith after Chadwick et al. (2000) is introduced to differentiate from Peninsular gneiss as the rocks are more granitic than gneissic. In most of the areas, the contact between the Dharwar Batholith and the schist belt is blind. Since, Kadiri is the northern extension of Kolar schist belt it is logical to accept the same mechanism of rifting associated with mantle-plume as suggested for Kolar schist belt (Rajamani et al. 1985; Rajamani 1990). However, the mantle plume related extension and rifting perhaps

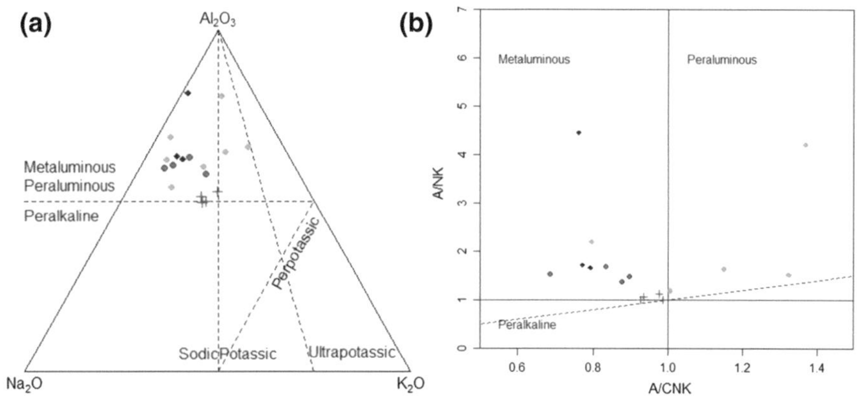

Fig. 18 **a** Molar Na_2O–Al_2O_3–K_2O ternary plot to distinguish meta/peraluminous from peralkaline rocks as well as potassic, sodic and ultrapotassic suites. **b** $Al_2O_3/(CaO+Na_2O+K_2O)$ versus $Al_2O_3/(Na_2O+K_2O)$ (mol%) i.e., A/CNK versus A/NK plot of Shand (1943) discriminating metaluminous, peraluminous and peralkaline compositions

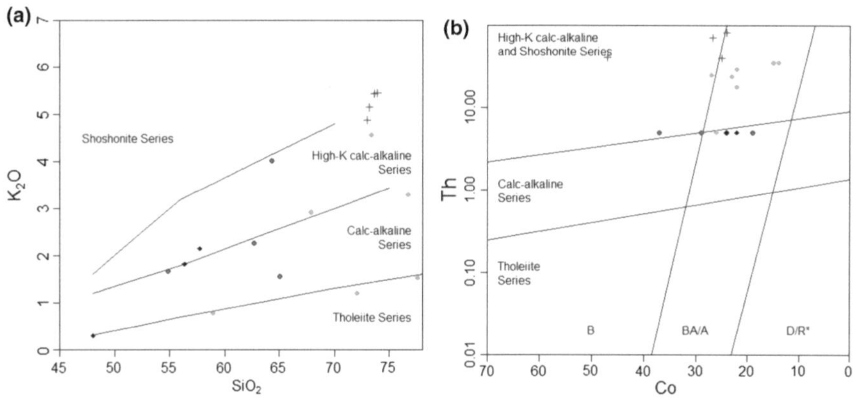

Fig. 19 **a** SiO_2 versus K_2O binary plot (after Peccerillo and Taylor 1976) to distinguish various series of tholeiitic, calc-alkaline and shoshonitic rocks. **b** Confirmation of magma series using less mobile elements, Co versus Th plot (after Hastie et al. 2007)

made back arc basin environment for the deposition of volcano-sedimentary sequence. According to Jayananda et al. (2013a, b), an ultrahot orogen involving two-stage accretion (2600–2700 Ma and 2520–2580 Ma) along W-dipping subduction zones in individual greenstone belts, with later development of plumes could explain extensive granite magmatism. Chadwick et al. (2000) reviewed the isotopic and geochemical data and reports of earlier workers (viz, Friend 1983; Friend and Nutman 1991; Oak 1990; Jayananda and Mahabaleshwar 1991) and indicated predominance of granites derived by partial melting of older gneisses (>2900 Ma) in the W portion of the batholith in contrast with juvenile granites and related suites, which predominate at E. Therefore, based on available literature it is logical to expect that high-grade terranes are the deep crustal equivalents of granite-greenstone terranes. However, local field observation of the present study area is insufficient to justify accurately such a broad interpretation.

7 Tectonic Evolution

According to Dey et al. (2014), the East Dharwar craton have Mesoarchaean and Palaeoarchaean crust in the E. In fact, the typical BADR trend consists of tholeiitic basalts, high Mg andesites and dacites-rhyolites (Dey et al. 2013) of the Kadiri schist belt is the signature volcanic rock suite of a convergent margin (Goswami et al. 2017). Based on the discussion on the different factors it can be said that the Kadiri greenstone granite terrain rocks indicate convergent margin magmatism affected by a number of variables, each of which have diversified magma composition (Fig. 20a). The eroded volcanic arc materials deposited as volcaniclastics in the back arc basin. More specifically, in this present context the entire study have given clue towards back arc basin environment for greenstone belt development, which is also discussed in case of the Ramagiri schist belt in the W to the Kadiri terrain (Goswami et al. 2016). The precipitation of chert and BIF atop the volcano-sedimentary succession can be interpreted from the model. In fact, the hydrothermal vents released elements in submarine environment and the free oxygen in the atmosphere started developing during that time by photo dissociation and some Archaean blue-green algae ($\sim$2.6 Ga). Thus the BIF precipitated in the oxidation-reduction interface (Fig. 20b). The cartoon (Fig. 20c) of the arc and back arc basin, proposed in this work, shows a two-stage evolution starting with volcanic eruption followed by erosion. The local extension in the back arc basin also erupted lava and surrounding areas too experienced characteristic phases of batholithic intrusions. Small pockets of plumes might have generated later magmatic eruptions and thus younger granitic intrusions enriched with large ion lithophile elements (LILE) like uranium. Uranium anomalies in younger granites are located from the study area (Fig. 1) during this work. The younger granitic apophyses affected the schist belt volcanics and late shear and tectonics brecciated the rocks. Thus, after formation and infilling of back arc basin followed by granitic intrusions, the long-term erosion allowed the upper chert and BIF-rich portion to get transported by surface agents when the Cuddapah basin started rifting (Fig. 21). This chert, jasper and BIFs are the representative of significant time of erosion in the form of basal conglomerate clast indicating the Eparchean nonconformity. Therefore, it can be concluded that submarine hydrothermal vent activity in the back arc basin is the source of iron and silica and thus volcanic hot springs released hot waters containing dissolved iron, which probably precipitated after oxydation and interbedded with volcanic rocks, graywackes etc. Subsequently, emplacement of dykes and quartz reefs along extension mode cracks possibly during initiation of rifting and Cuddapah basin development is another episode of tectonic history. The provinance

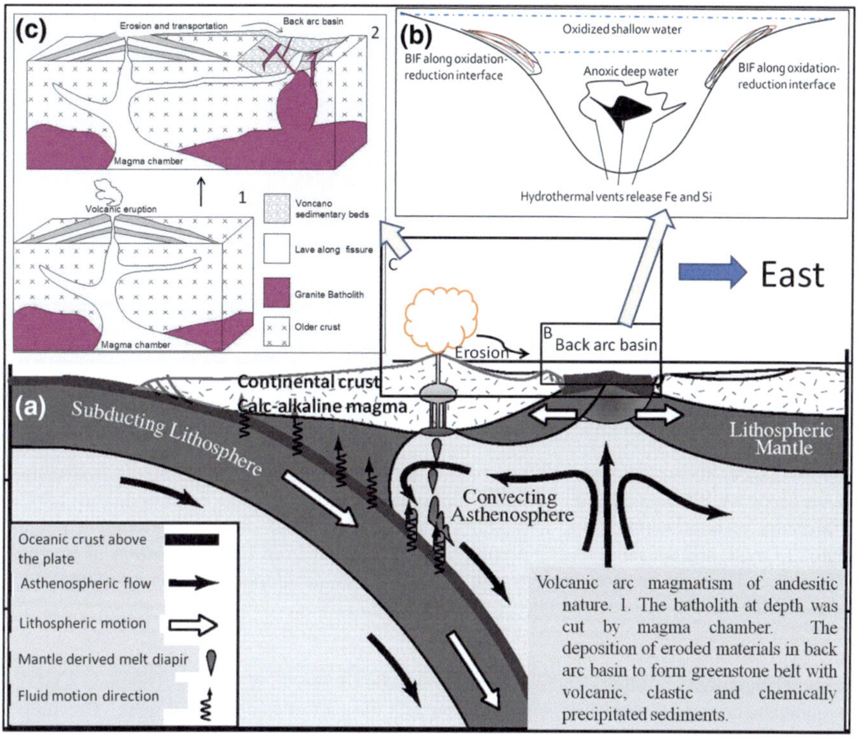

Fig. 20 Tectonic model showing ocean-continent convergence setting of greenstone belt in which the back arc basin environment is more specifically the position of succession during first phase (Episode from ~2.8 to 2.1 Ga). **a** The regional tectonic setting indicates subducting oceanic lithosphere below lighter continental plate and associated volcanic arc development. The deposition of eroded arc material in the back arc basin along with dominantly felsic back arc magmatism must have provided volcano—sedimentary greenstone belt succession. **b** Cartoon depicting a detail zoom in back arc basin and BIF, chert precipitation at the final stage. **c** Cartoon represents step wise evolution of the convergence setting in which stage 1 shows arc magmatism followed by erosion and deposition of arc materials in to the back arc basin with simultaneous back arc basin volcanism

erosion for the newly generated rift supplied initial basin fill and thus conglomerate with clasts of granite, jasper, chert, quartz veins and BIFs are derived from basement.

The SIMS U–Pb age (2556 ± 13 Ma) of the felsic volcanics from the Kadiri greenstone belt (Jayananda et al. 2013b) is corresponding to the Kenoran orogeny (>2500 Ma). Emplacement of later developed plume-related granitic magmatism of ~2520 Ma (Jayananda et al. 2013a) is reasonably related to the Hudsonian orogeny. In fact, the Hudsonian orogeny (~2500–1750 Ma) is an event resulting extention by granitic plutonism, emplacements of dykes and veins along fractures, rifting and associated opening of the Cuddapah basin.

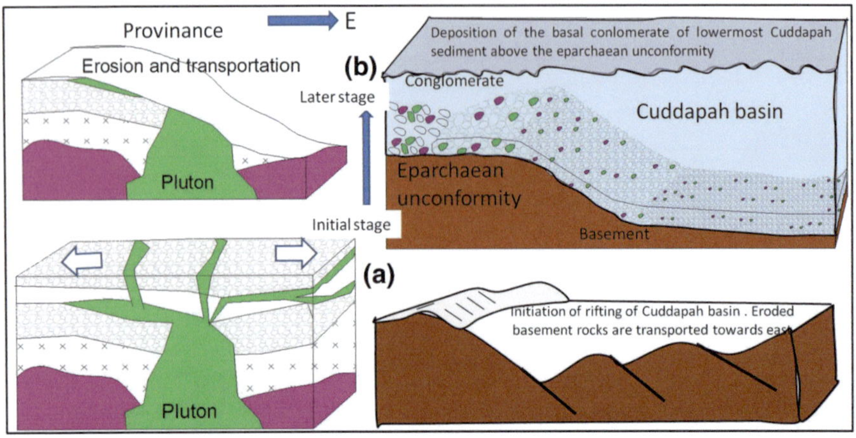

Fig. 21 Model shows the second phase of the tectonic evolution in which a switch over from convergence to divergence after a stable phase in between (episode from ~2.0 to 1.9 Ga). **a** Initiation of continental rifting and associated extension mode cracks development possibly related to dyke and quartz reef emplacement. **b** The erosion from topmost portion of basement provenance (i.e., granite-greenstone terrain) of the newly developing basin are redeposited as conglomerate-sandstone beds of Gulcheru Formation

Acknowledgements Authors express sincere gratitude to Shri L. K. Nanda, Director, AMD for encouragement and infrastructure support to publish the part of the assigned work. The support and help extended by Dr. Syed Zakaulla (RD/SR, Bangalore), Shri. A. K. Bhatt (Dy. RD/SR, Bangalore) and Shri. V. Natarajan (SO/H) are thankfully acknowledged. We thank Soumyajit Mukherjee (IIT Bombay) for reviewing and handling this manuscript. The Springer team is thanked for proof preparation and other assistance. Vide Vanik et al. (2018) for paleostressanalyses procedure, and Tripathy et al. (2019) and Mukherjee et al. (in press) for other tectonic updates of the Cuddapah basin. Mukherjee (2019) summarizes this work.

References

Batchelor RA, Bowden P (1985) Petrogenetic interpretation of granitoid rock series using multicationic parameters. Chemical Geology 48, 43–55

Bowen NL (1928) The evolution of igneous rocks. Princeton University Press, Princeton NJ, 332pp

Cas RAF, Wright JV (1988) Volcanic succesions, modern and ancient. Chapman & Hall, London, 528pp

Chadwick B, Vasudev VN, Hegde GV (2000) The Dharwar craton, southern India, interpreted as the result of late Archaean oblique convergence. Precambrian Research 99, 91–101

Chadwick B, Vasudev V, Hegde GV, Nutman AP (2007) Structure and SHRIMP U/Pb zircon ages of granites adjacent to the Chitradurga schist belt: implications for Neoarchean convergence in the Dharwar craton, southern India. Journal of Geological Society of India 69, 5–24

Chardon D, Jayananda M (2008) Three-dimensional field perspective on deformation, flow, and growth of the lower continental crust (Dharwar craton, India). Tectonics 27 TC 1014

Chardon D, Jayananda M, Peucat JJ (2011) Lateral constrictional flow of hot orogenic crust: insights from the Neoarchean of South India, geological and geophysical implications for orogenic plateaux. Geochemistry, Geophysics, Geosystems 12, Q02005

Dasgupta S, Mukherjee S (2017) Brittle shear tectonics in a narrow continental rift: asymmetric non-volcanic Barmer basin (Rajasthan, India). The Journal of Geology 125, 561–591

Davies GF (1992) On the emergence of plate tectonics. Geology 20, 963–966

De La Roche H, Leterrier J, Grandclaude P, Marchal M (1980) A classification of volcanic and plutonic rocks using R1R2-diagram and major element analyses – its relationships with current nomenclature. Chemical Geology 29, 183–210

Dey S, Nandy J, Choudhary AK, Liu Y, Zong K (2013) Neoarchaean crustal growth by combined arc–plume action: evidences from the Kadiri greenstone belt, eastern Dharwar craton, India. In: Roberts N, van Kranendonk M, Parman S, Shirey S, Clift P (eds) Continent formation through time, vol 389. Geological Society of London, Special Publications, pp 135–163

Dey S, Nandy J, Choudhary AK, Liu Y, Zong K (2014) Origin and evolution of granitoids associated with the Kadiri greenstone belt, eastern Dharwar craton: A history of orogenic to anorogenic magmatism. Precambrian Research 246, 64–90

Ducea M (2001) The California arc: thick granitic batholiths, eclogitic residues, lithospheric—scale thrusting, and magmatic fl are—ups. GSA Today 11, 4–10

Ernst WG (2007) Speculations on evolution of the terrestrial lithosphere—asthenosphere system—plumes and plates. Gondwana Research 11, 38–49

Fisher RV, Schmincke HU (1984) Pyroclastic rocks. Springer, Heidelberg, 474pp

Fliedner MM, Klemperer SL, Christensen NI (2000) Three-dimensional seismic model of the Sierra Nevada arc, California, and its implications for crustal and upper mantle compositions. Journal of Geophysical Research 105, 10899–10921

French JE, Heaman LM (2010) Precise U/Pb dating of Paleoproterozoic mafic dyke swarms of the Dharwar Craton, India: implications for the existence of the Neoarchean supercraton Sclavia. Precambrian Research 183, 416–441

Friend CRL (1983) The link between granite production and the formation of charnockites: evidence from Kabbaldurga, Karnataka. In: Atherton MP, Gribble CD (eds) Migmatities, melting and metamorphism. Shiva Press, Nantwich, pp 264–276

Friend CRL, Nutman AP (1991) SHRIMP U-Pb geochronology of the closepet Granite and Peninsular Gneiss, Karnataka, South India. Journal of Geological Society of India 38, 357–368

Gill JB (1981) Orogenic andesites and plate tectonics. Springer-Verlag, New York, p 390

Goodwin AM, Ambrose JW, Ayres LD et al (1972) The superior province. Geological Association of Canada Special Papers 11, 527–624

Goswami S, Bhattacharjee P, Bhagat S, Kumar S, Zakaulla S (2015) Petrography of chert nodules in stromatolitic dolostone of Vempalle Formation along Tummalapalle - Motnutalapalle, Cuddapah Basin, India. Indian J Geosci 69:13–24

Goswami S, Sivasubramaniam R, Bhagat S, Kumar Suresh, Sarbajna C (2016) Algoma type BIF and associated submarine volcano-sedimentary sequence in Ramagiri granite-greenstone terrain, Andhra Pradesh, India. Journal of Applied Geochemistry 18, 155–169

Goswami S, Upadhayay PK, Bhattacharjee P, Murugan MG (2017) Tectonic setting of the Kadiri schist belt, Andhra Pradesh, India. Acta Geologica Sinica-english edition. 91(6):1992–2006

Goswami S, Upadhyay PK, Bhagat S, Zakaulla S, Bhatt AK, Natarajan V, Dey S (2018a) An approach of understanding acid volcanics and tuffaceous volcaniclastics from field studies: a case from Tadpatri Formation, Proterozoic Cuddapah basin, Andhra Pradesh, India. J Earth Syst Sci 127:20. https://doi.org/10.1007/s12040-018-0929-0

Goswami S, Dey S (2018b) Facies analysis of tuffaceous volcaniclastics and felsic volcanics of Tadpatri Formation, Cuddapah basin, Andhra Pradesh, India. Int J Earth Sci (Geol Rundsch). https://doi.org/10.1007/s00531-018-1620-z

Grove TL, Kinzler RJ (1986) Petrogenesis of andesites. Annual Review of Earth and Planetary Sciences 14, 417–454

Halls HC, Kumar A, Srinivasan R, Hamilton MA (2007) Palaeomagnetic and U-Pb geochronology of easterly trending dykes in the Dharwar Craton, India: feldspar clouding, radiating dyke swarms and the position of India at 2.37 Ga. Precambrian Research 155, 47–68
Harker A (1909) The natural history of igneous rocks. McMillan Publishers, New York, 384pp
Hastie AR, Kerr AC, Pearce JA, Mitchell SF (2007) Classification of altered volcanic island arc rocks using immobile trace elements: development of the Th, Co discrimination diagram. Journal of Petrology 48, 2341–2357
Hawkesworth CJ, Gallagher K, Hergt JM, Keynes M (1993) Mantle and slab contributions in arc magmas. Annual Review of Earth and Planetary Sciences 21, 175–204
Irvine TN, Barager WRA (1971) A guide to the chemical classification of the common volcanic rocks. Canadian Journal of Earth Sciences 8, 523–548
Jakes P, White AJR (1972) Major and trace element abundances in volcanic rocks of orogenic areas. Bulletin of Geological Society of America 83, 29–40
Jayananda M, Mahabaleshwar B (1991) Relationship between shear zones and igneous activity: the Closepet graninte of southern India. Indian Academy of Sciences (Earth and Planetary Sciences) Proceedings 100, 31–36
Jayananda M, Tsutsumi Y, Miyasaki T, Gireesh RV, Kapfo Kowe-u, Tushipokla Hiroshi Hidaka, Kano T (2013a) Geochronological constraints on Meso- and Neoarchean regional metamorphism and magmatism in the Dharwar craton, southern India. Journal of Asian Earth Sciences 78, 18–38
Jayananda M, Peucat JJ, Chardon D, Krishna Rao B, Fanning CM, Corfu F (2013b) Neoarchean greenstone volcanism and continental growth, Dharwar craton, southern India: Constraints from SIMS U-Pb zircon geochronology and Nd isotopes. Precambrian Research 227, 55–76
Maniar PD, Piccoli PM (1989) Tectonic discriminations of granitoids. Geological Society of America Bulletin 101, 635–643
Mcphie J, Doyle M, Allen R (1993) Volcanic Textures. A guide to the interpretation of textures in volcanic rocks. Tasmanian Government Printing, Office, Tasmania, p 196pp
Misra AA, Mukherjee S (2015) Tectonic inheritance in continental rifts and passive margins. Springerbriefs in Earth Sciences. ISBN 978-3-319-20576-2
Misra AA, Mukherjee S (2018) Atlas of structural geological interpretation from seismic images. Wiley Blackwell. ISBN: 978-1-119-15832-5
Misra AA, Bhattacharya G, Mukherjee S, Bose N (2014) Near N-S paleo-extension in the western Deccan region in India: does it link strike-slip tectonics with India-Seychelles rifting? International Journal of Earth Sciences 103, 1645–1680
Misra AA, Sinha N, Mukherjee S (2015) Repeat ridge jumps and microcontinent separation: insights from NE Arabian Sea. Marine and Petroleum Geology 59, 406–428
Misra AA, Sinha N, Mukherjee S (2018a) The gop rift: a paleo slow spreading centre, Offshore Gujarat, India. In: Misra AA, Mukherjee S (eds) Atlas of structural geological interpretation from seismic images. Wiley Blackwell, pp 208-212 ISBN: 978-1-119-15832-5
Misra AA, Sinha N, Mukherjee S (2018b) The Ratnagiri fracture zone: a paleo oceanic-fracture-zone in the Mumbai-Ratnagiri Offshore Region, West India. In: Misra AA, Mukherjee S (eds) Atlas of structural geological interpretation from seismic images. Wiley Blackwell, pp. 195-199. ISBN: 978-1-119-15832-5
Miyashiro A (1974) Volcanic rock series in island arcs and active continental margins. American Journal of Science 274, 321–355
Mukherje S (2015) Atlas of structural geology. Elsevier, Amsterdam
Mukherjee S (2010a) Structures in Meso- and Micro-scales in the Sutlej section of the Higher Himalayan shear zone, Indian Himalaya. e-Terra 7, 1–27
Mukherjee S (2010b) Microstructures of the Zanskar shear zone. Earth Science India 3, 9–27
Mukherjee S (2011a) Flanking Microstructures from the Zanskar shear zone, NW Indian Himalaya. YES Bulletin 1, 21–29
Mukherjee S (2011b) Mineral Fish: their morphological classification, usefulness as shear sense indicators and genesis. International Journal of Earth Sciences 100, 1303–1314

Mukherjee S (2012) Tectonic implications and morphology of trapezoidal mica grains from the Sutlej section of the Higher Himalayan shear zone, Indian Himalaya. Journal of Geology 120, 575–590

Mukherjee S (2013) Deformation microstructures in rocks. Springer Geochemistry/Mineralogy, Berlin, pp 1–111. ISBN 978-3-642-25608-0

Mukherjee S (2014a) Mica inclusions inside host mica grains from the Sutlej section of the Higher Himalayan Crystallines, India- morphology and constrains in genesis. Acta Geologica Sinica 88, 1729–1741

Mukherjee S (2014b) Review of flanking structures in Meso- and Micro-scales. Geological Magazine 151, 957–974

Mukherjee S (2019) Introduction to "Tectonics and Structural Geology: Indian Context". In: Mukherjee S (ed) Tectonics and structural geology: Indian context. Springer International Publishing AG, Cham, pp 1–5. ISBN: 978-3-319-99340-9

Mukherjee S, Chakraborty R (2007) Pull-apart micro-structures and associated passive folds. In: Aho J (ed) Annual transactions of the nordic rheology society 15, pp 247–252. 16th Nordic Rheology Conference, Stavanger, Norway, 13–15 June 2007

Mukherjee S, Koyi HA (2009) Flanking microstructures. Geological Magazine 146, 517–526

Mukherjee S, Koyi HA (2010a) Higher Himalayan Shear Zone, Zanskar Section-microstructural studies & extrusion mechanism by a combination of simple shear & channel flow. International Journal of Earth Sciences 99, 1083–1110

Mukherjee S, Koyi HA (2010b) Higher Himalayan Shear Zone, Sutlej Section-structural geology & extrusion mechanism by various combinations of simple shear, pure shear & channel flow in shifting modes. International Journal of Earth Sciences 99, 1267–1303

Mukherjee S, Goswami S, Mukherjee A (In press) Structures and their tectonic implications form the southern part of the Cuddapah basin, Andhra Pradesh, India. Iranian J Sci Technol Trans A: Sci. https://doi.org/10.1007/s40995-018-0566-0

Nandy J, Dey S (2013) The mechanism of Neoarchaean granitoid formation: evidence from Eastern Dharwar Craton, Southern India. American International Journal of Research in Formal, Applied and Natural Sciences (AIJRFANS) 3, 105–109

Oak KA (1990) The geology and geochemistry of the Closepet granite, Karnataka, South India. Unpublished Ph.D. thesis, Council for National Academic Awards, Oxford Polytechnic, UK

Pearce JA (1982) Trace element characteristics of lavas from destructive plate boundaries. In: Thorpe RS (ed) Andesites: orogenic andesites and related rocks. Wiley, Chichester, pp 525–548. ISBN 0 471 28034 8

Pearce JA, Peate DW (1995) Tectonic implications of the composition of volcanic arc magmas. Annual Review of Earth and Planetary Sciences 23, 251–285

Peccerillo A, Taylor SR (1976) Geochemistry of Eocene calc-alkaline volcanic rocks from the Kastamonu area, Northern Turkey. Contributions to Mineralogy and Petrology 58, 63–81

Perfit MR, Gust DA, Bence AE, Arculus RJ, Taylor SR (1980) Chemical characteristics of island-arc basalts: implications for mantle sources. Chemical Geology 30, 227–256

Rajamani V (1990) Petrogenesis of Metabasites from the schist belts of Dharwar Craton: implications to Archean Mafic Magmatism. Journal of Geological Society of India 36, 565–587

Rajamani V, Shivkumar K, Hanson GN, Shirey SB (1985) Geochemistry and petrogenesis of amphibolites, Kolar Schist Belt, South India; evidence for komatiitic magma derived by low percent of melting of the mantle. Journal of Petrology 96, 92–123

Rowland SK, Walker GPL (1990) Pahoehoe and aa in Hawaii: volumetric flow rate controls the lava structure. Bulletin of Volcanology 52, 615–628

Satyanarayana K, Siddilingam J, Jetty J (2000) Geochemistry of Archean Metavolcanic Rocks from Kadiri Schist Belt, Andhra Pradesh, India. Gondwana Research 3, 235–244

Shackleton RM (1976) Shallow and deep-level exposures of the Archean crust in India and Africa. In: Windley BF (ed) The Early History of the Earth. Wiley, New York, pp 317–322

Shand SJ (1943) Eruptive rocks. Their genesis, composition, classification, and their relation to ore-deposits with a chapter on meteorite. Wiley, New York

Sreenivasulu P, Padmasree P, Hanumanthu PC (2014) Granitoids adjoining Kadiri schist belt, Andhra Pradesh, South India: field and petrographic implications. International Journal of Geology, Earth and Environmental Sciences 4, 244–258

Stern RJ (2008) Modern style plate tectonics began in Neoproterozoic time: an alternative interpretation of Earth' s tectonic history. In: Condie KC and Pease V (eds) When did plate tectonics begin on planet earth? Geological Society of America Special Paper 440, pp 265–280

Tripathy V, Satyapal, Mitra SK, Sai BBS (2019) Fold-thrust belt architecture and structural evolution of the Northern part of the Nallamalai Fold Belt, Cuddapah basin, Andhra Pradesh, India. In: Mukherjee S (ed) Tectonics and structural geology: Indian context. Springer International Publishing AG, Cham, pp 219–252. ISBN 978-3-319-99340-9

Wohletz KH, Mcqueen RG (1984) Volcanic and stratospheric dust like particles produced by experimental water-melt interactions. Geology 12, 591–594

Vanik N, Shaikh H, Mukherjee S, Maurya DM, Chamyal LS (2018) Post-deccan trap stress reorientation under transpression: evidence from fault slip analyses from SW Saurashtra, Western India. Journal of Geodynamics 121, 9–19

Basement Tectonics and Shear Zones in Cauvery Basin (India): Implications in Hydrocarbon Exploration

S. Mazumder, Blecy Tep, K. K. S. Pangtey and D. S. Mitra

1 Introduction

Structural and tectonic modelling plays a major role in providing solutions in different phases and aspects of hydrocarbon exploration. Different tectonic phases and their associated structures play a significant role right from basin evolution up to prospect generation and exploitation/development strategies. This influence of tectonics and structural analysis have more impact in case of basement exploration where different phases of Pre-Cambrian to Phanerozoic basement tectonics control various elements of exploration like prospective areas, well placement, reservoir capacity etc. These elements are governed by the parameters of basement prospects like formation of basement highs, fracture pattern and densities, generation of secondary porosities by neotectonic deformations and reactivations of earlier structures.

In the Indian context, basement exploration is not a rather new concept with the fractured basement in Borholla field of South Assam Shelf in Assam and Assam Arakan Basin being under production for almost five decades. However, with gradual decrease in large and easy-to-find oil pools, there is a shift in the focus of exploration from conventional sedimentary reservoirs to unconventional hydrocarbon accumulations. Basement oil pools fit into the latter category with present efforts for basement exploration being underway in the basins of Cambay, Mumbai High, Cauvery, Krishna-Godavari and Assam Shelf of India.

The Cauvery Basin forms a Category I sedimentary basin or an oil province with an established commercial production. Most of its established petroleum systems are limited to Cretaceous and Paleocene sequences with an estimated hydrocarbon resource of 700 MMT of oil and oil equivalent gas (DGH India 2015). However, in

S. Mazumder (✉) · B. Tep · K. K. S. Pangtey · D. S. Mitra
ONGC, Dehradun, India
e-mail: subhom007@gmail.com

S. Mukherjee (ed.), *Tectonics and Structural Geology: Indian Context*,
Springer Geology, https://doi.org/10.1007/978-3-319-99341-6_9

the recent years (around 2012), hydrocarbon discoveries in the basement at depths around 1400 m from a few fields have indicated the existence of an additional basement play which may have a significant role in the overall hydrocarbon exploration scenario. The present study based on a basement exploration perspective has been made to decipher different evidences of structural deformation of basement in outcrops and subcrops in different scales pertaining to the various episodes of tectonism affecting the basin. These structural signatures derived from independent G&G studies were subsequently integrated in a GIS platform to build a conceptual model for basement exploration and thus delineate optimal areas where the basement can act as a promising hydrocarbon reservoir.

The Cauvery Basin is a peri-cratonic rift basin in the southern part of India hosting a sedimentary sequence ranging in age from Late Jurassic to Recent (Mukherjee 2015a, b). The basin is bounded at west by exposures of the Pre-Cambrian South Indian Shield that rests unconformably below the sedimentary layers forming the basement to this basin. A number of NE oriented basement highs divide the basin into alternating ridge and sub basin areas, each of which form discrete fault-bound blocks (Fig. 1a). In the north of the basin, another basement high known as the Chingelput High (Mazumder et al. 2013a, b) separates the Cauvery Basin from the adjacent Palar Basin whereas in the eastern part, the Cauvery Basin continues underneath the Bay of Bengal as the Cauvery offshore.

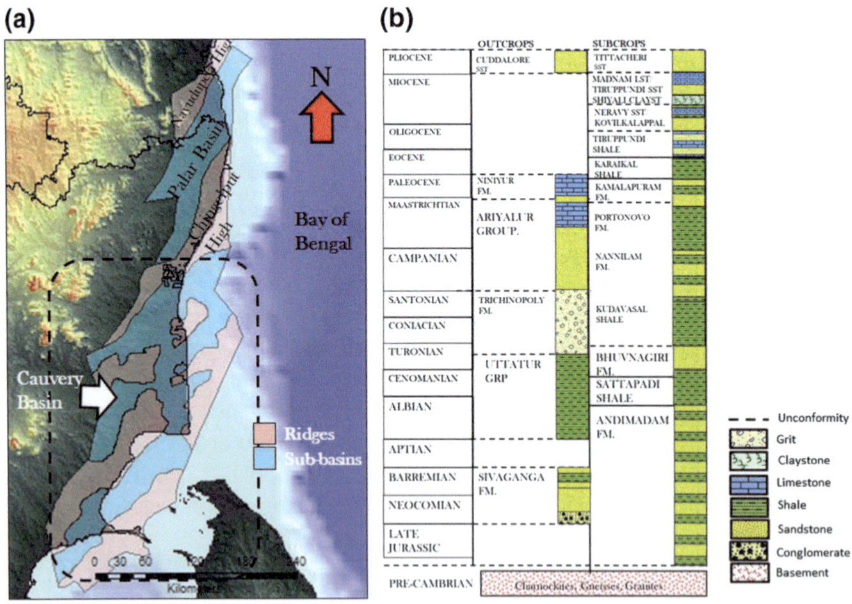

Fig. 1 **a** General tectonic & geographical set up of the Cauvery Basin showing different sub-basins and their dividing ridges **b** Generalized stratigraphy encountered in the Cauvery Basin in outcrops and subcrops depicting the different source rock sequences (After Rangaraju et al. 1993; dghindia.com 2011; Nagendra et al. 2017)

The basinal area in onshore part is covered by a thick layer of alluvium that conceals structural and geological information.

1.1 Basement Tectonics

The sedimentary sequence of the Cauvery Basin along with its basement (Fig. 1b) has suffered multiple episodes of tectonism. These different episodes of tectonism and their effects in the basement are described below.

1.1.1 Precambrian Basement Tectonics

The Southern Granulite Terrane (SGT) of Late Archean to Neo-Proterozoic age (Meert et al. 2010) effectively forms the basement of the entire Cauvery Basin and is exposed in the western part beyond the basin margin. The SGT is effectively constituted of polydeformed, poly-metamorphosed (Bhutani et al. 2007) and exhumed granulite facies lower-crustal rocks e.g. charnockites, khondalites and other high-grade metamorphic rocks with intrusions of granitic rocks (Ravindra Kumar 2005; Plavsa 2014). Genetically SGT is believed to have been formed by the accretion of number of high-grade metamorphic blocks (Meert et al. 2010). The boundaries of each of these juxtaposed crustal provinces are marked by Neoproterozoic shear zones (Meert et al. 2010) most of which are associated with deep seated faults (Rao and Prasad 2006; Harinarayana et al. 2006). One of the most significant tectonic feature in the SGT is the Cauvery Shear Zone (CSZ) that forms an crustal scale E-W trending zone of intense planar fabrics that are exposed in outcrop areas beyond basin margin and separate the northern Late Archaean granulite block from the Proterozoic Madurai granulite block (Chetty 2015). The Madurai block is itself separated by an isotopic boundary into a southern and northern Madurai Block (SMB and NMB) (Plavsa 2014). This boundary correlates with the surface extension of prominent southward dipping reflector in the upper and the middle crust (Rajendra Prasad et al. 2007) suggesting it to be a major terrane boundary. Studies by Gupta et al. (2015) conceptualized the extension of this boundary below Cauvery Basin parallel to the CSZ.

The CSZ comprises a network of anastomosing broadly E-W shear zones ranging from 1 to 10 km width, which represent belts of high strain separated by domains of relatively low strain (Santosh et al. 2009). It is bound by the Moyar Bhawani Shear Zone (MBSZ) in the north and the Palghat Cauvery Shear Zone (PCSZ) (Janardhan 1999; Rao et al. 2006) to the south (Chetty and Bhaskar Rao 2006) with its eastern extension disappearing in the Bay of Bengal. Overall, it is considered a product of two deformation phases, D1 during the Archean-Proterozoic boundary and D2 during Proterozoic-Cambrian boundary (Santosh et al. 2012). The D1 episode is associated with large-scale north verging thrusts and N-S shortening, probably a result of subduction between different blocks, one of them probably the southern margin of the Dharwar craton (Chetty and Santosh 2013). The D2-episode is characterized by transpressional tectonic regime with extensive dextral shear and migmatisation (Chetty et al. 2003). Here it is to be noted that in general shear fabrics in shear zones are developed ubiquitously in migmatites (Mukherjee and Koyi 2010a; Mukherjee 2013, 2014a, 2015a, b). These E-W individual shear zones are presumably rooted to a decollement zone in the upper mantle forming a crustal-scale 'flower structure' (Chetty and Bhaskar Rao 2006; Chetty 2015) implying that individual shear zones branch out of a deeper and thicker zone: vide fig A5 in the Appendix/Repository of Dasgupta and Mukherjee (2017) for a meso-scale flower structure from a different terrain. Of these shear zones, the MBSZ in the northern boundary of CSZ dip steeply southward whereas the other constituent shear zones dip to north with moderate values.

The shear zones of CSZ have been mapped in the outcrop, but their disposition below the Cauvery Basin plays a major role controlling the basement morphology and guiding later development of structures in the northern part of the Cauvery Basin (Murty et al. 2002). As such, CSZ influences basement exploration in the areas of Ariyalur-Pondicherry Sub Basin, Kumbakonam-Madanam ridge, Tranquebar-Tanjore Sub Basin, Karaikal ridge and Nagapattinam Sub Basin (Fig. 2).

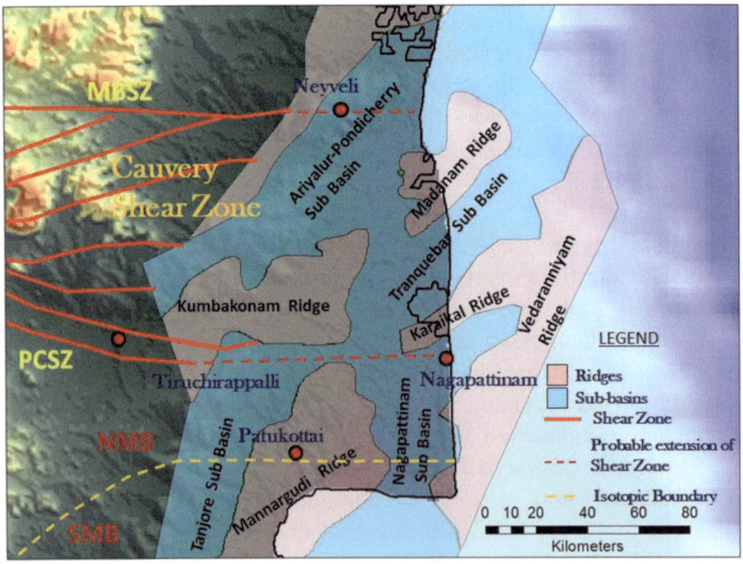

Fig. 2 Layout of the Cauvery Shear Zone (after Chetty 2015; Chardon et al. 2008; Plavsa 2014; Gupta et al. 2015) consisting of broadly E-W oriented shear zones with most of it impacting basement tectonics of Ariyalur-Pondicherry Sub Basin, Kumbakonam-Madanam Ridge, Tranquebar Sub Basin, Karaikal-Vedaranniyam Ridge and Nagapattinam Sub Basin of Cauvery Basin. *MBSZ* Moyar Bhawani Shear Zone, *PCSZ* Palghat Cauvery Shear Zone, *NMB* Northern Madurai Block and *SMB* Southern Madurai Block

1.1.2 Phanerozoic Basement Tectonics

Other than the formation of shear zones during Precambrian, the basement of Cauvery Basin has also undergone three major episodes of tectonism that governed the development of basement structures.

Formation of N-S Faults

One of the probable episodes of tectonism affecting the basement is the formation of N-S faults during Lower Gondwana (Late Carboniferous—Permian). Evidences of these tectonic phase exists as a number of N-S morphotectonic faults, major N-S lineaments (Subrahmanyam et al. 1995; Saravanavel and Ramasamy 2003; Prabaharan et al. 2013) and magnetic lineaments mapped in the onland Cauvery Basin and offshore (Subrahmanyam et al. 1995). These structural trends are sympathetic to the southern part of the adjacent Palar rift that hosts Ongur Formation (Fig. 3a) dated to be Late Carboniferous to Early Permian (Rangaraju et al. 1993; Vairavan 1993). In the Cauvery Basin, well P-2 drilled in Kumbakonam Ridge along the N-S trend has also yielded palynofossil assemblage equivalent to Ongur Formation (Prasad and Phor 2009). This implies the existence of N-S faults/rifts continuing from Palar Basin in the north and affecting at least the northern part of basement of Cauvery Basin (Fig. 3a).

Formation of NE-SW Faults

The principal event that initiated and shaped the morphology of Cauvery Basin is the NE trending rifting associated with the separation of India-Sri Lanka system from Antarctica and Australia in Late Jurassic to Early Cretaceous (Veevers and Tewari 1995; Gaina et al. 2007; Veevers 2009; Lal et al. 2009; Nemcok et al. 2013). The continental separation associated with rifting began at Late Jurassic (160 ma) and produced genetically related rift systems of Cauvery, Palar, Pennar and Krishna-Godavari. In Cauvery and Palar Basins the separation probably occurred along pre-existing NE-SW oriented weaknesses in the Pre Cambrian basement (linkages between shear zones) that may have guided later structures (Lal et al. 2009; Dasgupta and Mukherjee 2017). Such reactivation along existing fabric is called "tectonic inheritance" (Misra and Mukherjee 2015). This led to the formation of a series of major NE-SW horst and graben features in the basement dividing the basin into a number of fault-bound ridges and sub-basin areas (Lal et al. 2009).

Formation of NW-SE Faults

In Early Cretaceous, transpression along the junction of Antarctica and Sri Lanka-India developed a N-S strike slip fault that had been variously interpreted as

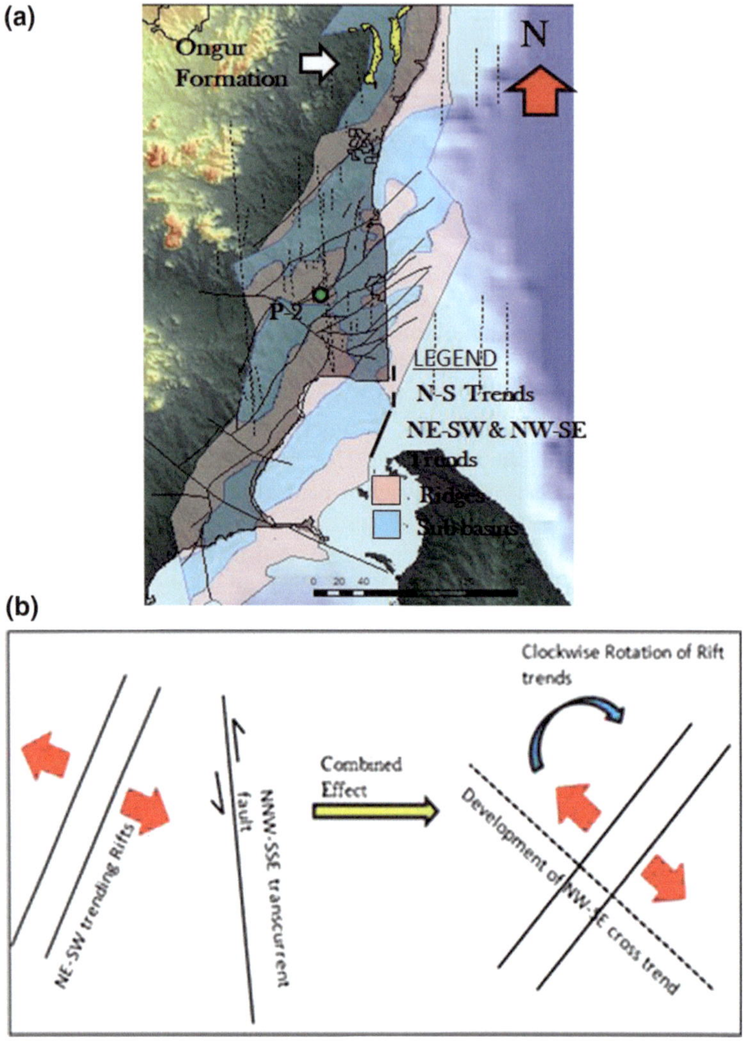

Fig. 3 **a** N-S oriented morphotectonic faults and lineaments (Subrahmanyam et al. 1995; Saravanavel and Ramasamy 2003; Prabaharan et al. 2013) probably formed during Lower Gondwana (Lower Carboniferous—Permian) that are sympathetic to the Ongur rift in Palar Basin and also host well P-2 in Thanjavur Sub Basin that yielded palynoflora equivalent to Ongur Formation. **b** Formation of NW-SE cross trends (Saravanavel and Ramasamy 2003; Prabaharan et al. 2013) that had affected the basement as a result of simultaneous activity of the NE-SW rifting and N-S strike slip faulting (Mazumder et al. 2013)

a transcurrent fault (Lal et al. 2009) or a dextral strike-slip transform fault (Nemcok et al. 2013) along which Antarctica moved southward. The combined and simultaneous effect of the NE-SW rift trend and NNW trending strike slip fault probably led to the development of NW-SE cross trends (Mazumder et al. 2013a, b). This also resulted in a rotation of the earlier rift trends (Mazumder et al. 2013a, b) causing a wide zone of fracturing along the rotational axis (Fig. 3a, b).

These tectonic phases in their respective times have its individual sympathetic structural signatures that constitute part of the fabric of the basin, controlling its evolution, sedimentation pattern to even the distribution of hydrocarbon systems. Results of these phases of tectonism may also have led to parts of the basement acting as a hydrocarbon reservoir as observed from the results of a number of wells drilled by ONGC in the basin with depths varying from 2000 to 3500 m.

1.2 Basement Character as Hydrocarbon Reservoir

Now, in order to act as an effective reservoir rock, the basement should have the following characteristics:

(a) Most basement hydrocarbons are hosted in structural highs of varying but generally moderate to large elevation. The highs are formed by fault-controlled blocks, often in rift settings like upthrown sides of faulted block, horsts within a graben (Chung 1982) or by palaeo-hills buried unconformably below sedimentary cover (Gutmanis 2009).
(b) Younger sediments, which act as hydrocarbon sources, either flank or directly overlie the basement (Sircar 2004). Charging mostly takes place laterally or by up-dip migration from adjacent structural lows, however, downward migration may also take place from overlying source rocks into underlying basement (Gutmanis 2009).
(c) Most basement rocks are characterized with low matrix porosity (typically <0.5%) and permeability. As a result, reservoir quality in basement rocks depends on the development of secondary porosity and the presence of a connected open fracture network. A number of such fractured reservoirs in basement exist worldwide e.g. Aerxia Oil Field in China with well-developed joints, faults and fractures in phyllite, slate and meta-sandstone (Sircar 2004), White Tiger field, Vietnam constituted of fractured granite basement (Dang et al. 2011), Dayung Field in South Sumatra with fractured basement comprising Permian carbonates intruded by granites (Darmadi et al. 2013) and fractured basement reservoir in Zeit Bay oil field, Gulf of Suez, Egypt composed of pegmatic granites (Zahran and Askary 1988). This secondary porosity consists of results of structural deformation like joints, faults, fractures ranging in scale from microfractures to seismic scale faults and their damage zones (Sircar 2004; Satyanaryana et al. 2010). As a result, the porosity and permeability values in a basement reservoir are irregularly distributed laterally as well as in depth, In case of White Tiger field, the largest field of Vietnam for basement oil, maximum porosity is found to be 4% with permeability occasionally reaching 0.8D (Huy et al. 2012). However, dense micro-cracks associated to major fault structures that may be formed as a result of multiple

episodes of tectonism can develop porosity up to 10% and permeability up to 0.1 D (Walter et al. 2017). Accordingly, areas of basement that suffer and bear the imprints of multiple episodes of tectonism and reactivation can be considered as optimal areas that can act as basement reservoir (Trice 2014).

(d) Both up-dip and lateral charging occur in periods of tectonic activity when active faults act as open conduits for the migration of fluids with overlying non-breached relatively tight and low permeability sedimentary units acting as seals (Sircar 2004; Gutmanis 2009). This is aided by the episodic and repeated mechanisms of seismic pumping, associated with reactivations and earthquake processes (Gutmanis 2009) depending upon the state of stress acting on the region. Here seismic pumping refers to the intermittent re-distribution of fluids including hydrocarbons mainly in and around fault zones during fault activity. Seismic faulting acts as a pumping mechanism whereby release of seismogenic stress during earthquakes creates a fluid potential difference between upper and lower part in a fault and thus acts as an important force resulting in rapid pore fluid migration (Sibson et al. 1975; Cao et al. 2011; Hua 1995). Basement prospects that are oriented parallel/sub parallel to the maximum in situ stresses in the region are considered as more favourable targets compared to those at high angles to these (Vasudevan et al. 2012).

(e) Analysis of production performance in basement reservoirs indicate high flow rates concentrated in late faults that are reactivated closely before the oil migration time (Huy et al. 2012).

The above mentioned salient features that qualify a basement to be a good reservoir indicates the significance of the different episodes of tectonism and their corresponding structural deformations in controlling basement exploration parameters like optimal reservoir area, areas of maximum fracturing, conduits for migration etc. In the rest of the manuscript, different studies like remote sensing and geomorphology, geological fieldworks, seismic data, electrologs, gravity and magnetics have been used to analyse the basement of Cauvery Basin to detect different structural signatures of tectonic episodes. Based on the signatures, the above parameters are integrated to delineate the promising areas of basement exploration in the basin.

2 Analysis and Methodology

The present study presumes that signatures of structural deformation resulting from a particular episode of tectonics are similar in pattern but may vary in their scale depending upon the tools and perspective used to view them. Based on this premise, pattern of signatures of a particular phase of tectonism observed in outcrops may be repeated in larger scale in subsurface seismic data or in further regional scale in remote sensing images. In accordance to the above assumption, structural patterns

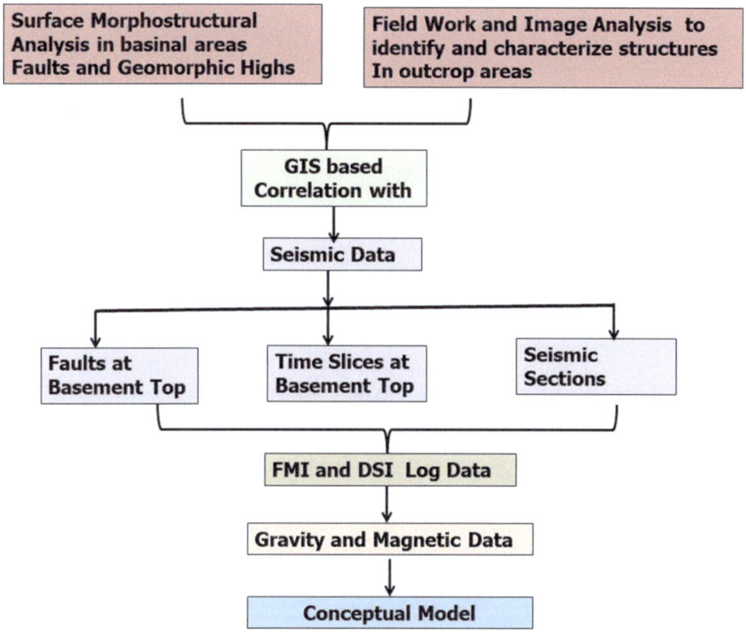

Fig. 4 Flowchart of analysis for deriving conceptual model of basement exploration

in the basement below the Cauvery Basin are extracted from different geophysical and geological data sets. These are subsequently compared with field analogues and integrated to prepare a conceptual model for basement exploration (Fig. 4).

2.1 *Surface Analysis*

In this part of the study, the Cauvery Basin and the Precambrian basement outcrops beyond it are analysed to delineate structures exposed and manifested on the surface.

In the basinal area, a neotectonic based structural analysis/morphostructural analysis is carried out on principles discussed in Mazumder et al. (2011, 2013a, b, 2016) to detect surface manifestations of deep-seated/basement-related structures. These basement related structures include both faults and structural highs. Both of these elements reactivate repeatedly in the neotectonic regime and are manifested on the surface as anomalies in geomorphic elements like topography, soil and especially drainage, which form the most sensitive parameter. Such neotectonic manifestations like fluvial anomalies, seismogenic lineaments and seismicity are well distributed and well-studied in the Cauvery Basin (Murty et al. 2002; Kumaran and Ramasamy 2005; Ganapathy and Rajarathinam 2010; Prabaharan et al. 2013,

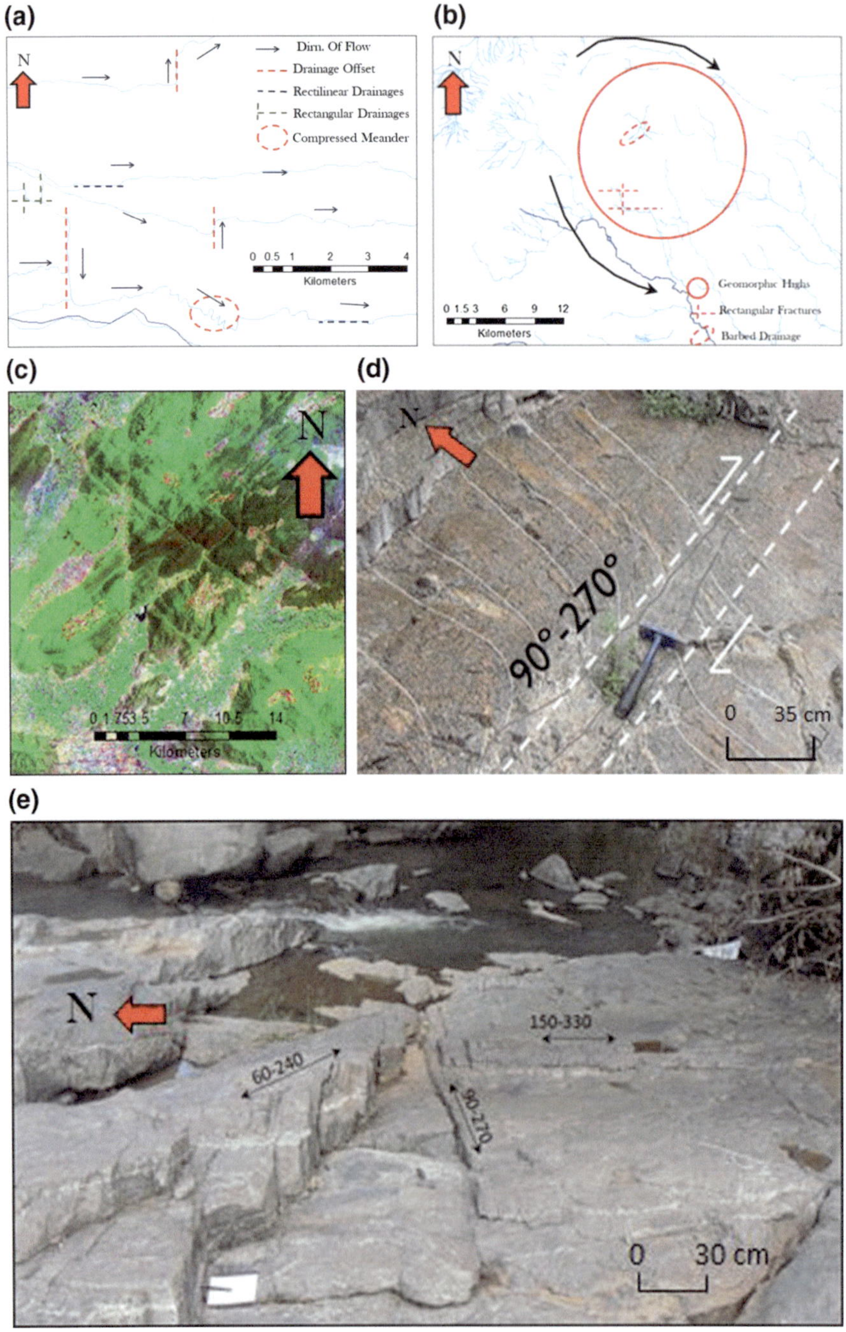
(a)
N
Dirn. Of Flow
Drainage Offset
Rectilinear Drainages
Rectangular Drainages
Compressed Meander
0 0.5 1 2 3 4
Kilometers
(b)
N
Geomorphic Highs
Rectangular Fractures
Barbed Drainage
0 1.5 3 6 9 12
Kilometers
(c)
N
Kilometers
(d)
N
90°-270°
0 35 cm
(e)
N
150-330
60-240
90-270
0 30 cm

◂**Fig. 5** **a** Drainage features used to interpret probable fault segments that may be manifestations of basement faults. **b** Geomorphic highs indicating sub surface structural highs interpreted based on peripheral drainage geometry. Associations with elements like barbed drainage and rectangular drainages imply a greater degree of confidence of representing sub surface structural highs (Trenchard 2007). **c** Evidences of folding and jointing in the exposed basement near Salem delineated from LANDSAT ETM+ band combination 7, 4 and 1. **d** Shear zone exposed in a river bed in Belur near Salem oriented E-W with earlier NE-SW foliations rotated parallel to sense of shearing. **e** Joints developed in Charnockite in the exposed basement. **f** Fault oriented NW-SE in the Cuddalore Formation near Neyveli. Geographical positions of the geomorphic, outcrop and image based features cited above marked in the Fig. 5**g**

Singh et al. 2016; Resmi et al. 2016 etc.). In this study, features in drainage geometry (e.g. drainage offsets, rectangular drainages, compressed meanders) (Fig. 5a), abrupt changes in surface slope and lineaments derived from tonal variations in images are used to delineate fault segments. Similarly, geomorphic highs forming surface expression of probable structural highs in basement are derived from features like radial and peripheral drainages. Associations with features like barbed drainages, rectangular drainage patterns and low moisture contents are used to build up and classify these geomorphic highs as per their degree of confidence based on their probabilities in representing basement structural highs (Fig. 5b). These morpho-structural interpretations are subsequently validated by extensive field checks where the mapped structures are checked for neotectonic faulting imprints and evidences like unpaired terracing, abnormal incisions and sudden appearance of braiding.

In areas beyond the basin margin of Cauvery Basin, extensive field work and image analysis have been carried out to identify structural features in the exposed basement rocks. LANDSAT ETM+ (EarthExplorer, USGS) band combinations of band 7, 4 and 1 have been pan-sharpened, contrast enhanced and edge filtered to identify different regional scale structural features (Fig. 5c). Abrupt termination of basement outcrops to delineate faults, instances of folding in basement lithology, prominent lineament patterns have been deciphered from separate image data sets. As a part of fieldwork in these outcrops, attitude of joint sets and fractures (Fig. 5e) developed in basement lithology were measured, characterized and used to identify trends of their causative regional faults. Outcrop scale deformations like folding, faulting (Fig. 5f) and shearing (Fig. 5d) were also analysed in view of their geometry and inter-relationships for modelling of subsurface basement features.

These two sets of surface structural data pertaining to the alluvial covered basinal area and the exposed basement area were correlated and joined in a GIS platform to understand the disposition of basement structures in the basinal area (Fig. 6).

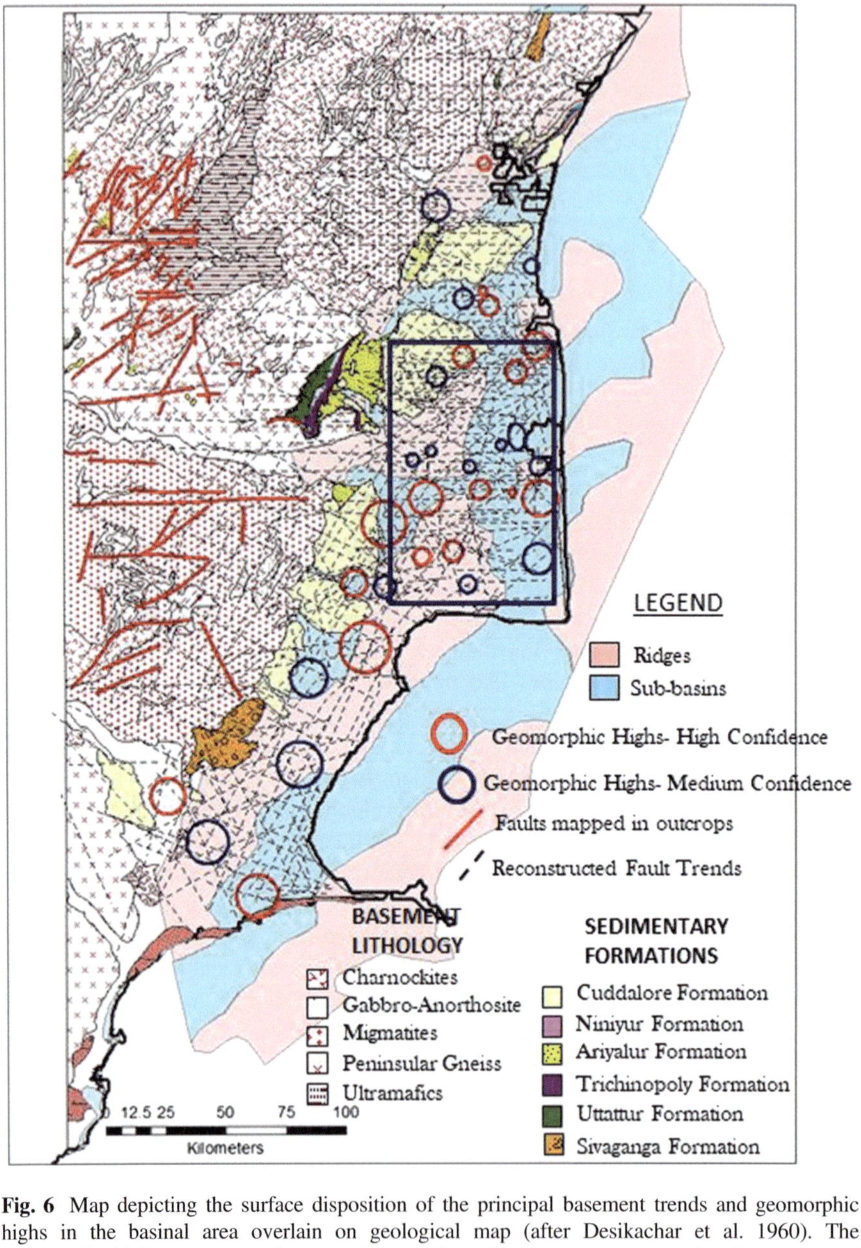

Fig. 6 Map depicting the surface disposition of the principal basement trends and geomorphic highs in the basinal area overlain on geological map (after Desikachar et al. 1960). The geomorphic highs are arranged as per degree of confidence to represent subsurface structural highs. The surface trends in the rectangular area has been correlated with faults at basement top in Fig. 7

2.2 Correlation with Seismic Data

3D Pre-Stack Time Migrated (PSTM) seismic data used for imaging subsurface datasets exhibiting complex geological structures recorded in the Cauvery Basin area has been used to validate and establish the continuity of the surface morphostructural interpretations and to identify the structural effects of different phases of tectonism.

An overlay analysis of faults mapped on basement top from seismic data in Kali-Kuthalam area (Fig. 6) with surface basement trends and morphostructural faults shows a positive correlation. This probably implies the basement related subsurface faults may have cut through Mesozoic and Tertiary sections as a result of multiple reactivations of them and have been manifested on the surface due to present day neotectonism affecting Cauvery Basin (Fig. 7). This also suggests the interpreted surface trends represent the basement faults.

In a further attempt to correlation, seismic time slices acquired at 2000 ms time equivalent depth (within basement, near basement top) in Madanam and Kuthalam area (Fig. 8a) had been digitally filtered in an image processing software to highlight and extract lines of high seismic amplitude. Here it is to be noted that though crystalline formations are associated with lack of acoustic contrast, however structures on or near basement top can be identified (Gutmanis 2009). These extracted lines appear to define a similar geometry to the E-W oriented folds observed in image data (Fig. 8c) and in outcrops (Fig. 8d) and also display an analogous association with the E-W faults as observed in those areas. As such, it is possible that these features are sheath folds defined as highly non-cylindrical folds (Reber et al. 2013) that may be produced as a result of pronounced ductile shear (Mukherjee 2014b; Mukherjee et al. 2015). Similar sheath folds have been mapped by Chetty et al. (2012) in Mahadevi Hills in the axial zone of Cauvery Shear Zone that are also characterized by E-W axial traces (Fig. 9). The E-W trends probably act as manifestations of the shear zones in basement with the sheath folds forming contemporaneously with the dextral shearing along them. Fold amplifies during shearing with the fold axes rotating towards shear direction. At the highest strain level, the fold axes becomes sub-parallel to shear direction (Burg 2017a, b) and existing foliations of the country rock are found to be deflected parallel to the E-W trends (Fig. 5d).

In addition, seismic sections are correlated with surface interpretations to validate the morpho-structural manifestations of basement top morphology (Fig. 5b). The correlation indicates that faults affecting basement top when extrapolated up-dip match with the surface fault trends re-iterating the fact that they are reactivation related manifestations of the basement faults (Fig. 9e). Also, most of the geomorphic highs spatially coincide with the areas of basement highs associated with a high amount of structural relief. Since most of the interpreted geomorphic highs are found to be associated with NE trending surface features, probably the NE faults create basement highs with a higher structural relief (Fig. 9a). Other than these structural highs, the basement top is also found to depict a series of smaller undulations with much smaller variation in basement relief compared to the structural highs. These small undulations

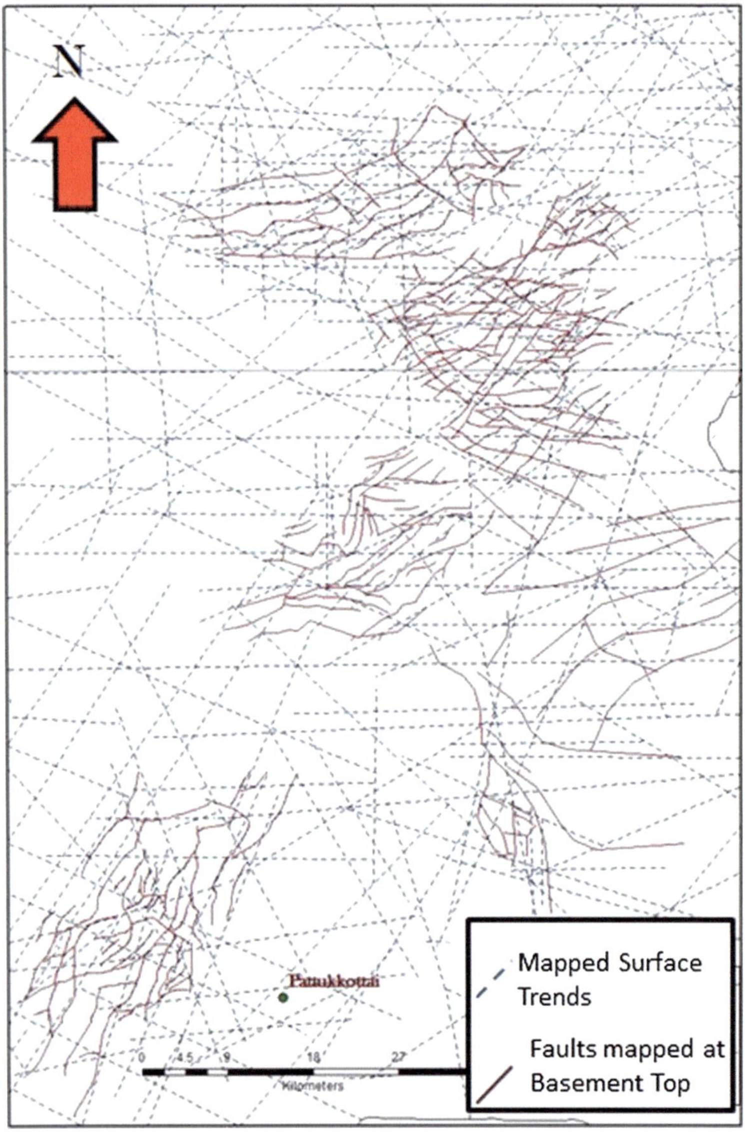

Fig. 7 Correlation of faults mapped using seismic data at basement top with surface basement trends corresponding to the area depicted in Fig. 6

are found to correlate with the sheath folded layers delineated from time slices with individual separated by the E-W trending faults (Fig. 9b). Since both the sheath folded layers and the E-W trending faults are associated with shearing, it indicates that the series of small undulations at the basement top are representative of shear zones in a sectional view. Analogous signatures are also observed in sectional views of shear zones exposed beyond the basin margin (Fig. 9c) as well as other shear zones reported in public domain (Fig. 9d).

Since the shear zones (Mukherjee and Mulchrone 2015) associate with E-W surface trends, their spatial disposition provides an estimate of the basinal extent affected by shearing. The results indicate that most of the trends are concentrated in the area north of the Cauvery Basin (Fig. 10a). Seismic sections near Ramnathapuram area (Fig. 10b) do not show any effects of the shear zones in basement like undulations or sheath folds suggesting no such Pre-Cambrian basement deformations in the area. However, basement related structural highs exist and are manifested as geomorphic highs (Fig. 10b).

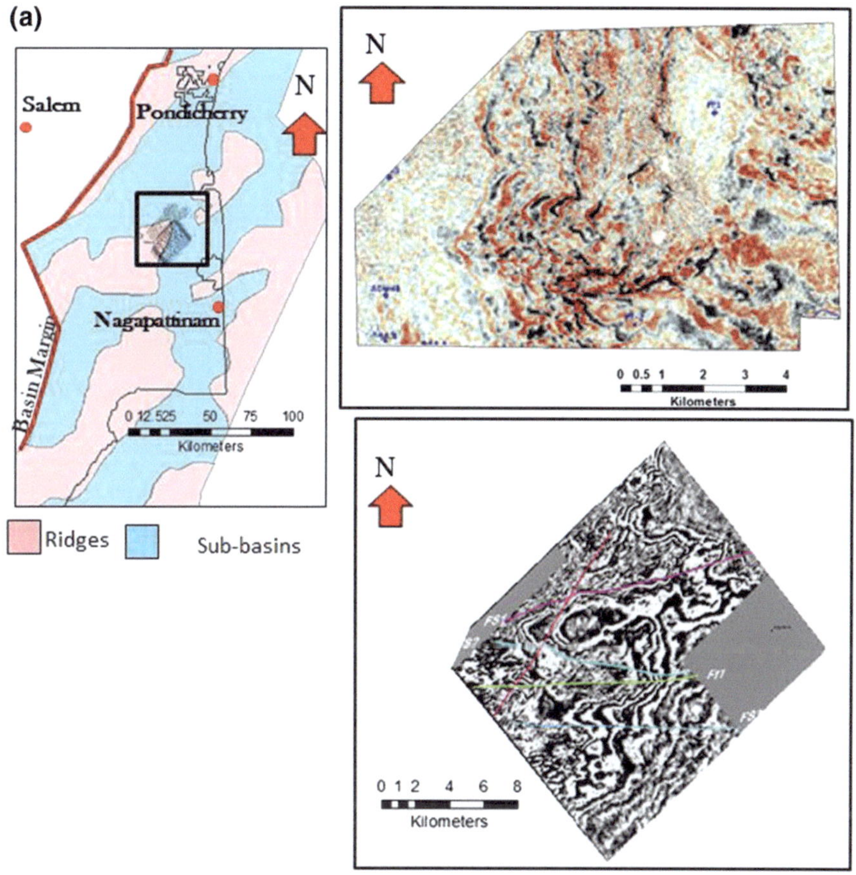

Fig. 8 **a** Seismic time slices in Madanam and Kuthalam area at 2000 ms time equivalent depth (within basement, near basement top) as shown in the index map. **b** Lines of high seismic amplitude extracted from the basement time slices appear to define a folded pattern with an axial trace that is parallel to the E-W basement trends mapped at the surface. These patterns are similar to the E-W trending sheath folds defined in image. **c** (LANDSAT ETM+ band 8) and outcrop scale (8d) near Salem

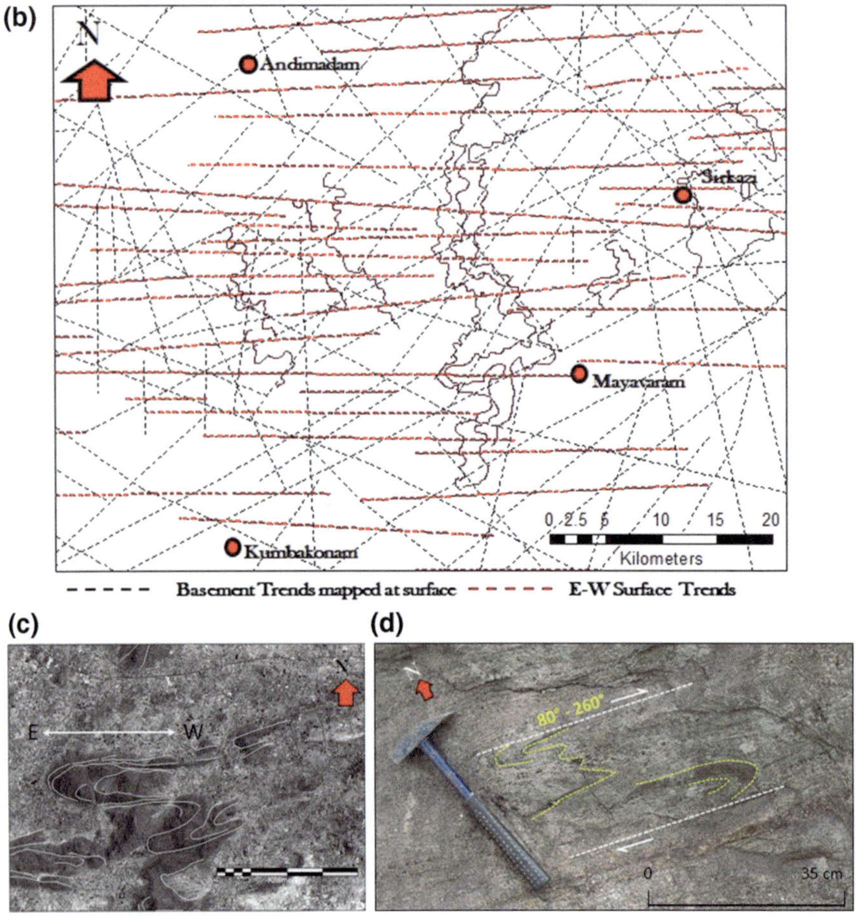

Fig. 8 (continued)

2.3 Correlation with Image Log Data

Formation Microimager logs (FMI) recorded in deep wells M-5 and M-6 drilled by ONGC along the E-W faults were analysed to determine the fracture patterns and porosity associated with the shear zones. An orientation analysis of fracture azimuths derived from FMI logs for every 50 m interval from the basement top indicates that E-W fractures occur in zones of 200–300 m thickness. These fractures are mostly regularly spaced and found to be oriented along E-W foliations (Fig. 11a, b and d). The E-W foliated and fractured zones represent the signatures of shear zones. Observations from FMI data indicate that these zones are separated by areas of un-sheared rock where other tectonic trends are dominant. Similar features are observed in cores cut in deep wells drilled by ONGC along the E-W surface trends. A 3 m cored interval recovered from basement in well MT-9 indicates parallel quartzo-feldspathic and biotite rich layers exhibiting mylonitic foliation with fracture planes developed sub-parallel to them. These fractures show a steep structural dip (60°) towards north (from FMI log). The fracturing or brittle plane generation parallel to mylonitic foliation is commonly observed in some other shear zones as well (Mukherjee and Koyi 2010b; Mukherjee 2012). Dipole Shear Sonic Imager logs (DSI log) carried out for fracture characterization of another well P-6 also associated with E-W surface trends indicate that the E-W trending fractures associated with the foliations are open and deep (Fig. 11c). Field analogs from outcrops of basement across basin margin also indicate foliation parallel fracturing or jointing with open fracture sets (Fig. 11d).

2.4 Gravity and Magnetic Data

Both gravity and magnetic data helps to build a regional representation of basement configuration. Deep-seated regional scale structural elements affecting basement are manifested as gravity or magnetic signatures. Based on this premise, E-W signatures were derived from gravity derivative (Kumar et al. 2009) and reconnaissance scale aeromagnetic data over Cauvery Basin (Rajaram and Anand 2014). Also since these E-W shear zones are believed to extend offshore (Murty et al. 2002), the E-W trending faults mapped by Subrahmanyam et al. 1995 had also been extracted from the magnetic map in offshore areas.

An analysis of the different independent surface and subsurface data sets thus provided a number of E-W basement related trends that can be regarded as components of shear zone. These are (a) E-W surface trends derived from geomorphic and image data (b) Shear zones and faults mapped in field (Chardon et al. 2008; Desikachar et al. 1960) (c) E-W trending faults from seismic maps at top of basement level in both onshore and offshore Cauvery Basin d) E-W trends extracted from gravity derivative maps (Kumar et al. 2009) and (e) E-W trends derived from aero-magnetic data (Rajaram and Anand 2014) and offshore magnetic data (Subrahmanyam et al. 1995). All these components signifying parts of a shear zone are joined in a GIS platform to define a probable shear zone network in Cauvery Basin (Fig. 12) maintaining protocols like thickness and spacing constraints, relationship with sheath folded layers derived from basement time slices and available well control in terms of both FMI logs and well data.

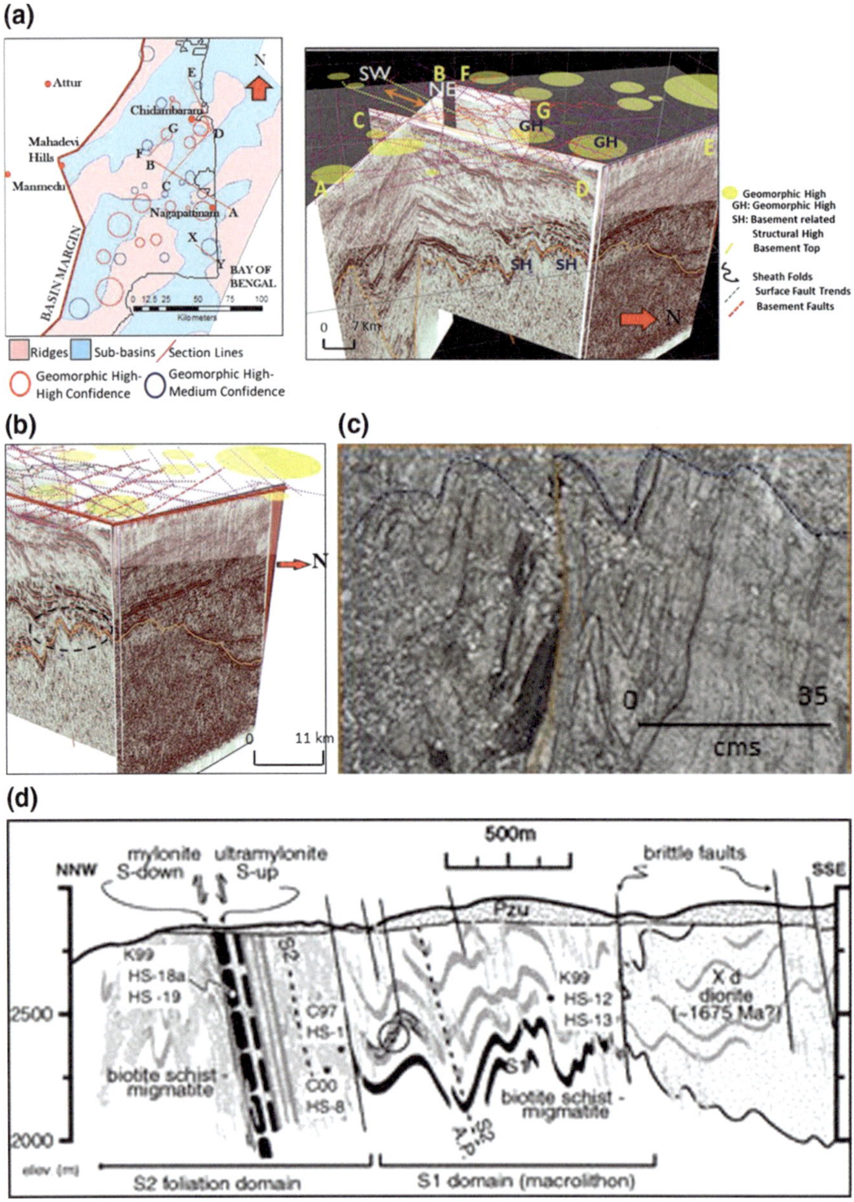
(a)
Attur
Chidambaram
Mahadevi Hills
Manmedu
Nagapattinam
BAY OF BENGAL
BASIN MARGIN
Kilometers
Ridges
Sub-basins
Section Lines
Geomorphic High-High Confidence
Geomorphic High-Medium Confidence
SW
NE
GH
SH
7 Km
Geomorphic High
GH: Geomorphic High
SH: Basement related Structural High
Basement Top
Sheath Folds
Surface Fault Trends
Basement Faults
(b)
11 km
(c)
cms
(d)
NNW
SSE
mylonite S-down
ultramylonite S-up
500m
brittle faults
Pzu
K99
HS-18a
HS-19
C97
HS-1
C00
HS-8
HS-12
HS-13
X d diorite (~1675 Ma?)
biotite schist - migmatite
2500
2000
elev (m)
S2 foliation domain
S1 domain (macrolithon)

◂**Fig. 9** **a** Seismic sections along AB, CDE, FG and XY (as in inset map) correlated with surface faults and geomorphic highs. Most of the geomorphic highs in the surface are found to overlie areas of basement high that are associated with a high amount of structural relief. **b** Basement top in seismic section also depicts a series of smaller undulations characterized with a much smaller variation in relief. These small undulations (encircled) correlate with the sheath folded layers extracted from time slice and are found to be separated by the E-W trending faults (in red). **c** Shear zone exposed in sectional view in Manmedu near Tiruchurapalli displays similar such undulation. **d** Undulations also observed in the cross section of Homestake Shear Zone, Hornsilver Ridge, Colorado similar to those interpreted in seismic data (Shaw et al. 2002). **e** Faults affecting basement top when extrapolated up-dip are found to correlate with interpreted surface fault trends (especially oriented E-W) implying that these surface trends are reactivation related manifestations of the basement faults

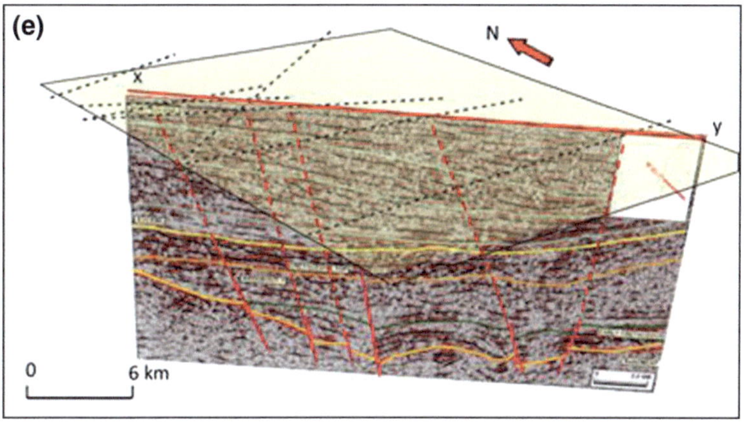

Fig. 9 (continued)

3 Discussions

3.1 *Fracturing and Layout of Shear Zones*

The Precambrian shear zones are considered products of deep and ductile deformation with the concentration of large strain. However, with decrease in depth and associated decrease in temperature and pressure, wider zones of ductile displacement are transformed into discrete planes and narrow zones of brittle displacement (Burg 2017a, b). The areas affected by the shearing show higher fracture density than the host rock, with the highest fracture densities restricted to fine-grained and strongly foliated sections (Laws et al. 2003).

In the present case, dextral shearing (D2) leads to the formation of a shear foliation and sheath folding aligned along the E-W shear zones. Most of these shear zones are continuous from the basement exposures to the basinal part and show

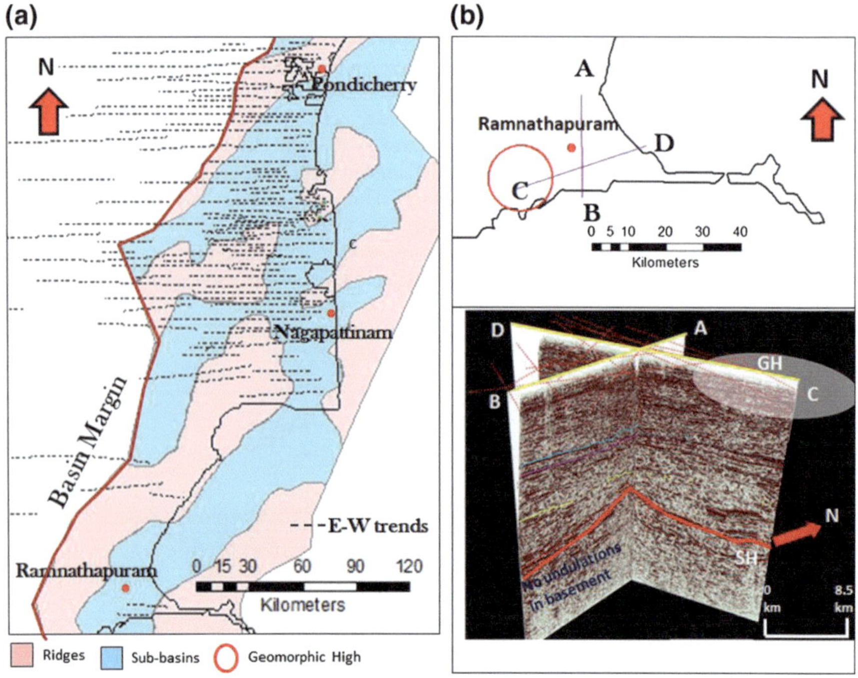

Fig. 10 **a** E-W trends appear to be concentrated in the northern part of the Cauvery Basin indicating that the effect of shear zone on basement is restricted to solely that area. **b** Seismic sections along lines AB and CD in southern part of Cauvery Basin showing no indications of shearing (GH stands for Geomorphic Highs)

similar type of sheath folding and foliation properties both in outcrops and subcrops (Fig. 13c). However, upon reaching a critical length, the shear zones link obliquely along step over areas (Schrank et al. 2008) forming NE-SW oriented linkages between the adjacent shear belts. The shear foliations along the E-W shear belts exposed at the basement top level changes to a weakly bonded planar fabric. Deformations occurring subsequent to the formation of the shear zones reactivated these planes of weakness, both along the shear zones and their linkages eventually forming fractures oriented along the foliations. These areas of fracturing due to the formation of shear zones are about 250–300 m in orthogonal thickness and are associated with a moderate to steep dip. As such, these shear zones could have branched out of a thicker, gentler and a more ductile layer in the deeper part of basement (Fig. 13a, b). Hence it is possible that more than one shear zone could be encountered in a particular drilled well that are separated by an interval of un-sheared and less fractured rock thus increasing the probability of encountering a zone of high porosity (secondary).

The isotopic boundary probably forms the limit of the area where the basement is affected by shearing.

3.2 Areas of Maximum Fracture Porosity

Other than the Precambrian tectonism creating E-W shear zones, the basement is deformed by to the three episodes of Phanerozoic tectonism as discussed earlier. Each of these phases probably inherited and reactivated older fracture patterns within basement depending upon the state of stress prevalent at the time. Of these episodes, the NE-SW trending main rifting phase in Late Jurassic-Early Cretaceous had the most dominant effect on basement. Additionally, each deformation phase is instrumental in reactivation of faults/fractures that were created in an earlier phase.

The areas of intersection of Phanerozoic basement trends demarcated from their surface manifestations with the shear zones will form the areas of maximum fracture porosity in Cauvery Basin (Figs. 11a, d and 14). These are also the areas of maximum reactivation resulting in extensive fracturing along the shear zones with deep, open and connected fracture networks increasing secondary porosity in the area. The areas of intersections of the different sets of fractures form an interconnected fracture mesh increasing the areas of permeability in the region.

3.3 Reactivation and Migration from Source

The Cauvery Basin forms a neotectonically active area with number of earthquake epicentres and seismic prone lineaments mapped in its geographical extent (Saravanavel and Ramasamy 2003; Ganapathy and Rajarathinam 2010). Accordingly, most of its structural and tectonic elements can be considered to be neotectonically reactivated. The in situ tectonic stress acting on the northern part of the basin as deduced from different sources like Drilling Induced Fractures in different wells (Fig. 15a), Focal Plane Solutions (Murty et al. 2002), World Stress Maps (Heidbach et al. 2016) and works of Chatterjee and Mukhopadhyay (2008), Subrahmanyam (2014) indicate S_{Hmax} direction oriented ENE-WSW to NW-SE.

In the present analyses, a trend based correlation of morphostructural faults at the surface with faults mapped at the top of basement from seismic data in the Madanam area indicates that the E-W trends are more reflected in the surface manifestations (Fig. 15b). This specifies that under the present stress regime the E-W faults are the most reactivated. Consequently, the E-W shear zones turn out to be the most neotectonically reactivated structures in the basement. This implies that the shear zones are associated with a greater damage zone constituted of a cluster of incipient fractures with the damage zone width increasing with each phase of reactivation associated displacement. This further enhances the reservoir character of the shear zone area.

Although the shear zone affected areas of basement can effectively act as reservoir with secondary porosity, it needs to be associated with a younger sedimentary source for charging. These areas of basement either need to directly overlie the hydrocarbon source horizon or needs to be connected to them by faults acting as

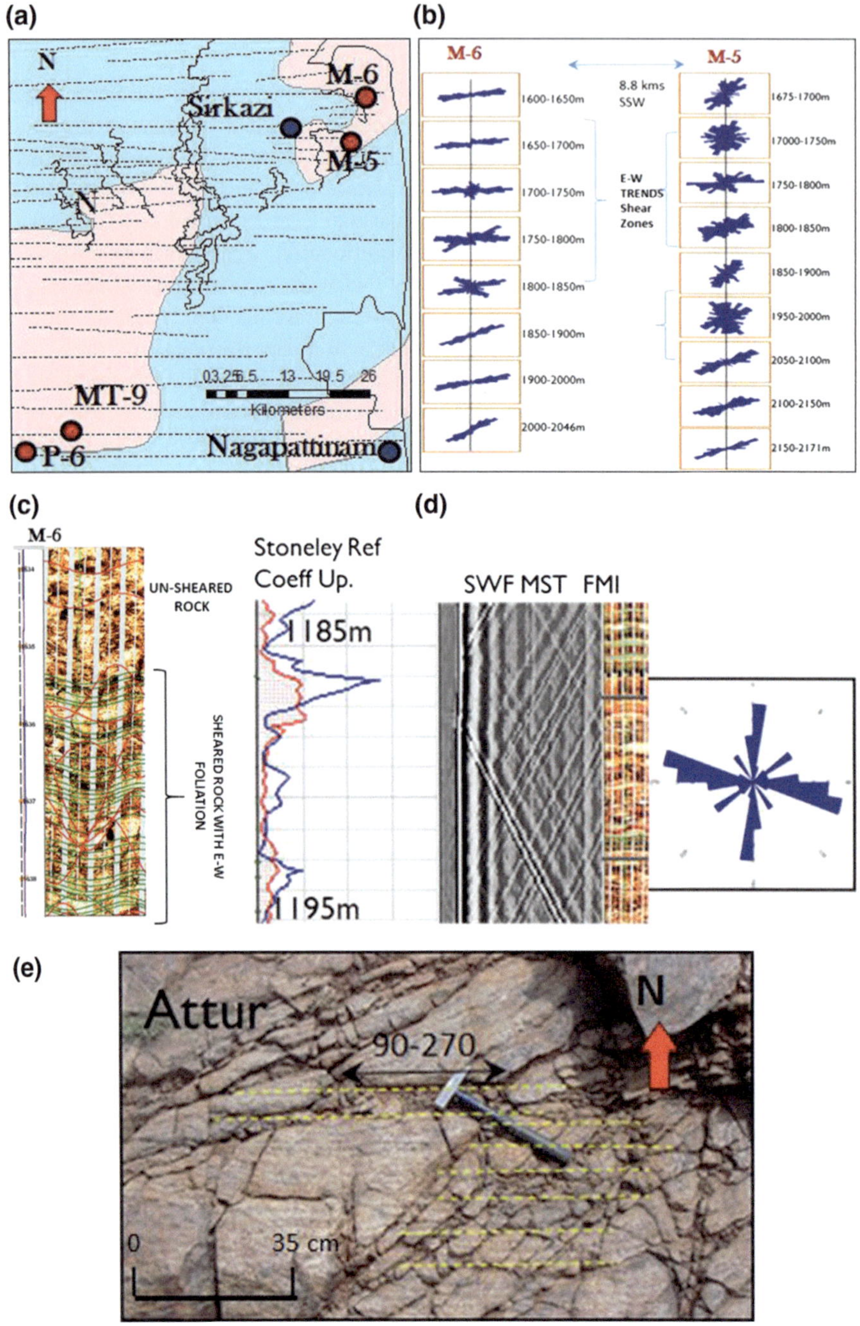
(a)
N
M-6
Sirkazi
M-5
N
MT-9
0 3.25 6.5 13 19.5 26
Kilometers
P-6
Nagapattinam
(b)
M-6
M-5
8.8 kms SSW
1600-1650m
1650-1700m
1700-1750m
1750-1800m
1800-1850m
1850-1900m
1900-2000m
2000-2046m
E-W TRENDS Shear Zones
1675-1700m
17000-1750m
1750-1800m
1800-1850m
1850-1900m
1950-2000m
2050-2100m
2100-2150m
2150-2171m
(c)
M-6
UN-SHEARED ROCK
SHEARED ROCK WITH E-W FOLIATION
Stoneley Ref Coeff Up.
1185m
1195m
(d)
SWF MST FMI
(e)
Attur
N
90-270
0
35 cm

◂**Fig. 11** **a** Trend analysis of fractures of wells M-5 and M-6 at every 50 m interval from basement top indicates 200–250 m thick zones with dominance of E-W fracture sets. **b** FMI of well M-6 shows regularly spaced E-W fractures oriented along a zone of E-W foliations. **c** DSI logs of well P-6 correlated with FMI indicate that the E-W oriented fractures are open and deep. **d** Field analog showing E-W oriented open fractures associated with shear foliations in Attur (Fig. 9). Imprints of fracture sets oriented NE-SW (along linkages) and E-W lead to an increase of fracture porosity in the area

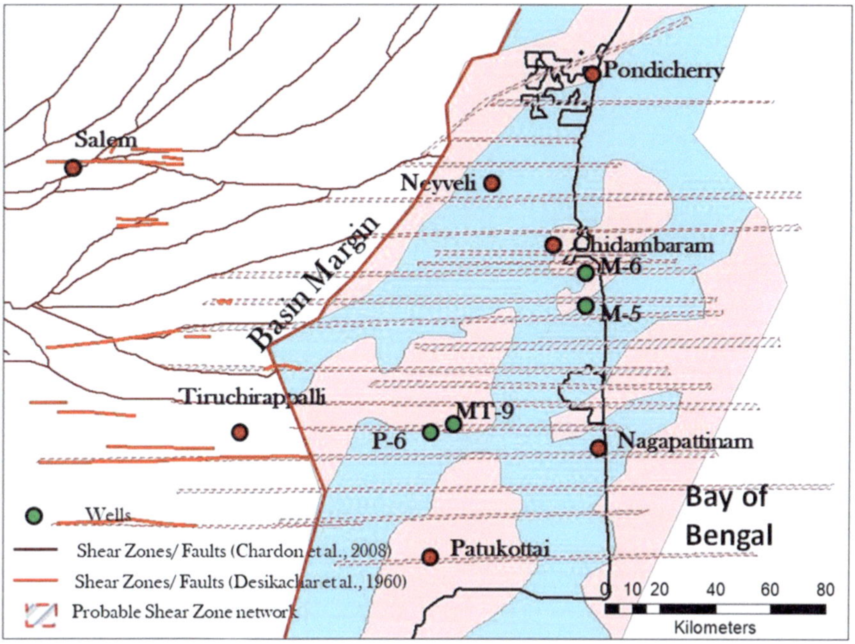

Fig. 12 Probable shear zone network derived from different components in the Cauvery Basin (pink indicates ridges and blue indicates sub-basins)

conduits. For Cauvery Basin, shales within Andimadam Formation and Sattapadi Shales of Early Cretaceous are considered the main source kitchen for the area. The critical moment or the time of maximum hydrocarbon generation prior to migration is observed to be around 65.5 Ma (Phaye et al. 2011). Since after Early Cretaceous, no major tectonic event occurred in the basin other than subsidence and tilting (Nagendra et al. 2017; DGH India), the in situ stress direction is considered similar to that prevailing at the time of hydrocarbon migration. With this assumption, based on principles in Zhang et al. (2011), E-W faults being parallel to the in situ stress direction tend to remain open and act as conduits of migration in Eocene-Oligocene. As a result they play a significant role in charging basement reservoirs by updip/lateral migration from adjacent kitchen areas into structural highs (Chandrasekhar et al. 2015). This is corroborated by adsorbed gas anomalies in onland areas and natural hydrocarbon seepages in offshore (Dave et al. 2012) both of which are considered as manifestations of migration/remigration conduits that show a positive correlation with E-W shear zones (Fig. 15c). This suggests that the E-W shear zones and associated faults are most instrumental in migration/

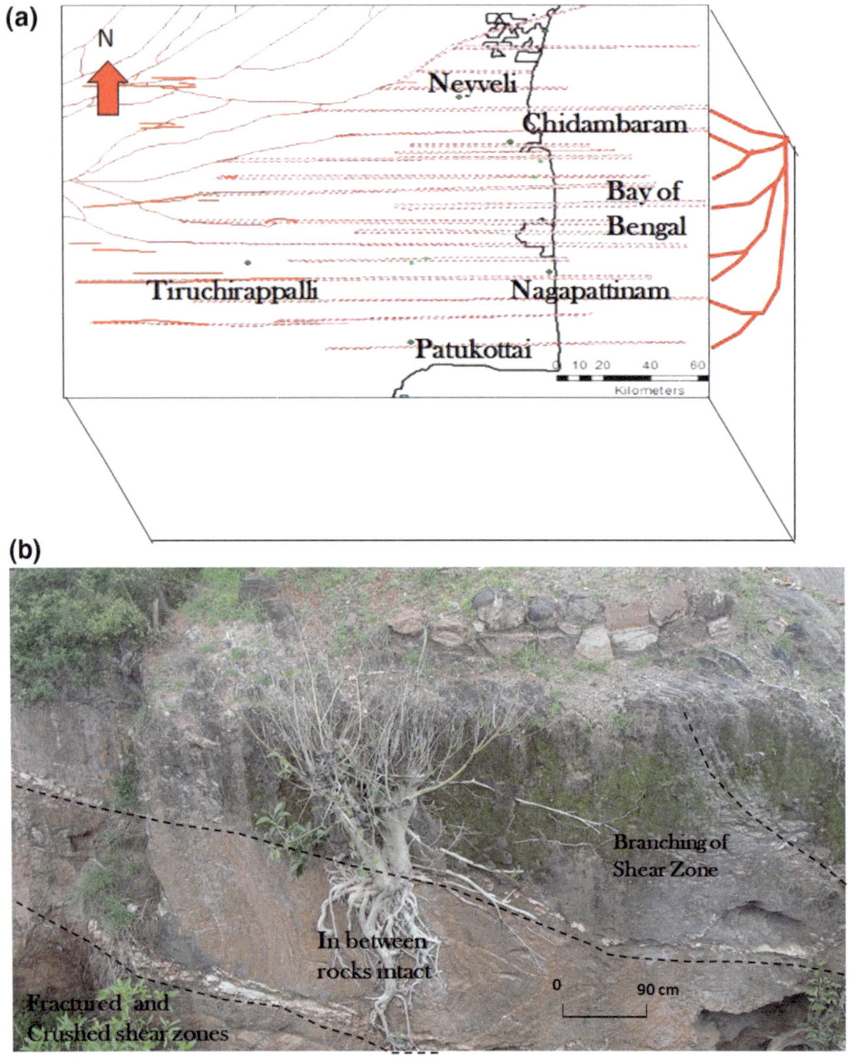

Fig. 13 **a** Probable model of arrangement of the shear zones branching off from a thicker and deeper shear zone **b** Field analog from Manmedu showing shear zones in a cliff section branching from each other enveloping unsheared rock in between **c** E-W oriented shear zones found to be continuous from the basement exposures to the basinal part, showing similar type of fracturing properties in outcrop in Manamedu and in FMI of subcrop in well M-5 (pink indicates ridges and blue indicates sub-basins)

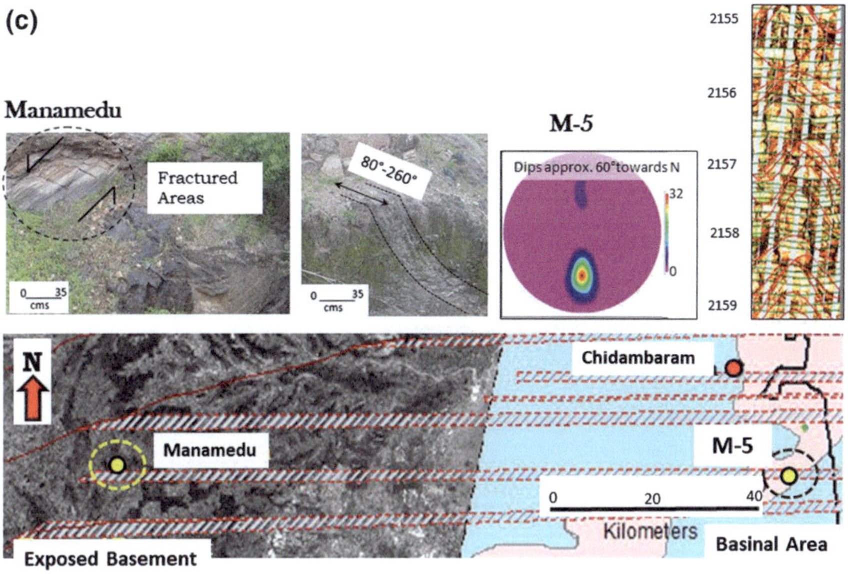

Fig. 13 (continued)

remigration of hydrocarbon in northern part of Cauvery basin thus forming an essential element in formation of a basement prospect.

3.4 Delineation of Promising Areas of Basement Exploration

The target areas for hydrocarbon exploration in the Cauvery Basin are the areas of overlap of (1) structural highs (2) areas of maximum fracture porosity overlying shear zones and (3) source pods of hydrocarbon generation that either overlie these regions or are connected with them by E-W faults formed by reactivations of shear zones. The areas of structural highs are manifested on the surface as geomorphic highs and are classified as per their degree of confidence. Geomorphic highs of high and medium confidence in the ridge areas are used to represent structural highs. Areas of positive correlation between all the elements in a GIS platform indicate the promising areas of basement exploration in Cauvery Basin (Fig. 16a, b).

The structural highs are mostly products of the main rift tectonics of Late Jurassic-Early Cretaceous that uplifted the basement with high structural relief. Areas of basement, affected by shear zones are also uplifted in these basement highs and form zones of high fracture porosity as described earlier. The post rift Cretaceous shales in Portonovo, Kudavasal, Sattapadi and Synrift Andimadam Formation form the potential source rocks (Singh et al. 2007) that are mapped as the

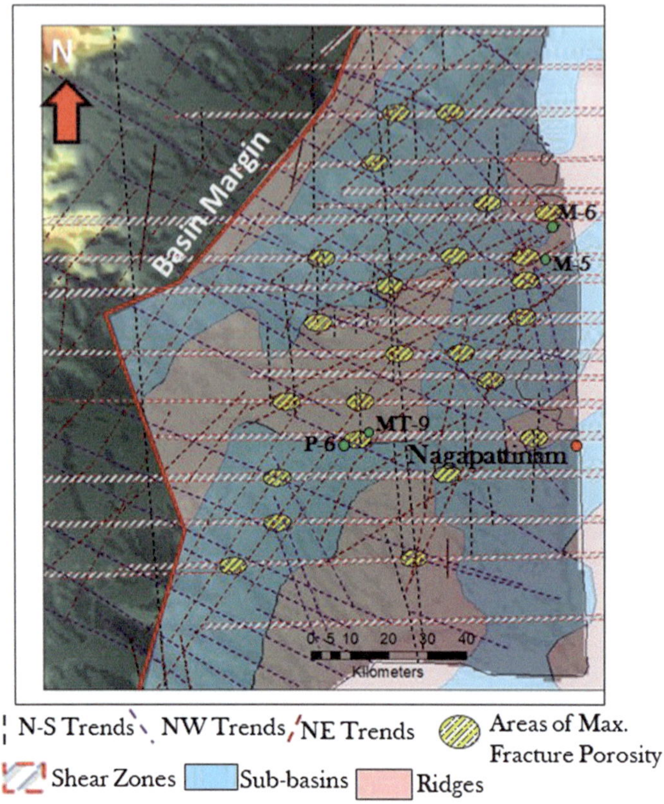

Fig. 14 Areas of maximum fracture porosity delineated from Phanerozoic basement trends associated with shear zones in the Cauvery Basin. Wells P-6, MT-9, M-5 and M-6 are hydrocarbon bearing wells producing from basement in these areas

source pods. These are deposited in the lows adjacent to the basement highs or directly overlie the basement highs. Shales within Kudavasal and Kamlapuram Formation act as seals. On attaining critical moment, the hydrocarbon expulsed migrate updip or migrate laterally along E-W reactivated faults and charge the areas of fractured basement in the highs. In case of these source rocks directly overlying the basement highs, downward migration resulting from hydrocarbon expelled during compaction of the source sediments may also charge the reservoir areas (Sircar 2004). The fractures serve as essential porosity for the hydrocarbon that may move downdip along them and form accumulations (Fig. 16b).

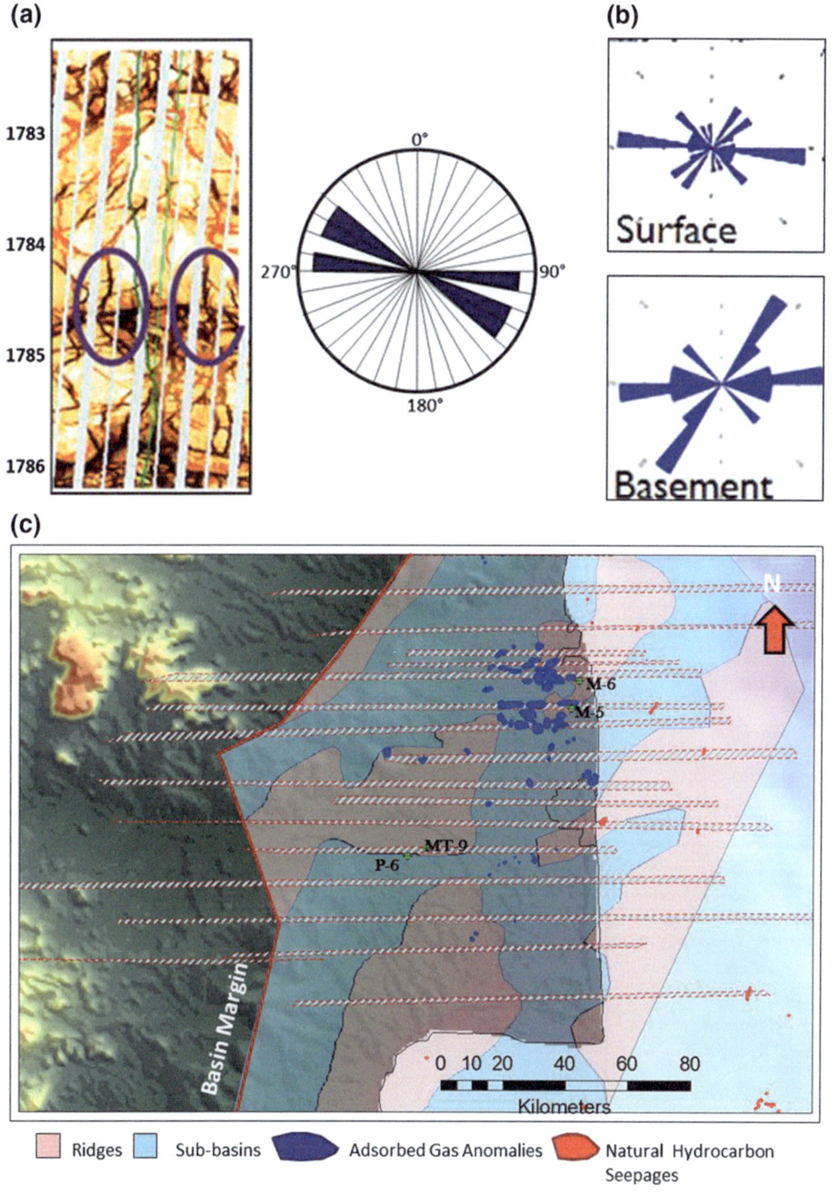

Fig. 15 **a** Drilling Induced Fractures (DIFs) interpreted from well M-5 indicating SHmax oriented E-W to NW-SE. **b** Comparison of rose diagrams of surface morphostructural trends and faults mapped at the basement top in Madanam field (Fig. 7) indicate that the E-W trends are the most reactivated. **c** Adsorbed gas anomalies in onshore Cauvery Basin and natural hydrocarbon seepages in offshore areas (Dave et al. 2012) associated with E-W oriented shear zones indicating that the E-W trending faults act as conduits of migration/remigration under present stress regime

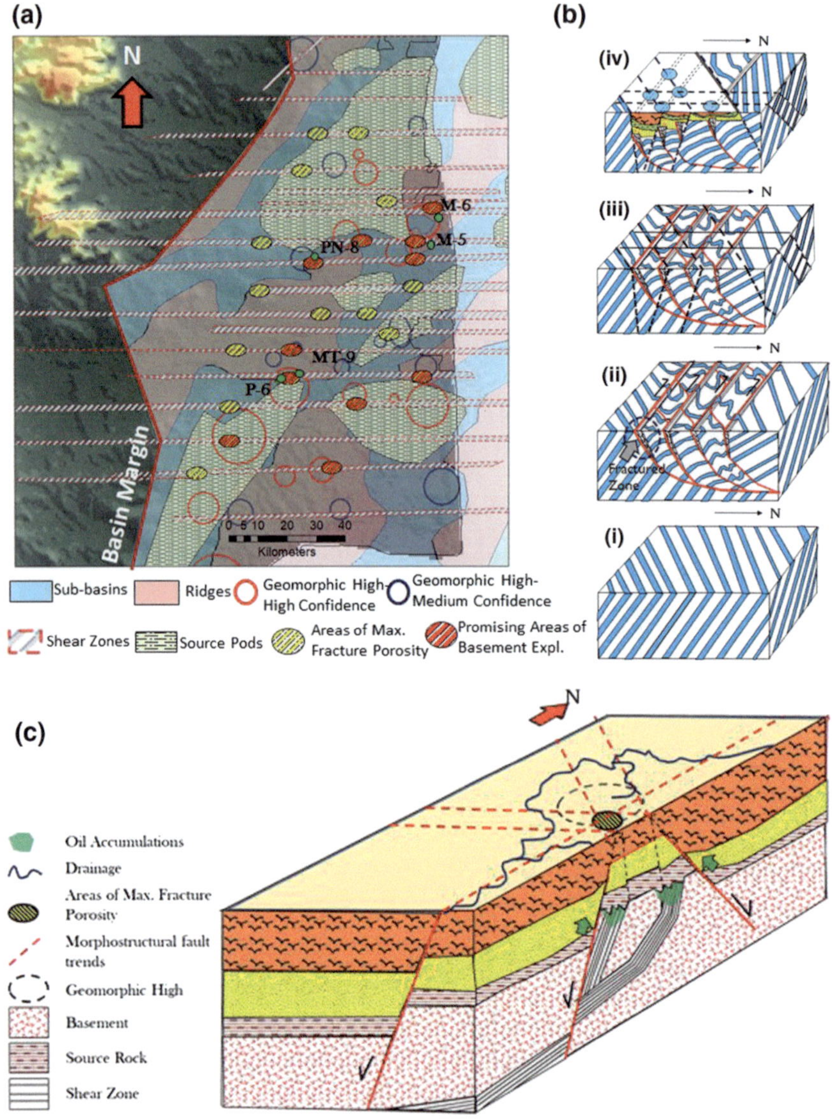

Fig. 16 **a** Promising areas of basement exploration extracted from the spatial correlation of the areas of maximum porosity in shear zones, geomorphic highs and source pods that spatially overlap each other. Wells P-6, MT-9, M-5, M-6 and P-9 are hydrocarbon bearing wells producing from basement in these areas. **b** Schematic model (not to scale) of (i) Initial basement rock with probable NE-SW trending foliations (ii) Formation of E-W trending shear zones (in red lines) associated with dextral shearing resulting in re-orientation of earlier foliations, formation of E-W fractures at basement top and NE-SW oriented linkages connecting the shear zones (iii) Probable formation of N-S trending faults resulting in reactivations of earlier fractures (iv) Main phase of rifting due to NE-SW trending faults forming a number of basement highs and lows. Basement highs associated with shear fractures and fault intersections act as promising areas **c** Block diagram of conceptual model for basement exploration

4 Conclusions

The study emphasizes on the use of tectonics and structures in basement exploration in Cauvery Basin. A GIS based correlation of different surface and subsurface geological & geophysical datasets using outcrop structural analogues model the disposition of shear zones in the northern part of Cauvery Basin. The mapped shear zones occur as individual bands of 250–400 m thick spaced apart by 500 m to a few km and dipping moderately to steeply near the basement top thereby fracturing extensively basement top areas. These shear zone fractures develop in an E-W orientation parallel to the shear foliations that opened due to repeated reactivations. Imprints of subsequent basement tectonism on these zones increases fracture porosity and creates connectivity with different fracture sets thus enhancing the reservoir character of the basement. Structural highs created during the main phase of rifting and associated with fractured basement forms promising areas of basement exploration. Neotectonic reactivations due to changes in in-situ stress regime results in hydrocarbon charging into the basement reservoirs thus creating hydrocarbon prospects. A similar procedure may be effective delineating basement hydrocarbon in other similar basins that have undergone multiple episodes of tectonism.

Acknowledgements The authors express their gratitude to Soumyajit Mukherjee (IIT Bombay) for handing and reviewing this manuscript. The other anonymous reviewer is also thanked. Mukherjee (2019) summarizes this work.

The views expressed by the authors in this manuscript are not necessarily similar to the views of the organization they represent.

References

Bhutani R, Balakrishnan S, Nevin CG, Jeyabal S (2007) Sm–Nd isochron ages from Southern Granulite Terrain, South India: Age of protolith and metamorphism. In: Goldschmidt conference abstracts 2007, A89

Burg JP (2017a) Ductile Faults: Shear Zones. http://www.files.ethz.ch/structuralgeology/JPB/files/English/13shearzones.pdf

Burg JP (2017b). Folds. http://www.files.ethz.ch/structuralgeology/JPB/files/English/8folds.pdf

Cao J, Liu S, He X (2011) Earthquake-induced secondary hydrocarbon migration and accumulation in Longmanshan region, China. In: American Geophysical Union, fall meeting 2011, abstract #T42B-04

Chandrasekhar N, Mane PH, Rajappan P (2015) Basement hydrocarbon exploration-overview of basement exploration. In: Multicomponent seismic analysis, identification of fractures and status of Indian basin exploration. Geohorizons July, pp 36–46

Chardon D, Jayananda M, Chetty TRK, Peucat JJ (2008) Precambrian continental strain and shear zone patterns: South Indian case. Journal of Geophysical Research 113, B08402

Chatterjee R, Mukhopadhyay M (2008) In-situ stress results from east coast sedimentary basins of India. In: Oral session VI: stress field interpretation at regional scale, 3rd world stress map conference 15–17. Potsdam

Chetty TRK, Bhaskar Rao YJ (2006) The Cauvery Shear Zone, Southern Granulite Terrain, India: A crustal-scale flower structure. Gondwana Res 10(1–2):77–85

Chetty TRK, Santosh M (2013) Proterozoic orogens in southern Peninsular India: contiguities and complexities. Journal of Asian Earth Sciences 78, 39–53

Chetty TRK (2015) The Cauvery suture zone: map of structural architecture and recent advances. Journal of Geological Society of India 85, 37–44

Chetty TRK, Bhaskar Rao YJ, Narayana BL (2003) A structural cross section along Krishnagiri–Palani Corridor, Southern Granulite Terrain of India. Memoir Geological Society of India 50, 255–277

Chetty TRK, Yellappa T, Mohanty DP, Nagesh P, Sivappa VV, Santosh M, Tsunogae T (2012) Mega sheath fold of the Mahadevi Hills, Cauvery Suture Zone, Southern India: implication for accretionary tectonics. Journal Geological Society of India 80, 747–758

Chung HP (1982) Petroleum in basement rocks. American Association of Petroleum Geologists 66, 1597–1643

Dang CTQ, Chen Z, Nguyen NTB, Bae W, Phung TH (2011) Improved oil recovery for fractured granite basement reservoirs: historical lessons, successful application, and possibility for improvement. In: Society of petroleum engineers, SPE-144148-MS, SPE European formation damage conference, 7–10 June, Noordwijk, The Netherlands

Darmadi Y, Harahap A, Achidat R, Ginanjar M, Hughes J (2013) Reservoir characterization of fractured basement using seismic attributes, Dayung field case study, South Sumatra Indonesia. In: Proceedings, Indonesian petroleum association 37th annual convention and exhibition, May 2013

Dasgupta S, Mukherjee S (2017) Brittle shear tectonics in a narrow continental rift: asymmetric non-volcanic Barmer basin (Rajasthan, India). The Journal of Geology 125, 561–591

Dave HD, Mazumder S, Samal JK, Pangtey KKS, Mitra DS (2012) Mapping Hydrocarbon seepages using satellite SAR data in Eastern offshore—essential input in oil exploration, P-137. In: Proceedings of 9th Biennial international conference and exposition on petroleum geophysics, SPG Hyderabad, 2012

Desikachar SV, Ramanathan S, Babu PVLP (1960) Progress report of geological exploration of Cauvery Basin. 1959–60. ONGC Unpublished report

DGH India (2015) Estimated resources of crude oil & natural gas, exp.about.resources2015.pdf. https://www.ndrdgh.gov.in/NDR/?page_id=1251

EarthExplorer, USGS. https://earthexplorer.usgs.gov/

Gaina C, Müller RD, Brown B et al (2007) Breakup and early seafloor spreading between India and Antarctica. Geophysical Journal International, 170, 151–169

Ganapathy GP, Rajarathinam (2010) Use of remote sensing and seismotectonic parameters to identify seismogenic sources of Tamil Nadu State. International Journal of Applied Engineering Research 1, 59–76

Gupta P, Rathore SS, Raza S, Uniyal GC (2015) Radiometric dating and textural characterization of Basement rocks of East Coast, KG and Cauvery Basins, ONGC report, unpublished

Gutmanis J (2009) Basement reservoirs—a review of their geological and production characteristics. In: International petroleum technology conference, IPTC 13156

Harinarayana T, Naganjaneyulu K, Patro BPK (2006) Detection of a collision zone in south Indian shield region from magnetotelluric studies, Gondwana Research 10, 48–56

Heidbach O, Rajabi M, Reiter K, Ziegler M, Team WSM (2016) World stress map database release 2016. GFZ Data Services. https://doi.org/10.5880/WSM.2016.001

Hua B (1995) Stress field, seismic pumping and oil-gas migration. Acta Sedimentological Sinica 13, 77–85

Huy XN, Bae W, San TN, Xuan VT, SungMin J, Kim DY (2012) Fractured basement reservoirs and oil displacement mechanism in white tiger field. In: Offshore Vietnam, APG Search and Discovery Article #90155©2012 AAPG international conference & exhibition, Singapore, 16–19 September

Janardhan AS (1999) Southern Granulite Terrain, south of the Palghat Cauvery Shear Zone: implications for India Madagascar connection. Gondwana Research 2, 463–469

Kumar N, Singh AP, Singh B, (2009) Structural fabric of the Southern Indian shield as defined by gravity trends. Journal of Asian Earth Sciences 34, 577–585

Kumaran CJ, Ramasamy SM, (2005) Fluvial anomalies and Neotectonics of parts of Western Ghats, Tamil Nadu, India. In: Ramaswami SM (ed) Remote Sensing in Geomorphology, Chapter: Chapter-7, Publisher: New India Publishing Agency. pp 81–88

Lal NK, Siawal A, Kaul AK (2009) Evolution of east Coast of India—a plate tectonic reconstruction. Journal of Geological Society of India 73, 249–260

Laws S, Eberhardt E, Loew S, Descoeudres F (2003) Geomechanical properties of shear zones in the Eastern Aar Massif, Switzerland and their Implication on Tunnelling. Rock Mechanics and Rock Engineering 36, 271–303

Mazumder S, Dave HD, Samal JK, Mitra DS (2011) Delineation of shallow structures as exploratory input in Ganga Basin based on Morphotectonic studies. ONGC Bulletin 46, 190–198

Mazumder S, Pangtey KKS, Mitra DS (2013a) Delineation of a possible subsurface ridge in Onshore Palar Basin based on morphotectonic studies and its implications. In: Proceedings of 10th Biennial international conference & exposition, SPG, Kochi

Mazumder S, Tep B, Pangtey KKS, Mitra DS (2013b) A Morphotectonic based approach to derive tectonic framework and possible areas of hydrocarbon accumulation in onshore Palar Basin. ONGC Bulletin 48, 113–123

Mazumder S, Adhikari K, Mitra DS, Mahapatra S, Pangtey KKS (2016) A Neotectonic based geomorphic analysis using remote sensing data to delineate potential areas of hydrocarbon exploration: Cachar Area, Assam. Journal Geological Society of India 88, 87–97

Meert JG, Pandit MK, Pradhan VR, Banks J, Sirianni R, Stroud M, Newstead B, Gifford J (2010) Precambrian crustal evolution of Peninsular India: A 3.0 billion year odyssey, Journal of Asian Earth Sciences 39, 483–515

Misra AA, Mukherjee S (2015) Tectonic Inheritance in Continental Rifts and Passive Margins. Springerbriefs in Earth Sciences. ISBN 978-3-319-20576-2

Mukherjee S (2012) Tectonic implications and morphology of trapezoidal mica grains from the Sutlej section of the Higher Himalayan Shear Zone, Indian Himalaya. The Journal of Geology 120, 575–590

Mukherjee S (2013) Deformation microstructures in rocks. Springer Geochemistry/Mineralogy, Berlin, pp 1–111. ISBN 978-3-642-25608-0

Mukherjee S (2014a) Atlas of shear zone structures in meso-scale. Springer Geology, Cham, pp 1–124. ISBN 978-3-319-0088-6

Mukherjee S (2014b) Review of flanking structures in meso- and micro-scales. Geological Magazine 151, 957–974

Mukherjee S (2015a) Atlas of structural geology. Elsevier, Amsterdam. ISBN: 978-0-12-420152-1

Mukherjee S (2015b) Petroleum geosciences: Indian contexts. Springer Geology. ISBN 978-3-319-03119-4

Mukherjee S (2019) Introduction to "Tectonics and Structural Geology: Indian Context". In: Mukherjee S (ed) Tectonics and structural geology: Indian context. Springer International Publishing AG, Cham, pp 1–5. ISBN: 978-3-319-99340-9

Mukherjee S, Koyi HA (2010a) Higher Himalayan Shear Zone, Zanskar section-microstructural studies & extrusion mechanism by a combination of simple shear & channel flow. International Journal of Earth Sciences 99, 1083–1110

Mukherjee S, Koyi HA (2010b) Higher Himalayan Shear Zone, Sutlej section- structural geology & extrusion mechanism by various combinations of simple shear, pure shear & channel flow in shifting modes. International Journal of Earth Sciences 99, 1267–1303

Mukherjee S, Mulchrone KF (2015) Ductile Shear Zones: From Micro-to Macro-scales. Wiley Blackwell. ISBN: 978-1-118-84496-0

Mukherjee S, Punekar J, Mahadani T, Mukherjee R (2015) A review on intrafolial folds and their morphologies from the detachments of the western Indian Higher Himalaya. In: Mukherjee S, Mulchrone KF (eds) Ductile shear zones: from micro- to macro-scales. Wiley

Murty GPS, Subrahmaniyam AS, Murthy KSR, Sarma KVLNS (2002) Evidence of fault reactivation off Pondicherry coast from marine geophysical data. Current Science 83, 1446–1449

Nagendra R, Reddy AN (2017) Major geologic events of the Cauvery Basin, India and their correlation with global signatures a review. Journal of Palaeogeography 6, 69–83

Nemcok M, Sinha ST, Stuart CJ et al (2013) East Indian margin evolution and crustal architecture: integration of deep reflection seismic interpretation and gravity modelling. In: Mohriak WU, Danforth A, Post PJ et al (eds) Conjugate divergent margins. Special Publications 369, The Geological Society, London, pp 477–496

Phaye DK, Nambiar MV, Srivastava DK (2011) Evaluation of petroleum systems of Ariyalur-Pondicherry Sub-basin (Bhuvangiri area) of Cauvery Basin, India: a two dimensional (2-D) Basin modeling study. In: AAPG search and discovery Article #90118

Plavsa D (2014) The tectonic evolution of the Southern Granulite Terrane of India and its role in the amalgamation of Gondwana, Ph.D. Thesis, University of Adelaide

Prabaharan S, Subramani T, Manonmani R, Ramalingam M (2013) Satellite lineaments and Subtle structures in Cauvery Basin, Tamil Nadu. International Journal of Remote Sensing & Geoscience 2, 8–11

Prasad B, Phor L (2009) Palynostratigraphy of the subsurface Gondwana and post-Gondwana Mesozoics of the Cauvery Basin, India. Journal of the Palaeontological Society of India 54, 41–71

Rajaram M, Anand SP (2014) Aeromagnetic signatures of Precambrian shield and suture zones of Peninsular India. Geoscience Frontiers 5, 3–14

Rajendra Prasad B, Rao GK, Mall DM, Rao PK, Raju S, Reddy MS, Rao GSP, Sridher V, Prasad ASSSRS (2007) Tectonic implications of seismic reflectivity pattern observed over the Precambrian Southern Granulite Terrain, India. Precambrian Research 153, 1–10

Rangaraju MK, Agrawal A, Prabhakar KN (1993) Tectono-Stratigraphy, structural styles, evolutionary model and hydrocarbon habitat, Cauvery and Palar basins. In: Proceedings of second seminar on Petroliferous Basins of India, Vol I, pp 371–388

Rao VV, Prasad BR (2006) Structure and evolution of the Cauvery Shear Zone system, SouthernGranulite Terrain, India: evidence from deep seismic and other geophysical studies. Gondwana Research 10, 29–40

Rao VV, Sain K, Reddy PR, Mooney WD (2006) Crustal structure and tectonics of the northern part of the Southern Granulite Terrane, India. Earth and Planetary Science Letters 251, 90–103

Ravindra Kumar GR (2005) Lithology and metamorphic evolution of granulite-facies segments of Kerala, Southern India. Journal of Geological Society of India 66, 253–254

Reber JE, Dabrowski M, Galland O, Schmid DW (2013) Sheath fold morphology in simple shear. Journal of Structural Geology 53, 15–26

Resmi MR, Achyuthan H, Jaiswal MK (2016) Middle to late Holocene paleochannels and migration of the Palar River, Tamil Nadu: implications of neotectonic activity, Quaternary International 443, 211–222

Santosh M, Maruyama S, Sato K (2009) Anatomy of a Cambrian suture in Gondwana: Pacific-type orogeny in southern India? Gondwana Research 16, 321–34

Santosh M, Xiao WJ, Tsunogaed T, Chetty TRK, Yellappae T (2012) The Neoproterozoic subduction complex in southern India: SIMS zircon U–Pb ages and implications for Gondwana assembly. Precambrian Research 192–195, 190–208

Saravanavel J, Ramasamy SM (2003) Probable Seismo Tectonic Lineaments in South India. In: Proceedings of annual convection of ISRS and national conference on remote sensing, CESS, At Trivandrum

Satyanaryana P, Sinha PK, Gupta DK, Sathe AV, Katiyar GC (2010). Hydrocarbon prospectivity of the Basement of Mumbai High Field, P-374. In: Proceedings of 8th Biennial international conference and exposition on petroleum geophysics, SPG 2010, Hyderabad

Schrank CE, Handy MR, Fusseis F (2008) Multiscaling of shear zones and the evolution of the brittle-to-viscous transition in continental crust. Journal of Geophysical Research 113, B01407

Shaw CA, Karlstrom KE, McCoy A, Williams ML, Jercinovic MJ, Dueker K (2002) Proterozoic Shear Zones in the Colorado Rocky Mountains: From Continental Assembly to Intracontinental Reactivation. In: Lageson D (ed) Science at the highest level. Geological Society of America Field Guide 3, 102–117

Sibson RH, Moore JMM, Rankin AH (1975) Seismic pumping—a hydrothermal fluid transport mechanism. Journal of the Geological Society 131, 653–659

Singh H, Goswami BG, Pahari S, Prasad IVSV, Singh RR (2007) Petroleum Geochemistry of the Cauvery Basin, India. In: AAPG Search and Discover Article #90063©2007 AAPG Annual Convention, Long Beach, California

Singh Y, John B, Ganapathy GP, George A, Harisanth S, Divyalakshmi KS, Kesavan S (2016) Geomorphic observations from southwestern terminus of Palghat Gap, south India and their tectonic implications. Journal of Earth System Sciences 125, 821–839

Sircar A (2004) Hydrocarbon production from fractured basement formations. Current Science 87, 147–151

Subrahmanyam AS, Lakshminarayana S, Chandrasekhar DV, Murthy KSR, Rao TCS (1995) Offshore Structural Trends from Magnetic data over Cauvery Basin, East Coast of India. Journal of Geological Society of India 46, 269–273

Subrahmanyam DS (2014) Stress Provinces of India–contribution to world stress map. International Journal of Advanced Earth Science and Engineering 3, 108–113

Trenchard S (2007) Geomorphology as applied to oil exploration in Northwest Kansas. http://www.docstoc.com/docs/96655562/Geomorphology-as-Applied-to-Oil-Exploration-in-Northwest-Kansas

Trice R (2014) Fractured basement reservoirs: a new play for the UK, Offshore Engineer. http://www.oedigital.com/energy/item/5363-fractured-basement-reservoirs-a-new-p-lay-for-the-uk

Vairavan V (1993) Tectonic Hisyory and Hydrocarbon prospects of Palar and Pennar Basins, India. In: Proceedings of second seminar on Petroliferous Basins of India Vol I, 308–396

Vasudevan K, Rao PH, Vairavan V (2012) Role of tectonics in development of fractured basement reservoir in Mumbai High Field, Western Offshore Basin, India. In: 9th Biennial International Conference and Exposition on Petroleum Geophysics, p 482

Veevers JJ (2009) Palinspastic (pre-rift and -drift) fit of India and conjugate Antarctica and geological connections across the suture. Gondwana Research 16, 90–108

Zahran I, Askary S (1988) Basement reservoir in Zeit Bay oil field, Gulf of Suez. https://www.researchgate.net/publication/236487721_Basement_reservoir_in_Zeit_Bay_oil_field_Gulf_of_Suez

Zhang L, Luo X, Guy V, Yu C, Yang W, Lei Y, Song C, Yu L, Yan J (2011) Evaluation of geological factors in characterizing fault connectivity during hydrocarbon migration: Application to Bohai Bay basin, Marine and Petroleum Geology, 28, 1634–1647

Implication of Transfer Zones in Rift Fault Propagation: Example from Cauvery Basin, Indian East Coast

Swagato Dasgupta

1 Introduction

The Cauvery basin, a rift zone located in the southern part of the eastern India, records a poly-phase rift-drift transition (Veevers and Tewari 1995; Veevers 2009; Lal et al. 2009; Sinha et al. 2010; Nemčok et al. 2013; Mukherjee 2018). The basin evolved by Late Jurassic-Early Cretaceous rifting of the Indian plate from the Antarctic plate (Sastri et al. 1981; Veevers 2009; Sinha et al. 2010). It was a part of the rift system that included Palar–Pennar, Krishna–Godavari and Mahanadi rift zones, and dextral Coromondal, North Vizag and Konark transform faults (Veevers 2009; Sinha et al. 2010). The Cauvery rift zone evolved as a system of rift units controlled by NE striking normal faults, having a trend similar to that of Krishna–Godavari and Mahanadi rift zones. The Cauvery basin, in the southern part, initiated as India separated from Sri Lanka due to India–Africa break–up (Katz 1978; Chari Narasimha et al. 1995; Gaina et al. 2007; Veevers 2009). Thus the basin experienced a poly phase rifting episode which ended by dextral strike-slip and oblique-slip separation of Antarctic plate along the Coromondal transform margin (Veevers 2009; Sinha et al. 2010; Nemčok et al. 2013). This ~NNW trending transform fault (~200–300 km length) comprise of horse-tail structure, which overprints the previously developed normal fault pattern of the Cauvery offshore basin in the southern part of India (Nemčok et al. 2013). The northern horse-tail structure (~NNE-NE trend, ~150–200 km length) of the NNW striking Coromondal transform margin got linked with the Krishna–Godavari rift zone (Sinha et al. 2010; Nemčok et al. 2013). Some small-scale pull-apart basins exist along the Coromondal transfer zone, immediately N of the Cauvery basin (Dasgupta 2018b).

S. Dasgupta (✉)
301-Platinum Aura, Sector-20, Roadpali, Navi Mumbai 410218, India
e-mail: swagato.dg@gmail.com

S. Mukherjee (ed.), *Tectonics and Structural Geology: Indian Context*,
Springer Geology, https://doi.org/10.1007/978-3-319-99341-6_10

Some authors document that the Cauvery basin is a failed arm rift between the Peninsular India in NW and the Sri Lankan landmass towards SE with structural trend sub-parallel to the adjacent Precambrian Eastern Ghat lineaments (Katz 1978; Chari Narasimha et al. 1995; Lal et al. 2009; see Misra and Mukherjee 2015 on inheritance issue). Precisely this basin, having different rift-drift transition history, can be termed as a pericratonic rift basin considering its setting within the Indian carton (Biswas et al. 1993).

Like other extensional rift settings, relay structures and transfer zones play important roles in rift fault propagation in the Cauvery basin (Dasgupta and Maitra 2018; Dasgupta 2018a). In a rifted setting the strain transfer occurs along adjacent faults. This transfer of displacement along a set of faults is normally gradual, a typical characteristic of a relay structure or accommodation zone (Rosendahl 1987; Morley et al. 1990; Morley 1999a). Pre-existing basement lineations further govern the formation of transfer zones and segment the rift basin into different sub-basins (Morley 1999b; Morley et al. 2004; Fossen and Rotevatn 2016; Misra and Mukherjee 2015; Dasgupta and Mukherjee 2017).

2 Tectonics and Geology

The Cauvery basin is subdivided into a number of fault-controlled horst and graben structures with sediment thickness ranging from 1 to >6 km (Chari Narasimha et al. 1995). The basement rock of the Cauvery basin comprise of Archean–Proterozoic granite gneiss complex, a part of the Southern Granulite Terrain (Janardhan 1999; Rajaram and Anand 2003; Rao et al. 2006; Mazumder et al. 2019). The major rift faults orient NE. The ~NW trending extension initiated the rifting between the southern part of the Indian peninsular and Sri Lanka along pre-existing Precambrian ~E trending weak zones (Rao et al. 2006; Chetty 2015; Mazumder et al. 2019). Further, extension along NW sub-divided the Cauvery basin into five sub–basins corroborated by a number of NW trending transverse faults, which played a key role in the basin evolution. Rift extension and subsequent anticlockwise rotation of the peninsular India and Sri Lanka (Katz 1978; Chari Narasimha et al. 1995; Lal et al. 2009) developed the Gulf of Mannar basin towards south of the Cauvery basin.

In the onland portion and towards south, the major depressions are the Ariyalur–Pondicherry sub-basin, the Tranquebar sub-basin, the Tanjavur/Tanjore sub-basin, the Nagapattinam sub-basin and the Ramnad-Palk bay sub-basin (Fig. 1; Sastri et al. 1979; Chari Narasimha et al. 1995; Murthy et al. 2008; Chaudhuri et al. 2010). The Mandapam-Delft Ridge separates it in the south from the Gulf of Mannar Basin. The onshore and offshore portions of the Cauvery basin are separated by three major horsts, which include the Madanam/Portonovo Ridge in north, Karaikal Ridge and the Vedaranniyam Terrace towards south (Fig. 1; Chari Narasimha et al. 1995; Murthy et al. 2008). These basement highs and sub-basins, shear zones correlate well with the gravity anomaly map derived from the Bouguer gravity

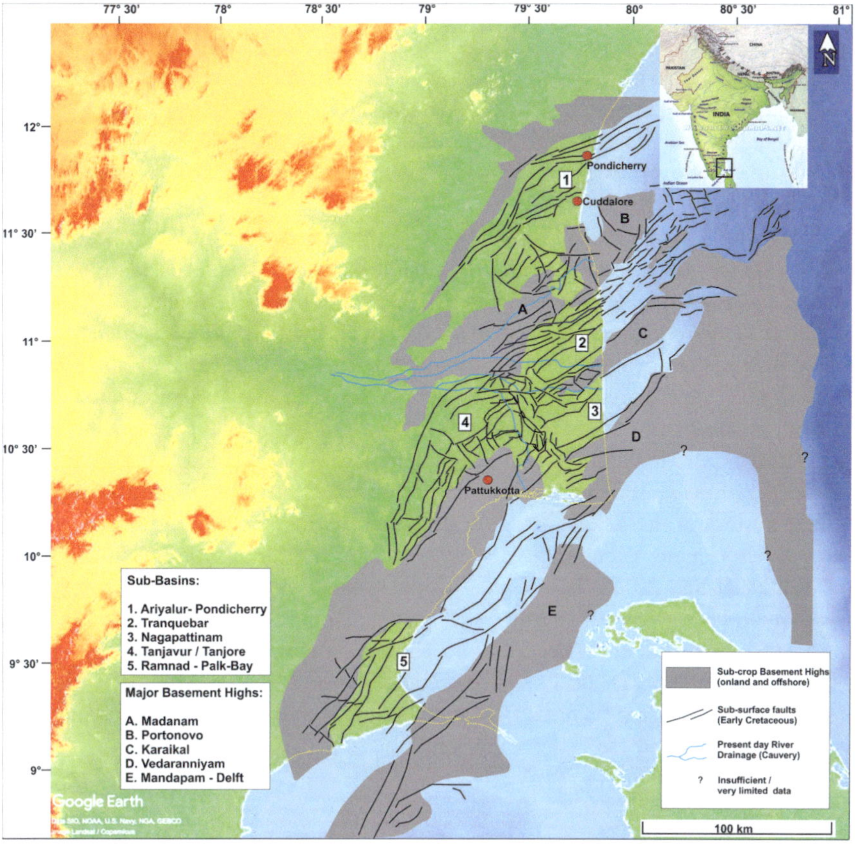

Fig. 1 Tectonic map of the Cauvery basin. Basement highs and sub-basins associated with various rift fault geometries shown that are overlaid on the present day physiographic map of southern India. Modified from Ram Babu and Lakshmi (2004), Murthy et al. (2008), Radhakrishna et al. (2012), Bastia and Radhakrishna (2012), Nemčok et al. (2013), Dasgupta and Maitra (2018)

anomaly and the free-air anomaly data (Verma et al. 1993; Ram Babu and Lakshmi 2004; Bastia and Radhakrishna 2012; Twinkle et al. 2016). The aeromagnetic map derived from aeromagnetic survey (by Geological Survey of India), also depicts similar type of basement morphology (Rajaram and Anand 2003). Interpretation of subsurface seismic data from offshore area also reveal the ~NE trending of rift grabens and fault pattern of the Cauvery basin (Sinha et al. 2010; Nemčok et al. 2013; Dasgupta and Maitra. 2018; Misra and Dasgupta 2018).

The Ramnad and the Mannar sub-basin appear to have come into being in the earliest stage of basin development (Katz 1978; Chari Narasimha et al. 1995). During the early-mid rift stage, the northern grabens of the Cauvery basin e.g., the Tranquebar, the Tanjavur, the Nagapattinam and the Ariyalur–Pondicherry remained nascent due to shear and rifting that deepened the basin. These sub-basins probably started with the deposition of ill-sorted Early Cretaceous sediments at the base. The oldest sediments of the Upper Gondwana expose in the western margin of the Cauvery basin known as the Sivaganga bed/Therani Formation comprising of coarse gritty conglomerates with occasionally argillaceous sandstones and shales (Roy Moulik and Prasad 2007; Watkinson et al. 2007; Kale 2010). The oldest unit of the sub-crop stratigraphy is of Jurassic age drilled in few sub-basins (Chaudhuri et al. 2010).

This tectonic phase was followed by continued deepening of the grabens mainly in response to the tensional forces and partly to thermal subsidence (Sag Phase: Roy Moulik and Prasad 2007; Nagendra et al. 2011). A cumulative thickness of ~ 3 km of Lower Cretaceous sediments in the northern part of the block suggests a long period of basin deepening followed by a differentiated sedimentation history (Raju et al. 2005; Nagendra et al. 2011). During the Cenomanian–Turonian (~ 100–89 Ma) as Madagascar was getting separated from the Indian west-coast, it tilted the Cauvery basin towards SE (Chari Narasimha et al. 1995; Storey et al. 1995; Roy Moulik and Prasad 2007; Reeves 2013). This generated a set of prograding channelized system in the deep-waters. Widespread marine sedimentation happened in the Late Cretaceous (Rangaraju et al. 1993; Roy Moulik and Prasad 2007; Bastia and Radhakrishna 2012). It essentially marks the development of the shelf-slope system with the outer shelf sinking or tilting relatively faster (Nagendra et al. 2011). Differential compaction (Mukherjee and Kumar submitted for mechanism) and fast rate of sedimentation in the Late Cretaceous formed a polygonal fault system that are mainly fluid escape features (Dasgupta and Misra 2018; also see Dasgupta and Mukherjee submitted for general concepts).

3 Transfer Zone Nomenclature

Transfer zone terminologies were initially applied to explain the structures formed in between the overlapping tips of the thrusts (Dahlstrom 1970). These terms were used subsequently in extensional settings as well to define the zone of accommodation between the two overlapping normal fault tips (Rosendahl 1987). Accommodation zones are areas enclosed by terminating segments of overlapping faults that dip in same or opposite direction (Rosendahl 1987; Faulds and Varga 1998). The amount of overlap between the normal faults controls the geometry of the accommodation zone (Rosendahl 1987). The term ‘transfer zone’ or ‘relay structure’ is preferred by many authors over accommodation zone in extensional regimes (Morley et al. 1990; Gawthorpe and Hurst 1993; Morley 1999a; Fossen 2016; Fossen and Rotevatn 2016). These transfer zones preserve both the strain amount as well as the elevation difference between the adjacent footwalls (Morley et al. 1990).

Transfer faults have been defined as transverse to oblique faults with a strong strike-slip component (Gibbs 1984). Strain transfer facilitates in extensional regimes from one normal fault to another along these transfer faults (Gibbs 1984;

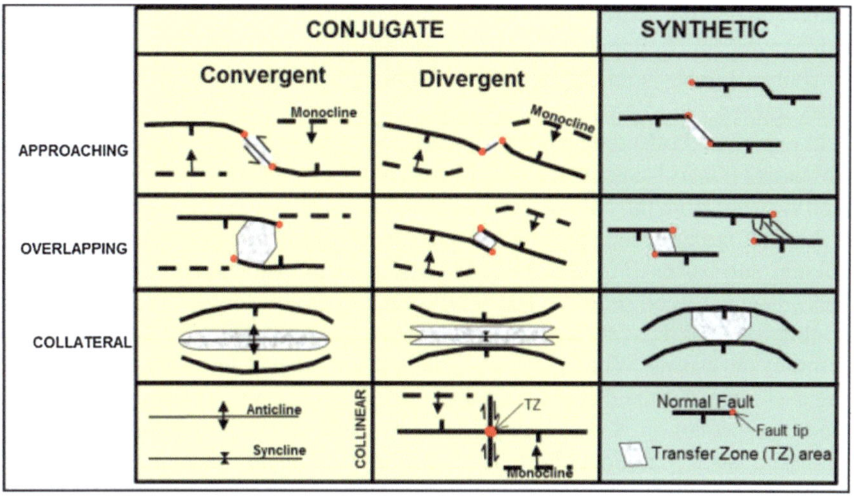

Fig. 2 Transfer zone nomenclature. Modified from Morley et al. (1990), Morley (1999a)

Morley et al. 1990). The dip of the normal faults on either side of a transfer fault are generally in same direction, although opposite dips are also common (e.g., Morley et al. 1990; Morley 1999a).

The classification of transfer zones by Morley et al. (1990; Fig. 2) is relatively robust than that of other authors. It encompasses both extents of overlap as well as changes in dip direction of the adjoining set of normal faults. The primary subdivision is based on the direction of throw of the two overlapping normal faults; (a) conjugate—boundary faults having opposite dips and (b) synthetic—boundary faults with similar dip directions. Conjugate transfer zones can be sub-divided: (a1) convergent—boundary faults dipping towards each other; and (a2) divergent—boundary faults dipping oppositely. Transfer zone comprising of approaching normal fault tips usually associate with transfer faults or transverse faults across which the slip transfers. This nomenclature of transfer zone as proposed by Morley et al. (1990) has been incorporated in the present work.

4 Discussions

In a typical rift system, the rate of occurrence and dip direction of secondary faults might change on successive dip lines leading formation of different type of transfer zones (Morley et al. 1990; Morley 1999a). This is well observed in the Cauvery offshore basin (as in Dasgupta and Maitra 2018). Switch of polarity of the rift-related faults separated the syn-rift grabens by basement highs (Morley et al. 1990; Withjack et al. 2002; Fossen 2016; Fossen and Rotevatn 2016). Variation in dip of the secondary faults is complemented by change in polarity of the major faults. The polarity switch over zones of the secondary faults is normally associated with convergent overlapping transfer zones (Morley et al. 1990). The transfer faults generally originate at much later stage of rifting, even after the overlapping transfer zones formed.

In onland and offshore areas of Cauvery basin, transfer faults segregate different sub- basins (Chari Narasimha et al. 1995; Twinkle et al. 2016). The overall rift fault trend is ~NE from the onland to the offshore areas. However, the rose diagram plot of the fault strike depicts that the rift fault trends has certain variations across the different sub-basins (Fig. 3). Also, there is a major variation of the transfer or transverse fault trend (Fig. 3). The WNW trend of the transfer faults is prevalent in the Ramnad-Palk Bay, the Ariyalur–Pondicherry and the NE offshore (Tranquebar offshore) sub-basins. While the NW trend is the dominant one in the Tanjavur, the Nagapattinum and the Tranquebar sub-basins (Figs. 1 and 3). There is also some good proportion of NW trending transfer faults in the Ariyalur–Pondicherry sub-basin. Such variation of rift fault geometry is mainly due to transfer zones guided by pre-existing Precambrian basement shear fabrics and basement rheology (Withjack and Jamison 1986; Morley 1999b; Withjack et al. 2002; Morley et al. 2004; Robertson et al. 2016; Dasgupta and Mukherjee 2017; Dasgupta and Mukherjee, in press). Few of the ~E trending faults (Fig. 3) were probably affected by the pre-existing basement lineations (Mazumder et al. 2019). These transfer faults also alter the basin depo-centers in the onland and the offshore parts.

The Ariyalur–Pondicherry, the Tranquebar and the Nagapattinam sub-basins extend into the deep-water areas towards NE. The Nagapattinam sub-basin terminates against the Karaikal high in NE (Fig. 1). The Portonovo high dies out towards NE through a set of relay faults associated with some ~NW trending transfer faults. This provides the sediment entry points to the deeper basins. These transfer faults probably associate with strike slip or oblique slip motion (e.g. Morley 1999a, b; Morley et al. 2004). The basin opening as well as deepening is influenced by the transfer zones guided rift faults. In each of the sub-basins, the ~NW trending extension separates the basement highs. Such separation commonly associate with a set of secondary fault systems across which slip transfers through different transfer zones (Figs. 4 and 5). Figure 4 explains this as a cartoon, which depicts that the basin opens towards NE as explained by the dip lines 1–3 (Fig. 5). Transfer zones can be easily identified and named in Fig. 4 (highlighted by yellow; also see Fig. 2).

The major Mesozoic discoveries and producing fields of the basin are in vicinity to the basement highs where the transfer zones are present to a great extent (Fig. 1; Chakraborty et al. 2011; www.dghindia.org). This indicates that the transfer zones act as good fluid migration pathway as well as create hydrocarbon entrapments. Near the basin margin area the rock physical parameters indicate a mixed type lithology with considerable variation of porosity and elastic properties (Chatterjee et al. 2013). This is obvious as the sediments undergo short distance of transport, dominated by transverse drainage, thereby having poor textural and mineralogical maturity (Ravnås and Steel 1998). Reservoir quality is a key concern in such areas. Thus in order to get a better reservoir quality (of Early Cretaceous units) and considerable hydrocarbon pool, the relatively basin interior areas over/adjacent to

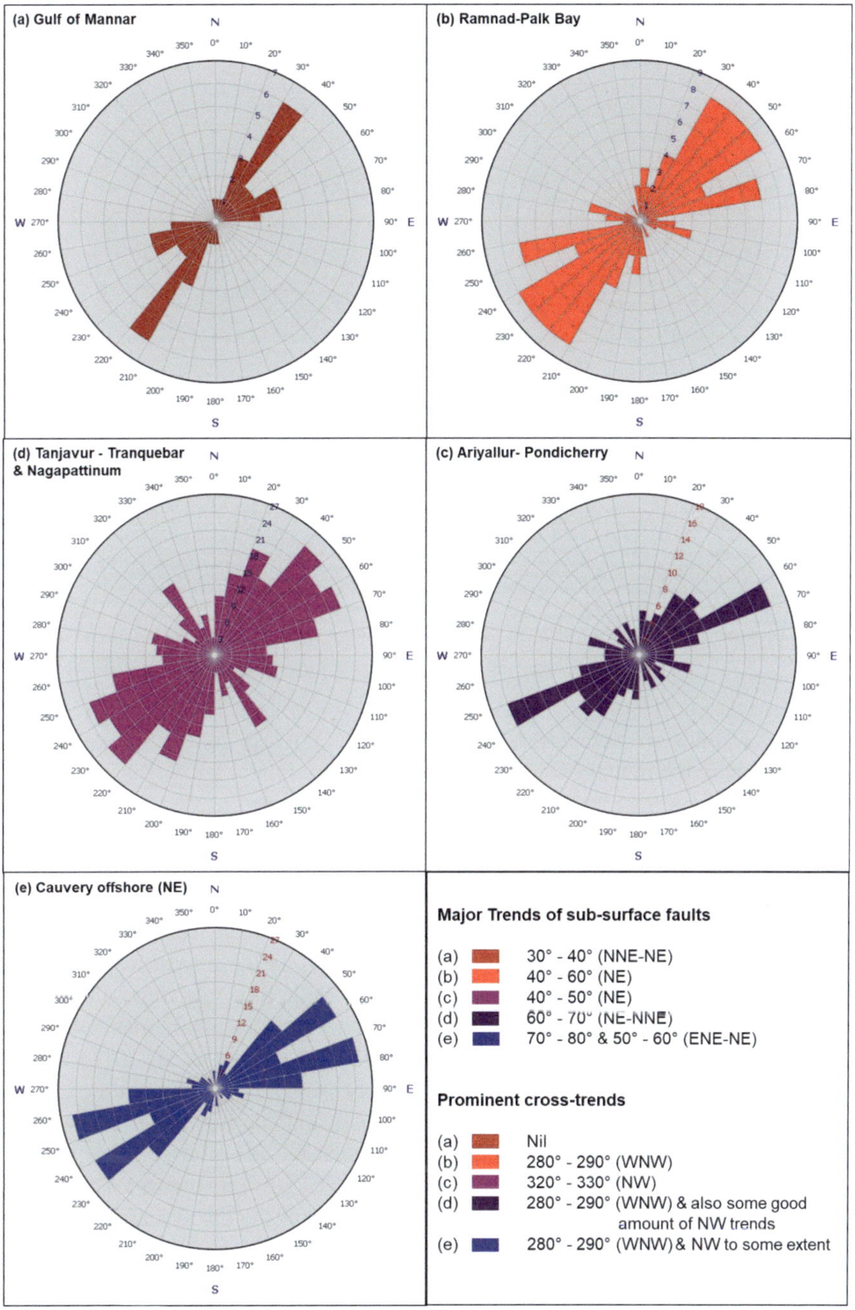

Fig. 3 Rose diagram plot of rift fault trends of different sub basins in Cauvery basin. Note the variation among the trends of the rift faults as well as the transfer faults across different sub basins. Data gathered from Murthy et al. (2008), Nemčok et al. (2013), Dasgupta and Maitra (2018)

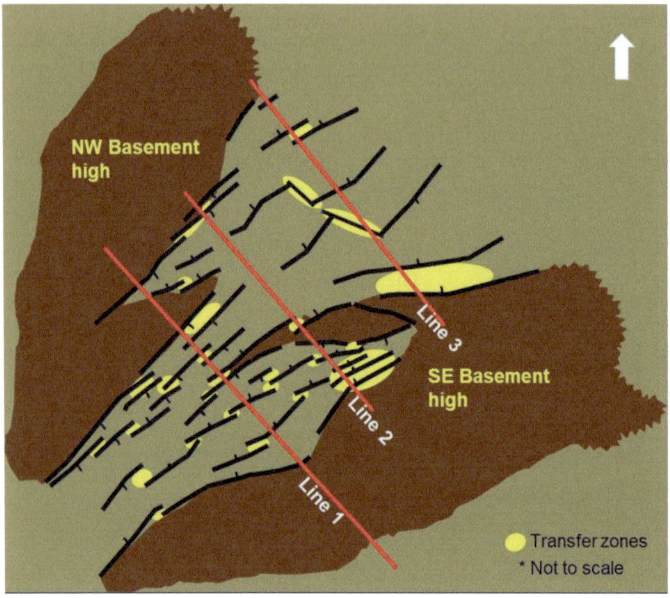

Fig. 4 Tectonic map (cartoon) of a rift system, depicting a typical basin opening process due to extension guided by a set of transfer zones

the transfer zones are potential targets. Such areas possess good structural entrapment guided by the transfer zones. Also, prominent sediment fairway leads to better reservoir quality.

The crustal architecture varies from the proximal to the distal margin of the basin. At the proximal margin, the sediment accommodation space in the half-grabens is controlled by high-angle faults having steeply inclined shear (Fig. 6). Thus, the syn-rift sediments can be quite thick (e.g., sedimentary fills of the Ariyalur-Pondicherry and the Ramnad sub-basins). In these areas, especially the Nagapattinum, the Tranquebar and the Ramnad–Palk Bay sub basins, rift faults interacted with the transfer faults producing stress locking geometries leading to inverted structures (Fig. 6; Murthy et al. 2008; Chakraborty et al. 2011). At the distal margin, the rifting geometries are relatively younger and are governed by the low-angle detachments (Nemčok et al. 2013; Misra and Dasgupta 2018). The dip of the normal rift faults becomes more listric towards the deep water offshore areas (Fig. 7; Mukherjee and Agarwal, in press, for listric fault kinematics). Differential stretching of the continental crust from the proximal to the distal margin may thin the crust and exhume partially the the continental lithosphere (Sinha et al. 2010; Nemčok et al. 2013).

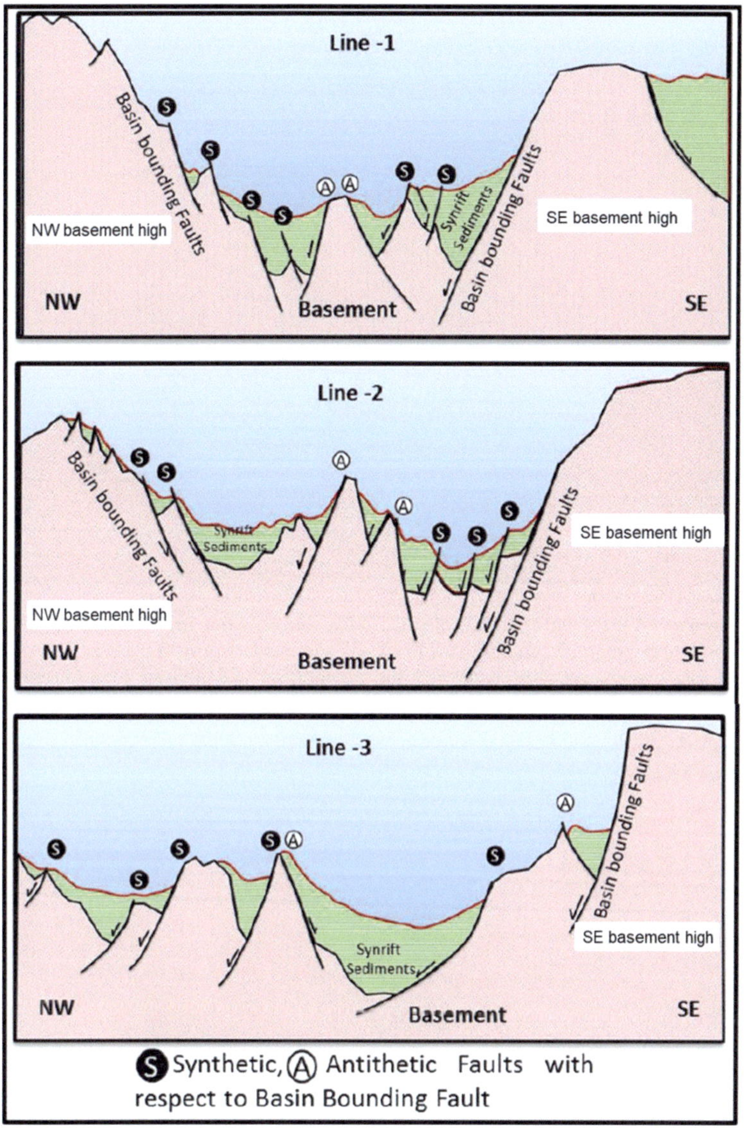

Fig. 5 Schematic dip sections of three lines from SW to NE (shown in Fig. 4) in the direction of basin opening

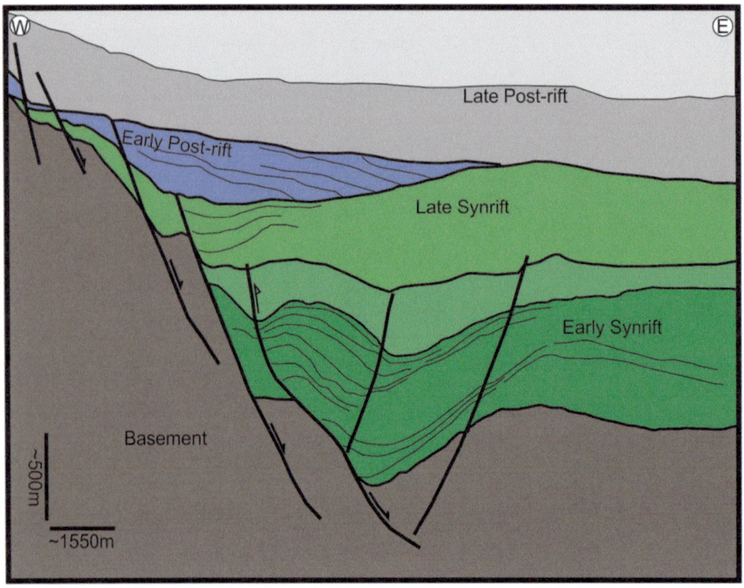

Fig. 6 Around 14 km long seismo-geological cross-section from the Nagapattinum sub-basin showing high-angle rift faults associated with syn-rift sedimentary growth features (line drawing in early and late syn-rift section). Note that the rift-induced and transfer fault-related inverted structures. Modified from Murthy et al. (2008)

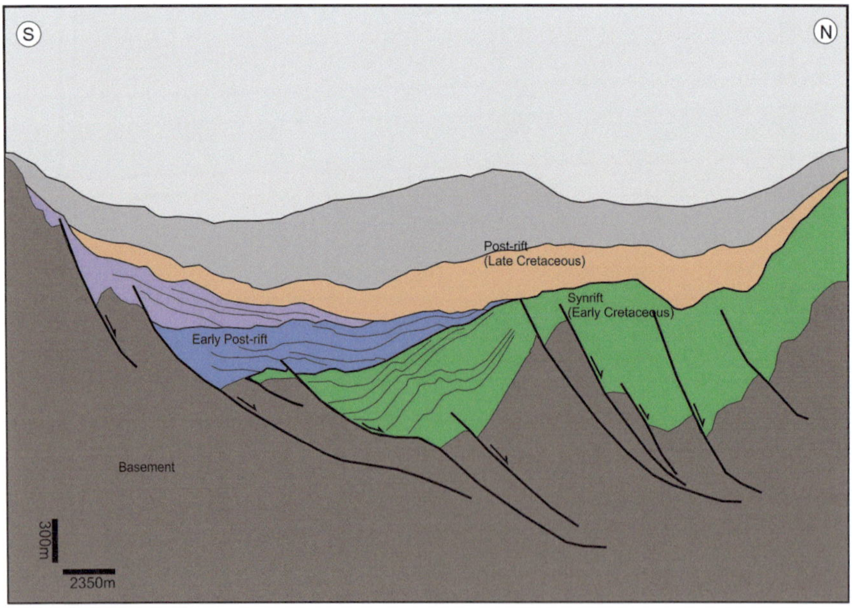

Fig. 7 Around 30 km long seismo-geological cross-section from offshore deep-water area of the Cauvery basin adjacent to the outboard part of the Karaikal high. Note the low-angle listric faults with typical syn-sedimentary growth features. Modified from Dasgupta and Maitra (2018)

The vertical movements of rifted fault blocks and isostatic uplifts (Mukherjee 2017 for mechanisms) in consequence to continental break-up (Nemčok et al. 2013; Dasgupta and Mukherjee 2017) control the distribution of trap, reservoir and source rock elements. Thus, understanding the driving dynamics plays an important role in petroleum system evaluation at rifted and sheared continental margin segments like those of the Cauvery basin.

5 Conclusions

It is observed that the basin expands towards NE associated with change in dip of basin bounding faults. Rift fault propagation is guided by the transfer zones. Switching of polarity of the basin-bounding rift faults link with the transfer zone geometry. The transfer faults present in the basin are possibly influenced by the older basement lineaments. This is well observed in the Cauvery rift basin where the Southern Granulite Terrain is the basement rock. The transfer faults occur at late-stage of rifting. In a typical rift system, transfer zone and transfer faults associate with changes in slip form one rift fault to another. These act as major sediment entry as well as barrier zones. Thus a variation in the transfer zone geometry can affect the sediment depositional system and the fluid migration pathway. The transfer zone also plays a key role in forming structural- and stratigraphic traps.

Ackowledgements SDG thanks Soumyajit Mukherjee for handling this article and provide two rounds of review comments. SDG thanks Abhimanyu Maitra for helping in preparation of Fig. 5 in this article. The work is solely based on published literature survey. No data has been taken from any companies/organizations. Mukherjee (2019) summarized this work.

References

Bastia R, Radhakrishna M (2012) Basin evolution and petroleum prospectivity of the continental margins of India, vol 59. Elsevier, p 11–200. ISBN: 978-0-444-53604-4

Biswas SK, Bhasin AL, Ram J (1993) Classification of Indian sedimentary basins in the framework of plate tectonics. In: Biswas SK, Pomdery J, Dave A, Maithani A, Garg P, Thomas NJ (eds) Proceedings of the 2nd seminar on petroliferous basins of India, vol 1. Indian Petroleum Publishers, Dehradun, p 1–46

Chakraborty C, Srivastava DK, Rana MS, Saxena VM, Sati GC (2011) Paradigm shift in exploration strategy: identification of prospective corridors in Ramnad-Palk Bay sub basin, Cauvery basin, India. In: The 2nd South Asian geoscience conference and exhibition, GEO India

Chari Narasimha MV, Sahu JN, Banerjee B, Zutshi PL, Chandra K (1995) Evolution of the Cauvery basin. India from subsidence modelling, Marine and Petroleum Geology 12, 667–675

Chatterjee R, Manoharan K, Mukhopadhyay M (2013) Petrophysical and mechanical properties of cretaceous sedimentary rocks of Cauvery basin, eastern continental margin of India. Journal of the Indian Geophysical Union 17, 349–359

Chaudhuri A, Rao MV, Dobriyal JP, Saha GC, Chidambaram L, Mehta AK, Ramana LV, Murthy KS (2010) Prospectivity of Cauvery Basin in deep syn-rift sequences, SE India; Search and Discovery Article #10232, Adapted from oral presentation at AAPG Annual Convention and Exhibition, Denver, Colorado, USA, June 7–10

Chetty TRK (2015) The Cauvery suture zone: map of structural architecture and recent advances. Journal of Geological Society of India 85, 37–44

Dahlstrom CDA (1970) Structural geology in the eastern margin of the Canadian Rocky Mountains. Bulletin of Canadian Petroleum Geology 18, 332–406

Dasgupta S (2018a) Sedimentary deformation features produced by differential compaction and buoyancy in the offshore Palar Basin, East Coast of India (Chapter 24). In: Misra AA, Mukherjee S (eds) Atlas of structural geological interpretation from seismic images. Wiley Blackwell, Hoboken, pp 131–133. ISBN: 978-1-119-15832-5

Dasgupta S (2018b) Pull-apart basin in the offshore Cauvery–Palar Basin, India (Chapter 23). In: Misra AA, Mukherjee S (eds) Atlas of structural geological interpretation from seismic images. Wiley Blackwell, Hoboken, pp 127–129. ISBN: 978-1-119-15832-5

Dasgupta S, Maitra A (2018) Transfer zone geometry in the offshore Cauvery Basin, India. In: Misra AA, Mukherjee S (eds) Atlas of structural geological interpretation from seismic images. Wiley Blackwell, Hoboken, pp 117–120. ISBN: 978-1-119-15832-5

Dasgupta S, Misra AA (2018) Polygonal fault system in late cretaceous sediments of Cauvery deepwater Basin, India. In: Misra AA, Mukherjee S (eds) Atlas of structural geological interpretation from seismic images. Wiley Blackwell, Hoboken, pp 113–116. ISBN: 978-1-119-15832-5

Dasgupta S, Mukherjee S (2017) Brittle shear tectonics in a narrow continental rift: asymmetric non-volcanic Barmer basin (Rajasthan, India). The Journal of Geology 125, 561–591

Dasgupta S, Mukherjee S (in press) Remote sensing in lineament identification: examples from western India. In: Billi A, Fagereng A (eds) Problems and solutions in structural geology and tectonics. Developments in structural geology and tectonics book series. Series Editor: Mukherjee S, Elsevier ISSN: 2542-9000

Dasgupta T, Mukherjee S (Submitted) Sediment compaction and applications in petroleum geoscience. Springer. Series: Advances in oil and gas exploration & production. Series Editor: Swenner R. ISSN: 2509-372X

Faulds JE, Varga RJ (1998) The role of accommodation zones and transfer zones in the regional segmentation of extended terranes. Geological Society of America, Special Paper, p 323

Fossen H (2016) Structural geology, 2nd edn. Cambridge University Press, Cambridge. ISBN: 9781107057647

Fossen H, Rotevatn A (2016) Fault linkage and relay structures in extensional settings—a review. Earth-Science Reviews 154, 14–28

Gaina C, Müller RD, Brown B, Ishihara T, Ivanov S (2007) Breakup and early seafloor spreading between India and Antarctica. Geophysical Journal International 170, 151–169

Gawthorpe RL, Hurst JM (1993) Transfer zones in extensional basins: their structural style and influence on drainage development and stratigraphy. Geological Society of London 150, 1137–1152

Gibbs AD (1984) Structural evolution of extensional basin margins. Geological Society of London 141, 609–620

Janardhan AS (1999) Southern granulite terrain, south of the Palghat-Cauvery shear zone: implications for India-Madagascar connection. Gondwana Research 2, 463–469

Kale A (2010) Comments on 'Sequence surfaces and paleobathymetric trends in Albian to Maastrichtian sediments of Ariyalur area, Cauvery Basin, India' from Nagendra, Kannan, Sen, Gilbert, Bakkiaraj, Reddy, and Jaiprakash. Mar Pet Geol 28:1252–1259

Katz MB (1978) Sri Lanka in Gondwanaland and the evolution of the Indian Ocean. Geological Magazine 115, 237–244

Lal NK, Siwal A, Kaul AK (2009) Evolution of East Coast of India—a plate tectonic reconstruction. Journal Geological Society of India 73, 249–260

Mazumder S, Tep B, Pangtey KKS et al (2019) Basement tectonics and shear zones in Cauvery basin: implications in hydrocarbon exploration. In: Mukherjee S (ed) Tectonics and structural geology: Indian context. Springer International Publishing AG, Cham, pp 279–311. ISBN 978-3-319-99340-9
Misra AA, Dasgupta S (2018) Shallow detachment along a transform margin in Cauvery Basin, India. In: Misra AA, Mukherjee S (eds) Atlas of structural geological interpretation from seismic images. Wiley Blackwell, Hoboken, pp 101–105. ISBN: 978-1-119-15832-5
Misra AA, Mukherjee S (2015) Tectonic inheritance in continental rifts and passive margins. Springer briefs in Earth Sciences, Berlin. ISBN 978-3-319-20576-2
Morley CK (1999a) Aspects of transfer zone geometry and evolution in East African rifts. In: Morley CK (ed) Geoscience of rift systems—evolution of East Africa, AAPG Studies in Geology No. 44, p 161–171
Morley CK (1999b) How successful are analogue models in addressing the influence of pre-existing fabrics on rift structure? Journal of Structural Geology 21, 1267–1274
Morley CK, Nelson RA, Patton TL, Munn SG (1990) Transfer zones in the East African rift systems and their relevance to hydrocarbon exploration in rifts. AAPG Bulletin 74, 1234–1253
Morley CK, Haranya C, Phoosongsee W, Pongwappe S, Kornsawan A, Wonganan N (2004) Activation of rift oblique and rift parallel pre-existing fabrics during extension and their effect on deformation style: examples from the rifts of Thailand. Journal of Structural Geology 26, 1803–1829
Mukherjee S (2017) Airy's isostatic model: a proposal for a realistic case. Arabian Journal of Geosciences 10, 268
Mukherjee S (2019) Introduction to "Tectonics and Structural Geology: Indian Context". In: Mukherjee S (ed) Tectonics and structural geology: Indian context. Springer International Publishing AG, Cham, pp 1–5. ISBN: 978-3-319-99340-9
Mukherjee S, Agarwal I (in press) Shear heat model for gouge free dip-slip listric normal faults. Marine and petroleum geology. https://doi.org/10.1016/j.marpetgeo.2018.09.004
Murthy KS, Chaudhuri A, Ramana LV, Rao MV, Dobriyal JP (2008) Hydrocarbon exploration of syn rift sediments in Nagapattinam Sub Basin, Cauvery Basin—a case study. In: 7th Biennial international conference and exposition on petroleum geophysics, society of petroleum geophysicists (SPG), p 443
Nagendra R, Kamalak Kannan BV, Sen G, Gilbert H, Bakkiaraj D, Reddy AN, Jaiprakash BC (2011) Sequence surfaces and paleobathymetric trends in Albian to Maastrichtian sediments of Ariyalur area, Cauvery Basin, India. Marine and Petroleum Geology 28, 895–905
Nemčok M, Sinha ST, Stuart CJ et al (2013) East Indian margin evolution and crustal architecture: integration of deep reflection seismic interpretation and gravity modelling. In: Mohriak WU, Danforth A, Post PJ et al (eds) Conjugate divergent margins, special publications, vol 369. Geological Society London, pp 477–496
Radhakrishna M, Twinkle D, Nayak S, Bastia R, Rao S (2012) Crystal structure and rift architecture across the Krishna–Godavari basin in the central Eastern Continental Margin of India based on analysis of gravity and seismic data. Marine and Petroleum Geology 37, 129–146
Rajaram M, Anand SP (2003) Crustal structure of south India from aeromagnetic data. Journal of the Virtual Explorer 12, 72–82
Raju DSN, Ramesh P, Mohan SKG, Uppal S (2005) Sequence and bio-chronostratigraphic subdivisions of the Cretaceous and Cenozoic of India with notes on pre-Cretaceous events: an overview. In: Raju DSN, Peters J, Shankar R, Kumar G (eds) Association of petroleum geologists special publication No. 1, pp 97–103
Ram Babu HV, Lakshmi MP (2004) A reappraisal of the structure and tectonics of the Cauvery Basin (India) from aeromagnetics and gravity. In: 5th conference & exposition on petroleum geophysics. SPG Hyderabad, India, pp 19–23
Rangaraju MK, Aggarwal A, Prabhakar KN (1993) Tectono-stratigraphy, structural styles, evolutionary model and hydrocarbon prospects of Cauvery and Palar Basins, India. In: Biswas SK, Alak D, Garg P, Pandey J, Maithani A, Thomas NJ (eds) Proceedings of the 2nd

seminar on petroliferous basins of India, vol 1. Indian Petroleum Publishers, Dehradun, India, pp 331–354
Rao VV, Sain K, Reddy PR, Mooney WD (2006) Crustal structure and tectonics of the northern part of the Southern Granulite Terrane, India. Earth and Planetary Science Letters 251, 90–103
Ravnås R, Steel RJ (1998) Architecture of marine rift-basin successions. AAPG Bulletin 82(1), 110–146
Reeves CV (2013) The global tectonics of the Indian Ocean and its relevance to India's western margin. J Geophys 34:87–94
Robertson EAM, Biggs J, Cashman KV, Floyd MA, Vye-Brown C (2016) Influence of regional tectonics and pre-existing structures on the formation of elliptical calderas in the Kenyan Rift. In: Wright TJ, Ayele A, Ferguson DJ, Kidane T, Vye-Brown C (eds) Magmatic rifting and active volcanism, vol 420. Geological Society of London Special Publications, pp 43–67
Rosendahl BR (1987) Architecture of continental rifts with special reference to East Africa. Annual review of Earth and Planetory Science Letters 15, 445–504
Roy Moulik SK, Prasad GK (2007) Seismic expression of the canyon fill facies and its geological significance—a case study from Ariyalur-Pondicherry Sub-basin, Cauvery Basin, India. AAPG Annual Convention, Long Beach, California
Sastri VV, Raju DSN, Venkatachala BS, Acharyya SK (1979) Sedimentary basin map of India, stratigraphic correlation between sedimentary basins of the ESCAP region. Mineral Resource Development Series, United Nations, New York, p 45
Sastri VV, Venkatachala BS, Narayanan V (1981) The evolution of the east coast of India. Palaeogeography, Palaeoclimatology, Palaeoecology 36, 23–54
Sinha ST, Nemčok M, Choudhuri M, Misra AA, Sharma SP, Sinha N, Venkatraman S (2010) The crustal architecture and continental break up of East India Passive Margin: an integrated study of deep reflection seismic interpretation and gravity modeling; Search and Discovery Article 40611
Storey M, Mahoney JJ, Saunders AD, Duncan RA, Kelley SP, Coffin MF (1995) Timing of hot-spot related volcanism and the breakup of Madagascar and India. Sci 267:852–855
Twinkle D, Rao GS, Radhakrishna M, Murthy KSR (2016) Crustal structure and rift tectonics across the Cauvery–Palar basin, eastern continental margin of India based on seismic and potential field modelling. Journal of Earth System Science 125, 329–342
Veevers JJ (2009) Palinspastic (pre-rift and -drift) fit of India and conjugate Antarctica and geological connections across the suture. Gondwana Research 16, 90–108
Veevers JJ, Tewari RC (1995) Gondwana master basin of peninsular india between Tethys and the interior of the Gondwanaland Province of Pangea. Geological Society of America Memoir, pp 72–187
Verma RK, Rao SC, Satyanarayana Y (1993) Gravity field and evolution of Krishna–Godavari and Cauvery basins of India. Indian Journal of Petroleum Geology 2, 39–52
Watkinson MP, Hart MB, Joshi A (2007) Cretaceous tectonostratigraphy and the development of the Cauvery Basin, Southeast India. Petroleum Geoscience 13, 181–191
Withjack MO, Jamison WR (1986) Deformation produced by oblique rifting. Tectonophysics 126, 99–124
Withjack MO, Schlische RW, Olsen PE (2002) Rift-basin structure and its influence on sedimentary systems. In: Renaut RW, Ashley GM (eds) Sedimentation of continental rifts. SEPM Special Publication 73, pp 57–81

Remote Sensing, Structural and Rock Magnetic Analyses of the Ramgarh Structure of SE Rajasthan, Central India-Further Clues to Its Impact Origin and Time of Genesis

Saumitra Misra, Pankaj Kumar Srivastava and Md. Arif

1 Introduction

The Ramgarh structure (centered at 25° 20′N, 76° 38′E) of SE Rajasthan, central India (Figs. 1 and 2a), is a ring-shaped structure of an outer diameter of ≤ 4 km, and is situated within the Vindhyan Supergroup of sedimentary rocks of Meso- to Neoproterozoic age (Sharma 1973; Prasad 1984; Ramasamy 1987). Although, it is a potential candidate of asteroid impact crater for last many decades (e.g., Crawford 1972; Grieve et al. 1988; Sisodia et al. 2006a, b), the supporting evidences in favour of this idea are slowly accumulating. A re-investigation of its origin is thus important since it is the only known potential candidate for asteroid impact crater in India excavated in the sedimentary target rocks. The other comparable terrestrial asteroid impact craters on sedimentary target-rocks of equivalent diameter are: B. P. structure, Libya (25° 19′N, 24° 20′E); Goyder Crater, northern Australia (13° 9′S, 135° 2′E), and Gusev Crater, Russia (48° 26′N, 40° 32′E) (Earth Impact database, http://www.unb.ca/passc/ImpactDatabase).

The Ramgarh structure has many of the physical attributes of an impact crater with a raised rim (Fig. 2a), as described below, but its origin has been debated and more than one hypotheses have been proposed. Sharma (1973) first suggested that

S. Misra (✉)
Discipline of Geological Sciences, SAEES, University of KwaZulu-Natal,
Durban 4000, South Africa
e-mail: misras@ukzn.ac.za

P. K. Srivastava
Department of Petroleum Engineering and Earth Sciences,
University of Petroleum and Energy Studies, Dehradun 248007, India

Md. Arif
Birbal Sahni Institute of Palaeosciences, Lucknow 226007, India

S. Mukherjee (ed.), *Tectonics and Structural Geology: Indian Context*,
Springer Geology, https://doi.org/10.1007/978-3-319-99341-6_11

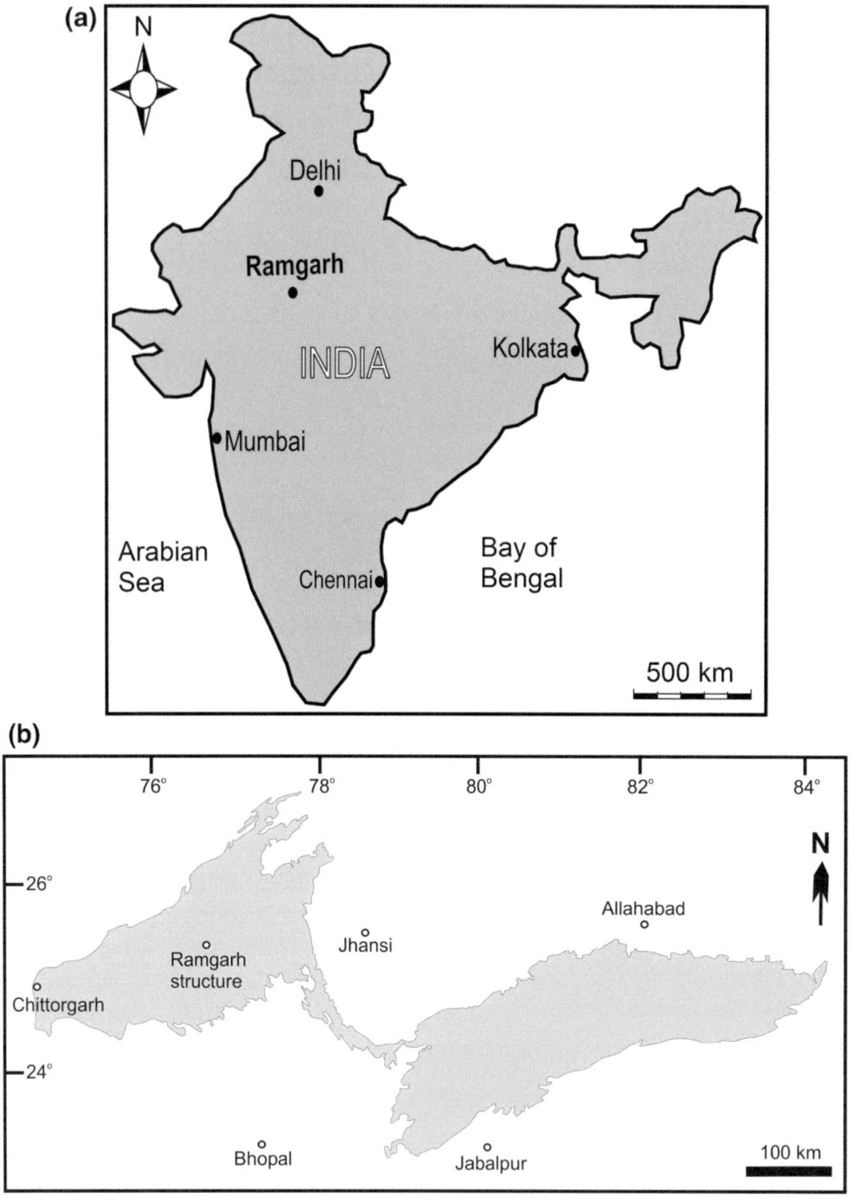

Fig. 1 Sketch map of **a** India, and **b** Vindhyan basin (in light grey shade), central India, showing the position of the Ramgarh structure

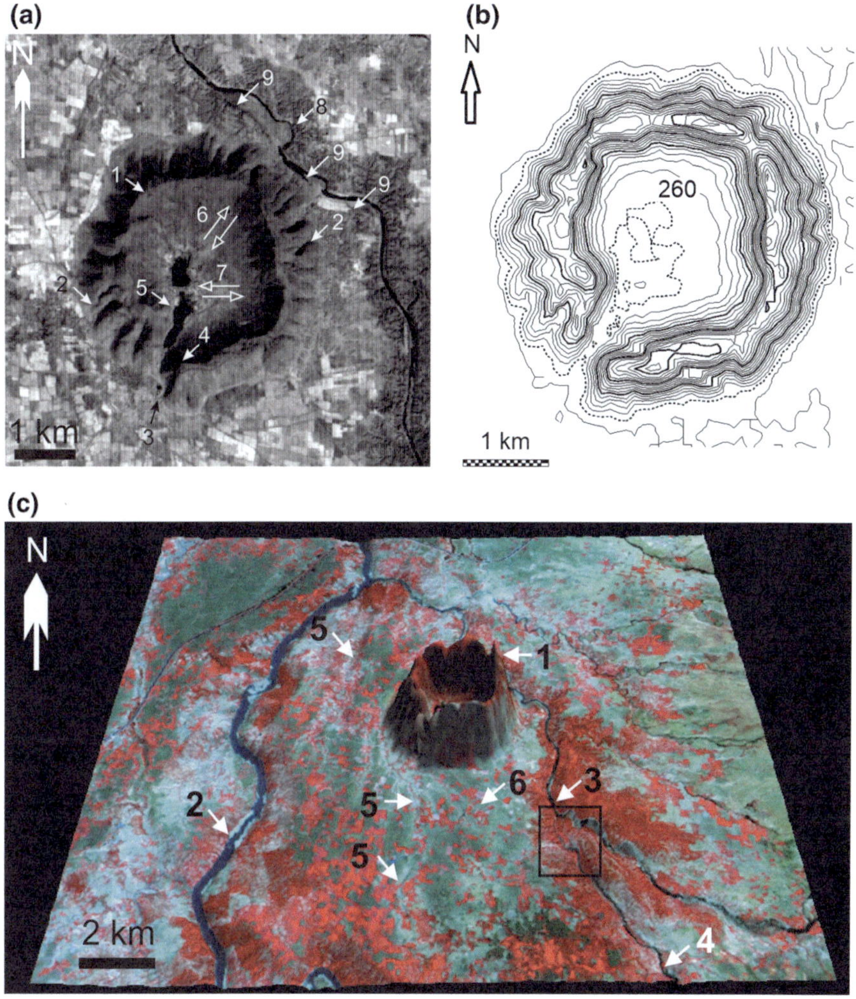

Fig. 2 **a** Grey-scale ASTER image at 15 m resolution of Ramgarh structure, note: roughly rectangular shape of the structure's rim (1), numerous gullies cross-cutting structure's rim (2), a narrow deep-seated gully to SW making only avenue to enter into the structure (3), the structure's rim is discontinuous towards SW (4) and its counter-part to W displaced towards N (5) due to a major dextral NE-SW slip (6), the southern half of Ramgarh structure moved marginally towards E perhaps due to sinistral faulting (7), the Kul river channel displaced towards NE (8), and palaeo-channel of Kul river (9). **b** Topographic map of Ramgarh structure prepared from SRTM-DEM with a contour interval of 10 m. **c** False-coloured SRTM-DEM image (special resolution 90 m) of Ramgarh structure, note: steeply standing structure's rim on the ground (1), present channel of Parvati River to W (2) and Kul River to E (3), a tributary of Kul River (4), palaeo-channel of Parvati River (5), and the box showing where tributary (4) merging with Kul River at present leaving behind its palaeo-channel flowing towards NW (6)

this structure could have been resulted due to magmatic activity. However, this hypothesis can safely be discarded because no igneous plug/activity has been described within this structure, including in borehole studies reaching to ~450 m depth and in surrounding areas (Ramasamy 1987). Furthermore, the regional aeromagnetic anomaly map does not support the presence of any magnetic high below the structure (Kumar et al. 2011) suggestive of any magmatism. Alternatively, it has been suggested that the Ramgarh structure formed due to tectonic activity by 'rheid' flow of shales and associated horizontal compression in subordinate scale (Ramasamy 1987). But there is no evidence of structures dictated by compressive forces in the bed rocks surrounding the Ramgarh structure. This model, therefore, fails to explain the origin of this unique structure within the undeformed, flat-lying sedimentary country rocks.

The third upcoming hypothesis is that this structure could have been formed by asteroid impact. Some of the important observations favouring the impact hypothesis are (a) presence of a prominent conical central peak (Grieve et al. 1988; Master and Pandit 1999), (b) shatter-cone like structure in colluvium near its center (Crawford 1972); (c) presence of closed spaced Planar Fractures (PF) and granulation in quartz grains in shocked sandstones from this structure (Balasundaram and Dube 1973; Das et al. 2011), and (d) occurrences of impact spherule-looking materials, diaplectic glasses, and possible Planer Deformation Fracture (PDF) in quartz grains from this structure (Sisodia et al. 2006a, b; Purohot and Sisodia 2013; Rana and Agrawal 2016), although the true nature of these PDFs remain questionable (Reimold et al. 2006; also cf. Vernooij and Langenhorst 2005; French and Koeberl 2010).

The iron-silica rich mm-sized particles/spherules from within the alluvium inside the Ramgarh structure have been primarily investigated in Misra et al. (2008b). The Back Scattered Electron (BSE) images of these magnetic particles are similar to impact-induced accretionary lapilli reported from other established impact structures (cf. Graup 1981; Koeberl et al. 2007; Newsom et al. 2016). These particles/spherules have cores consisting of coarse grained quartz fragments surrounded by multiple concentric layers of fine-grained materials with smaller quartz fragments (Misra et al. 2008b; Das et al. 2011). Our studies further suggest that these iron-silica rich particles have very high Natural Remnant Magnetization (NRM) (~2–19 Am^{-1}) and high REM (after Yu 2006 and references therein) (~7–145%) that could indicate the presence of a high magnetic field during their formation, much higher than the ambient Earth magnetic field (~40 Am^{-1}) (Das et al. 2009). The exact reason of this local high magnetic field in and around the Ramgarh structure is not clearly known, but it could be resulted by asteroid impact (cf. Crawford and Schultz 1988, 1999). More recently, a few fragments of iron-rich Ca–Al silicate glassy pieces, recovered from the soil inside the Ramgarh structure and nearby outside the western rim of the structure, contain dendritic magnetite with occasional inclusions of relict native iron containing high proportions of Co (~350–3000 ppm), Ni (~200–4000 ppm) and Cu (~2200–7000 ppm) (with similar Co/Ni ratios), and these iron globules are thought to be extra-terrestrial in origin (Misra et al. 2013).

Hence, the emerging geological evidences favour more towards the asteroid impact origin of the Ramgarh structure. The only lacuna is that the bedrock sandstones from the rim of this structure do not show any confirming PDF under the transmitted light polarizing microscope satisfying the definition described in French and Koeberl (2010). However, the general absence of PDF in quartz grains within the Ramgarh structure sandstones does not preclude the asteroid impact origin of this structure. This is because this microstructure could also be formed by terrestrial lightning strikes (Chen et al. 2017; Melosh 2017), and hence cannot be considered as a confirming criterion for identification of an impact crater. On the other hand, although some geochemical clues on possible impact origin of the Ramgarh structure have been documented (Ahmed et al. 1974; Sisodia and Lashkari 2003; Sisodia et al. 2006a; Misra et al. 2008b, 2013), no detail information on the structural geology and/or rock magnetic properties on this structure in favour or against of its asteroid impact origin presently exist. In the present study, we thus examine the available remote sensing images, and structural data of the Ramgarh structure in more detail (cf. Misra et al. 2008a). The rock magnetic properties of a few glassy samples from this structure (cf. Das et al. 2009) have also been examined to re-evaluate the asteroid impact origin of the Ramgarh structure.

2 Geology of Ramgarh Structure

The Vindhyan Supergroup, into which the Ramgarh structure is situated, exposes as a sickle-shaped, east-west elongated outcrop in the central India (Oldham 1856; Mallet 1869; Auden 1933), and occupies $\sim$178,000 km^2 (Tandon et al. 1991) with a thickness up to $\sim$4.5 km (Ahmad 1958) (Fig. 1). This Supergroup of rocks are unmetamorphosed and only mildly deformed. Geophysical investigations revealed that the rifting across E-W faults initiated the formation of the Vindhyan basin (Bose et al. 2001). The N-S rifting was accompanied by a dextral shear and as a result the basin segmented into several NW-SE elongated sub-basins. Between the two major sectors of the Vindhyan Supergroup, the Ramgarh structure occurs in the western (Rajasthan) sector of this huge sedimentary outcrop (Ramasamy 1987). This structure is excavated in the flat-lying sandstone, shale, and minor limestone horizons of the Lower Bhander Group of the Vindhyan Supergroup (Prasad 1984; Ramasamy 1987). The oldest to youngest stratigraphic sequences of the Lower Bhander Group that are exposed within the Ramgarh structure include the Ganurgarh Shale (thickness $\sim$630 m), Lower Bhander Limestone ($\sim$26 m), Samaria Shale ($\sim$52 m) and Lower Bhander sandstone ($\sim$400 m) (cf. Fig. 3 in Ramasamy 1987).

The ring-like, prominently rectangular shaped Ramgarh structure has a rim-to-rim diameter of $\sim$2.37 and 2.48 km along E-W and N-S respectively (measured from Google Earth Imagery 2017), and a more or less continuous raised rim along its periphery except to the southwest, which rises $\sim$200 m above the surrounding undisturbed, flat-lying plain (cf. Sisodia et al. 2006a; Misra et al. 2008b) (Fig. 2a). Earlier reports indicate the presence of a central conical peak within this structure (Grieve et al. 1988; Master and Pandit 1999; Sisodia et al. 2006a). The Upper Bhander sandstone, constituting the youngest Formation of the Vindhyan Supergroup, has been dated as 1000–1070 Ma (U-Pb zircon age, Malone et al 2008), which could be the upper age limit of formation of the Ramgarh structure.

The sandstone exposures on the rim of the Ramgarh structure have dips between ~45° and 78°, which gradually reduce to ~20°–30° at the lower outer slope, and to ~5°–15° at the foot of the outer rim (Prasad 1984). The rim of the structure is only discontinuous at the SW corner and displaced by a set of NE-SW dextral strike slip faults (Prasad 1984; Ramasamy 1987; Sisodia et al. 2006a; Misra et al. 2008a). Additional joint/fault systems that cross-cut the Ramgarh structure trend E-W, N-S and NW-SE (Ramasamy 1987; Misra et al. 2008a). Geomorphologically, the Ramgarh structure appears to have more degraded state (i.e. modified by erosion) as compared to those of the Arizona Crater, USA, and Lonar Crater, India (Grant 1999). This conclusion is well supported by the presence of reworked materials (i.e. materials derived during weathering of the rim crest rocks of the structure) within a very limited zone (~770 m) in the outer flank of the structure rim (Misra et al. 2008a). Although the Ramgarh structure is eroded, the sandstone exposed on its rim shows quaquaversal dips (Sharma 1973; Ramasamy 1987) like other impact craters (e.g., Lonar Crater, Misra et al. 2010).

3 Analytical Techniques

The rock magnetic characterization experiments were carried out on a dark reddish coloured, glassy silicate sample (RJ-26) of a maximum size of 3 cm (Fig. 9 in Appendix), which was recovered from within the rocky debris derived from the rim crest rocks and deposited on the outer slope of the western rim of the Ramgarh structure (Fig. 3). The glassy piece was magnetic, and under transmitted light, polarizing, optical microscope it was looking similar to the glassy sample R-82 reported in Misra et al. (2013).

Our experimental studies include measurements of Natural Remanent Magnetization (NRM), REM (=NRM/SIRM expressed in %, where SIRM—Saturation Isothermal Remanent Magnetization) (after Yu 2006 and references therein), Isothermal Remanent Magnetization (IRM) acquisition and backfield SIRM dc demagnetization, and Lowrie-Fuller test (methods and techniques of our experiments after McElhinny and McFadden 2000). All these measurements were carried out at the Palaeomagnetism Laboratory, Birbal Sahni Institute of Palaeosciences (Lucknow, India). Four mm-sized sub-samples were cut from the original glassy silicate piece for our present measurements. The NRM was measured by an Agico JR-6 spinner magnetometer. The IRM was imparted in progressively increasing magnetic fields up to 1 T and their back field application to the saturation IRM for evaluating remanence coercive force (Hcr) by an ASC scientific IM-10–30 impulse magnetizer; magnetic remanence after each IRM step

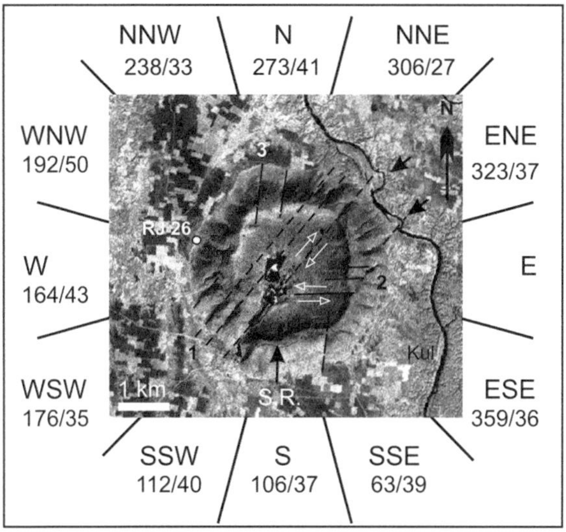

Fig. 3 The panchromatic band Landsat-7 grey shed image of Ramgarh structure (source GLCF 2007) showing three prominent sets of lineament transecting the structure's rim, abbreviations: S.R.-structure's rim, Kul- Kul River, 1, 2, 3-lineaments cross-cutting rim of Ramgarh structure, NE-SW dextral slip and E-W sinistral slip shown by arrows. The rim and adjoining area of Ramgarh structure are segmented into twelve geographic sectors, average orientations of sedimentary country rocks (sandstones) on structure's rim in each sector shown for reference (read as follows: strike in degree/dip amount in degree, right-hand rule should be followed to interpret dip direction), location of sample RJ 26 also shown. Black arrows at north eastern part of the map indicating displacement of Kul River channel along NE-SW lineament

was measured by an Agico JR-6 spinner magnetometer. The domain structure of magnetic minerals and origin of the magnetization of representative samples were investigated using the Lowrie-Fuller test (Lowrie and Fuller 1971). This test is based on a comparison of the AF demagnetization spectra of NRM and SIRM. First, the NRM was stepwise AF demagnetized up to 100 mT using ASC scientific D2000 AF demagnetizer; magnetic remanence after each AF demagnetization step was measured by an Agico JR-6 spinner magnetometer. After that, an isothermal remanent magnetization (IRM) was induced in a 1 T field along the z-direction and progressively AF demagnetized with the same scheme.

4 Analyses

4.1 Shape and Structural Analyses of Ramgarh Structure from Remote Sensing Images

A multispectral Advanced Spaceborn Thermal Emission and Reflection Radiometer (ASTER) image (for technical details see Misra et al. 2010) (Fig. 2a), and a high spatial resolution ($\sim$15 m) panchromatic band of Landsat 7 imagery [acquired on October 22, 1999 (GLCF 2007)] (Fig. 3) were used to evaluate the morphology of the Ramgarh structure and its surroundings in the present study. A contour map of the Ramgarh structure with maximum and minimum contours of 260 and 450 m respectively was also produced using Shuttle Radar Topographic Mission (SRTM) data of 90 m special resolution with a contour interval of $\sim$10 m (after Misra et al. 2010) (Fig. 2b). The SRTM-DEM (DEM-Digital Elevation Model) was also draped over the panchromatic image to reconstruct 3D view of the Ramgarh structure (Fig. 2c).

The ASTER and Landsat 7 images, and the contour map reconfirm that the Ramgarh structure is a rectangular-shaped morphologic feature with a steeply raised rim along its periphery except to the southwest (Figs. 2a, b and 3). The analysis of the SRTM-DEM image further confirms that the rim of the structure is steeply standing on the ground (Fig. 2c). The highest elevation of the rim is $\sim$450 m, observed in the ENE sector of the structure (Fig. 2b). There are several furrow-like features on the rim (Fig. 2a), and many of them appear to be continuous and cross-cut the Ramgarh structure rim especially in NE-SW direction (Fig. 3, lineament 1). The displacement of the NNW-SSE trending Kul River channel (situated to E of the Ramgarh structure) towards NE is evident at least at two places along the NE-SW lineament (Figs. 2a, and 3). The SW rim crest appears to have been detached from its continuation towards the east and have moved dextrally along NE-SW lineament. There are two more sets of lineament trending E-W and N-S those cross-cut the Ramgarh structure rim (Fig. 3, lineaments 2 and 3 respectively). The southern half of the structure appears to have been displaced marginally towards the east along the E-W lineament (Fig. 2a).

4.2 *Field Geology and Field Structural Analyses of Ramgarh Structure*

The sedimentary country rocks surrounding the Ramgarh structure are essentially flat (Fig. 4a). Our observations in the Parvati and Kul River sections to the W and NNE respectively of the Ramgarh structure (Fig. 2), and around $\sim$19.4 km south of this structure show wide-spaced ($\sim$2 m) cross-cutting fractures trending NE-SW and NW-SE (Figs. 4b and 6a). The cross-cutting field relationship at the Parvati River section suggests that the NE-SW fracture could have developed early in sequence, followed by the formation of NW-SE fracture.

The bedding plane (S_o) in sandstone on the rim crest of the Ramgarh structure show quaquaversal dips at moderate angle (Fig. 4c) (cf. Sharma 1973; Ramasamy 1987). In places, the rim sandstones also dip $\sim$vertical (Fig. 4d). The sandstones examined on the rim crest (Fig. 4e, f), the structure wall (Fig. 4g), and the shale exposed at the base inside the structure (Fig. 4h) are extremely fractured. The sequence of fracture development within the Ramgarh structure appears to be extremely complicated. Besides the NE-SW and NW-SE cross-cutting orthogonal fracture sets, the sandstones on the Ramgarh structure rim also show two additional fracture sets trending N-S and E-W (Fig. 4f). Although our observation on the flat-lying sandstone in the Parvati River section suggests an early NE-SW fracture (Fig. 4b), the same fracture set presents in the SW and SE Ramgarh structure rim, however, show reverse time relationship (Fig. 4i). It appears from our present study in field and on satellite imageries (Figs. 2a and 3) that the E-W fracture was the youngest in the sequence, preceded by the slip along NE-SW brittle plane. The dextral slip along the NE-SW plane displaced the Kul River section towards NE leaving its palaeo-channel close to the Ramgarh structure (Figs. 2a and 3, also see Google Earth imagery 2017), and the transformation of the sandstone in the river section into gneissic rocks (Fig. 4j). Intense activity of Mn-rich solution (as confirmed by our microprobe analyses) is observed within the Ramgarh structure, which injected into the bedding planes and some fracture planes cross-cutting the sandstones (Fig. 4k–m).

On the inner wall of the Ramgarh structure towards NW, we observed in one case, a surficial feature on the sandstone looking similar to shatter cone observed in many asteroid impact craters (Fig. 4n, an orthogonal cross-sectional view could not be photographed due to steepness of valley section) (cf. French 1998; French and Koeberl 2010). At the central part of the structure, a small conical peak is noticed that has a height of around 6 m at present (Fig. 4o) (cf. Grieve et al. 1988). This peak consists of sandstone having >1 sets of fracture, among which the NE-SW fracture, dipping moderately towards the southeast, is the most common (Fig. 4p).

The quaquaversal dips of the bedding planes (S_o) in sandstones exposed on the rim of the Ramgarh structure were described in literature (Sharma 1973; Ramasamy 1987). However, no detail study exists to document any possible variations in the dips of the rim crest sandstones in different geographic sectors. Hence, for a systematic structural analyses on the variation in dips of the rim sandstones, the Ramgarh structure rim and adjoining area were segmented into twelve arbitrary sectors based on the geographic directions (Fig. 3) (after Misra et al. 2010). The detail of our structural analyses show that the rim sandstones in the northern sector have average dip of 273°/41°N (i.e., strike/dip amount, the right hand rule is followed to express dip direction) (Fig. 5a). The rim sandstones on the NNE and ENE sectors gradually vary in dip directions from N to E (Fig. 5b, c). In the E-sector, the rim crest sandstones dip at moderate angle either towards N or E (Fig. 5d). In the ESE sector, the rim sandstones show dips towards E with an average of 359°/36°E (Fig. 5e). In the SSE sector, the general dip of the crater rim sandstone is towards SE with an average of 63°/39°E (Fig. 5f). In the S and SSW sectors, the rim sandstones show dip towards S or SW with averages of 106°/37°S and 112°/40°S respectively (Fig. 5g, h). In the WSW and W sectors, the dips of the rim sandstones are towards W with averages of 176°/35°W and 164°/43°W respectively (Fig. 5i, j). In the WNW sector, the dip directions of the rim sandstones are extremely variable between SW and W with an average of 192°/50°W (Fig. 5k). In the NNW-sector, both the components of westerly and northerly dipping sandstones are present with an average of 238°/33°NW (Fig. 5l).

(a)
R S
lst+ sst+ shale
sst
(b)
NW-SE
NE-SW
(c)
(d)
(e)
(f)
(g)
(h)

◂**Fig. 4** **a** Cross sectional view of flat-lying Vindhyan Supergroup of sedimentary rocks including limestone (lst), sandstone (sst) and shale at Parvati River section at ~4 km W of Ramgarh structure [width of photograph is ~30 m, Ramgarh structure (RS) to E in background], **b** plan view of flat-lying Vindhyan Supergroup of sedimentary rocks at Parvati River section showing development of well-spaced (~2 m) orthogonal fractures where NW-SE fracture cross-cuts NE-SW fracture (length of hammer ~27.9 cm), **c** quaquaversal dip of bedding plane (S_o) in sandstone on the western rim crest of Ramgarh structure, white arrow indicates towards the center of structure to E. **d** sub-vertical orientation of S_o in sandstone on NNE rim crest [hammer (~27.9 cm length) placed parallel to S_o]. **e** plan view of N-S subvertical fracture (shown by arrow) and cross fracture on sandstones exposed at SSW sector of rim crest (hammer's head points to N). **f** plan view of sandstone exposed on SSW rim crest showing development of N-S, E-W and NE-SW trending fractures intersecting each other (hammer's head pointed towards E). **g** cross-sectional view of inner wall of Ramgarh structure at the southern sector of rim crest showing highly fractured sandstone with flowage of Mn-rich solution along few selected fractures (shown by arrow). **h** plan view of Ganurgarh Shale exposed at the base inside the Ramgarh structure towards SW close to the entrance showing development of extensive cross-cutting fractures. **i** plan view of sandstone on SSW structure rim showing NE-SW fracture crosscuts NW-SE fracture, both fractures are cross-cut by E-W fracture. **j** a view of rock sample at Kul River section towards NE of Ramgarh structure showing gneissic foliation (diameter of coin shown as scale is 25 mm). **k** cross-sectional view of inner wall of Ramgarh structure at southern sector showing flowage of Mn-rich solution along few fracture planes (shown by arrow) (diameter of coin shown as scale is 23 mm). **l** plan view on northern rim crest showing injection of Mn-solution along S_o plane of tilted sandstone (shown by arrow). **m** plan view on SE rim crest showing injection of Mn-solution along S_o plane of sandstone and fracture planes (shown by arrow, width of photograph ~50 cm). **n** a shatter cone-like structure on the inner wall of Ramgarh structure at NW rim crest. **o** view of ~6 m high central uplift with an ancient temple at top (shown by arrow), and. **p** cross sectional view of extremely fractured sandstone in central uplift (width of photograph ~20 cm)

Unlike the surrounding undeformed sedimentary country rocks (Fig. 6a), the rocks inside the Ramgarh structure show extreme development of fractures (Fig. 6b). These fractures, including both the joints and faults, could primarily be classified into two types: vertical and moderately dipping fractures. The former fractures trend in all directions, while the dipping fractures trend ~NE-SW and NW-SE. In the N-NNE-ENE sectors of the Ramgarh structure (Fig. 3), the dipping fractures trend ~NE-SW and dip at moderate angle both towards NW and SE (Fig. 6c, d). Two subordinate sets of fractures are also identified having (1) trend of NW-SE and dipping moderately towards SW, and (2) trend E-W and dipping moderately towards S. The second important sector, where sufficient data were available, is the S-SSW-WSW sectors (Fig. 3). Here, two sets of fractures are noticed, e.g. the NE-SW trending fracture dipping towards NW, and the NW-SE trending fracture dipping towards NE (Fig. 6g, h). In the W-WNW-NNW sectors, we have the NE-SW trending fracture dipping both towards NW and SE, and the NW-SE trending fracture dips towards NE. There is also an E-W trending fracture dips towards the south.

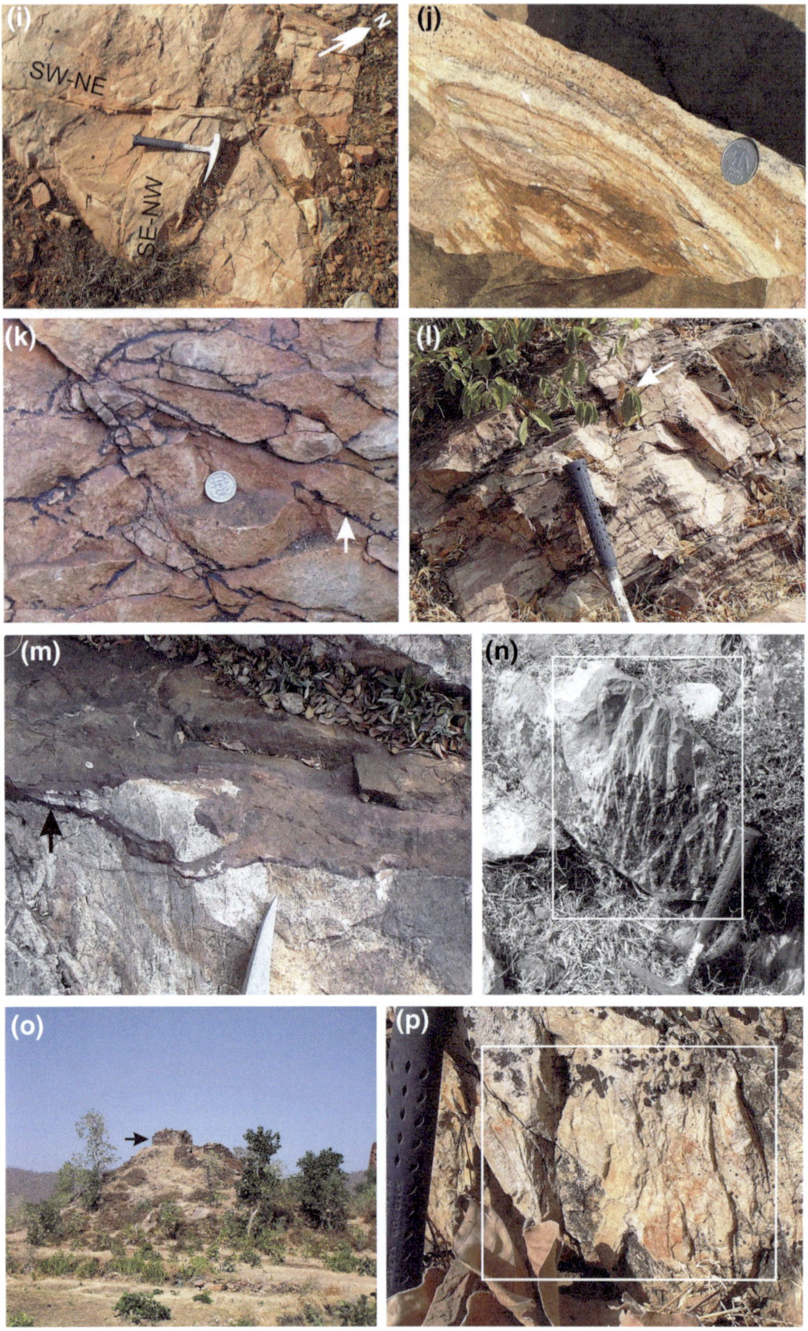

Fig. 4 (continued)

4.3 Remote Sensing Images in and Around Ramgarh Structure

An examination of the ASTER image of the greater areal extent suggests that the Ramgarh structure is most likely situated on the palaeo-channel of the Parvati River, which is indicated by the presence of a NNW-SSE trending abundant channel of this river to the south and west of this structure (Fig. 7). The present course of Parvati River to the west of the Ramgarh structure is extremely irregular compared to the rest of the course of this river. The SRTM-DEM image further shows that the river marked '4' was previously a tributary of Parvati River (Fig. 2c). Presently this river changes its course leaving its palaeo-channel to the south of Ramgarh structure (channel marked 6) and merges with the Kul River.

4.4 Rock Magnetic Properties of Glassy Samples from Ramgarh Structure

The NRM of all four sub-samples from sample RJ26 (Fig. 3) ranges between 9 and 19 Am^{-1}, and their REM (i.e., NRM/SIRM ratio in %) varies between 40 and 78%. In all sub-samples, the NRM decay curves are above the SIRM decay curves for the lower range of AF demagnetization fields, whereas there are always overlap between these two curves in the higher AF fields (Fig. 8a–d) indicating a mixture of Single Domain (SD), Pseudo-SD and Multi Domain (MD) grains in them. The IRM acquisition curves of all our sub-samples seem to saturate at higher fields of 500 mT indicating high coercivity magnetic mineral as the remanence carrier (Fig. 8e). Their IRM backfield curves indicated a range of coercivity of remanence (Hcr) between 30 and 45 mT (Fig. 8f).

5 Discussion

The Vindhyan Supergroup of sedimentary rocks studied in the Chittorgarh area, ~208 km WSW of Ramgarh structure (Fig. 1b), contain four sets of fractures (joints), e.g., E-W/vertical (or 80°N); N-S/vertical (or 45–65°W); NE-SW/vertical and NNW-SSE/vertical (Prasad 1984). The area surrounding the Ramgarh structure, however, shows only a set of orthogonal, widely spaced (~2 m), vertical fractures trending NE-SW and NW-SE (Figs. 4b and 6a).

On the other hand, the sedimentary country rocks within the Ramgarh structure show evidences of extensive brittle deformations. For examples, the vertical radial fractures present within the Ramgarh structure have trend in numerous orientations (Fig. 6b). The NE-SW and NW-SE fractures dip moderately within the Ramgarh structure (Fig. 6c–j) in contrast to their vertical dip as observed in Chittorgarh

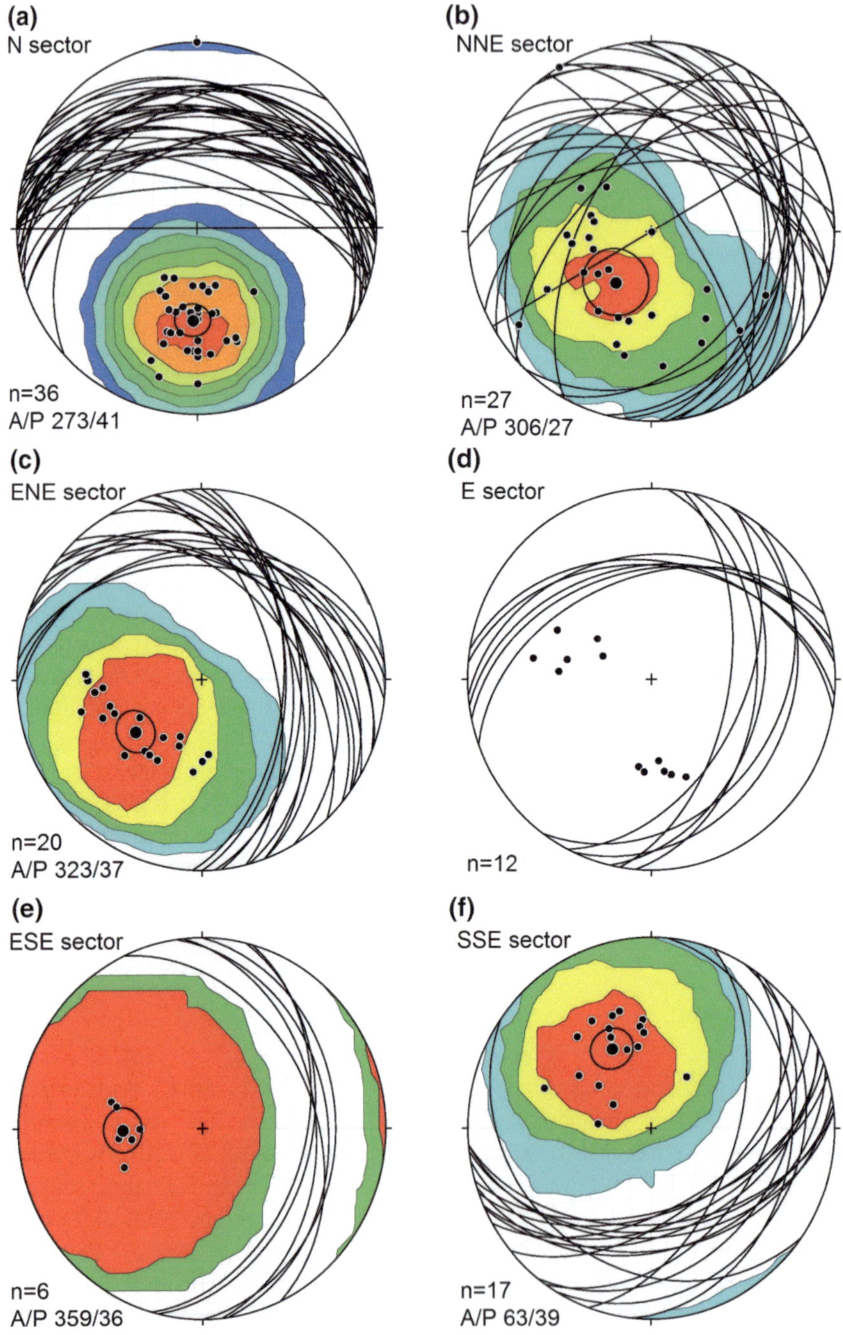

Fig. 5 Stereographic projections of bedding planes (S_o) in sandstones exposed along the rim crest of the Ramgarh structure, the sectors selected along the rim (Fig. 3) is purely geographical after the convention described in Misra et al. (2010), for detail see text

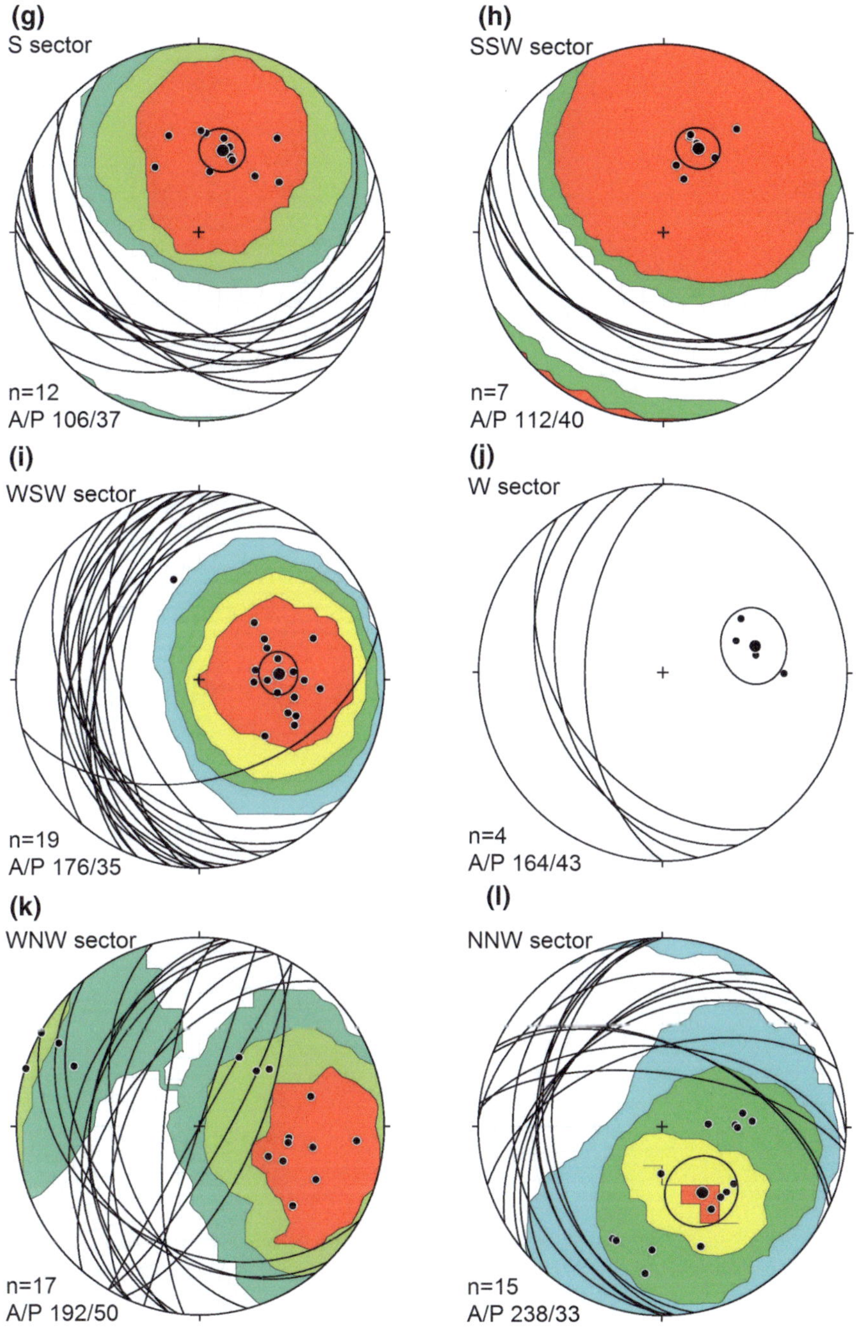

Fig. 5 (continued)

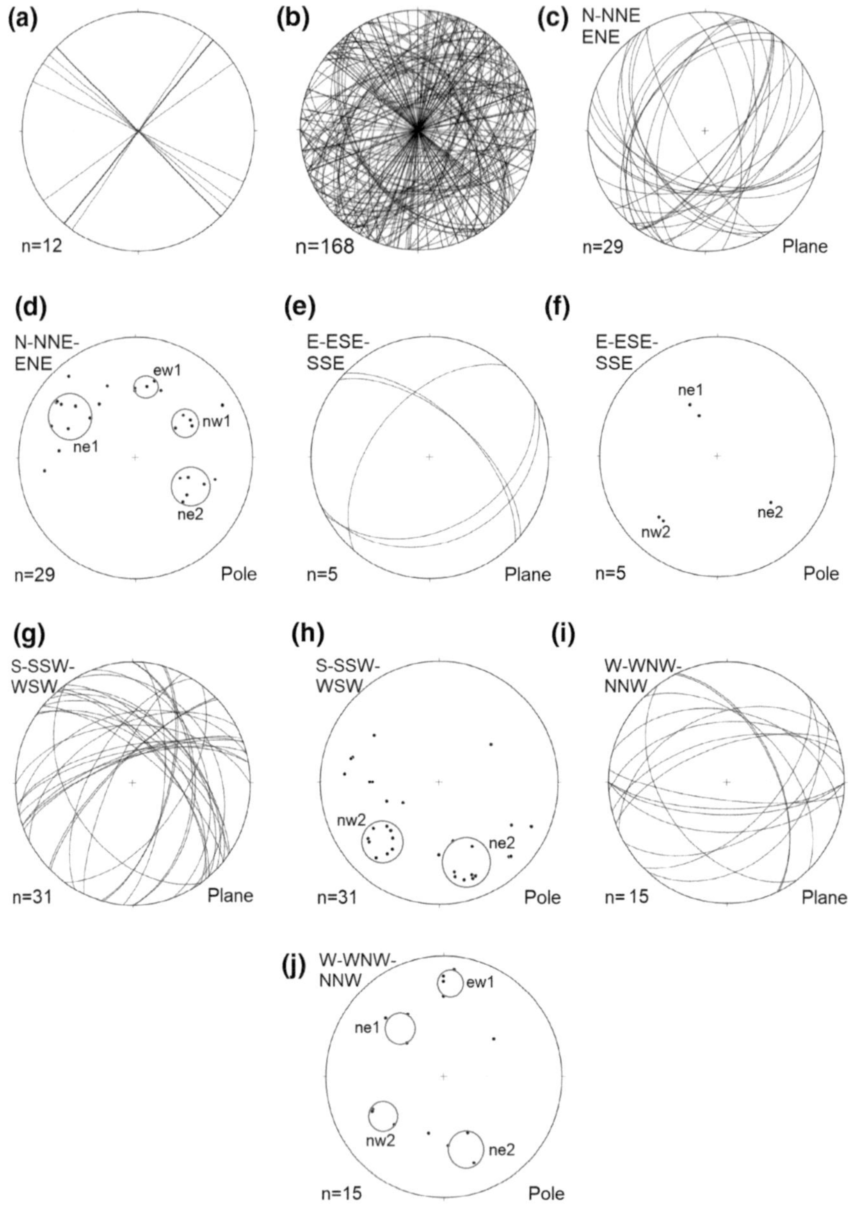

Fig. 6 Stereographic projections of **a** fracture planes in flat-lying sedimentary country rock(s) around Ramgarh structure, and **b**, **c**, **e**, **g**, **i** fracture planes developed in the rim crest sandstones of Ramgarh structure, and **d**, **f**, **h**, **j** projections of their respective poles, which are grouped according to their strike directions, e.g., northeast (ne), northwest (nw) and east-west (ew), note Ramgarh structure is associated with numerous fractures, which are both vertical and dipping, different geographic sectors of Ramgarh structure rim after Fig. 3

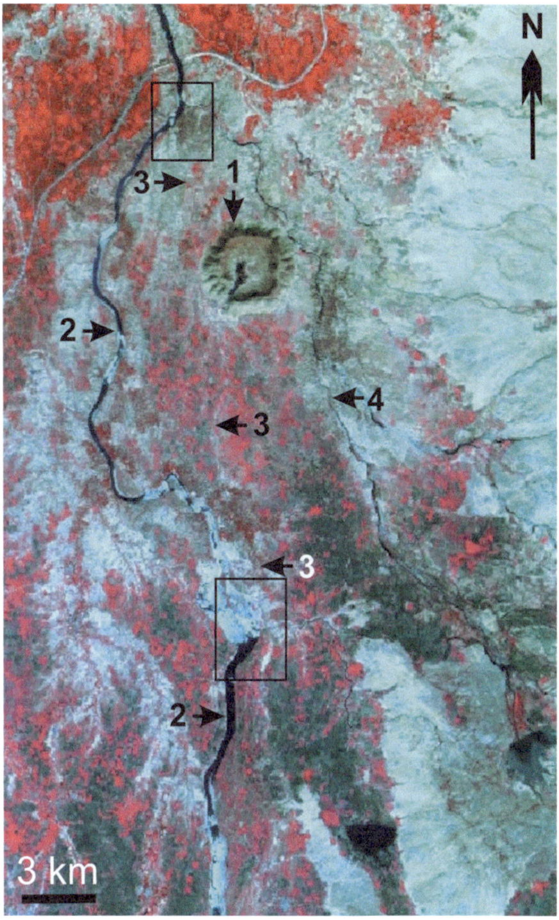

Fig. 7 False colour ASTER image (resolution ~15 m) showing modification of drainage pattern adjacent to Ramgarh structure at present, abbreviations: 1—Ramgarh structure, 2—present course of Parvati River, 3—palaeo-channel of Parvati River, 4—present course of Kul River, note shift of Parvati River from a relatively smooth palaeo-channel adjacent to Ramgarh structure to an irregular channel to W at present, the points where palaeo-channel of Parvati River meets with its present course are shown in box

(Prasad 1984), and in area surrounding the Ramgarh structure (Figs. 4b and 6a). The evidence of extensive dextral shear along the NE-SW fracture plane cross-cutting the Ramgarh structure, and following minor sinistral shear along the E-W fracture planes are only seen within the Ramgarh structure (Figs. 2a and 3). All these brittle deformation features suggest that the sedimentary country rocks within the Ramgarh structure could have experienced a sudden shock that generated new fractures and reactivated the pre-existing fractures promoting limited slip.

The quaquaversal dips of the country rock sandstone along the rim of the Ramgarh structure (Fig. 5), along with its extreme brittle deformations resemble in several aspects to those observed in the Lonar impact Crater, India (Kumar 2005; Misra et al. 2010). The Ramgarh structure is ~ rectangular (Figs. 2a and 3), which is similar in shape to the Arizona Crater, USA (Grieve et al. 1988). The present diameter/depth ratio of the Ramgarh structure is ~12 (measured from Google Earth Imagery 2017), which is within the range (10–20) of the terrestrial complex asteroid

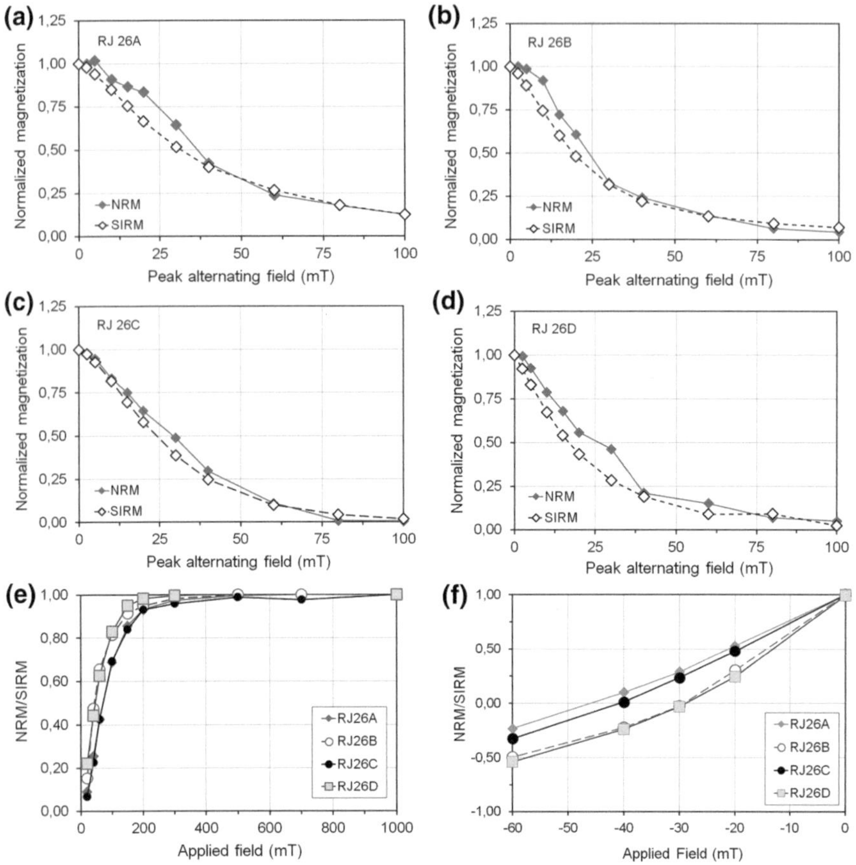

Fig. 8 **a–d** Alternating field demagnetization of NRM and SIRM of sub-samples from RJ26 glassy sample recovered from Ramgarh structure (Fig. 3), note position of NRM decay curves above SIRM decay curve for lower range of AF demagnetization field. **e** IRM acquisition curves of all four glassy samples showing saturation at higher field of ~500 mT. **f** IRM backfield curves of glassy sub-samples

impact craters (cf. Smith 1971; Koeberl and Sharpton 2017). Like other terrestrial complex impact crater, the Ramgarh structure has a prominent conical central peak (Fig. 4o) (also Grieve et al. 1988; Master and Pandit 1999), which consists of sandstone and has a present height of ~6 m (Misra et al. 2008a). The subsurface presence of this central peak below the present alluvium is also modelled based on borehole data (Ramasamy 1987). Hence it can be modelled that the extreme brittle deformation of the country rocks observed within the Ramgarh structure, the development of shatter cone like structure on the wall of the Ramgarh structure (Fig. 4n) and central uplift within the structure (Fig. 4o) were resulted due to a sudden shock effect caused by asteroid impact (cf. French 1998).

Our previous study on magnetic particles recovered from the Ramgarh structure (Das et al. 2009) and our present study on four small (mm-size) sub-samples from a cm-size magnetic glassy rock, recovered from the rocky debris lying on the outer slope of western rim of this crater (Figs. 3 and 8), confirm that the glassy sample/particles from this structure have very high NRM ($\sim$2–19 Am^{-1}) and high REM ($\sim$7–145%) indicating a high magnetic field during their formation. The REM values provide an estimation of the paleomagnetic field (Kletetschka et al. 2003) and a ratio close to 1.5% indicates an Earth-magnetic field (Gattacceca and Rochette 2004; Kletetschka et al. 2004; Yu 2006). The low-field processes other than thermal remanent magnetization, e.g., viscous or chemical remanent magnetization, yield lower REM values for the same paleofield (Fuller et al. 1988), whereas the high-field processes, e.g., lightning-induced or plasma-induced magnetization, yield REM values above 10% (Wasilewski and Dickinson 2000). As the lightening alone cannot produce the observed ring-like shape of the Ramgarh structure and its brittle deformations, the increase of initial NRM intensity of the Ramgarh magnetic particles (Das et al. 2009, and Fig. 8a–d) could only be acquired by any secondary magnetization process such as shock metamorphism by asteroid impact (cf. Gold and Soter 1976; Schultz and Srnka 1980). This impact also led to brittle deformations of sedimentary country rocks. The role of the impact was to create a high magnetic field during the formation of the Ramgarh structure, which perhaps covers an area $>\sim$20 km^2 during its formation (cf. Crawford and Schultz 1988, 1999). Hence, our present observations on structural geology and rock magnetics, and previous observations on petrography (Misra et al. 2008b; Das et al. 2011; Purohit and Sisodia, 2013) and geochemistry (Misra et al. 2008b; 2013) have enhanced the possibility of an asteroid impact origin for the Ramgarh structure.

One of the against-hypothesis on the asteroid impact origin for the Ramgarh structure could be the general absence of PDF in the sandstones in this structure (cf. Reimold et al. 2006). However, a detailed observation under the optical microscope shows that the sandstones from this structure is extremely fine grained (grain size $\leq$200 μm) (Fig. 10 in Appendix), and such fine-grained rocks would inhibit the formation of PDF than their coarse-grained counterparts (cf. French and Koeberl 2010 and references therein). The absence of any PDF under the microscope, therefore, does not negate the asteroid impact origin of the Ramgarh structure (cf. Chen et al. 2017; Melosh 2017). In fact, a few of quartz grains within the bed rock sandstone from the rim of the Ramgarh structure show one set very closed-spaced ($\leq$20 μm) fractures (PFs) (Das et al. 2011), which could favour the asteroid impact origin of this structure (cf. French 1998). Further, more recent observation of Purohot and Sisodia (2013) on the universal stage measurements probably identified a few relict PDFs within the bed rock sandstone from this structure.

The quaquaversal dips of the rim crest sandstones are extremely disturbed in the NNE, ENE and E sectors, and in the WNW and NNW sectors (Fig. 5). In the former three sectors, the dips of S_o planes in sandstones vary from N to E, while in the latter two sectors it varies in wide range from SW to NNE. The deviation of dips of the rim crest sandstones from its typical quaquaversal dips in the NNE, ENE and E sectors is definitely due to the effect of the dextral NE-SW faulting that transected and partly displaced the structure rim (Figs. 2a and 3) (cf. Sharma 1973). However, the reason behind the same type of deviation of dips for the rim sandstone in the WNW and NNW sectors is not clearly understood in the present study. It could be due to another NW-SE dextral faulting that was early in sequence to that of the major NE-SW fault and cross-cut the Ramgarh structure (Ramasamy 1987). Hence the rectangular shape of the Ramgarh structure (Figs. 2a and 3) is a combined result of slip along dextral NW-SE, dextral NE-SW and sinistral E-W faults in sequence. It appears that the NE-SW and NW-SE fracture sets within the surrounding Vindhyan Supergroup were reactivated within the Ramgarh structure due to the shock waves generated by the impact (cf. French 1998). A similar example could be the Arizona Crater, USA. The rectangular shape of this crater was resulted due to impact on sedimentary target rocks with pre-existing joint sets and faults (Poelchau et al. 2009).

The 'rheid flow of shale' model of Ramasamy (1987) to explain the origin of the Ramgarh structure was mainly based on the morphology of the folds observed in shale-siltstone sequence in central borehole between 20 and 210 m depth. According to his description, these shale and siltstone layers show intricate disharmonic folds. The limbs and axial surfaces of the folds are parallel and vertical with either sharp or rounded hinges. These folds are refolded with sub-horizontal axial surfaces. To explain this folded sequence, Ramasamy (1987) introduced the concept of vertical injection of rheid shales in the axial portion of the Ramgarh structure. However, no mechanism is suggested to explain why this vertical injection of solid rocks took place at the central part of this structure. It is understood from the impact cratering process that during the end part of excavation stage and the following modification stage of evolution of complex impact crater, the solid materials flow vertically upward close to the central uplift of the impact crater (French 1998). This type of material flow could thus explain the vertical folding developed in the shale and siltstone in the central borehole within the Ramgarh structure.

The Ramgarh structure is situated on the palaeo-channel of Parvati River (Figs. 2c and 7) and hence its formation must has post-dated the formation of this river. It appears that the impact has resulted the shift of river courses of Parvati and its tributary adjacent to Ramgarh structure. Impact on river channel, hence, enhances the possibility of involvement of surface water during the evolution of this structure. Our previous microprobe analyses on spherules recovered from this structure consistently show total major oxide around 80 wt% supporting possible involvement of volatiles during the formation of Ramgarh structure (Misra et al. 2008b). Therefore, based on this observation along with our present observation on intense activity of Mn-rich hydrothermal solution along the bedding planes of sandstones and some fracture planes cross-cutting these sandstones within the Ramgarh structure (Fig. 4k–m), it can be modelled that the Ramgarh impact cratering took place essentially under hydrous environment.

6 Conclusions

(a) The flat-lying, nearly undeformed Vindhyan Supergroup of sedimentary rocks surrounding to the Ramgarh structure show only two sets of vertical fracture trending NE-SW and NW-SE. In contrast, the sedimentary country rocks within and adjacent to the Ramgarh structure show extreme brittle deformation including vertical radial fracture in numerous directions, and moderately dipping fractures trending NE-SW and NW-SE.
(b) The ring-like shape of the Ramgarh structure with a central conical peak, the quaquaversal dips of its rim crest sandstones, along with its present diameter/depth ratio ($\sim$12) suggest that this structure could have formed by asteroid impact. The impact-induced shock was responsible for the extreme brittle deformation of the sedimentary country rocks within and adjacent to this structure and reactivation of some pre-impact fracture plane in the country rocks.
(c) The present rock magnetic studies on sub-samples from a glassy silicate piece, along with our previous observation on magnetic particles recovered from the alluvium of the Ramgarh structure (Das et al. 2009) suggest that these magnetic substances have very high NRM ($\sim$2–19 Am^{-1}) and REM ($\sim$7–145%) indicating the presence of a high magnetic field during their formation, much higher than the ambient Earth's magnetic field.
(d) A natural phenomenon that could generate a unique ring-shaped deformation structure on a monotonously flat-lying, undeformed sedimentary country rock as well as a high magnetic field in and around this structure is a hypervelocity asteroid impact. Many previous observations on petrography (Das et al. 2011; Purohot and Sisodia 2013) and geochemistry (Misra et al. 2008b; 2013) published in the form of extended abstracts also support this conclusion.
(e) The rectangular shape of the Ramgarh structure, which is similar to that of the Arizona Crater, USA, was resulted due to post-impact dextral movement along NW-SE fault, followed by dextral movement along NE-SW fracture, and sinistral movement along E-W fracture in minor scale.

(f) Observation on remote sensing images confirms that the asteroid impact took place on the palaeo-channel of Parvati River, which is now displaced towards W.

Acknowledgements The first author (S. M.) is grateful to PLANEX, Indian Space Research Organization, and NRF, South Africa (grant no. 91089) for supporting this research work. Special thanks to Anand Dube of India, for helpful guidance to the first author during the field work, to Horton Newsom of USA for his continuous encouragement during the progress of this research, and to Tesfaye Kidane for helping in stereoplot software. We are indebted to Dr. Soumyajit Mukherjee and an anonymous reviewer for their constructive comments on the early version of the manuscript.

Appendix

See Figs. 9 and 10.

Fig. 9 Hand specimen photograph of cm-sized silicate glassy piece recovered from within the rocky debris deposited on the outer wall of western rim of Ramgarh structure (diameter of coin shown as scale is 22 mm)

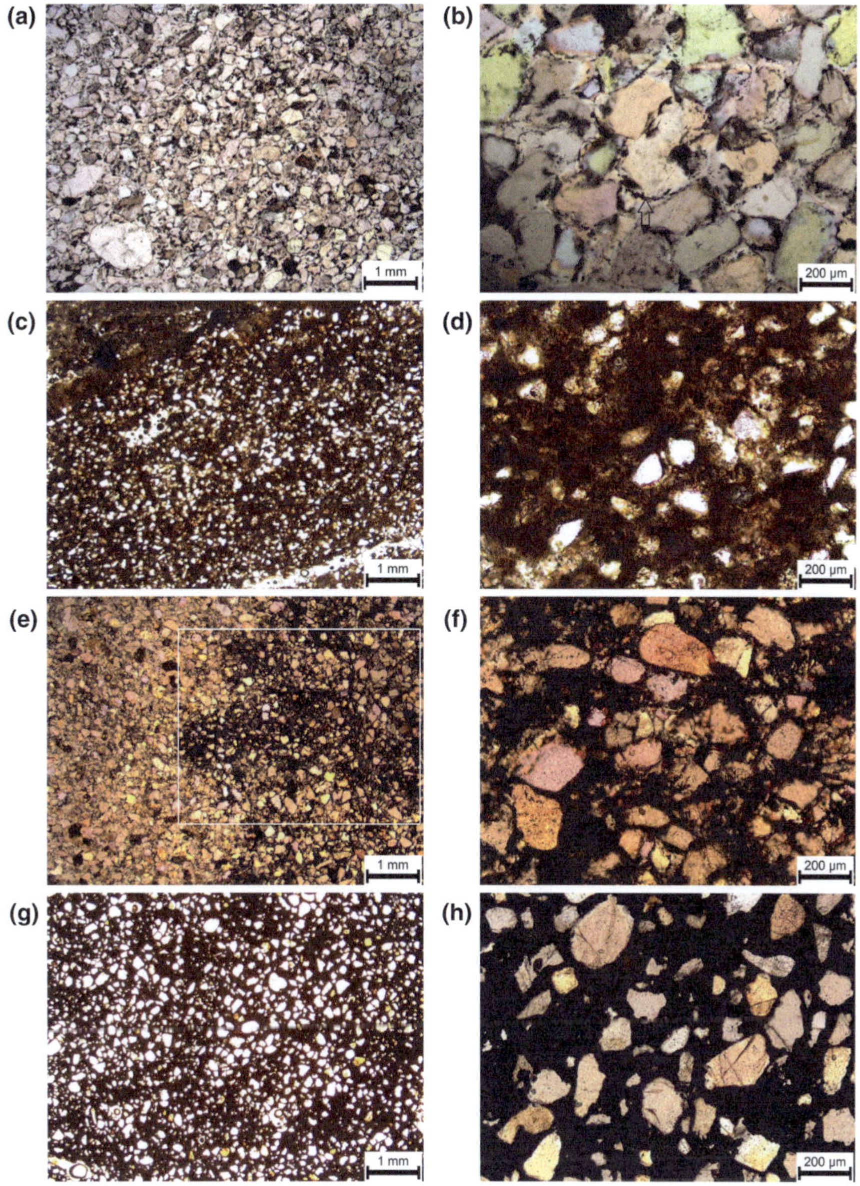

Fig. 10 **a**, **b** Sandstone sample from Kul River section N of Ramgarh structure (location R 38), the dark matters among the grains (shown by arrow in **b**) are manganese encrustation. **c, d** impure limestone from Ramgarh structure (R 18) containing numerous fine grain quartz sand particles. **e**, **f** percolation of manganese solution within the sandstone (shown in box) sample from the rim of Ramgarh structure (R 11). **g**, **h** manganese encrustation containing numerous sub-angular to sub-rounded quartz grains (R 15). GPS locations of samples: R 11: 25° 19.378′N, 76° 37.933′E; R 15: 25° 19.582′N, 76° 37.570′E; R 18: 25° 19.406′N, 76° 37.071′E; R 38: 25° 22.929′N, 76° 36.961′E. Note fineness of grain size of sandstones occurring on rim crest of Ramgarh structure

References

Ahmad K (1958) Paleogeography of Central India in the Vindhyan period. Geological Suvey of India Records 87, 531–548

Ahmed N, Bhardwaj BD, Sajid HA, Hasnain I (1974) Ramgarh meteorite crater. Current Science 43, 598

Auden JB (1933) Vindhyan sedimentation in the Son valley, Mirzapur district. Geological Survey of India Memoirs 62, 141–250

Balasundaram MS, Dube A (1973) Ramgarh structure, India. Nature 242, 40

Bose PK, Sarkar S, Chakrabarty S, Banerjee S (2001) Overview of the Meso- to Neoproterozoic evolution of the Vindhyan basin, central India. Sedimentary Geology 41–142, 395–419

Chen J, Elmi C, Goldsby DL, Gieré R (2017) Generation of shock lamellae and melting in rocks by lightning-induced shock waves and electrical heating. Geophysical Research Letters 44, 8757–8768

Crawford AR (1972) Possible impact structure in India. Nature 237, 96

Crawford DA, Schultz PH (1988) Laboratory observations of impact–generated magnetic fields. Nature 336, 50–52

Crawford DA, Schultz PH (1999) Electromagnetic properties of impact-generated plasma, vapor and debris. International Journal of Impact Engineering 23, 169–180

Das PK, Misra S, Basavaiah N, Newsom H, Dube A (2009). Rock magnetic evidence of asteroid impact origin of Ramgarh structure. India. In: 40th Lunar and planetary science conference. Abstract no. 1466

Das PK, Misra S, Newsom HE, Sisodia MS (2011) Possible planer fractures, coesite, and accretionary lapilli from Ramgarh structure, India: new evidence suggesting an impact origin of the crater. In: 42nd Lunar and planetary science conference. Abstract no. 1294

French BM (1998) Traces of catastrophe: a handbook of shock-metamorphic effects in terrestrial meteorite impact structures. In: Lunar and planetary institute contribution series. vol 954, p 120

French BM, Koeberl C (2010) The convincing identification of terrestrial meteorite impact structures: what works, what doesn't, and why. Earth-Science Reviews 98, 123–170

Fuller M, Cisowski S, Hart M, Haston R, Schmidtke E, Jarrard R (1988) NRM:IRM(S) demagnetization plots; an aid to the interpretation of natural remanent magnetization. Geophysical Research Letters 15, 518–521

Gattacceca J, Rochette P (2004) Toward a robust normalized magnetic paleointensity method applied to meteorites. Earth and Planetary Science Letters 227, 377–393

GLCF (2007) Global land cover facility. University of Maryland, USA. Available online at http://glcf.umiacs.umd.edu

Gold T, Soter S (1976) Cometary impact and the magnetization of the Moon. Planetary and Space Science 24, 45–54

Grant JA (1999) Evaluating the evolution of process specific degradation signatures around impact craters. International Journal of Impact Engineering 23, 331–340

Graup G (1981) Terrestrial chondrules, glass spherules and accretionary lapilli from the suevite, Ries Crater, Germany. Earth and Planetary Science Letters 55, 407–418

Grieve RAF, Wood CA, Garvin JB, Mclaughlin G, McHone JF (1988) Astronaut's guide to terrestrial impact craters. Lunar and planetary institute technical report 88-03, Houston, p 89

Kletetschka G, Kohout T, Wasilewski PJ (2003) Magnetic remanence in the Murchison meteorite. Meteoritics and Planetary Science 38, 399–405

Kletetschka G, Acuna MH, Kohout T, Wasilewski PJ, Connerney JEP (2004) An empirical scaling law for acquisition of thermoremanent magnetization. Earth and Planetary Science Letters 226, 521–528

Koeberl C, Sharpton VL (2017) Terrestrial impact craters, 2nd edn. www.lpi.usra.edu/publications/slidesets/craters

Koeberl C, Brandstätter F, Glass BP, Hecht L, Mader D, Reimold WU (2007) Uppermost impact fall back layer in the Bosumtwi crater (Ghana): mineralogy, geochemistry, and comparison with ivory coast tektites. Meteoritics and Planetary Science 42, 709–729

Kumar PS (2005) Structural effects of meteorite impact on basalt: evidence from lonar crater. Journal of Geophysical Research 110, B12402

Kumar J, Negi MS, Sharma R, Saha D, Mayor S, Asthana M (2011) Ramgarh magnetic anomaly in the Chambal valley sector of Vindhyan basin: a possible meteorite impact structure and its implications in hydrocarbon exploration. In: American Association of Petroleum Geologists, Search and Discovery. Article #80145

Lowrie W, Fuller M (1971) On the alternating field demagnetization characteristics of multidomain thermoremanent magnetization in magnetite. Journal of Geophysical Research 76, 6339–6349

Mallet FR (1869) On the Vindhyan series, as exhibited in the North-western and Central Province of India. Memoir of Geological Survey of India 7(Part 1), 129

Malone SJ, Meert JG, Banerjee DM, Pandit MK, Tamrat E, Kamenov GD, Pradhan VR, Sohl LE (2008) Paleomagnetism and detrital zircon geochronology of the Upper Vindhyan sequence, Son valley and Rajasthan, India: a ca. 1000 Ma closure age for the Purana basins? Precambrian Research 164, 137–159

Master S, Pandit MK (1999) New evidence for an impact origin of the Ramgarh structure. Meteoritics and Planetary Science 34, 4

McElhinny MW, McFadden PL (2000) Paleomagnetism: continents and oceans. In: International geophysics series, vol 73. Academic Press, San Diego, CA, p 386. ISBN: 0124833551

Melosh HJ (2017) Impact geologists, beware! Geophys Res Lett 44, 8873–8874

Misra S, Dube A, Srivastava PK, Newsom HE (2008a) Time of formation of Ramgarh crater, India-constraints from geological structures. In: 39th Lunar and planetary science conference. Abstract. no. 1502

Misra S, Lashkari G, Panda D, Dube A, Sisodia MS, Newsom HE, Sengupta D (2008b) Geochemical evidence for the meteorite impact origin of Ramgarh structure, India. 39th Lunar and planetary science conference. Abstract. no. 1499

Misra S, Arif M, Basavaiah N, Srivastava PK, Dube A (2010) Structural and anisotropy of magnetic susceptibility (AMS) evidence for oblique impact on terrestrial basalt flows: Lonar crater, India. Bulletin of Geological Society of America 122, 563–574

Misra S, Panda D, Ray D, Newsom H, Dube A, Sisodia MS (2013) Geochemistry of glassy rocks from Ramgarh structure, India. In: 44th Lunar and planetary science conference. Abstract. no. 1020

Newsom H, Gasnault O, Le Mouelic S, Mangold N, Le Deit L, Wiens R, Anderson R, Edgar L, Herkenhoff K, Johnson JR, Bridges N, Grotzinger JP, Gupta S, Jacob S (2016) Long distance observation with the ChemCam remote micro-imager: mount sharp and related deposits on gale crater floor? Geological Society of America, Denver, 25–28 Sept 2016

Oldham T (1856) Remarks on the classification of the rocks of central India resulting from the investigation of the Geological Survey. Journal the Asiatic Society, Calcutta 25, 224–256

Poelchau MH, Kenkmann T, Kring DA (2009) Rim uplift and crater shape in meteor crater: effects of target heterogeneities and trajectory obliquity. Journal of Geophysical Research 114, E01006

Prasad B (1984) Geology, sedimentation and palageography of the Vindhyan Supergroup, Southeast Rajasthan. Geological Survey of India Memoirs 116, 1–107

Purohot V, Sisodia MS (2013) Universal-stage measurements of planar deformation features in shocked quartz grains recovered from Ramgarh structure. In: 44th Lunar and planetary science conference. Abstract. no. 1151

Ramasamy SM (1987) Evolution of Ramgarh dome, Rajasthan: India. Records of Geological Survey of India 113, 13–22

Rana S, Agrawal V (2016) Microscopic evidences for the impact origin of Ramgarh structure, Rajasthan, India. Journal of Indian Geophysical Union 20, 544–550

Reimold WU, Trepmann C, Simonson B (2006) Comment on "Impact origin of the Ramgarh structure, Rajasthan: some new evidences by M. S. Sisodia, G. Lashkari and N. Bhandari. Journal of Geological Society of India, v. 67, pp. 423–431". Journal of Geological Society of India 68, 561–563

Schultz PH, Srnka LJ (1980) Cometary collision on the moon and mercury. Nature 284, 22–26

Sharma HS (1973) Ramgarh structure, India. Nature 242, 39–40

Sisodia MS, Lashkari G (2003) Ramgarh structure, Rajasthan, India: meteorite impact evidences. Workshop on Impact cratering: bridging the gap between modeling and observations, Houston. Abstract no. 8008

Sisodia MS, Lashkari G, Bhandari N (2006a) Impact origin of the Ramgarh structure, Rajasthan: Some new evidences. Journal of Geological Society of India 67, 423–431

Sisodia MS, Lashkari G, Bhandari N (2006b) Reply to "The comment on Impact origin of the Ramgarh structure, Rajasthan: Some new evidences by W. U. Reimold, C. Trepmann and B. Simonson". Journal of Geological Society of India 68, 561–563

Smith EI (1971) Determination of origin of small lunar and terrestrial craters by depth diameter ratio. Journal of Geophysical Research, 76, 5683–5689

Tandon SK, Pant CC, Casshyap SM (1991) Sedimentary basins of India-Tectonic context. Gyanodaya Prakashan, Nainital

Vernooij MGC, Langenhorst F (2005) Experimental reproduction of tectonic deformation lamellae in quartz and comparison to shock-induced planar deformation features. Meteoritics and Planetary Science 40, 1353–1361

Wasilewski P, Dickinson T (2000) Aspects of the validation of magnetic remanence in meteorites. Meteoritics and Planetary Science 35, 537–544

Yu YJ (2006) How accurately can NRM/SIRM determine the ancient planetary magnetic field intensity? Earth and Planetary Science Letters 250, 27–37

Geology, Structural Architecture and Tectonic Framework of the Rocks of Southern Lalitpur District Uttar Pradesh, India: An Epitome of the Indian Peninsular Shield

G. K. Dinkar, A. R. Bhattacharya, A. K. Verma and Pankaj Sharma

1 Introduction

The southern part of Uttar Pradesh is geologically known for exposing a part of the Indian peninsular shield. Both basement and cover rocks are exposed at several areas. The Lalitpur district of this region typically exposes several geological elements of the peninsular shield. The southern part of Lalitpur district (Fig. 1) epitomizes the peninsular shield to a larger extent. This area has been studied by various workers mainly from the viewpoints of geology, rock types, and mineral resources. The early studies were confined to identifying the major rock groups. Later studies focused on identifying various rock types together with the associated metasedimentary sequences. Recent studies focus amongst others, on mineralogical, geochemical and geochronological studies to understand the evolution of the ancient crust and crustal growth in the context of the peninsular shield. Despite geological studies on several aspects, detailed lithostratigraphic and structural-tectonic studies of south Lalitpur district are scanty. This paper fills up this gap.

2 Geological Set up

The geological setup of southern Lalitpur district is intimately related to that of southern U.P. The area has been studied by various workers mainly from the viewpoints of geology, rock types, and mineral resources. Our knowledge on the geology of south Lalitpur developed in several phases during the last 5 or 6 decades.

G. K. Dinkar · A. R. Bhattacharya (✉) · A. K. Verma · P. Sharma
Centre of Advanced Study in Geology, University of Lucknow, Lucknow 226007, India
e-mail: arb65k@gmail.com

S. Mukherjee (ed.), *Tectonics and Structural Geology: Indian Context*,
Springer Geology, https://doi.org/10.1007/978-3-319-99341-6_12

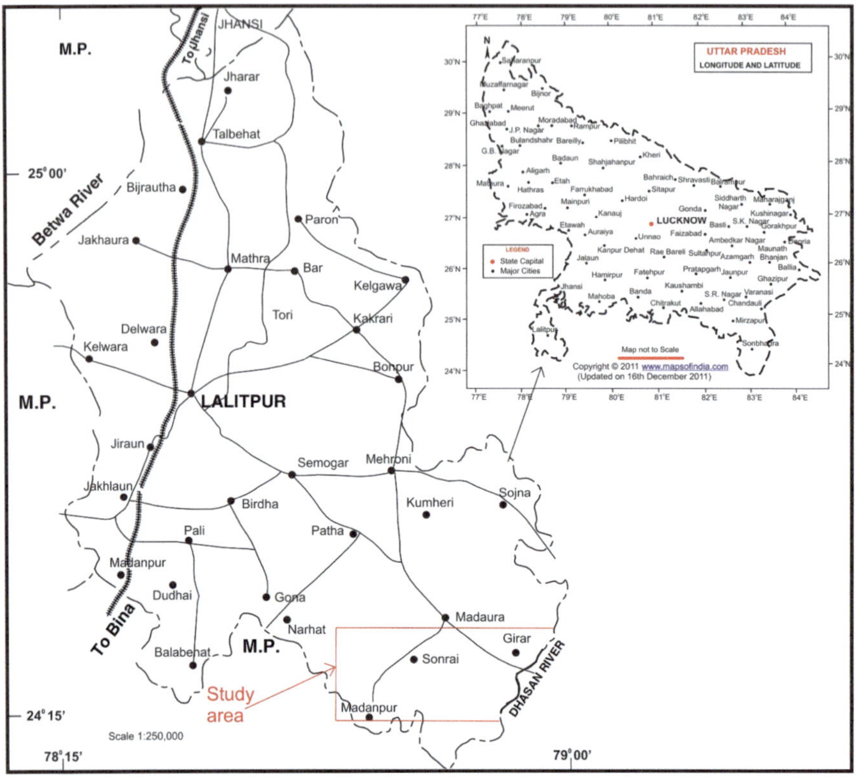

Fig. 1 Map showing the location of the study area. The inset shows the state Uttar Pradesh

On a regional basis, the area has been referred to, directly or indirectly, by several workers. Medlicott (1859) was possibly the first to make a geological study of the rocks of the Bundelkhand region. He made a detailed study of the Banded Iron Formation (Baraitha) and various schistose rocks from the southwestern part of the Bundelkhand region and grouped them as crystallines of Bundelkhand. The metasedimentary belt around Gwalior was studied by Hacket (1870). Much later, Heron (1935) studied the crystalline rocks from several areas of the Bundelkhand region and he referred these rocks as "Bundelkhand Granite".

During the above period, study remained confined only to some general aspects of the Bundelkhand region. Thereafter, the trend of study gradually shifted to some specific aspects of the region. Srivastava (1951–1952) was one of the earlier workers who carried out detailed geological and mineralogical investigations in several areas of the Bundelkhand region and published his new findings on geological and petrological characters of the Bundelkhand granites. He also reported pyrophyllite deposits in some parts of the Bundelkhand massif. By that time, the granitic rocks of the Bundelkhand massif became the major subject matter of study mainly because a variety of granitic rocks were reported till then. Jhingran (1958) identified two major types of granites: pink granite and grey granite and he suggested that the former (pink granite) is younger than the latter. He also identified ten types of granites within the great batholithic massif on the basis of grain size, colour of feldspar and presence or absence of ferromagnesian minerals. In addition to these, he also reported gneisses, quartz–reefs, basic dikesand tuffaceous serpentinite rocks within the massif. Saxena (1961) suggested a lithostratigraphic correlation of the metasediments of the Bundelkhand massif with the middle Dharwarian rocks of South India and the Bundelkhand granitic activity as equivalent of the Closepet granite.

During the nineteen hundred seventies, the different rock types of the Bundelkhand region including lithostratigraphic sequences of some areas came to limelight. Prakash et al. (1975) carried out a detailed lithostratigraphic study of the southwestern part of Bundelkhand massif and distinguished three tectonic blocks in this massif: a northern block (Jhansi–Lalitpur) consisting of various types of granitoids with enclaves of metasedimentary rocks, a southern block (south of Lalitpur) comprising of Bundelkhand granitoids including the low grade metasediments (Mehroni Group) together with the overlying Vindhyan sediments, and a central block mainly exposing the pink, massive granites. Misra and Sharma (1975) identified metasedimentary rocks within the Bundelkhand complex and suggested that the Bundelkhand rocks can be divided into four formations; from lower to upper these are Kuraicha Formation of high grade metamorphic rocks, Palar Formation with low grade metamorphic rocks, Bundelkhand Granite, and Bundelkhand basic intrusives.

The nineteen hundred eighties witnessed a multifarious progress on several yet untouched aspects on the geology of the region. In this connection, the contributions of Basu (1986, 2001, 2007, 2010) deserve special mention. He carried out extensive field investigations and mapped almost the entire area of the Bundelkhand Craton. He proposed an evolutionary model for the Precambrian crust on the basis of petrological, structural and chemical characteristics of various types of granitoids and basic rocks. He suggested that the metallogenic and tectonometamorphic events of the massif were regionally dominated by plutonic and hypabyssal rocks.

Despite the available information on several aspects of the geology of the region, studies from structural angles were almost lacking. Bhattacharya (1985) estimated tectonic strain for the rocks of the Bundelkhand massif and has shown that the mineralogical behaviour of the rocks has controlled the deformation patterns of the massif. He (Bhattacharya 1986) developed a semi-quantitative method for identifying different fold episodes of deformed terrains and on applying this method to the rocks of the Bundelkhand massif, he identified five different folding episodes for the massif.

On a much larger perspective, Radhakrishna (1989) identified a "Central Indian Tectonic Zone" encompassing the SONATA Belt that marks the junction between the Bundelkhand block in the north and the Peninsular block in the south. Sarkar et al. (1989, 1995, 1997) on the basis of the geochronology and petrology of various rock types of the Bundelkhand complex suggested that the granitic rocks belong to three distinct intrusive phases. They also suggested that the trondhjemitic gneiss containing Rb-Sr isotopes may be considered as 3.5 Ga old (Baghora TTG suite). The older crust of the region (Archaean to Palaeoproterozoic) is comprised of amphibolites, quartzites, BIF, schists, marbles and calc–silicate rocks and that these rocks are intruded by 3.3 Ga old syn-to-late tectonic Na-rich Tonalite-Trondjhemite-Granodiorite (TTG) rocks. The oldest metamorphic events of the Bundelkhand region have been dated at 3.3 Ga, while the oldest relics of younger geneisses may be 2.7–2.5 Ga old (Mondal et al. 2002). Attempts have been made to work out the crustal evolutionary pattern of the Bundelkhand region (Singh et al. 2007; Singh and Dwivedi 2009; Singh 2012).

Our knowledge on the hitherto unknown crustal shear systems of the Bundelkhand massif is definitely a new addition as came into light by the joint work of A. R. Bhattacharya and S. P. Singh (Singh and Bhattacharya 2010, 2017; Bhattacharya and Singh 2013). They have identified the following four different crustal shear systems in the Bundelkhand massif. (1) The E-W shear system: It is mainly confined to the central parts of the massif and is represented by mylonites showing vertical foliation together with a greenstone belt developed at several places with dimensions ranging from about 10 × 30 m to 30 × 50 m. This is an example of crustal scale vertical ductile shear zone and is possibly the first report of vertical crustal shear zone in the Indian subcontinent. (2) The NE-SW shear system: It is represented by the quartz reefs that constitute the most spectacular structure in the Bundelkhand terrain. The quartz reefs constitute strike-slip dominated vertical to subvertical shear zones with dominant sinistral sense of shear and they represent Proterozoic crustal scale extensional events in the Bundelkhand massif. (3) The NW-SE shear system: These are in the form of a number of minor basic igneous dikes emplaced in the fractures developed due to extensional processes followed by strike-slip shearing. (4) N-S shear system: This is a major tectono–magmatic event of the Bundelkhand craton and is represented by N-S trending quartz and quarto-feldspathic veins. This is the youngest shear system that cross-cuts all the earlier (E-W, NE-SW and NW-SE) shear systems.

In the context of vertical shear zones of the Bundelkhand massif, it may be mentioned here that elsewhere in the Indian subcontinent, shear zones with sub-vertical foliations are reported from Karakoram Shear Zone (Mukherjee 2011) and vertical brittle shear zones from trap rocks near Mumbai (Misra et a. 2014).

Recently, Dinkar (2016) carried out a detailed study on the structure and deformation patterns of the rocks of the area around Sonrai. He presented a geological and a structural map of the area both on the 1:50,000 scale and worked out the structure and structural evolution of the rocks of the area

In a nutshell, south Lalitpur district is characterized by three different geological subdivisions. The northern part is represented by the Bundelkhand Gneissic Complex, the central part by the rocks of the Bijawar Group and the southern part by the Vindhyan Supergroup. In addition, there are a few minor rock units also that fall in any of these three major subdivisions. The geological information brought to light through this paper, and the results arrived at, should thus advance our knowledge on the Indian peninsular shield.

3 The Present Work

About 460 km^2 has been mapped on 1:50,000 scale covering the area around Digwar in the northwest, Balla in the northeast, Madanpur in the southwest and Kurrat in the southeast. The study area (Fig. 1) encompasses the river valley areas of the Rohini River in the west, Bandai River in the central part and Dhasan River in the east.

4 Physiography

The physiography of the southern Lalitpur district is rather typical from the remaining parts of the southern Uttar Pradesh. The main reason is that in this part the physiography appears to be related to the geology of the area. The general surface level of flat lying regions is 400 m above the mean sea level with hillocks rising up to about 566 m from the surrounding areas. The southern part (Fig. 2) of the area represents plateaus and ridges while the vast expanse of the central (Fig. 2) and northern parts is mostly alluvial exposing scattered hillocks. In general, we identify three physiographic subdivisions (Fig. 2). The northern subdivision is mostly alluvial exposing scattered hillocks. The central subdivision mainly includes E-W trending arcuate to elongated hills. The southern part of the area is occupied by plateau with E-W trending ridges rising up to about 566 m.

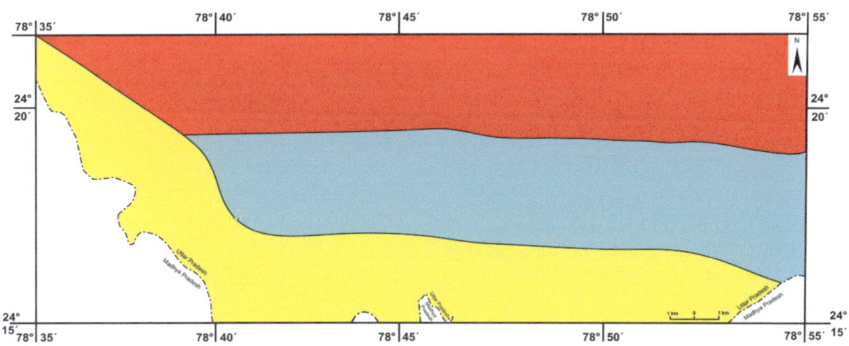

Fig. 2 Photographs showing the physiographic features of the southern Lalitpur district. Our studies suggest that the area shows three different geological subdivisions: the northern subdivision exposing the Bundelkhand Gneissic Complex, the central subdivision exposing a sedimentary/metasedimentary belt consisting of the Bijawar Group of rocks while the southern one exposes the rocks of the Vindhyan Supergroup. Each of these geological subdivisions has been observed to show its typical physiography. The northern subdivision is mostly alluvial exposing scattered hillocks. The central subdivision mainly includes E-W trending arcuate to elongated, discontinuous hills. The southern one shows plateau with long, E-W trending ridges. The photographs shown here are representative of the latter two types of physiography, i.e. those shown by the rocks of the Bijawar Group (**a**) and those of the Vindhyan Supergroup (**b**)

The above mentioned three physiographic features seem to be related to the geology of the area. Thus the northern part exposes Bundelkhand Gneissic Complex, the central part exposes a sedimentary/metasedimentary belt consisting mainly of the Bijawar Group of rocks and the southern part exposes the rocks of the Vindhyan Supergroup.

5 Lithostratigraphy

From the above discussion, we identify three distinct belts in southern Lalitpur district as described above. The basement rocks are represented by the BGC, the metasedimentary sequence by the Bijawar Group while the cover sediments by the Vindhyan Supergroup. In addition, a few outcrops of Deccan Trap also expose at the southern parts of the area. A generalized lithostratigraphic succession has been presented in Table 1. Detailed lithostratigraphic description and the various rock types of the study area are given below. The lithological characteristics and lithostratigraphic details of the rocks of the area are described here: (A) Bundelkhand Gneissic Complex (BGC), (B) Bijawar Group, and (C) Vindhyan Supergroup (Fig. 3).

Table 1 Generalized lithostratigraphic succession of southern Lalitpur district

Deccan Trap	
〜〜〜	
Vindhyan Supergroup (Sandstone and Shale)	
〜〜〜	
Bijawar Group	Solda Formation (Carbonate, Shale and Sandstone)
	Sonrai Formation (Ferruginous quartzite, Shale and Sandstone)
—— Thrust ——	
Mahrauni Group (Banded hematite quartzite, Conglomirate, Fuchsite quartzite, Grey to green Chlorite schists)	
—— Thrust ——	
Bundelkhand Gneissic Complex (Gneisses, Granite, Amphibolite schist)	

Fig. 3 Photographs showing the physiographic features of the study area. **a** Central part with east-west trending ridges, **b** southern part dominated by plateau. These physiographic subdivisions are characterized by their typical geological attributes (see text)

(A) **Bundelkhand Gneissic Complex**

The rocks of the Bundelkhand massif comprise of a variety of granitoids including pink coloured and fine to coarse grained porphyritic granite along with a variety of metamorphic rocks including gneisses, mylonitised granite gneisses, migmatites and amphibolite schists. The rocks of the BGC commonly show well developed dominantly planar banding represented by melanosomes and leucosomes of variable thicknesses. In the study area, the Bundelkhand massif represents the type area of 'Mehroni Schist Belt' of Pascoe (1950) and the Archaean Mehroni Formation. In addition, basic and ultrabasic rocks together with metavolcanic rocks are also present. Doleritic dykes trending NW-SE are noticed at a few places. Quartz reefs constitute one of the most spectacular features of the Bundelkhand terrain. These reefs occur as linear ridges and typically show NE-SW to ENE-WSW alignments. The Bundelkhand granitic and mafic rocks are broadly considered as intrusives into the basement of the overlying metasediments. In the study area the Bundelkhand massif is cut by a series of younger intrusive rocks, some of which are post-Bijawars (Prakash et al. 1975; Jha et al. 2012). The rocks of the BGC are generally metamorphosed twice. A prograde metamorphism was high grade with the the presence of garnet, silliminite, perthite, diopside, hornblende and anthophyllite and with the absence of muscovite, chlorite, kyanite, actinolite, tremolite, talc and epidote minerals. This was followed by a retrograde metamorphism and is mainly characterized by tonalite-trondhjemit—granodiorite suite of rocks (Fig. 4).

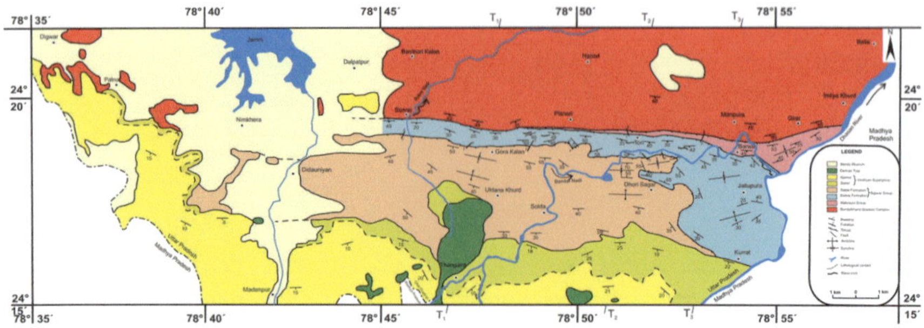

Fig. 4 Geological map of the southern Lalitpur a district, Uttar Pradesh. Modified after Prakash et al. (1975) and Sharma (1982)

In the course of our field study, we have identified a few lithologic associations within this complex as described below.

(i) **Bundelkhand Gneisses**

The Bundelkhand Gneisses (Fig. 5) are exposed in the northern parts of the study area. These rocks are dominantly constituted of moderate to high grade metamorphic rocks. The major rock types include granite-gneiss, migmatite, biotite-gneiss, tonlite-trondhjemite–granodiorite (TTG), hornblende-biotite-gneiss, garnet-sillimanite-gneiss, garnet-sileimanite–cordierite-gneiss and amphibolites. The foliation of the gneissic rocks generally trend ENE-WSW to E-W with dips of 30°–50° southwards.

Fig. 5 Field photograph showing outcrop of gneissic rock. Loc. North of Balla

(ii) Bundelkhand Granites

Bundelkhand granites are exposed to the north of the Sonrai-Girar thrust zone. These rocks mainly include fine-grained granite, coarse-grained porphyritic granite, biotite granite and hornblende granite.

(a) Coarse Grained Pink Granite

Pinkish red coloured coarse-grained granite (Fig. 6) is commonly exposed as a few small patches to the north of Girar and Manpura.

(b) Medium Grained Pink Granite

A few small exposures of medium grained pink granite (Fig. 7) are commonly noticed to the NE of Imalia Khurd, east of Girar. This rock appears to be intrusive into the greygranite (Fig. 7). This rock is a medium grained and is pink to pinkish white in colour. The rock is characterized by small phenocrysts of feldspar.

(c) Grey Granite

A small outcrop of grey quartz (Fig. 8) has been observed north of Sonrai. It is coarse grained and is grey to pinkish grey in colour. The rock is mainly constituted of K-feldspar and biotite.

(d) Leucogranite

Exposures of massive, medium to coarse grained, leucogranite (Fig. 9) are noticed near Girar, Imliya Khurd and at a few places in the western flanks of the Dhasan River.

Fig. 6 Field photograph of the coarse grained pink granite. The diameter of the coin is 2.5 cm. Loc. Near Manpura

Fig. 7 Field photograph of the medium grained pink granite. The diameter of the coin is 2.5 cm. Loc. 200 m east of Imaliya Khurd

Fig. 8 Field photograph of the grey granite. The diameter of the coin is 2.5 cm. Loc. Near Balla

Fig. 9 Field photograph of the leucogranite Loc. Imliya Khurd

(e) **Pegmatites**

5–12 cm wide and 1–10 m long pegmatite veins have been noticed at several places. Most of them trend N-S, though variations are also there at many places. The pegmatite commonly shows large crystal of orthoclase and quartz, e.g. north of Sonrai (Fig. 10). Occasionally, the veins show large crystals of biotite and

Fig. 10 Field photograph of pegmatite showing large crystals of orthoclase and quartz. The diameter of the coin is 2.5 cm. Loc. North of Sonrai

muscovite, e.g. near Nathikhera. It appears that the pegmatites have intruded into the preexisting rocks in phases as based on cross-cutting relations. Basu (1986) has reported three generations of pegmatite veins in the Bundelkhand massif.

(iii) **Mafic and Ultramafic Suite (Metavolcanic Suite)**

A variety of mafic and ultramafic (Fig. 11a, b) rocks are exposed in various parts of the BGC. We assign these rocks to constitute a metavolcanic suite that is mainly represented by metaperidotite, metapyroxinite and coarse to medium grained gabbro. This suite of rocks is intrusive into different types of granites as well as gabbroic rocks. Earlier Prakash et al. (1975) and Sharma (1982) have reported a suite of ultrabasic and basic rocks around madaura and Sonrai. Basu (1986) reported medium grained and dark grey gabbro associated with these ultrabasics to the east of Madaura.

Fig. 11 Field photographs of metavolcanic suite of rocks. The dashed lines show the boundaries of a metavolcanic rock (in each photo) intruded into the host granitic rocks. Loc. NW of Balla

(iv) Quartz Reef

In the Bundelkhand massif, the quartz reefs constitute a spectacular physiographic feature. These reefs constitute the NE-SW trending ridges of quartz. In the study area, the approximate length of these ridges vary from 50 to 200 m. Approximately 400 quartz reefs (individual exposures) are exposed in BGC covering the part of southern Uttar Pradesh. Quartz reefs are sharply exposed as linear ridges/hills with varying thickness from a few meters to the maximum height of 175 m above ground level and thus form prominent physiographic features. The ridges have been observed near Girar, Hansra-Hansri, north of Manpura, Balla and adjoining areas. Commonly the quartz reefs are pinched and swelled. The rocks exposures occasionally show brecciation and shattering such as near Girar and north of Manpura.

The quartz reefs have been interpreted in several ways. Pascoe (1950) described these as quartz veins. Jhingran (1958) describes these as cherty small offshoots having cataclastic and granulated nature. Misra (1969) considers these reefs as product of recrystallisation of quartzites. Saxena (1961) opines the quartz reefs constitute a part of the older rocks of the Bundelkhand massif. Prakash et al. (1975) considered these as dense sheared veins of quartz. The reefs exhibit sharp contacts with the enclosing granites. These reefs are variedly described as a monomineralic rock (Sharma 1982), as mega-veins of quartz occasionally mildly sheared by (Basu 2004) and as giant quartz veins by Pati et al. (2007) and Mondal (2010). Roday et al. (1995) presented a kinemetic model of emplacement of the quartz reefs and subsequent deformation pattern of these reefs in Central Indian Bundelkhand batholiths. Pati et al. (2007) described the geology and geochemisty of these reefs. Bhattacharya and Singh (2013) described these quartz reefs as shear indicators of the Proterozoic crust in the Bundelkhand massif, and suggested that these are associated with the NE-SW strike-slip dominated vertical to sub-vertical shear zones.

(v) Mafic Dike Swarms

Occurrence of mafic dikes is another important physiographic feature of the Bundelkhand massif. These rocks occur as NW-SE trending dike swarms. Some of them also show ENE-WSW trends. In the study area, mafic dikes are noticed around Girar (Fig. 12).

(B) Mahrauni Group

Within the Bundelkhand Gneissic Complex some relics of higher grade metamorphism are occasionally noticed that have been assigned as the Mehroni Group (Prakash et al. 1975; Jha et al. 2012). Prakash et al. (1975) subdivided the Mehroni Group into a Lower Rajaula Formation, followed up by Barwar Formation and Madaura Formation. The Rajaula Formation is mainly constituted of quartz–feldspar biotite gneiss, chlorite schists, and meta-basalts. The Barwar Formation is constituted of banded hematite quartzite, conglomerate, fuchsite quartzite and grey to green chlorite schists. The Madaura Formation mainly includes ultrabasic rocks, gabbro, granites and pegmeatites. However, in the study area only the rocks of the

Fig. 12 Field photograph of a mafic dike. The yellow line demarcates the upper and lower boundaries of the dike. Loc. Near Girar

Barwar Formation represent the Mahrauni Group. In this work we name the Mehroni Group as Mahrauni Group following the recent toposheets. Occurrence of banded hematite quartzite seems to be a characteristic feature of the Barwar Formation in the study area and so also for the Mahrauni Group at least in the present area.

The Barwar Formation is exposed in the eastern part of the study area. Major rock types of the Barwar Formation include banded hematite quartzite (BHQ), conglomerate, fuchsite quartzite, grey to green chloritic schists. These rocks are typically exposed in the Girar and Barwar areas (Fig. 13). The banded ferruginous rocks with their parallel layering appear to be product of marine deposition. During sedimentation, the basin also witnessed sedimentation of clastic constituents in the form of arenaceous and politic sediments. Occurrence of chloritic schists and quartzites indicate that the rocks were subsequently subjected to low grade metamorphism. Folding in the BHQ layers represents deformation of the rocks of this Formation.

The earlier workers have considered the Mahrauni Group as part of the basement. In the present area, only a small part of the Barwar Formation (Mahrauni Group) is exposed to northeast. Both to the north (against BGC) and to the south (against Bijawars), the Barwar Formation bears thrust contact. As such, we consider the Barwar Formation to constitute a thrust slice.

Fig. 13 Field photograph showing outcrop of folded banded hematite quartzite of Barwar Formation. Loc. East of Barwar

(C) Bijawar Group

The Bijawar Group is exposed to the south of the Bundelkhand Gneissic Complex. The contact of this group with the BGC as well as with the Mahrauni Group has been suggested to be unconformable (Prakash et al. 1975; Swaroop and Saxena 1969; Jha et al. 2012). The Bijawar Group consists of a sequence of ferruginous rocks, various types of shale, sandstone and carbonate. This group is unconformably overlain by the rocks of the Vindhyan Supergroup.

On a regional basis, the Bijawar Group generally occurs with an E-W trend in the central part of the study area. The rocks of this group have been interpreted to suggest conditions of shallow carbonate platform with a clastic influx from the nearby landmass, typical for very shallow marine environment with several periods of an intensive tropical weathering (Chaudhuri et al. 1999; Bose et al. 2001; Chakraborty 2006; Banerjee 1982).

The Bijawars exposed in the area are variously classified as (i) the basal rocks of the Sonrai Formation named after the Sonrai village, (ii) the sandstone and pebble conglomerate overlain by carbonates rocks, shale and sandstone named as the Jamuni Carbonate Formation named after the Jamuni river section and (iii) an overlying sequence of shale, quartzite and ferruginous beds named as Solda Formation after the village Solda (Prakash et al. 1975; Swaroop and Saxena 1969; Jha et al. 2012). However in the study area, the Sonrai Formation and the Solda Formation are well represented.

Deposition of different parts of the sedimentary sequence of the Bijawar Group may have taken place directly over the crystalline basement. The unconformable contacts of some of the litho-units of this group suggest repeated periods of vertical movement, which eroded the uplifted areas on which overlying sequence was deposited (Kumar et al. 1990). The end of the evolution of the Bijawar Basin is marked by regional uplift and erosion. Subsequently, Lower Neoproterozoic sedimentary sequences of the Vindhyan Supergroup unconformably deposited over the Bijawar Group (Prakash et al. 1975; Swaroop and Saxena 1969).

On the basis of lithological characteristics, this Group has been subdivided in the present work into two formations; a lower Sonrai Formation and an upper Solda Formation.

(i) Sonrai Formation

The Sonrai Formation is mostly a carbonate sequence without tuffaceous material. The various rock types include carbonates (Fig. 14a), shales (Fig. 14b) and sandstones (Fig. 14c) with some basaltic and gabbroic intrusive. The rocks occasionally show abundant bedding-controlled sulphides. Prakash et al. (1972) have subdivided the Sonrai Formation into five members; from base to top, these members are Jamuni Carbonate Member, Gora Kalan Shale Member, Rohini Carbonate Member, Bandaui Sandstone Member, and Kurrat Member. However, in the study area, these members are not fully developed. As such, it is not possible to subdivide this formation into different members.

(ii) Solda Formation

The rocks of the Solda Formation are exposed with NW-SE to E-W trends and occasionally show steep dips. This formation is mainly constituted of ferruginous shale with quartz arenite interbands together with tuffaceous material and some sandstone (Fig. 14d). The various rock types of this formation include ferruginous shale, ferruginous quartzite, iron stone, calcareous sandstone and chloritic shale. Prasad et al. (1999) considered this formation to represent a volcano-sedimentary sequence.

Fig. 14 Field photographs showing outcrops of Bijawar Group of rocks. **a** Carbonate rocks of the Sonrai Formation. Loc. South of Tori. **b** Shales of Sonrai Formation. Loc. North of Dhori Sagar. **c** Sandstone of Sonrai Formation. Loc. Near Gora Kalan. **d** Ferruginous sandstone of Solda Formation. Loc. Near Solda

Fig. 15 Field photographs of the rocks of the Vindhyan Supergroup (Lower Vindhyan). **a** Kaimur sandstone. Loc. South of Uldana Khurd. **b** Fragile sandstone of Semri Group. Loc. Near Madanpur. **c** Bedded sandstone of Semri Group. Loc. South of Kurrat. **d** Contact of Semri and Kaimur Groups. Loc. Near Madanpur

(D) **Vindhyan Supergroup**

A part of the Vindhyan Supergroup is exposed in the southern part of the study area. The rocks of this unit occur as E-W and NE-SW trending continuous ridges and smaller hills. In the study area, this supergroup is represented by two groups: a Lower Semri Group and an Upper Kaimur Group. In the study area, the Semri Group (Fig. 15a, b) includes thin-bedded sandstones and shales together with locally developed conglomerate horizons. The Kaimur Group (Fig. 15c, d) is dominantly represented by massive sandstone, though thin-bedded sandstones are also locally noticed.

(E) **Deccan Traps**

A few outcrops of Deccan Trap occur in the southern part of the area. Of these only one outcrop is relatively large in size; it trend almost N-S with a length of about 8 km and a width of ~4 km. The other outcrops are smaller and extend for a few tens of meters only. The basalt is usually light to dark grey, fine to medium grained rock. The Deccan Trap occurs as flat cap rocks in the area. In the study area, the Deccan Trap is mainly constituted of basaltic rocks. Most of the basaltic rocks show highly weathered and irregular surfaces as well as lateritic soil and black alluvium soil cover. Recently, a number of papers have been published on the structural geology and tectonics of the Deccan Trap (Misra et al. 2014, 2015; Misra and Mukherjee 2015; Mukherjee et al. 2017).

6 Structural Architecture

(A) **Structures**

The large-scale structure of the rocks of the area has been observed to vary with different lithostratigraahic units as well as, to a larger extent, on the overall physiographic set-up. As mentioned above, the area can be subdivided into three physiographic–lithostratigraphic subdivisions, we can describe the structure under three domains as described below.

Structure of Domain-I

The Domain-I is considered here including the BGC. Structurally, the rocks of this domain do not show any visible large-scale structure excepting the dominantly south-dipping foliation. The rocks however occasionally show complications in their internal domain as reflected in the mesoscopic structures described below.

Structure of Domain-II

Structurally, the central part of the study area has been identified as Domain-II. This domain is mainly represented by the rocks of the Bijawar Group. In the study area, the rocks of the Bijawar Group are folded into an anticline to the south and a corresponding syncline to the north. The axes of these folds trend approximately E-W to ENE-WSW and are slightly curved in nature with gentle plunges towards east. These folds are traversed by a number of roughly N-S trending steep to vertical faults of varying dimensions.

Structure of Domain-III

Domain-III is constituted of the rocks of the Vindhyan Supergroup. This lithounit includes a massive succession of sandstones shales. Structurally, the Vindhyan rocks show a simple dipping structure with southward dipping strata. The Vindhyan rocks are believed to rest unconformably over the Bijawar rocks. At some places, the Vindhyan and Bijawar strata show some effects of deformation along their contacts in the form of some small scale folds and faults.

A prominent outcrop of Deccan Trap also occurs in the southwestern part of the area. This outcrop trends almost N-S and is about 8 km N-S and about 4 km E-W. In addition to this, four scattered outcrops of Deccan Trap also occur at the southern fringe of the study area. These outcrops extend for a few tens meters only. The Deccan Trap occurs as flat cap rocks and thus adds a special feature to the physiography of the area. The Deccan Trap rocks are constituted of basaltic rocks. As such, we consider these rocks as discordant igneous bodies and possibly represent the last igneous (volcanic) event in the study area. The Deccan Trap rocks apparently do not show any structure in their internal domain.

(B) **Mesoscopic Structures**

The rocks of the area show a variety of mesoscopic structures that are distributed in the area rather sporadically. Most of these structures occur in the BGC followed by the Bijawar Group. In the rocks of the Vindhyan Supergroup, occurrence of mesocopic structures is very rare excepting near the Vindhyan-Bijawar contact zone. The mesoscopic structures of the area are in the form of minor folds, planar structures and linear structures.

Minor Folds

Minor folds developed in the area range in size from a few centimeters to a few meters. The study of fold geometry and their attitudes indicate that the rocks of the area have undergone at least three phases of deformation. Each phase is represented by a set of minor folds showing their typical fold geometry and attitudes as described below (Fig. 16).

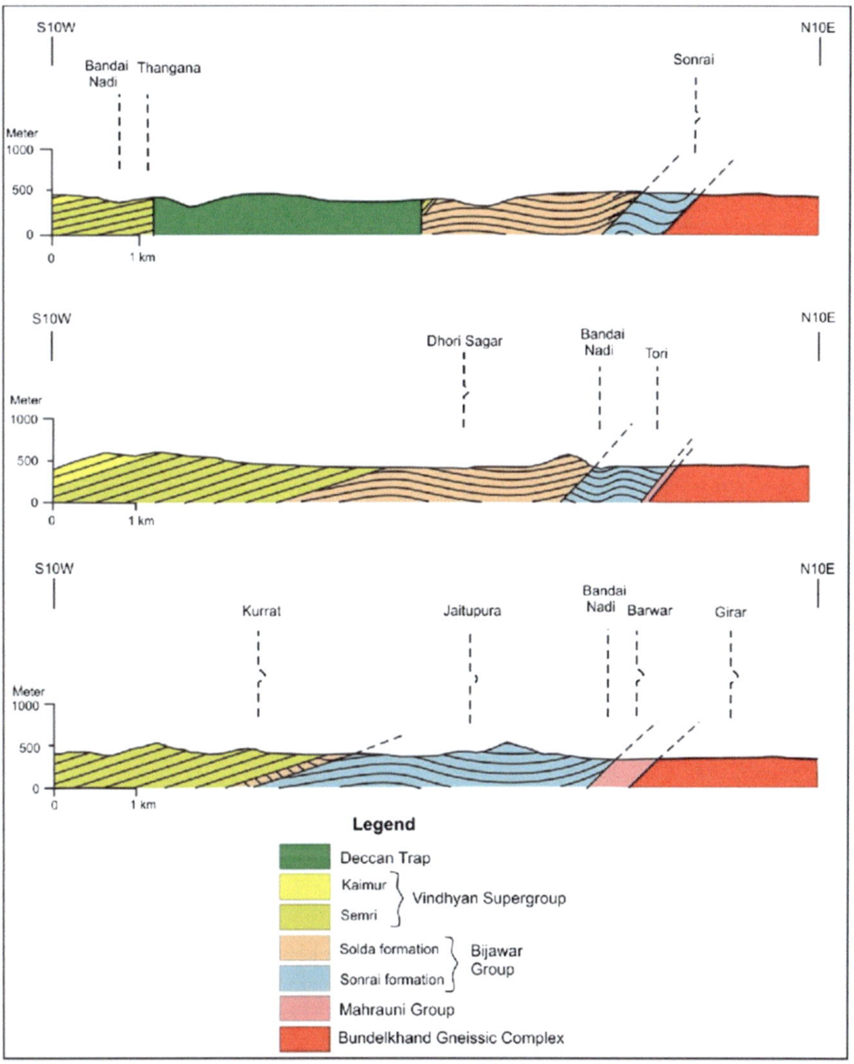

Fig. 16 Geological cross-sections of the rocks of the study area in three traverses (marked in Fig. 4) across the strike of the rock formations: T_1: upper diagram, T_2: middle diagram, T_3: lower diagram

(i) **F_1 Folds**

The first phase of deformation structures are represented by the F_1 folds (Fig. 17a, b). These folds are well developed in the quartzite (banded hematite quartzite) of Mahrauni Group. In the BGC, these folds are commonly traced by silica bands of the gneissic rocks. The F_1 folds are generally tight recumbent isoclinal with interlimb angle of about 20°–30°. They commonly show thickened hinge zones and thin limbs. The F_1 folds vary in size from a few centimeters to a few meters and are characterized by high amplitude/wavelength ratio which is up to about 4 at some places. The axes commonly trend between NW-SE and E-W but occasionally show minor variations due to the effect of later deformations. In general, the higher degree of flattening at the hinge zone is associated with higher thinning of the limbs. The F_1 folds are occasionally associated with an axial plane foliation designated as S_1 which is parallel to F_1 fold axes. However, S_1 transects bedding at the closures of the F_1 folds.

(ii) **F_2 Folds**

The second generation folds (F_2) are commonly developed in the banded hematite quartzite (Fig. 17c) and in the banded hematite jasper of Barwar Formation. The F_2 folds are generally coaxial to F_1. In comparison with the F_1 folds, the F_2 folds are more open, asymmetrical, and show low amplitude/wavelength ratio and higher interlimb angle up to 60°. These folds generally show almost constant orthogonal thickness in the competent folded layers.

Fig. 17 Photograph showing minor folds of the study area. **a**, **b** F_1 folds in the banded hematite quartzite affected by ductile shearing showing thinning of limb. Loc.1.8 km west of Girar. **c** F_2 folds in banded hematite jasper, 1.8 km southwest of Girar. **d** F_3 folds in carbonate rocks together with ferruginous layering. Loc. 0.5 km south of Gorakalan

(iii) **F_3 Folds**

The third generation (F_3) folds (Fig. 17d) are relatively much less common in the area. These are open folds with rounded hinges. Generally these are parallel folds thus showing class 1B of Ramsay (1967). The amplitude/wavelength ratios are low as compared with the F_1 and F_2 folds. In the schistose rocks, these folds commonly occur as crenulation folds. The F_3 folds commonly show vertical axial planes striking NNE-SSW to NNW-SSE direction. These folds are noticed in the basic dikes as well as in some rocks of the Bijawar Group. The F_3 folds possibly represent signatures of some large scale open folds in the form of warps that developed a curvilinear pattern to the some of the ridges of the area.

7 Tectonic Framework

Our knowledge on the tectonic framework of the rocks of the area is as yet meager. The main reason is that the nature of contact of a few lithostratigraphic units (e.g. Sonrai-Solda, Bijawar-Vindhyan) is either not known properly or is debatable (e.g. Jha et al. 2012; Kumar et al. 1990; Prakash et al. 1972). In the course of our present work, we have made efforts to study the nature of contact of important lithostratigraphic units and the related structural features, if any.

(i) **BGC-Bijawar contact**

The contact of crystalline rocks with sedimentary/metasedimentary (cover) rocks in any area is always of fundamental significance. The main reason, amongst others, is that the study of the contact zone helps us to unravel the geological and palaeogeograhical conditions of the past. In this context, the presence of the contact of the BGC and the Bijawars is also of great significance not only to the present area but also to the peninsular shield in general. When we talk of the contact of basement rocks with cover rocks, there appears to be two possibilities: the contact is either unconformable or tectonic. In the first case, i.e. unconformable contact, commonly the occurrence of conglomerates is expected in which the conglomerate horizon should contain pebbles, cobbles, etc. derived from the basement rocks. In the second case, i.e. tectonic contact, development of structural features indicating the presence of a fault or thrust is expected.

In the study area (Fig. 4), the BGC-Bijawar contact is exposed in the northern parts of the area with almost an E-W trend. It passes from the south of Sonrai in the west through the south of Pisnari in the central part, and south of Girar in the eastern part of the area. Unfortunately the direct contact of these two rock units is exposed only at a few localities. Also, in most cases the contact zone is rather vaguely exposed. We have carried out detailed study of the contact zone at Sonrai, Pisnari, Tori, Barwar, Girar and their adjoining areas. So far, we have not observed any conglomerate horizon at the contact. We thus rule out any unconformable contact between the BGC and the Bijawars. We have instead observed the occurrence of tectonic structures practically in all the localities where the BGC-Bijawar contact is exposed.

Along the BGC-Bijawar contact we have observed structures indicating ductile deformation of rocks, e.g. thick-hinge folds showing asymmetry of their axial planes with respect to the surrounding foliation. In the light of these features, we suggest a thrust contact between the BGC and the Bijawar Group. It may be mentioned here that earlier Prakash et al. (1975) had also suggested a thrust contact between the BGC and the Bijawars though no distinct structures or structural features to support this view had been described. Following Prakash et al. (1975) we have retained the name of this contact as *Sonrai-Girar thrust*. Several mafic and ultra-mafic bodies are intruded along the BGC-Bijawar contact.

(ii) **Bijawar-Vindhyan Contact**

The contact of the Bijawar Group with the overlying Vindhyan Supergroup is believed to be unconformable. However in the study area (Fig. 4), this is not so. Our field studies and mapping indicates that there are several evidences that lead to the otherwise. In the area, the Vindhyan Supergroup is exposed in the southern parts and is represented by the Semri Group and the Kaimur Group. Both these groups are exposed as almost E-W trending continuous outcrops. The outcrops of the Semri Group however do not maintain uniform width in different parts of the area and show variations from about 2 km to about 200 m. At the same time, the dip of the strata maintains almost uniform values, i.e. from 15° to 25° roughly towards S. We assign this variation in width of the outcrops to the existence of a fault due to which a part of the Semri Group has been concealed. Occurrence of breccia is noticed at a few places. Further, at some places, e.g. about 3 km north of Madanpur, the Semri Group is missing and, as a result, the Bijawar Group is directly juxtaposed against the Kaimur Group. We therefore consider a faulted contact between the Bijawar Group and the overlying Vindhyan Supergroup at least in the study area. We name this contact as *Patna–Kurrat fault*.

(iii) **Tectonic status of the Mahrauni Group**

The Mahrauni Group is exposed in the northeastern parts of the study area. This rock unit thins in the west. In this work, we have mapped the Mahrauni Group as a single lithostratigraphic unit (Fig. 4) because the Rajaula and the Barwar Formations are not mappable on the scale of our map. Along the BGC-Mahrauni contact we have not observed any evidence of unconformity. Rather, development of minor shear zones has been observed at one or two localities. It therefore seems possible that the BGC-Mahrauni contact is tectonic, possibly thrusted, in the study area. As a matter of fact, the entire contact zone is not well exposed. Therefore on the basis of our interpretation of a thrust contact, we can extend this generalization for the entire BGC-Mahraunu contact. Since this contact is in direct continuity of the Sonrai-Girar thrust, we therefore extend the Sonrai-Girar thrust for the BGC-Mahrauni contact also.

(iv) **The Overall Tectonic Framework**

All the major rock units of south Lalitpur district bear tectonic contacts. The BGC-Bijawar contact is a thrust named as the *Sonrai-Girar thrust* while the Birawar-Vindhyan contact is a fault named as the *Patna-Kurrat fault*. If we go back to the existing concepts, the Bijawars are believed to have deposited unconformably over the crystalline basement (the BGC in this case). Likewise, the Vindhyans are also believed to have deposited unconformably over the Bijawars. But in the south lalitpur district, the situation differs Here the contacts of practically all the major lithostratigraphic units are tectonic. This suggests that initially the BGC-Bijawar and Bijawar-Vindhyan contacts may have been unconformable. Later, the contacts were disturbed tectonically. We suggest that the differences in the overall lithologic, and so rheologic, characters of the individual lithostratigraphic units were the major factors due to which the individual lithostratigraphic units behaved differently to the deforming stresses. This possibly explains why is the overall tectonic framework of the lithostratigraphic units of the study area not exactly what is elsewhere in the peninsular shield. Another possibility could be that this part of the shield may have experienced tectonic activities in due course of time due to some crustal disturbances that did not affect other parts of the shield. And, further, even if other parts of the shield were affected by tectonic disturbances, the effects may have been feeble so that the degree of tectonization did not leave well developed and recognizable signatures today, as we see them in the present study area.

8 Discussion and Conclusions

- South Lalitpur district exposes some typical rock units of the Indian peninsular shield.The area shows three distinct physiographic features: scattered hillocks to the northern parts, nearly continuous hills in the central parts and plateau in the southern parts. This physiographic set-up is closely related to the geological setup of the area. The northern part exposes crystalline-metamorphic rocks of the Bundelkhand Gneissic Complex (BGC), the central part exposes Bijawar Group and the southern part exposes rocks of the Vindhyan Supergroup.
- The BGC consists of a variety of granitoids, gneisses and schists together with mylonitised granite gneisses, migmatites and amphibolite schists. Of the granitoids, pink and grey granites are the dominating types. In the study area, we have identified the following lithologic associations: gneisses designated here as Bundelkhand Gneisses, Bundelkhand Granites that include coarse grained pink porphyritic granite, medium grained pink porphyritic granite, grey granite, leucogranite, aplites and pegmatites. A few NE-SW trending quartz reefs and NW-SE trending mafic dikes are also present.
- South of the BGC, a group of schistose rocks is exposed. This group was earlier referred to as Mehroni Group, which we rename as Mahrauni Group following

recent toposheets. The Mehroni Group was divided into four Formations. However our work reveals that in the study area only one formation, i.e. Barwar Formation, is exposed. This Formation is constituted of banded hematite quartzite, conglomerate, fuchsite quartzite and grey to green chlorite schists.

- South of the BGC occurs the Bijawar Group that extends almost E-W in the central parts of the area. This group includes a sequence of sedimentary rocks that are affected by very low metamorphism. In the study area, this group is represented by two Formations: a Lower Sonrai Formation and an Upper Solda Formation. The Sonrai Formation includes carbonates, shales and sandstones with some basaltic and gabbroic intrusive while the Solda Formation is mainly constituted of ferruginous shales with quartz arenite interbands together with tuffaceous material and some sandstone.
- The southern part of the area is mostly occupied by the rocks of the Vindhyan Supergroup and is represented by only two formations, viz. Semri and Kaimur Formations.
- In addition to the above rock units, a few outcrops of Deccan Trap also occur in the southern parts of the area. The Trap is constituted of fine to medium grained basaltic rocks that are light to dark grey in colour.
- The structure of the area seems to be related to the individual lithostratigraphic units. Thus we can identify three different structural domains towards south. The Domain-I is assigned to the BGC, which apparently does not show any visible large-scale structure excepting a dominantly south-dipping foliation. Domain-II is represented mainly by the Bijawar Group folded into an anticline to the south and a corresponding syncline to the north. The axes of these folds trend approximately E-W to ENE-WSW. Domain-III mainly includes the Vindhyan Supergroup. The rocks of this rock units show a simple dipping structure with southward dipping strata.
- The rocks of the area show a variety of mesoscopic structures rather irregularly distributed in different parts. We have identified a variety of minor folds, planar structures and linear structures. Study of minor folds indicates that the rocks of the area have undergone at least three generations of folds. The first generation (F1) folds are commonly traced by quartz bands of the associated rocks and are generally tight recumbent isoclinal with interlimb angle of about 20°–30°. The second generation (F2) folds are generally co-axial to F1 but they differ in geometry from F1 folds in being more open, asymmetrical and occasionally greater interlimb angle up to 60°. The third generation (F3) folds are relatively less common in the area and occur as open folds with high wavelength-amplitude ratios.
- The tectonic framework of the rocks of the area is as yet not clearly known mainly because the nature of contact of the lithostratigraphic units is either not known properly or is debatable. One practical problem is that the contact of most of the major rock units is not clearly exposed. In the light of our observations, it appears that the BGC and the Bijawars bear a thrust contact while the Bijawars and the Vindhyans bear a faulted contact. The Mahrauni Group

appears to bear a thrust contact with the BGC to the north as well as with the Bijawars to the south. We thus consider this group to occur as a thrust slice.

- It seems possible that the contacts of the major lithotectonic units were tectonized due to tectonic disturbances in different parts of the peninsular shield possibly due to the differences in the overall lithologic, and so rheologic, characters of the individual lithostratigraphic units. As a result, the individual lithostratigraphic units behaved differently to the deforming stresses. Possibly because of this the overall tectonic framework of the lithostratigraphic units of the study area does not fully reflect the general framework elsewhere in the peninsular shield.

Acknowledgements We express our sincere thanks to the Head, Centre of Advanced Study in Geology, University of Lucknow, for providing working facilities. GKD extends his thanks to the Director, Directorate of Geology and Mining, U.P., for granting him permission to carry out this work as part of his doctorate thesis. We thank Prof. Soumyajit Mukherjee (IIT Bombay) for his editorial and reviewing works that significantly improved the quality of the manuscript. Mukherjee (2019) summarizes this work.

References

Banerjee DM (1982) Lithotectonics phosphate mineralization and regional correlation of Bijawar Group of rocks in Central India. In: Valdia KS, Bhatia SB, Gaur VK (eds) Geology of Vindhyanchal. Hindustan Publishing Corporation, pp 49–54

Basu AK (1986) Geology of parts of Bundelkhand Granitic Massif. Records of Geological Survey of India 117, 61–124

Basu AK (2001) Some characteristics of the Precambrian crust in the northern part of Central India. Geological Survey of India 55, 182–204

Basu AK (2004) Contemplations on the role of the Bundelkhand Massif on structural evolution and mineralization in the western Indian Cration, Rajasthan, India. Geological Survey of India Special Publication 72, 325–344

Basu AK (2007) Role of Bundelkhand Granite Massif and the Son-Narmada megafault in Precambrian crustal evolution and tectonism in Central and Western India. Journal of Geological Society of India 70, 745–770

Basu AK (2010) Precambrian geology of the Bundelkhand terrain, Central India and adjacent part of western India. Journal of Economic Geology and Georesource Management 7, 1–53

Bhattacharya AR (1985) Some unusual strain relations in elliptically deformed xenoliths and feldspar porphyroblasts. Zeitschrift für Geologische Wissenschaften 13, 689–697

Bhattacharya AR (1986) Wavelength-amplitude characteristics of polyphase folds in the Precambrian Bundelkhand Complex. Tectonophysics 128, 121–125

Bhattacharya AR, Singh SP (2013) Proterozoic crustal scale shearing in the Bundelkhand Massif with special reference to quartz reefs. Journal of Geological Society of India 82, 474–484

Bose PK, Sarkar S, Chakraborty S, Banerjee S (2001) Overview of the Meso- to Neoproterozoic evolution of the Vindhyan basin, Central India. Sedimentary Geology 141, 395–419

Chakraborty C (2006) Proterozoic intra continental basin: the Vindhyan example. Journal of Earth System Science 115, 3–22

Chaudhuri J, Chowdhury D, Maitra U (1999) Distinct functions of eukaryotic translation initiation factors eIF1A and eIF3 in the formation of the 40 S ribosomal preinitiation complex. Journal of Biological Chemistry 18, 17975–17980

Dinkar GK (2016) Deformation pattern of the rocks exposed around Sonrai, district Lalitpur, Uttar Pradesh. Unpublished Ph.D. thesis, University of Lucknow, Lucknow
Hacket CA (1870) Geology of Gwalior and vicinity. Records of Geological Survey of India 3:33–42
Heron AM (1935) Synopsis of the Pre-Vindhyan geology of Rajasthan, Rajputana. Transactions of the National Institute of Science 1, 17
Jha SK, Shrivastava JP, Bhairam CL (2012) Clay mineralogical studies on Bijawars of the Sonrai Basin: Palaeoenvironmental implecations and influences on the uranium mineralization. Journal of Geological Society of India 79, 117–134
Jhingran AG (1958) The problem of Bundelkhand granites and gneisses. In: Presidential Address, Proceedings of Indian Science Congress 45th session, Madras, pp 48–120
Kumar B, Srivastava RK, Jha DK, Pant NC, Bhandaru BK (1990) A revised stratigraphy of the rocks of type area of the Bijawar Group in Central India. Indian Minerals 44, 303–314
Misra RC (1969) The Vindhyan System. Presidential Address (Geology & Geography Section) 56th Session, Bombay
Misra RC, Sharma RP (1975) New data on the geology of Bundelkhand complex of Central India. In: Verma VK et al (ed) Recent research in geology. Hindustan Publishing Co., Delhi, vol 2, pp 311–346
Misra AA, Mukherjee S (2015) Tectonic inheritance in continental Rifts and Passive Margins. Springer-Briefs in Earth Sciences. Springer International Publishing, Berlin. ISBN 978-3-319-20576-2
Misra AA, Bhattacharya G, Mukherjee S, Bose N (2014) Near N-S paleo-extension in the western Deccan region in India: does it link strike-slip tectonics with India-seychelles rifting? International Journal of Earth Science 103, 1645–1680
Misra AA, Sinha N, Mukherjee S (2015) Repeat ridge jumps and microcontinent separation: insights from NE Arabian Sea. Marine and Petroleum Geology 59, 406–428
Medlicott HB (1859) Vindhyan rocks and their associates in Bundelkhand. Memoir of Geological Survey of India II, pp 1–95
Mondal MEA, Goswami JN, Deomurari MP, Sharma KK (2002) Ion microprobe ^{207}Pb/Pb^{206} ages of Zircons from Bundelkhand massif, northern India, implications for crustal evolution of the Bundelkhand-Aravalli Proto Continent. Precambrian Research 117, 85–110
Mondal MEA (2010) Geochemical evolution of the Archaean-paleoproterozoic Bundelkhand Craton, Central India shield. Journal of Economic Geology and Georesource Management 7, 69–80
Mukherjee S, Misra AA, Calvès G, Nemčok M (2017) Tectonics of the Deccan large igneous province: an introduction. In: Mukherjee S, Misra AA, Calvès G, Nemčok M (eds) Tectonics of the Deccan large igneous province. Geological Society, London, Special Publication 445, pp 1–9
Mukherjee S (2011) Mineral fish: their morphological classification, usefulness as shear sense indicators and genesis. International Journal of Earth sciences 100, 1303–1314
Mukherjee S (2019) Introduction to "Tectonics and Structural Geology: Indian Context". In: Mukherjee S (ed) Tectonics and structural geology: Indian context. Springer International Publishing AG, Cham, pp 1–5. ISBN: 978-3-319-99340-9
Pascoe EH (1950) A manual of the Geology of India and Burma. Geological Survey of India, Calcutta, 1
Pati JK, Patel SC, Pruseth KL, Malviya VP, Arima M, Raju S, Pati P, Prakash K (2007) Geology and geochemistry of gaint quartz veins from the Bundelkhand Craton, Central India and their implications. Journal of Earth System Science 116, 497–510
Prakash R, Srivastava RN, Saharma DP, Bhatt GD (1972) Progress of geological investigation on Sonarai copper belt, Mehroni Tehsil, Jhansi, Uttar Pradesh. DGM, U.P.
Prakash R, Swarup P, Srivastava RN (1975) Geology and mineralization in the southern parts of Bundelkhand in Lalitpur district, Uttar Pradesh. Journal of Geological Society of India 16, 143–156

Prasad MH, Hakim A, Rao BK (1999) Metavolcanic and Metasedimentary Inclusions in the Bundelkhand Granitic Complex in Tikamgarh District, Madhya Pradesh. Journal of Geological Society of India 54, 359–368

Radhakrishna BV (1989) Suspect Tectono-stratigraphic terrain elements in the Indian Subcontinent. Journal of Geological Society of India 34, 1–24

Ramsay JG (1967) Folding and fracturing of rocks. McGraw-Hill, New York

Roday PP, Diwan P, Singh S (1995) A kinematic model of emplacement of quartz reef and subsequent deformation pattern in Central Indian Bundelkhand batholiths. Proceedings of the Indian Academy of Science—Earth and Planetary Science 104, 465–488

Sarkar A, Bhalla JK, Bishui PK, Gupta SN, Srimal N (1989) Geochemistry and geochronology of the Early Proterozoic Bundelkhand granitic complex, Central India. In: Symposium on Precambrian, Granitoids, Helsinki, Finland, Geological Survey of Finland, p 117 (Abstract 8)

Sarkar A, Paul DK, Potts PJ (1995) Geochronology and geochemistry of the Mid-Archean, Torndhjemitic gneisses from the Bundelkhand Craton, Central India. In: Saha AK (ed) Recent researches in geology, vol 16, pp 76–92

Sarkar A, Ghosh S, Singhsi RK, Gupta SN (1997). Rb-Sr geochronology of the Dargawan sill: constraint on the age of the type Bijawar sequence of Central India. In: International conference on isotopes in solar system, pp 100–101 (Abstract pp 11–14)

Saxena MN (1961) Bundelkhand granites and associated rocks from Kabrai and Mauranipur area of Harnirpur and Jhasi district; U.P., India. Research Bulletin of Panjab University 12, 85–107

Sharma RP (1982) Lithostratigraphy, structure and petrology of Bundelkhand Group. In: Valdia KS, Bhatia SB, Gaur VK (eds) Geology of Vindhyanchal. Hindustan Publishing Corporation, pp 30–46

Singh SP (2012) Geochemical signatures of Archaean felsic volcanism in Central part of Bundelkhand Craton, Central India. International Journal of Advance Earth Sciences 1, 20–32

Singh SP, Bhattacharya AR (2017) N-S crustal shear system in the Bundelkhand massif: a unique crustal evolution signature in the northern Indian peninsula. Journal of Earth System Science 126, 121

Singh SP, Bhattacharya AR (2010) Signature of Archaean E-W crustal-scale shears in the Bundelkhand massif, Central India, an example of vertical ductile shearing. Earth Science India 3, 217–225

Singh SP, Singh MM, Srivastava GS, Basu AK (2007) Crustal evolution in Bundelkhand area, Central India. Journal of Himalayan Geology 28, 79–101

Singh SP, Dwivedi SB (2009) Garnet-silliminite-cordierite-quartz bearing assemblages from the early Archaean supracrustal rocks of Bundelkhand massif, Central India. Current Science 97, 103–107

Srivastava JP (1951–1952) Rocks of Probable Bijawar age overlying the Bundelkhand granite in Tikamgarh district. Geological Survey of India, Unpublished Progress Report

Swaroop P, Saxena AK (1969) A report on the preliminary investigation of iron-formations near Obra, District Mirzapur, Uttar Pradesh. Report of the Directorate of Geology and Mining, U.P.

Deformation in the Kangra Reentrant, Himachal Pradesh of NW-Sub Himalaya of India: A Paradox

Tejpal Singh and A. K. Awasthi

1 Introduction

The Himalayan orogeny is geologically very complex and heterogeneous along its entire length even though it consists of longitudinal curvilinear stripes of similar lithologies (Mukherjee 2013a). For the ease of correlation and understanding, broad sub-divisions have been made based on comparable litho-tectonic units (Fig. 1). These sub-divisions have been further verified through detailed lab (geochemical) and field studies (Valdiya 1984). Accordingly, the southern (outer)most litho-tectonic sub-division of the Himalaya is called the Sub-Himalayan belt. The Sub-Himalayan belt in the NW Himalaya has been studied in detail and classified based on energy sequences, which are equivalent to various rock types (Raiverman 2002). This belt is the youngest and most actively deforming tectonic sub-division of the Himalayan orogen that accommodates about 20% ($\sim$4–11 mm/yr) of the total India-Asia convergence (Larson et al. 1999; Lave and Avouac 2000; Bilham et al. 2001). Note that the Sub-Himalayan belt is assumed to be tectonically col-located with the external locked zone of the Himalayas (Bilham et al. 2001; Stevens and Avouac 2015). This locked zone has an approximate width of 100 km and tends to accumulate the potential tectonic slip almost entirely as elastic (>90%) rather than inelastic (<10%) strain, which would permanently deform the rock

T. Singh (✉)
CSIR-Central Scientific Instruments Organisation, Chandigarh 160030, India
e-mail: tejpal@csio.res.in; geotejpal@yahoo.co.in

A. K. Awasthi
Graphic Era (Deemed to be University), Dehradun 248002, India

A. K. Awasthi
Formerly at Department of Earth Sciences, Indian Institute of Technology
Roorkee, Roorkee 247667, India

S. Mukherjee (ed.), *Tectonics and Structural Geology: Indian Context*,
Springer Geology, https://doi.org/10.1007/978-3-319-99341-6_13

(Bilham et al. 2001). The cover sequences in this locked zone are variably coupled with the basement and the coupling tends to maximum near the deformation front i.e. Himalayan Frontal Thrust: HFT (Stevens and Avouac 2015).

The Sub-Himalayan belt is a typical example of a "cover sequence", which is faulted and folded over a rigid basement. The structures characterize the thin skin deformation of Fold-Thrust belts (FTBs). There has been a considerable progress in our understanding of thin-skin deformation of FTBs over the last few decades (Ruh et al. 2012) and as more and more data and new tools become available, refined interpretations are coming up (e.g., Rajendra Prasad et al. 2011; Carrillo et al. 2017). The initial models are commonly based on surface geology and structure. However, the current understanding largely draws from the theoretical and experimental models (Suppe 1983; Suppe and Medwedeff 1990; Stokmal et al. 2007). Most of these models are based on (i) different material properties of the detachment e.g., frictional vs. viscous detachments (Cotton and Koyi 2000; Bahroudi and Koyi 2003), (ii) inherent properties of the cover sequences such as cohesion and pore pressure, and (iii) interaction between the detachment and cover sequences such as their coupling (Withjack and Callaway 2000; Paul and Mitra 2015; Borderie et al. 2018). The results are commonly interpreted in terms of evolution of (i) structural geometries such as imbricate thrusts, back thrusts, open folds etc., (ii) topographic slope (α), and (iii) detachment dip (β) (Singh et al. 2012). In most cases, the models are tuned compatible with the available seismic sections, well logs, structural- and topographic maps.

The width of the Sub-Himalayan belt varies along the length of the orogen. It reaches to a maximum of over 300 km in the western sections of Pakistan whereas

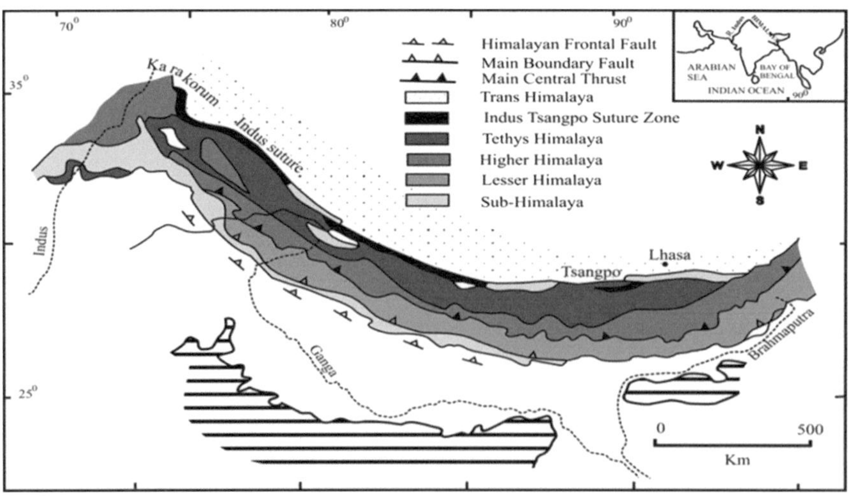

Fig. 1 Broad litho-tectonic subdivisions of the 'Himalaya' (after Valdiya 1984, 1998)

it thins to <10 km in the eastern sections, especially east of Nepal. The variable width to a large extent is attributed to the original basin geometry and rheology of the basement. The basins of deposition towards the west have been extremely wide and sediments have been deformed over rheologically weak basement in salt (Davis and Lillie 1994). This has not been the case eastwards where, the rheologically weak salt is not found in the basement and the belts are narrower. However, there is an interesting paradox in the Kanrga Reentrant of Western India where the width of the Sub-Himalayan belt is ~ 80 km: much wider than that at east. The topographic slope (α) of the Kangra Reenetrant is $\sim 1°$ and together with the basement dip (β) of 3°, it forms a wedge taper of around 4°. This wedge taper is comparable to those of Salt Range-Potwar Plateau in Pakistan (Davis and Lillie 1994) and Fars and Lorestan segments of the Zagros FTB in Iran (McQuarrie 2004). Notably the FTBs in Pakistan and Iran are extremely wide and have also evolved over a weak detachment in salt which justifies a low wedge taper. However, there have been no reports on the presence of salt at the surface or sub-surface in the Kangra Reentrant which could justify a relatively low wedge taper spread over a width of 80 km. This article focuses on this paradoxical reentrant to elaborate on the geological structure and to understand the reasons for its relatively larger width.

2 Litho-Tectonic Setting

The Sub-Himalayan belt, in general comprises the Tertiary rock units, which are classified into the Lower and the Upper Tertiaries. In western India, it consists of parautochthonous and the autochthonous units (Table 1). The former units include the Subathu and the Dagshai (=Lower Dharamshala) and the Kasauli (=Upper Dharamshala) Formation of Lower Tertiary age. The autochthonous unit includes the Siwalik Supergroup of Upper Tertiary to Quaternary age (Raiverman 2002). The Siwalik Supergroup is conventionally sub-divided into the Lower, Middle and Upper Siwalik Groups (Table 1). Both parautochthonous and autochthonous units are unconformably overlain by Quaternary terraces along the stream network (Singh et al. 2011). The Sub-Himalayan rock units thrust over the sediments of the Indo-Gangetic plains along the Himalayan Frontal Thrust (HFT). The HFT defines the present-day tectonic boundary of active tectonic convergence (Yeats et al. 1992; Powers et al. 1998; Delcaillau et al. 2006; Kumar et al. 2006; Singh and Jain 2009).

3 Surface Topography and Taper of the Wedge

Deformation style and kinematics vary considerably across the fold-thrust belts in orogenic systems (e.g., Mukherjee 2013b). These differences may be present at much finer scales but their first order manifestation is readily visible in the surface topography as seen in various modeling approaches (Koyi 1988; Cotton and

Table 1 Generalized stratigraphy of the Kangra Reentrant

	Formation/group		Description	Thickness (m)	Age
Post-Siwalik deposits	Alluvium/piedmont		Sand and silt		Recent
Siwalik Supergroup	Upper Siwalik		Predominantly conglomerate with sand and clay matrix	600–1500	Middle Miocene to Pleistocene
	Middle Siwalik		Predominantly sandstone and interbedded clays	1400–2000	
	Lower Siwalik		Alterations of sandstone, claystone and minor siltstone	1200–1600	
	Dharamshala	Upper Dharamshala	Sandstone with subordinate claystone	1100–1600	Lower–Upper Eocene to Middle Miocene
		Lower Dharamshala	Predominantly claystone with minor sandstone	1900–2300	
	Subathu		Shales, limestone and orthoquatzite	400–800	Late paleocene-middle Eocene
Haitus					
	Bilaspur/Bandla limestone		Limestone	450+ Base not exposed	Neo-Meso Proterozoic
Metamorphic/granitic basement			Granitic	–	Archean

Modified after Raiverman (2002) and Rajendra Prasad et al. (2011)

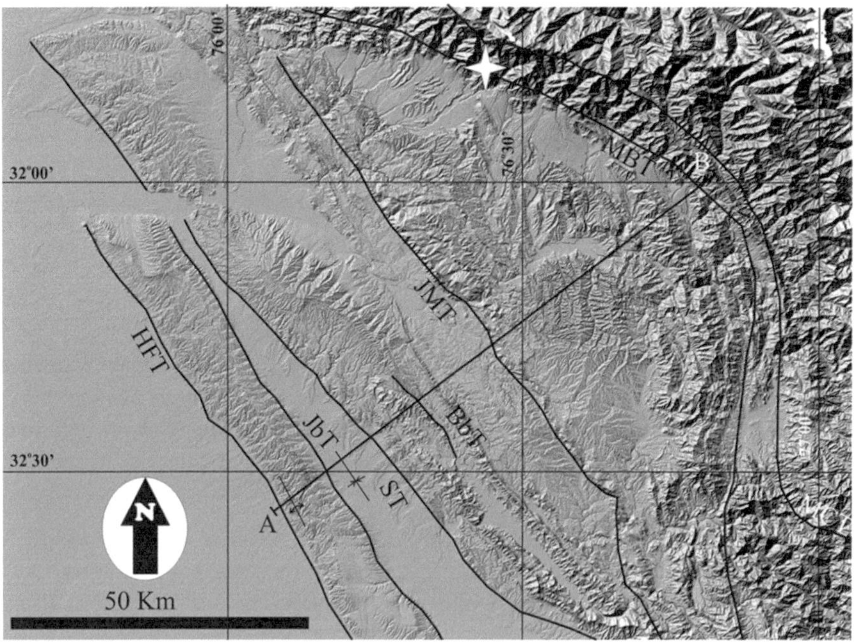

Fig. 2 Generalized surface topography of the Kangra Reentrant showing NW-SE elongated ridges separated by synclinal valleys. The main structures viz. HFT: Himalayan Frontal Thrust, JbT: Januari back Thrust, ST: Soan Thrust, BbT: Barsar back Thrust, JMT: Jwalamukhi Thrust, MBT: Main Boundary Thrust and MCT: Main Central Thrust are shown. The representative anticlinal and synclinal topography are marked. The star in the top middle panel represents the last major earthquake of 1905 around Kangra. The line of structural cross-section A–B shows location of Fig. 3

Koyi 2000; Costa and Vendeville 2002; Nilforoushan et al. 2008; Nilfouroushan et al. 2012; Ruh et al. 2012). In all these models, structural evolution readily compare with the large scale topographic features. Of course, surface processes and erosion play an equally important role in shaping the topography. Their role is to enhance the surface manifestations of geological structure, often modulated by lithology and climate (Singh 2008a, b; Singh and Jain 2009). Therefore surface topography forms a primary guide to the mechanical characters leading to the observed structural geometries and surface manifestations.

The present day surface topography of the Kangra Reentrant is characterized by a series of NW trending (sub)parallel ridges separated by wide and relatively flat valleys trending parallel to the ridges in between them (Fig. 2). This topographic arrangement is more apparent towards the southern part of the Fold-Thrust Belt (FTB) between the Himalayan Frontal Thrust (HFT) and the Jwalamukhi Thrust (JMT; Fig. 2). Here the ridges are more simply NW-SE elongated almost parallel to the HFT. North of JMT, although the same arrangement of ridges exists there is a curvature which most likely replicates the strike of the MBT. These form a different surface expression between the JMT and MBT that is more subdued by the cover of Quaternary alluvial fills and fans descending from the adjoining mountain ranges. Also, the surface expression in this part is complicated by the interaction of tectonic and surface processes strongly modulated by the climatic signal (Dey et al. 2016). The JMT forms an important tectonic boundary which internally constrains the Sub-Himalayan wedge, although its activity has been identified only through neotectonic studies and terrace dating (Singh et al. 2012; Mukherjee 2015; Dey et al. 2016).

The most important surface topographic parameter that contributes to the taper of the Sub-Himalayan wedge is the forelandward slope of the modern day topographic envelope (α). For the Kangra reentrant, the present day topographic slope α has been estimated to be $0.9° \pm 0.2°$ based on analysis of topographic maps and Digital Elevation Models (Singh et al. 2012). This too small alpha is comparable to many other wedges developed in the paleo-forelands of collisional settings over weak detachments in the Salt Range of Pakistan (Davis and Lillie 1994) and the Zagros FTB of Iran (McQuarrie 2004).

4 Detachment Geometry

Although uncertainties exist about the geometry and rheology of the detachment (Main Himalaya Thrust: MHT) in the NW Himalaya, it is generally considered to be a gently dipping single continuous sliding plane over which the cover sequences deform (Fig. 3). The dip of the detachment has been estimated $\sim 3°$in the external sector using well logs and seismic profiling carried out by the ONGC (Powers et al. 1998). The dip is more or less uniform under the entire Sub-Himalayan belt of Kangra reentrant. This is particularly true because any undulations in the geometry of the detachment would locally induce stress concentrations causing clustering of seismicity, which contradicts the seismic findings (Arora et al. 2009). However, in the down dip direction further north of the MBT, the dip of the detachment increases to about 10°–15°. This corresponds with the zone of intense microseismicity that defines the Himalayan mid-crustal ramp just north of the Main Central Thrust (MCT) (Figs. 1 and 2). Mechanically the basement is a rigid mass made-up of the Vindhayan rocks equivalent and followed by the metamorphic/granitic basement at deeper levels (Powers et al. 1998; Rajendra Prasad et al. 2011). The internal thrusts formed within the Sub-Himalayan belt are the result of continued contraction tectonics of the Himalaya (Singh et al. 2014). They emerge from the basement and deform the cover sequences into broad open folds (Powers et al. 1998). The anticlinal folds manifest in the surface topography as ridges followed by trailing synclines manifested as wide flat intermontane depressions (Fig. 2).

5 Cover Sequences

The overall cross-sectional taper of the wedge is intricately related to the competency contrast of the rocks in the wedge and its basal detachment (Davis et al. 1983). In the Kangra reentrant the basal detachment is resolved to be a rigid surface made of competent rocks. However, there are outstanding issues regarding the extent of the resolution and role of the cover rocks involved in deformation. The frictional strengths of a wide variety of rocks in the upper crust are similar (Davis and Engelder 1985). However, there are special considerations for clay and shales, which are relatively weak and their presence can significantly modify the deformation style if they are predominant in the cover sequence (Davis and Lillie 1994). They tend to behave similar to weak horizons as in the case of eastern Potwar plateau (Pakistan), which exhibits increased sedimentation rates of 0.46–0.53 m/kyr during the Middle Siwalik sedimentation towards the top of the sequence. The sequence is predominantly sandy with interbedded clays of the Middle Siwaliks (Pennock et al. 1989).

The cover sequences of the Sub-Himalayan FTB deform over a rigid basement forming foreland verging thrusts, fault-related folds and back-thrusts. The structural geometries are constrained through data from surface geology, 2D-seismic profiles, well logs and other ancillary data from geophysical techniques (Figs. 2 and 3; Powers et al. 1998; Mukhopadhyay and Mishra 1999; Jayangondaperumal et al. 2017). It is important to note the major thrusts towards from S to N are Himalayan

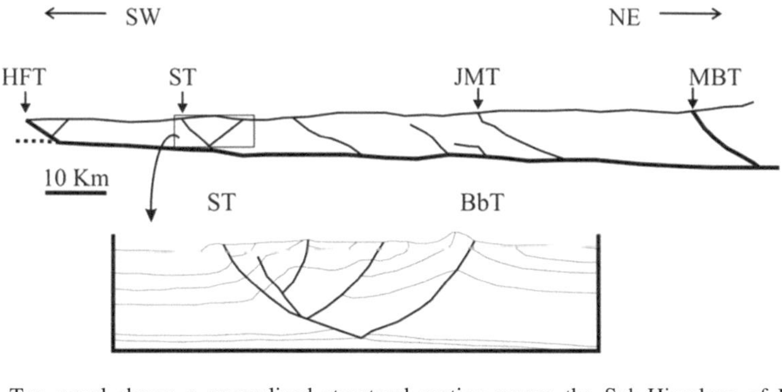

Fig. 3 Top panel shows a generalized structural section across the Sub-Himalaya of Kangra Reentrant (modified from Powers et al. 1998). For section line, refer to Fig. 2. Vertical and horizontal scales are the same. The bold line forming the base of the section represents the basement. The internal thrusts shown in the section emerge from the basement and deform the cover sequences into broad open folds (bottom panel). The bottom panel details the triangle zone structure formed between the Soan Thrust (ST) and the Barsar back Thrust (BbT). Regional drag folds (Mukherjee 2014) are also exemplified in this cross-section

Frontal Thrust (HFT), Soan Thrust (ST), Jwalamukhi Thrust (JWT), Palampur Thrust (PT) and the Main Boundary Thrust (MBT). The geometry of the folds i.e. Janauri Anticline (JA), Dera-Gopipur Anticline (DGA), Bahl Anticline (BA) and Paror Anticline (PA), is clearly attributed to the movement on associated thrust faults (Fig. 3). The spacing between the thrusts has controlled the development and preservation of fault-related folds in their hanging walls. The main back thrusts are Janauri back Thrust (JbT) and Barsar back Thrust (BbT).

In the tectonic wedge of the Kangra reentrant, the sedimentary cover rocks have been thrust southward over a series of thrust faults (Powers et al. 1998; Mukhopadhayay and Mishra 1999; Singh et al. 2012). There is almost no evidence of basement rocks involved in thrusting (Dubey 2004). The vergence of main structural elements i.e. HFT, ST, JT, PT and MBT is southward, towards the foreland. However, structures with opposite vergence are also common towards the frontal region e.g. JbT, BbT. The thrusts and back thrusts show active deformation consistent with the model of distributed deformation (Singh et al. 2012; Thakur et al. 2014; Dey et al. 2016; Jayangondaperumal et al. 2017; also Bose and Mukherjee, submitted-1, 2). The Tertiary formations thicken towards the northeast and the structural style becomes increasingly more complex towards the hinterland, as compared over the ~80 km length of the Sub-Himalayan wedge from SW towards NE (Powers et al. 1998; Mukhopadhyay and Mishra 1999).

6 Coupling Between the Cover Sequences and the Basement

There are a series of research contributions on the active tectonic deformation along thrust faults in the Kangra reentrant (Malik and Mohanty 2007; Singh et al. 2012; Thakur et al. 2014; Dey et al. 2016). The last major earthquake in this area occurred in 1905 (M = 7.8) near Kangra (Fig. 2). As this happened before the beginning of instrumental monitoring, there are no good records on the precise location, nature and extent of the rupture zone. However, there have been outstanding debates with few equivocal surface manifestations found (Szeliga and Bilham 2017). The 1905 Kangra earthquake is generally attributed to the recent slip along the Main Boundary Thrust (MBT). In the instrumental era since late 1950s when seismic monitoring through instruments began, there have been no major earthquakes in this area and most of the recorded smaller events are shallow (<10 km) with 3–5 magnitudes. Such seismic behaviour would also not suggest any involvement of the basement (Mukhopadhyay and Dasgupta 2015). Geologically this would mean that the cover rocks are poorly coupled with the basement which has a low estimated range of basal friction (μ_b) i.e. 0.26–0.36 in the Kangra Reentrant (Singh et al. 2012). Also range-normal convergence velocities across the Sub-Himalayan belt demonstrate reduced intensities connoting poor tectonic coupling between the cover and the basement (Schiffman et al. 2013). However, in the Kangra Reentrant it can be noted that the smoothed contours of range-normal convergence velocities follow the surface topography (Fig. 2). The spatial distribution of these topographic contours in the Kangra Reentrant suggests that complete coupling is only observed in the areas beyond the active deformation front (HFT).

7 Discussions

The Kangra Reenetrant exhibits a topographic slope (α) of $\sim 1°$ towards SW and a relatively high basement dip (β) of 3° towards NE (Fig. 4; in comparison to FTBs in Pakistan and Iran). The two parameters (α and β) contribute to the taper of the wedge, which is $\sim 4°$. This wedge taper is slightly higher than the Salt Range-Potwar Plateau (Pakistan) where it ranges 1°–3° (Davis and Lillie 1994) and Fars and Lorestan segments of the Zagros FTB (Iran) where it is 1.5° (McQuarrie 2004). The FTB's in Pakistan and Iran are comparable in age to the Sub-Himalayan belt but it is notable that both the FTBs in Pakistan and Iran have evolved over a weak detachment in salt. In fact, the wedge taper in the Kangra Reentrant is comparable to the Dezful embayment of the Zagros FTB of Iran which is around 4° (McQuarrie 2004). The most likely reason for the increased wedge taper of the Dezful embayment is its limited width of the FTB, which is controlled by the spatial distribution of the Hormuz salt (Mukherjee et al. 2010). Overall, for the Kangra Reentrant the high basement dip β is supporting a gentler topographic slope α (Davis et al. 1983; Dahlen 1990; Fig. 4). What is intriguing is that the topographic slope is similar to that observed over the weak basement in Pakistan and elsewhere (Fig. 4), whereas there is no report of such a basement rheology in the Kangra Reentrant. The topographic slope is extremely low ($\sim 1°$, almost similar to the FTBs in Pakistan and Iran) and forms a relatively narrow wedge taper over a width of about 80 km (largest in the Indian part). Such characters of a wedge which slides over a rigid basement give rise to a paradoxical situation that has been dealt in detail based on a synthesis of data presented in the earlier Sects. 3, 4, 5 and 6.

In general, assuming a particular rheology of the basement, the other important factors controlling the growth of the FTB and surface topographic slope (α) could be the mechanical stratigraphy, pore-fluid pressure of the cover rocks and climatic effects. Also in a particular area, under similar climatic conditions competent rocks would be able to retain a higher surface topography whereas incompetent rocks would not. Therefore in the Kangra Reentrant the basement is assumed to be a rigid competent mass over which the total thickness of incompetent clay-rich Paleogene–Neogene section is 4–8 km. This thickness is higher than similar sediments in the Potwar plateau (Davis and Lillie 1994). So, if a similar but thinner sediment layer could contribute to the weak coupling between the basement and cover sequence (Davis and Lillie 1994), it would be reasonable to believe that these horizons would definitely provide a reduced resistance to slippage along faults. Also these sediments would effectively couple weakly between the basement and cover rocks in the Kangra reentrant as well (Davis and Engelder 1985). In principle, this situation

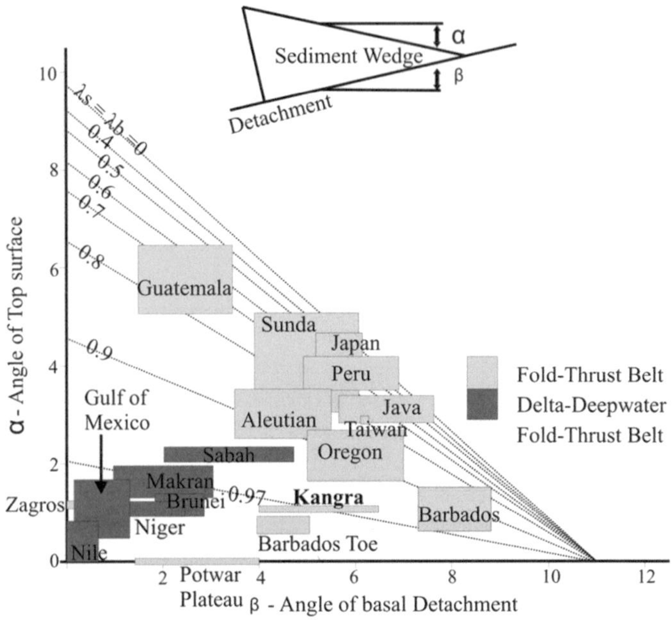

Fig. 4 Surface slope α versus dip of detachment β for natural FTBs and Delta-Deepwater FTBs (modified from Ford 2004; Tuitt et al. 2012)

is also analogous to many deep water FTBs, which contain detachments in shale (e.g. Niger delta; Fig. 4). In addition, the cover strata of the Kangra Reentrant are relatively overpressured (up to 1.5–2 times more than the hydrostatic pressure cf. Powers et al. 1998). This is also supported by high sedimentation rates in this oversized basin (Rajendra Prasad et al. 2011). Therefore the high pore-fluid pressures would efficiently act to reduce the effective normal stress in turn reducing the shear traction and contribute to the reduction in coupling between the cover strata and the basement. This is especially true and important because the presence of high pore-fluid pressure shales would act as an efficient detachment, although not as efficient as halite (Kreuger and Gilbert, 2009; also see Dasgupta and Mukherjee submitted). This would facilitate in generating a wider FTB such as in the Kangra Reentrant, similar to those that are formed also see over weak detachments in salt and over pressured shales elsewhere (Fig. 4).

The most interesting observation is that the strength of the wedge (its frictional character), which could be either due to the rheology or other characters such as pore pressure would strongly control the orientation of the principal stress axes. Accordingly, an increased pore-fluid pressure would reduce the internal friction of the wedge (μ). This in-turn would decrease Ψ_b as demonstrated by the Mohr-Coulomb criterion (Fig. 5). Ψ_b strongly controls the orientation of the principal stress axes and the style of deformation as illustrated in Fig. 6.

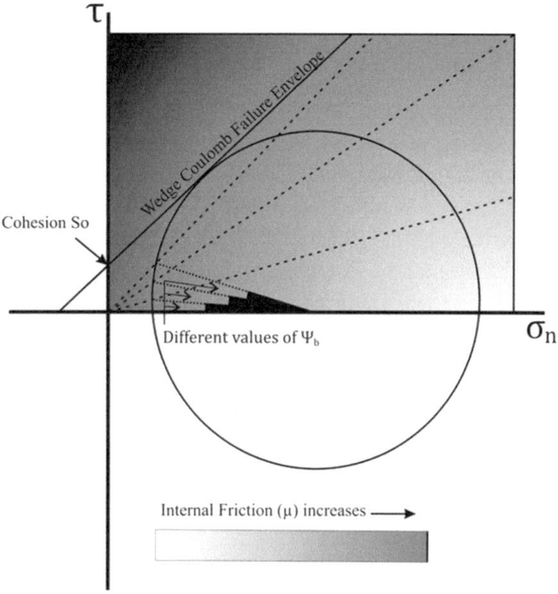

Fig. 5 Mohr-Coulomb failure criterion showing how different values of Ψ_b are related to the internal friction (μ) of the wedge. With the reduction of internal friction, the Ψ_b angle also reduces. In particular for wedges with basement in salt Ψ_b would be approximately 1° (modified after Davis and Engelder, 1985). The τ and σ represent shear and normal stress respectively

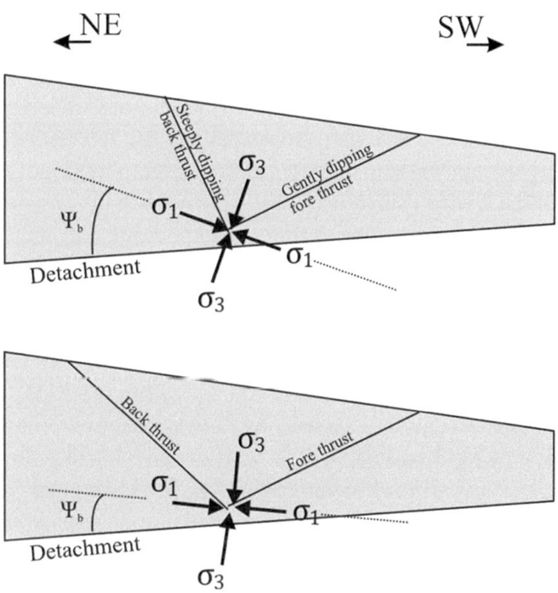

Fig. 6 The illustration shows the relationship between the angle Ψ_b (from Fig. 5) and the orientation of σ_1. **a** In a stronger wedge where the forethrust with a gentler dip is a preferred slip plane in comparison to the steeply dipping back thrust. **b** In a weaker wedge where the forethrust and back thrust are both preferred slip planes. The section is oriented NE–SW in accordance with the Kangra section of Fig. 3

The normal (σ) and shear stresses (τ) acting on a plane are evaluated using the Mohr-Coulomb failure criterion (Fig. 5). The normal to the plane is oriented at an angle Θ with the maximum principal stress, σ_1 direction. Accordingly, there are two planes oriented at an angle inclined on either side of the σ_1 axis. For frictional wedges the angle Ψ_b offsets the orientation of σ_1 in such a way that the σ_1 axis is highly inclined with respect to the detachment (Fig. 6a). In this case, the two slip planes formed have different orientations (vergence) with reference to the slope of the topographic wedge. The foreland verging plane, the fore-thrust, usually dips gently towards the hinterland whereas the hinterland verging plane, the back-thrust, dips steeply towards the foreland (Fig. 6a). Geometrically, the fore-thrust would be the preferred slip plane. Thus, these FTBs should have a preferred structural vergence.

On the contrary, for weaker wedges, Ψ_b tends to be quite small (Fig. 5) leading to a near horizontal σ_1. In such a situation there could be two slip planes on either side of the σ_1, similar to the frictional wedges. But these planes orient in such a way that they make almost equal angle with the surface topography. The two planes in this case also are foreland-verging fore-thrust and hinterland verging back-thrust but unlike the frictional wedges they have comparable dips. Therefore, both these planes i.e. the fore-thrust and the back-thrust would be equally preferred slip planes but with opposite vergence (Fig. 6b). The FTBs on such wedges have no preferred structural vergence as is the case with the Kangra Reentrant (Figs. 2 and 3). This is very well illustrated along the Una–Talmehra section exhibiting a triangle zone between the ST (fore-thrust) and the BbT (back-thrust) where the ST and BbT are separated by ~ 10 km on the surface (Fig. 3). Therefore, wedges with incompetent rocks in the cover strata and poor coupling between the basement and cover strata would allow the wedge to translate over a wide belt usually project outward (Figs. 2 and 3). Similar inferences have been drawn from the comparisons between the detachment strength to wedge strength (Stockmal 1983; Davis and Engelder 1985). As the wedge in the Kangra Reentrant would rapidly propagate forward it would incorporate the undeformed foreland into the wedge, towards the frontal part and spatially widen the wedge over the basement. This should significantly lead to the decrease in topographic slope (α) over time and reduce the taper of the wedge (Singh et al. 2012), as seen in the case of Kangra Reentrant.

In summary, the surface morphology of the Kangra Reentrant resembles deformation that is distributed in nature and exhibits poor coupling between the basement and cover units. Other attributes include: (i) presence of wide-spaced folds, (ii) abrupt changes in deformation style across the basin that usually correspond spatially to the weak layer distribution, (iii) narrow cross-sectional taper, and (iv) bimodal vergence of structures within the FTB. The structural evolution of the Kangra Reentrant is mainly by development of fault-related folds in the cover strata over a rigid basement. It is well understood that the basement structure and associated variations in stratigraphic thickness of cover rocks are potential controls for the localization and growth of frontal anticlines (McQuarrie 2004).

8 Conclusions

The structure of the Kangra reentrant is controlled by the geometry of the sedimentary basin (depocentre) that was present in front of the MBT, which largely follows the distribution of the thin Vindhyan basement as limited by the recognition of an older fabric beneath by Rajendra Prasad et al. (2011). This provides a larger accommodation space occupied by thick (4–8 km) incompetent clay rich Paleogene-Neogene section. Therefore the distribution of the excess pore fluid pressures at the depth appears to be stratigraphy-controlled.

The poor coupling between the cover strata together with high pore pressure and presence of thick incompetent rocks will not allow potential accumulation of shear stresses within the rocks of the Kangra area, which is characterized by the shallow and low seismicity (3–5 M) in the region since the last major event of 1905. In most cases there in no involvement of the basement in the India-Eurasia collision induced deformation process of the Sub-Himalayas in the Kangra Reentrant. This indicates that most of these earthquakes would occur by a dynamic weakening mechanism due to overpressured shales. However, further north of the JMT there are potential seismogenic zones, which are actively accumulating shear stresses over narrow localized zones. This may lead to a high magnitude seismicity, similar to the 1905 event. However, the chances of the rupture propagating towards the HFT seem to be limited.

Acknowledgements TPS thanks the Director CSIR-Central Scientific Instruments Organisation Chandigarh for providing infrastructural facilities. Mr. Sarabjeet Singh helped in meticulously drafting Figs. 2 and 4. Soumyajit Mukherjee reviewed this article. The work is summarized in Mukherjee (2019).

References

Arora BR, Gahalaut VK, Kumar N (2009) Structural control on along-strike variation in the seismicity of the northwest Himalaya. Journal of Asian Earth Sciences 57, 15–24

Bahroudi A, Koyi HA (2003) Effect of spatial distribution of Hormuz salt on deformation style in the Zagros fold and thrust belt: an analogue modelling approach. Journal of the Geological Society 160, 719–733

Bilham R, Gaur V, Molnar P (2001) Himalayan seismic hazard. Science 293, 1442–1444

Borderie S, Graveleau F, Witt C, Vendeville BC (2018) Impact of an interbedded viscous décollement on the structural and kinematic coupling in fold-and-thrust belts: insights from analogue modeling. Tectonophysics 722, 118–137

Bose N, Mukherjee S (Submitted-1) Field documentation and genesis of the back-structures from a part of the Garhwal Lesser Himalaya. Uttarakhand, India. Tectonic implications. In: Sharma, Villa IM, Kumar S (eds) Crustal architecture and evolution of the Himalaya-Karakoram-Tibet Orogen. Geological Society of London Special Publications

Bose N, Mukherjee S (Submitted-2) Field documentation and genesis of back-structures from the foreland part of a collisional orogen: Examples from the Darjeeling-Sikkim Lesser Himalaya, Sikkim, India. Gondwana Research

Carrillo E, Koyi HA, Nilfouroushan H (2017) Structural significance of an evaporite formation with lateral stratigraphic heterogeneities (Southeastern Pyrenean Basin, NE Spain). Marine and Petroleum Geology 86, 1310–1326

Costa E, Vendeville BC (2002) Experimental insights on the geometry and kinematics of fold-and-thrust belts above weak, viscous evaporitic decollement. Journal of Structural Geology 24, 1729–1739

Cotton JT, Koyi HA (2000) Modeling of thrust fronts above ductile and frictional detachments: application to structures in the Salt Range and Potwar Plateau, Pakistan. Geological Society of America Bulletin 112, 351–363
Dahlen FA (1990) Critical taper model of fold-and-thrust belts and accretionary wedges. Annual Reviews of Earth and Planetary Sciences 18, 55–99
Dasgupta T, Mukherjee S (submitted) Sediment compaction and applications in petroleum geoscience. In: Swenner R (ed) Springer series: advances in oil and gas exploration & production. ISSN: 2509-372X
Davis D, Engelder T (1985) The role of salt in fold-and-thrust belts. Tectonophysics 119, 67–88
Davis D, Lillie RJ (1994) Changing mechanical response during continental collision: active examples from the foreland thrust belts of Pakistan. Journal of Structural Geology 16, 21–34
Davis D, Suppe J, Dahlen FA (1983) Mechanics of fold-and-thrust belts and accretionary wedges. Journal Geophysical Research 88, 1153–1172
Delcaillau B, Carozza J-M, Laville E (2006) Recent fold growth and drainage development: the Janauri and Chandigarh anticlines in the Siwalik foothills, northwest India. Geomorphology 76, 241–256
Dey et al (2016) Climate-driven sediment aggradation and incision since the late Pleistocene in the NW Himalaya, India. Earth and Planetary Science Letters 449, 321–331
Dubey AK (2004) Formation of decollement upwarps during thrusting. Terra Nova 16, 91–94
Ford M (2004) Depositional wedge tops: interaction between low basal friction external orogenic wedges and flexural foreland basins. Basin Research 16, 361–375
Jayangondaperumal R, Kumahara Y, Thakur VC, Kumar A, Srivastava P, Dubey S, Joevivek V, Dubey AK (2017) Great earthquake surface ruptures along backthrust of the Janauri anticline, NW Himalaya. Journal of Asian Earth Sciences 133, 89–101
Koyi H (1988) Experimental modeling of role of gravity and lateral shortening in Zagros mountain belt. AAPG Bulletin 72, 1381–1394
Krueger A, Gilbert E (2009) Deepwater fold-thrust belts: not all the beasts are equal. In: American association of petroleum geologists international conference and exhibition. Cape Town, South Africa. AAPG Search and Discovery. Article 30085: http://www.searchanddiscovery.com/documents/2009/30085krueger
Kumar S, Wesnousky SG, Rockwell TK, Briggs RW, Thakur VC, Jayangondaperumal R (2006) Paleoseismic evidence of great surface rupture earthquakes along the Indian Himalaya. Journal Geophysical Research 111, B03304
Larson K, Burgmann R, Bilham R, Freymueller JT (1999) Kinematics of the India-Eurasia collision zone from GPS measurements. Journal Geophysical Research 104, 1077–1093
Lavé J, Avouac JP (2000) Active folding of fluvial terraces across the Siwalik Hills, Himalayas of central Nepal. Journal Geophysical Research 105, 5735–5770
Malik J, Mohanty C (2007) Active tectonic influence on the evolution of drainage and landscape: geomorphic signatures from frontal and hinterland areas along the Northwestern Himalaya, India. Journal of Asian Earth Sciences 29, 604–618
McQuarrie N (2004) Crustal scale geometry of the Zagros fold–thrust belt, Iran. Journal of Structural Geology 26, 519–535
Mukherjee S (2013a) Channel flow extrusion model to constrain dynamic viscosity and Prandtl number of the Higher Himalayan Shear Zone. International Journal of Earth Sciences 102, 1811–1835
Mukherjee S (2013b) Higher Himalaya in the Bhagirathi section (NW Himalaya, India): its structures, backthrusts and extrusion mechanism by both channel flow and critical taper mechanisms. International Journal of Earth Sciences 102, 1851–1870
Mukherjee S (2014) Review of flanking structures in meso- and micro-scales. Geological Magazine 151, 957–974
Mukherjee S (2015) A review on out-of-sequence deformation in the Himalaya. In: Mukherjee S, Carosi R, van der Beek P, Mukherjee BK, Robinson D (eds) Tectonics of the Himalaya, vol 412. Geological Society, London, Special Publications, pp 67–109

Mukherjee S (2019) Introduction to "Tectonics and Structural Geology: Indian Context". In: Mukherjee S (ed) Tectonics and structural geology: Indian context. Springer International Publishing AG, Cham, pp 1–5. ISBN: 978-3-319-99340-9

Mukherjee S, Talbot CJ, Koyi HA (2010) Viscosity estimates of salt in the Hormuz and Namakdan salt diapirs, Persian Gulf. Geological Magazine 147, 497–507

Mukhopadhayay DK, Mishra P (1999) A balanced cross section across the Himalayan foreland belt, the Punjab and Himachal foothills: a reinterpretation of structural styles and evolution. Proceedings of Indian Academy of Science 108, 189–205

Mukhopadhyay B, Dasgupta S (2015) Seismic hazard assessment of Kashmir and Kangra valley region, Western Himalaya, India. Geomatics, Natural Hazards and Risk 6, 149–183

Nilforoushan F, Koyi H, Swantesson J, Talbot CJ (2008) Effect of basal friction on surface and volumetric strain in models of convergent settings measured by laser scanner. Journal of Structural Geology 30, 366–379

Nilfouroushan F, Pysklywec R, Cruden A (2012) Sensitivity analysis of numerical scaled models of fold-and-thrust belts to granular material cohesion variation and comparison with analog experiments. Tectonophysics 526–529, 196–206

Paul D, Mitra S (2015) Role of salt in decoupling deformation and fault evolution in the Smørbukk area, Halten Terrace, offshore Mid-Norway. In: Annual convention and exhibition. Denver, Colorado, 31 May–3 June 2015

Pennock ES, Lillie RJ, Hamid Zaman AS, Yousaf M (1989) Structural interpretation of seismic reflection data from eastern salt range and Potwar Plateau, Pakistan. AAPG Bulletin 73, 841–857

Powers PM, Lillie RJ, Yeats RS (1998) Structure and shortening of the Kangra and Dehra Dun Reentrants, sub-Himalaya, India. Geological Society of America Bulletin 110, 1010–1027

Raiverman V (2002) Foreland sedimentation in Himalayan TectonicRegime. A relook at the orogenic process, Bishen Singh Mahendra Pal Singh Publisher, Dehra Dun, India, p 371

Rajendra Prasad B, Klemperer SL, VijayaRao V, Tewari HC, Khare P (2011) Crustal structure beneath the sub-Himalayan fold-thrustbelt, Kangra recess, north west India, from seismic reflection profiling: implications for Late Paleoproterozoicorogenesis and modern Earth hazard. Earth and Planetary Science Letters 308, 218–228

Ruh JB, Kaus BJP, Burg J-P (2012) Numerical investigation of deformation mechanics in fold-and-thrust belts: influence of rheology of single and multiple décollements. Tectonics 31, TC3005

Schiffman C, Bali BS, Szeliga W, Bilham R (2013) Seismic slip deficit in the Kashmir Himalaya from GPS observations. Geophysical Research Letters 40, 5642–5645

Singh T (2008a) Hypsometric analysis of watersheds developed on actively deforming Mohand Anticlinal ridge, NW Himalaya. Geocarto International 23, 417–427

Singh T (2008b) Tectonic implications of geomorphometric characterization of watersheds using spatial correlation: Mohand Ridge, NW Himalaya, India. Zeitschrift fur Geomorphologie 52, 489–501

Singh T, Jain V (2009) Tectonic constraints on watershed development on frontal ridges: Mohand Ridge, NW Himalaya, India. Geomorphology 106, 231–241

Singh T, Sharma U, Awasthi AK, Virdi NS, Kumar R (2011) Geomorphic and structural evidences of neotectonic activity in the Sub-Himalayan belt of Nahan Salient, India. Journal of Geological Society of India 77, 175–182

Singh T, Awasthi AK, Caputo R (2012) The Sub-Himalayan fold-thrust belt in the 1905-Kangra earthquake zone: a critical taper model perspective for Seismic Hazard Assessment. Tectonics 31, TC6002

Singh T, Awasthi AK, Paul D, Caputo R (2014) Role of internal thrusts in NW Sub-Himalaya, India, for SHA. Journal of Himalayan Earth Sciences 153

Stevens VL, Avouac J-P (2015) Interseismic coupling on the main Himalayan thrust. Geophysical Reseach Letters 42, 5828–5837

Stockmal GS (1983) Modeling of large-scale accretionary wedge deformation. Journal of Geophysical Research, 88, 8271–8287

Stokmal J, Beaumont C, Nguyen M, Lee B (2007) Mechanics of thin-skinned fold-and-thrust belts: insights from numerical models. In: Sears JW, Harms TA, Evenchick CA (eds) Whence the mountains? Inquiries into the evolution of orogenic systems: a volume in honor of Raymond. The Geological Society of America Special Paers, pp 63–98

Suppe J (1983) Geometry and kinematics of fault-bend folding. American Journal of Science 283, 684–721

Suppe J, Medwedeff DA (1990) Geometry and kinematics of fault-propagation folding. Eclogae Geologicae Helvetiae 83, 409–454

Szeliga W, Bilham R (2017) New Constraints on the mechanism and rupture area for the 1905 Mw 7.8 Kangra earthquake, Northwest Himalaya. Bulletin of the Seismological Society of America 107, 2467–2479

Thakur VC, Joshi M, Sahoo D, Suresh N, Jayangondaperumal R, Singh A (2014) Partitioning of convergence in Northwest Sub-Himalaya: estimation of late Quaternary uplift and convergence rates across the Kangra reentrant, North India. International Journal of Earth Sciences 103, 1037–1056

Tuitt A, King R, Hergert T, Tingay M, Hillis R (2012) Modelling of sediment wedge movement along low-angle detachments using ABAQUS. Geological Society of London Special Publication 367, 171–183

Valdiya KS (1984) Aspects of tectonics. Focus on south-central Asia. Tata McGraw-Hill, New Delhi, p 319

Valdiya KS (1998) Dynamic Himalaya. Universities Press, Hyderabad, p 186

Withjack MO, Callaway S (2000) Active normal faulting beneath a salt layer: an experimental study of deformation patterns in the cover sequence. AAPG Bulletin 84, 627–651

Yeats RS, Nakata T, Farah A, Fort M, Mirza MA, Pandey MR, Stein RS (1992) The Himalayan frontal fault system. Annals of Tectonicae 6, 85–98

Impact of Structural Damage Zones on Slope Stability: A Case Study from Mandakini Valley, Uttarakhand State (India)

Mohit Kumar, Ramesh Chander Joshi and Pitamber Dutt Pant

1 Introduction

A damage zone represents the volume of deformation confined in a zone around a fault surface (Cowie and Scholz 1992; McGrath and Davison 1995; Caine et al. 1996; Kim et al. 2004a; Childs et al. 2009; Mukherjee 2014; Choi et al. 2016 and Peacock et al. 2017). Several factors control slope stability viz., rainfall, earthquakes, joints and their geometries, geological material, slope angles, anthropogenic activities (Hoek and Bray 1981; Panikar and Subramanyam 1997; Paul and Mahajan 1999; Pant and Luirei 1999; Gupta and Bist 2004; Stead and Eberhardt 2013a; Puniya et al. 2013; Kumar et al. 2017; Sah et al. 2018). In this study, we are showing the role of structural damage zone on slope stability between Sonparyag and Kund.

2 Study Area and Geology

The Mandakini River originates from the Chorabari glaciers at an altitude of 3840 m altitude and is joined by Son Ganga, Kali Ganga and Madhyamaheshwar Ganga before it meets the Alaknanda River at Rudraprayag at 600 m altitude.

The study area is bound by the Main Central Thrust (MCT) from the south and Vaikrita Thrust (VT) from the north (Fig. 1), this it comes under the Greater

M. Kumar (✉)
National Geotechnical Facility, Survey of India, Dehradun, India
e-mail: puniyamohit@gmail.com

R. C. Joshi
Department of Geography, Kumaun University, DSB Campus, Nainital, India

P. D. Pant
Department of Geology, Kumaun University, Nainital, India

S. Mukherjee (ed.), *Tectonics and Structural Geology: Indian Context*,
Springer Geology, https://doi.org/10.1007/978-3-319-99341-6_14

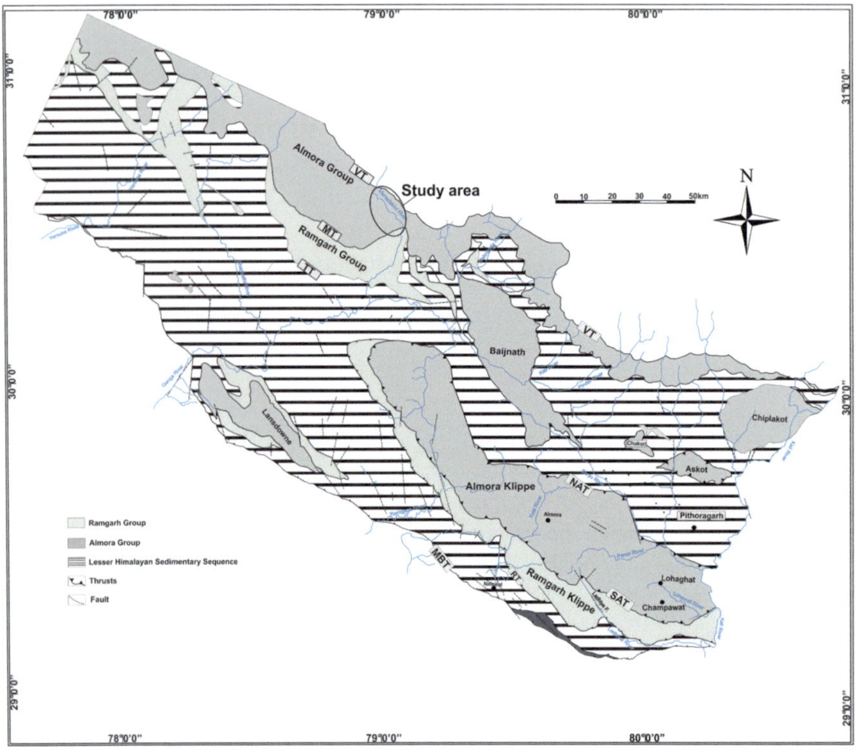

Fig. 1 Regional geological map of the area (Valdiya 1980)

Himalaya (see Mukherjee 2013a, 2015a; Mukherjee et al. 2012 etc.). MCT constitutes a wide zone near Kund (approximately 12 km downstream from Guptakashi), and marks the boundary between Munsiari Formation at north and Bhatwari Formation at south (Figs. 1 and 3). Tilwara Thrust (TT) separates the Ramgarh Group and Lesser Himalayan Metasedimentary Rocks (Fig. 1). Different shear zones rocks viz. mylonitic augen-gneiss, calc-silicates, biotite-gneiss and fine-grained gneiss (Fig. 2) are identified in the field during traverse mapping. The Bhatwari Formation consists of flaggy quartzite and basic intrusion in the form of amphibolites (Fig. 3).

3 Methodology

Detailed structural mapping has been done with the help of handheld GPS and available data have been collected such as foliation, different joint sets, folds, faults, shear sense indicators, slope angle and height. 1:10,000 scale map has been

Fig. 2 **a** mylonitic augen gneiss with symmetric clasts (Mukherjee 2017) at location-24 (30° 32′ 17.05′N, 79° 04′ 49.15″E); **b** a contact between clac-silicate and biotite gneiss dipping towards NNW (322°) with 48° dip amount (dash line represents the contact) at location-13 (30° 35′ 46.38′ N, 79° 01′ 20.14″E); **c** a xenoliths in fine grained gneiss at location-19 (30°34′41.99′N, 79° 02′ 38.18″E)

prepared along the National highway-109 parallel to Mandakini river valley from Sonparyag to Kund. Study area falls in the Survey of India toposheet numbers 53 N/3 and 53 N/4.

This study focuses on the structural damage zones and their relationship to the different rock mass classification methods, viz., rock mass rating (RMR), kinematics analysis of the slopes, slope mass rating (SMR). RMR has application for tunnels, slopes (for SMR calculation) and foundations (dam, bridges and other civil structures; Bieniawski 1989). Field data collection such as, the orientations of different discontinuities (joints, shear zones, faults etc.) and uniaxial compressive strength (UCS) were calculated using Schmidt hammer method as per the International Society for Rock Mechanics (ISRM) recommended methods (ISRM 1978, 1981, 2007; Bieniawski 1989; Brencich et al. 2013). Spacing of discontinuities, conditions of discontinuities (aperture, roughness, type of filling and weathering conditions), rock quality designation (RQD), groundwater condition, and orientations of slopes with dip angles have been collected (Appendix-I). All aforesaid parameters were used to calculate the RMR in this study. Kinematics analysis has been performed for the slopes to delineate the kind of failure. Internal

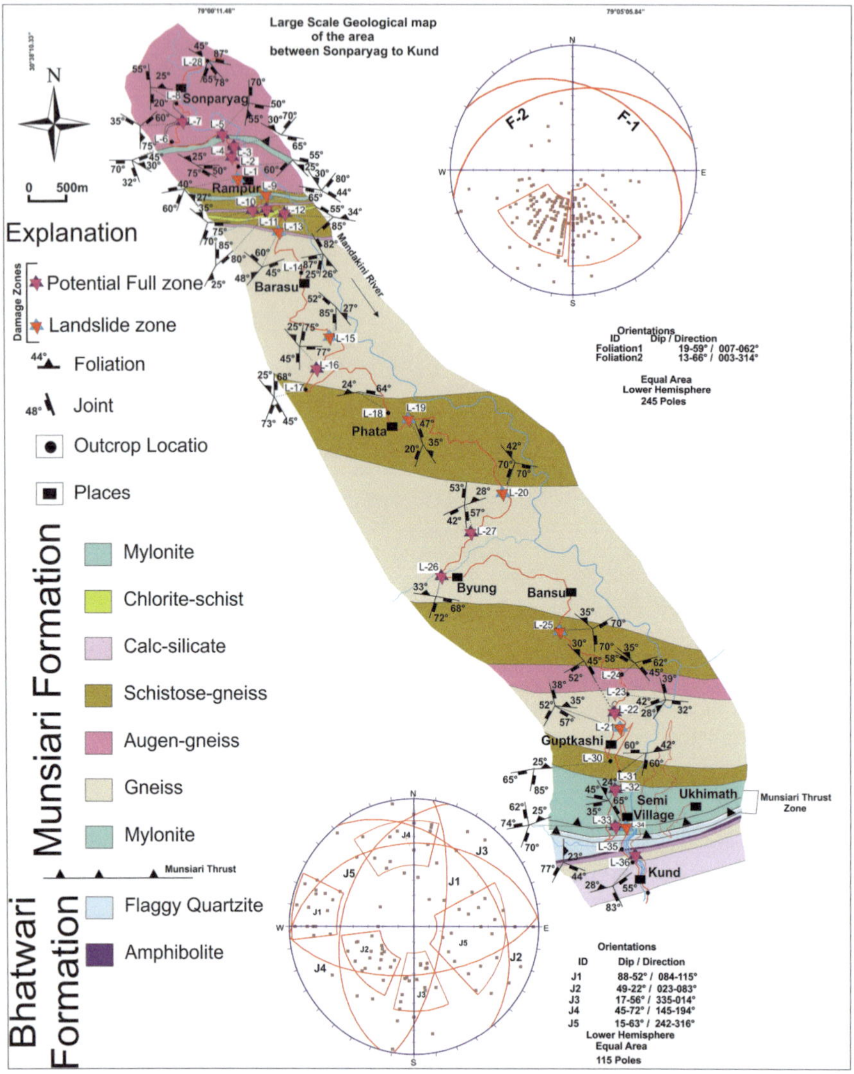

Fig. 3 Traverse map of the study area from Sonparyag to Kund

friction angle has been taken as 25°–35° for the slope according to Hoek and Bray (1981). The RMR is calculated as per Bieniawski (1989).

The slope mass rating (SMR) proposed by Romana (1985) and is obtained from RMR by:

$$\mathrm{SMR} = \mathrm{RMR} + (\mathrm{F1.F2.F3}) + \mathrm{F4} \tag{1}$$

Here

F1 depends on parallelism between joints and slope face. Its range is from 1.00 to 0.15.

$$\mathrm{F1} = (1 - \mathrm{SinA})^2 \tag{2}$$

A is angle between the dip directions of slope face and the joint for the planar and toppling failure. In wedge failure, A is the angle between trend and slope dip direction.

F2 refers to joint dip angle in the planar mode of failure.

$$\mathrm{F2} = \tan^2\mathrm{B} \tag{3}$$

B is dip angle of joint for planer failure and for wedge failure, B is plunge amount.

F3 represents the angle between the slope and joint dip in the planar failure. In wedge failure, F3 indicates the relationship between the slope face and plunge amount two joints.

F4 factor represents the excavation method and the rating depends on the method as:

$$\text{Natural slopes}: F4 = +15 \tag{4}$$

$$\text{Presplitting}: F4 = +10, \tag{5}$$

$$\text{Smooth blasting}: F4 = +8, \tag{6}$$

$$\text{Normal}\ blasting: F4 = 0, \tag{7}$$

$$\text{Deficient blasting}: F4 = -8, \tag{8}$$

and

$$\text{Mechanical excavation}: F4 = 0. \tag{9}$$

4 Results and Discussions

The study area is mapped along the Mandakini River for ~ 31 km length. The main lithology is mylonite, chloritic-schist, calc-silicates, schistose-gneiss, mylonitic augen gneiss, gneiss of the Munsiari Formation, and flaggy quartzite and

Fig. 4 Field photographs show: **a** contact between mylonite and protomylonite at location-5; **b** a ductile damage zone show shearing and folding at location-10; **c** sulphur leaching on the surface of sheared mylonitic gneiss at location-20; **d** ductile-brittle shearing show top to SW movement at location-22

amphibolites of the Bhatwari Formation (Fig. 3). The shear zone classification is performed in the field based on the proportion of matrix and grains. Based on the shear zone rocks and structure present at the field, we have identified ductile-shear zone, ductile-brittle shear zone and brittle shear zones (Ghosh 1993). Damage zones are deformed wall rock spatially associated with faults that form during the initiation and propagation of a fault, during the interaction of slip in fault linkages or jogs, or during the flexure of beds around a fault (Jamison and Stearns 1982; Chester and Logan 1986; Shipton and Cowie 2001, 2003; Flodin and Aydin 2004; Kim et al. 2004b). To delineate these damage zones in the field, we have collected several structures like rocks with mylonitic foliations (Figs. 3 and 4a) and asymmetric folding (Figs. 4b and 5b),

S-C structures (Fig. 4b; Mukherjee 2010a, b, 2011), phyllonites and mylonites (Figs. 4c and 5a; Bose et al. 2018), brittle-ductile S-C structures (Fig. 4d), fault stariations (Fig. 5c; Dasgupta and Mukherjee 2017; Vanik et al. 2018) and asymmetric boudinage (Fig. 5d; few examples in Mukherjee and Koyi 2010a, b; Mukherjee 2013b, 2015b).

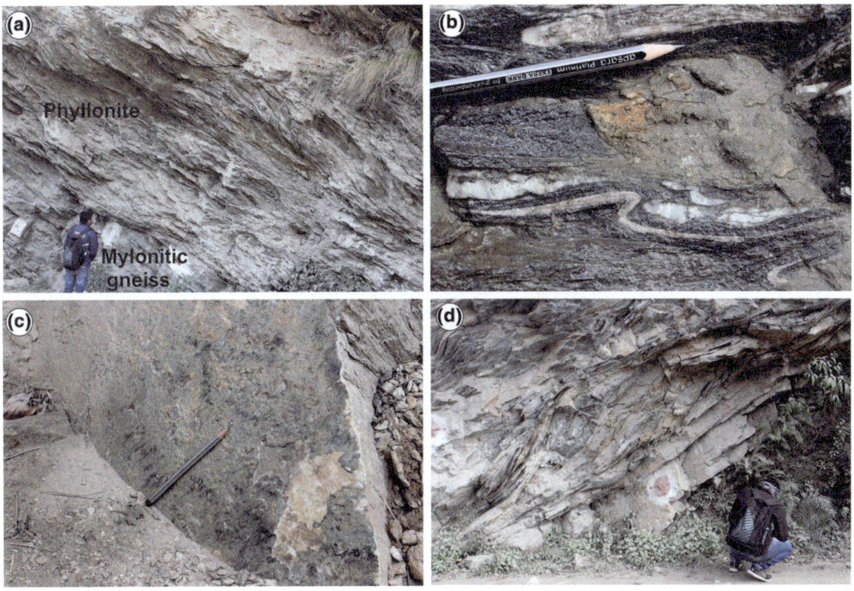

Fig. 5 Field photographs: **a** a contact between phyllonite and mylonitic gneiss at Main Central Thrust near Kund; **b** asymmetric folding, typically intrafolial: Mukherjee et al. (2015) shows top to NE shearing at location-34; **c** fault stariation show up dip movement at location-32; **d** a ductile-brittle shearing show top to SW movement at location-32

Based on aforesaid observations, we have marked 23 structural damage zones between Sonparyag to Kund (Fig. 3) in the form of landslide zones and potential full zone for landslide. The relationship of structural damage zones and slope stability is described by many workers around the world e.g. (Badger 2002; Ambrosi and Ceosta 2011; Brideau and Couture 2006; Stead and Eberhardt 2013b; Stead and Wolter 2015). Landslide zones are marked as active damage zones and potential full zones marked probable damage zones. Structural mapping identified regional synclinal folding in the area, with fold axis passing through Byung, foliation data of the northern limb dipping at 13°–66° towards NNE-NNW (003°–314°) while southern limb dip at 19°–59° towards NE (007–062°) (Fig. 3). Based on the orientation of the discontinuities, 5 sets of joints are identified (Fig. 3) viz., J1 dips at 88°–52° towards 084°–115°, J2 dipping 49°–22° dipping towards 023°–083°; J3 dips 17°–56° towards NNW-NNE (335°–014°); J4 with 45°–72° dip towards 145°–194° (SSE-SSW), and J5 at 15°–63° towards SW-NW (242°–316°).

Kinematics analysis has been performed for 23 damage zones (Fig. 5) and further divided into potential full landslide zone and landslide zone according to the Markland's test (Fig. 6). Kinematics analysis of the slope showed that 12 slopes in which locations L-7, L-9, L-11, L-13, L-15, L-16, L-19, L-20, L-25, L-26, L-27 and L-36 come under planar failure (Fig. 6; Table 1). Locations L-3, L-4A, L-5, L-10,

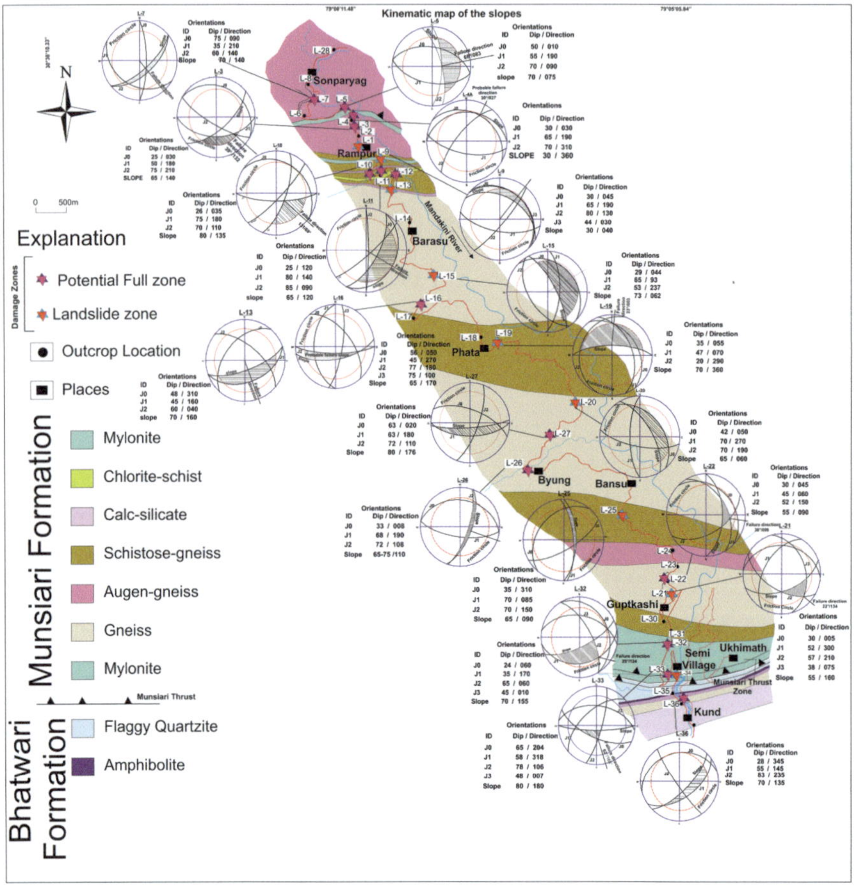

Fig. 6 Traverse map of the study area from Sonparyag to Kund

L-10, L-21, L-22, L-32 and L-33 show wedge failure. Two locations L-1 and L-12 are identified as anthropogenic induced landslides (Fig. 6; Table 1). Damage zones weaken the rocks so the slopes become more prone for failure (Fig. 7). Location-19 near the town Phata show planar failure. Here ductile shear zones are present along the foliation with 35° dip amount and dipping towards NE (055°). Rocks tend to fail due to these shear zones (Fig. 7; Table 1). At this location, the SMR values range 31.9–39.9 (Table 1). SMR values vary from 0 to 55.25 at L-3 and L-33, respectively (Table-1). SMR values also confirm the potential damage zone for landslid**es.** Similarly we have investigated all the locations and find that damage zone assessment and SMR confirms our observations.

Table 1 Data of geological investigation at different locations

Sr. No.	Location			Rock type	RMR (basic)		F1	F2	F3	F4	SMR		Structural condition
		Latitude	Longitude		Min	Max					Min	Max	
1	1	30° 37′ 05.08″	79° 00′ 35.48″	Mylonitic augen gneiss	54	55	0.15	1	−60	0	45	46	Due to anthropogenic activity
2	3	30° 37′ 15.49″	79° 00′ 32.12″	Mylonitic augen gneiss	53	55	0.85	1	−60	0	2	4	Wedge Failure/ ductile-brittle shear
3	4	30° 37′ 25.28″	79° 00′ 28.63″	Mylonitic augen gneiss	54	55	0.15	1	−50	0	46.5	47.5	Wedge failure/ ductile-brittle shear
4	4 (A)	30° 37′ 25.28″	79° 00′ 25.28″	Mylonitic gneiss	54	55	0.4	1	−25	0	44	45	Wedge failure/ ductile-brittle shear
5	5	30° 37′ 31.49″	79° 00′ 19.91″	Mylonitic augen gneiss	49	53	0.85	1	−50	0	6.5	10.5	Wedge Failure/ Ductile-brittle shear
6	7	30° 37′ 33.90″	78° 59′ 54.69″	Mylonitic gneiss	53	57	1	1	−50	0	3	7	Planar failure/mica bands
7	9	30° 36′ 55.11″	79° 00′ 55.52″	Mylonitic gneiss	47	56	1	1	−25	0	22	31	Planar failure/brittle shear
8	10	30° 36′ 43.82″	79° 00′ 45.01″	Biotite gneiss	52	57	0.85	1	−60	0	1	6	Wedge failure/ ductile-brittle shear
9	11	30° 36′ 45.23″	79° 01′ 05.63″	Biotite gneiss	51	51	0.85	1	−60	0	0	0	Planar failure/ductile shear
10	12	30° 36′ 32.11″	79° 01′ 05.98″	Augen gneiss	45	48	0.7	1	0	0	45	48	Lithological contact/Joint control
11	13	30° 36′ 14.22″	79° 01′ 18.34″	Biotite gneiss	54	55	0.4	1	−60	0	30	31	Planar failure/joint control
12	15	30° 35′ 34.07″	79° 01′ 35.81″	Granitic gneiss	30	35	0.15	1	0	0	30	35	Planar failure/ buctile-brittle shear

(continued)

Table 1 (continued)

Sr. No.	Location			Rock type	RMR (basic)		F1	F2	F3	F4	SMR		Structural condition
		Latitude	Longitude		Min	Max					Min	Max	
13	16	30° 35′ 21.07″	79° 01′ 44.92″	Biotite gneiss	46	47	0.85	1	0	0	46	47	Planar failure/joint control
14	19	30° 34′ 41.99″	79° 02′ 38.18″	Biotite gneiss	48	49	0.4	1	−60	0	24	25	Planar failure/ductile shear
15	20	30° 34′ 29.19″	79° 05′ 20.05″	Fine grained gneiss	42	43	0.15	1	−50	0	34.5	35.5	Planar failure/joint control
16	21	30° 31′ 43.85″	79° 05′ 05.84″	Fine grained schistose gneiss	46	47	0.4	1	−60	0	22	23	Wedge failure/joint control
17	22	30° 31′ 49.06″	79° 04′ 59.61″	Augen gneiss	53	57	0.85	1	−60	0	2	6	Wedge failure/ductile shear
18	25	30° 32′ 56.08″	79° 04′ 26.05″	Schistose gneiss	35	38	0.85	1	−6	0	29.9	32.9	Planar failure/ buctile-brittle shear
19	26	30° 33′ 13.60″	79° 02′ 57.91″	Gneiss	37	45	0.85	1	−6	0	31.9	39.9	Planar failure/joint control
20	27	30° 33′ 33.23″	79° 03′ 19.77″	Fine grained gneiss	37	48	0.4	1	−50	0	17	28	Planar failure/joint control
21	32	30° 31′ 23.10″	79° 04′ 58.70″	Mylonitic gneiss	52	53	0.4	1	−60	0	28	29	Wedge failure/ ductile-brittle shear
22	33	30° 31′ 03.80″	79° 04′ 58.30″	Biotite gneiss	58	59	0.15	1	−25	0	54.25	55.25	Wedge failure/lithological contact
23	36	30° 31′ 19.6″	79° 05′ 20.7″	Biotite gneiss/greenish gneiss	44	50	0.7	1	−60	0	2	8	Planar failure/lithological contact

Fig. 7 Location-19 near Phata town show planar failure along the foliation plane (J0), along J0 ductile shear zone inducing the landslide

5 Conclusions

23 structural damage zones are identified in the study area. 6 joint sets are identified that reduce the rock strength. Kinematics analyses confirm that 12 slopes show planar failure and 9 are wedge failure. SMR for different modes of failure ranges 0–55.25. This study confirms that structural damage zones are also responsible for the slope failure and are important to consider for the road construction in the hilly area. Structural damage zone assessment mapping also show relation with the slope kinematics analyses and the SMR.

Acknowledgements Authors are very thankful to the Department of Geography, DSB Campus Kumaun University Nainital to provide the necessary facility for the research. Authors are also very thanking to the NRDMS, Department of Science and Technology to the financial assistance in the project no. NRDMS/06/11/015 (G) Dt. 11.08.2015 "Large Scale Geological-Geomorphological Mapping of Rudraprayag-Sonprayag Area, Mandakini valley National highway No.109. Reviewing and editorial handling by Soumyajit Mukherjee. This work is summarized in Mukherjee (2019). The Springer team is thanked for assistance in proofreading.

References

Ambrosi C, Crosta GB (2011) Valley slope influence on deformation mechanisms of rock slopes. In: Jaboyedoff M (ed) Slope tectonics, vol 351. Geological Society, London, Special Publications, pp 215–233

Badger TC (2002) Fracturing within anticlines and its kinematic control on slope stability. Environmental & Engineering Geoscience 8, 19–33

Bieniawski ZT (1989) Engineering rock mass classification. Wiley, New York

Bose N, Dutta D, Mukherjee S (2018) Role of grain-size in phyllonitisation: Insights from mineralogy, microstructures, strain analyses and numerical modeling. Journal of Structural Geology 112, 39–52

Brencich A, Cassini G, Pera D, Riotto G (2013) Calibration and reliability of the rebound (Schmidt) hammer test. Civil Engineering and Architecture 1, 66–78

Brideau M-A, Stead D, Couture R (2006) Structural and engineering geology of the east gate landslide, Purcell mountains, British Columbia, Canada. Engineering Geology 84, 183–206

Caine JS, Evans JP, Forster CB (1996) Fault zone architecture and permeability structure. Geology 24, 1025–1028

Chester FM, Logan JM (1986) Implications for mechanical properties of brittle faults from observations of the Punchbowl fault zone, California. Pure and Applied Geophysics 124, 79–106

Childs C, Manzocchi T, Walsh JJ, Bonson CG, Nicol A, Schöpfer MPJ (2009) A geometric model of fault zone and fault rock thickness variations. Journal of Structural Geology 31, 117–127

Choi JH, Edwards P, Ko K, Kim YS (2016) Definition and classification of fault damage zones: a review and a new methodological approach. Earth-Science Reviews 152, 70–87

Cowie PA, Scholz CH (1992) Growth of faults by accumulation of seismic slip. Journal of Geophysical Research 97, 11085–11095

Dasgupta S, Mukherjee S (2017) Brittle shear tectonics in a narrow continental rift: asymmetric non-volcanic Barmer basin (Rajasthan, India). The Journal of Geology 125, 561–591

Flodin EA, Aydin A (2004) Evolution of a strike-slip fault network, Valley of Fire State Park, southern Nevada. Geological Society of America Bulletin 116, 42–59

Ghosh SK (1993) Structural geology, fundamentals and modern developments, vol 1. Pergamon Press, p 598

Gupta V, Bist KS (2004) The 23 September 2003 Varunavat Parvat landslide in Uttarkashi city, Uttaranchal. Current Science 87, 1600–1605

Hoek E, Bray J (1981) Rock slope engineering. The Institution of Mining and Metallurgy, London

International Society for Rock Mechanics (ISRM) (1978) Suggested methods for the quantitative description of discontinuities in a rock mass. International Journal of Rock Mechanics and Mining Sciences and Geomechanics Abstracts 6, 319–368

ISRM (1981) Suggested methods for determining hardness and abrasiveness of rocks. Rock characterization, testing and monitoring: ISRM suggested methods. Pergamon, Oxford, pp 95–96

ISRM (2007) The complete ISRM suggested methods for rock characterization, testing and monitoring: 1974–2006. Springer, Berlin

Jamison WR, Stearns DW (1982) Tectonic deformation of Wingate Sandstone, Colorado National Monument. American Association of Petroleum Geologists Bulletin 66, 2584–2608

Kim YS, Peacock DCP, Sanderson DJ (2004a) Fault damage zones. Structural Geology 26, 503–517

Kim YS, Peacock DCP, Sanderson DJ (2004b) Fault damage zones. Structural Geology 26, 503–517

Kumar M, Rana S, Pant PD, Patel RC (2017) Slope stability analysis of Balia Nala landslide, Kumaun Lesser Himalaya, Nainital, Uttarakhand, India. Journal of Rock Mechanics and Geotechnical Engineering 9, 150–158

McGrath AG, Davison I (1995) Damage zone geometry around fault tips. Structural Geology 17, 1011–1024
Mukherjee S (2010a) Structures in Meso- and Micro-scales in the Sutlej section of the Higher Himalayan Shear Zone, Indian Himalaya. e-Terra 7, 1–27
Mukherjee S (2010b) Microstructures of the Zanskar Shear Zone. E-journal: Earth Science India 3, 9–27
Mukherjee S (2011) Mineral fish: their morphological classification, usefulness as shear sense indicators and genesis. International Journal of Earth Sciences 100, 1303–1314
Mukherjee S (2013a) Channel flow extrusion model to constrain dynamic viscosity and Prandtl number of the Higher Himalayan Shear Zone. International Journal of Earth Sciences 102, 1811–1835
Mukherjee S (2013) Deformation microstructures in rocks. Springer Geochemistry/Mineralogy, Berlin, pp 1–111. ISBN 978-3-642-25608-0
Mukherjee S (2014) Atlas of shear zone structures in Meso-scale. Springer, Berlin, pp 1–128. https://doi.org/10.1007/978-3-319-00089-3
Mukherjee S (2015a) A review on out-of-sequence deformation in the Himalaya. In: Mukherjee S, Carosi R, van der Beek P, Mukherjee BK, Robinson D (eds) Tectonics of the Himalaya, vol 412. Geological Society, London, Special Publications, pp 67–109
Mukherjee S (2015) Atlas of structural geology. Elsevier. Amsterdam. ISBN: 978-0-12-420152-1
Mukherjee S (2017) Review on symmetric structures in ductile shear zones. International Journal of Earth Sciences 106, 1453–1468
Mukherjee S (2019) Introduction to "Tectonics and Structural Geology: Indian Context". In: Mukherjee S (ed) Tectonics and structural geology: Indian context. Springer International Publishing AG, Cham, pp 1–5. ISBN: 978-3-319-99340-9
Mukherjee S, Koyi HA (2010a) Higher Himalayan Shear Zone, Zanskar Section- microstructural studies & extrusion mechanism by a combination of simple shear & channel flow. International Journal of Earth Sciences 99, 1083–1110
Mukherjee S, Koyi HA (2010b) Higher Himalayan Shear Zone, Sutlej section- structural geology & extrusion mechanism by various combinations of simple shear, pure shear & channel flow in shifting modes. International Journal of Earth Sciences 99, 1267–1303
Mukherjee S, Koyi HA, Talbot CJ (2012) Implications of channel flow analogue models in extrusion of the Higher Himalayan Shear Zone with special reference to the out-of-sequence thrusting. International Journal of Earth Sciences 101, 253–272
Mukherjee S, Punekar J, Mahadani T, Mukherjee R (2015) A review on intrafolial folds and their morphologies from the detachments of the western Indian Higher Himalaya. In: Mukherjee S, Mulchrone KF (eds) Ductile Shear Zones: From Micro- to Macro-scales. Wiley Blackwell, pp 182–205
Panikkar SV, Subramanyan V (1997) Landslide hazard analysis of the area around DehraDun and Mussoorie, Uttar Pradesh. Current Science 73, 1117–1123
Pant PD, Luirei K (1999) Malpa rockfalls of 18 August 1998 in the Northeastern Kumaun Himalaya. Journal of the Geological Society of India 54, 415–420
Paul SK, Mahajan AK (1999) Malpa rockfall disaster, Kali valley, Kumaun Himalaya. Current Science 76, 485–487
Peacock DCP, Nixon CW, Rotevatn A, Sanderson DJ, Zuluaga LF (2017) Interacting faults. Journal of Structural Geology 97, 1–22
Puniya MK, Joshi P, Pant PD (2013) Geological investigation of Nainital Bypass: a special emphasis on slope stability analysis, Kumaun Lesser Himalaya. In: Himalayan vulnerability, Uttarakhand, 2013: learning for planning and action. Xpressions Print & Graphics Pvt. Ltd. pp 65–72
Romana M (1985) New adjustment ratings for application of Bieniawski classification to slopes. In: Proceedings of the international symposium on the role of rock mechanics in excavations for mining and civil works. International Society of Rock Mechanics, Zacatecas, pp 49–53
Sah N, Kumar M, Upadhyay R, Dutt S (2018) Hill slope instability of Nainital City. Kumaun Lesser Himalaya, Uttarakhand, India. https://doi.org/10.1016/j.jrmge.2017.09.011

Shipton ZK, Cowie PA (2001) Damage zone and slip-surface evolution over mm to km scales in high-porosity Navajo sandstone, Utah. Journal of Structural Geology 23, 1823–1844

Shipton ZK, Cowie PA (2003) A conceptual model for the origin of fault damage zone structures in high-porosity sandstone. Journal of Structural Geology 25, 333–344

Stead D, Eberhardt E (2013a) Understanding the mechanics of large landslides. Italian Journal of Engineering Geology and Environment Book Series (6), 85–108

Stead D, Eberhardt E (2013b) Understanding the mechanics of large landslides. Italian Journal of Engineering Geology and Environment, 85–109

Stead D, Wolter A (2015) A critical review of rock slope failure mechanism: the importance of structural geology. Journal of structural Geology 74, 1–23

Valdiya KS (1980) Geology of Kumaun Lesser Himalaya. Wadia Institute of Himalayan Geology, Dehradun

Vanik N, Shaikh H, Mukherjee S, Maurya DM, Chamyal LS (2018) Post-Deccan trap stress reorientation under transpression: evidence from fault slip analyses from SW Saurashtra, western India. Journal of Geodynamics, 121, 9–19

Documentation of Brittle Structures (Back Shear and Arc-Parallel Shear) from Sategal and Dhanaulti Regions of the Garhwal Lesser Himalaya (Uttarakhand, India)

Souradeep Mahato, Soumyajit Mukherjee and Narayan Bose

1 Introduction

We study the brittle shear tectonics from the Lesser Himalayan terrain in the Mussoorie, Dhanaulti and Satengal regions, Uttarakhand, India. The Lesser Himalaya is bound by the Main Boundary Thrust (MBT) in the south and the Main Central Thrust Zone (MCTZ) in the north. The Lesser Himalayan sequence is divided in this region as the Inner Lesser Himalaya and the Outer Lesser Himalaya (Célérier et al. 2009a, b). The Mussoorie- and the Garhwal synclines are the two major structures of the Outer Lesser Himalaya. The area around Dhanaulti (Fig. 1) constitutes a part of the northern limb of the Mussoorie syncline. The Satengal region (Fig. 1) lies at the core of the Mussoorie syncline. Several workers (e.g., Srivastava et al. 2011; Bose et al. submitted-1, 2; Bose et al. 2018) discuss major structural features and the deformation phases from the Garhwal Himalaya. However, a detail study remained due on its brittle deformations. This work describes brittle shear sense indicators from the terrain.

2 Previous Works

The first-hand rock description and regional mapping in this area have been done by Middlemiss (1887), Auden (1937), Jain (1972), Shankar and Ganesan (1973), Valdiya (1975), Saklani (1978). The Mussoorie Syncline mainly consists of sedimentary rocks and low-grade metamorphic rocks of Tal-, Krol-, Blaini-, Nagthat-, Chandpur-, and Mandhali Formation. Out of all these, the Mandhani, the Chandpur

S. Mahato · S. Mukherjee (✉) · N. Bose
Department of Earth Sciences, Indian Institute of Technology Bombay, Mumbai, India
e-mail: soumyajitm@gmail.com

S. Mukherjee (ed.), *Tectonics and Structural Geology: Indian Context*, Springer Geology, https://doi.org/10.1007/978-3-319-99341-6_15

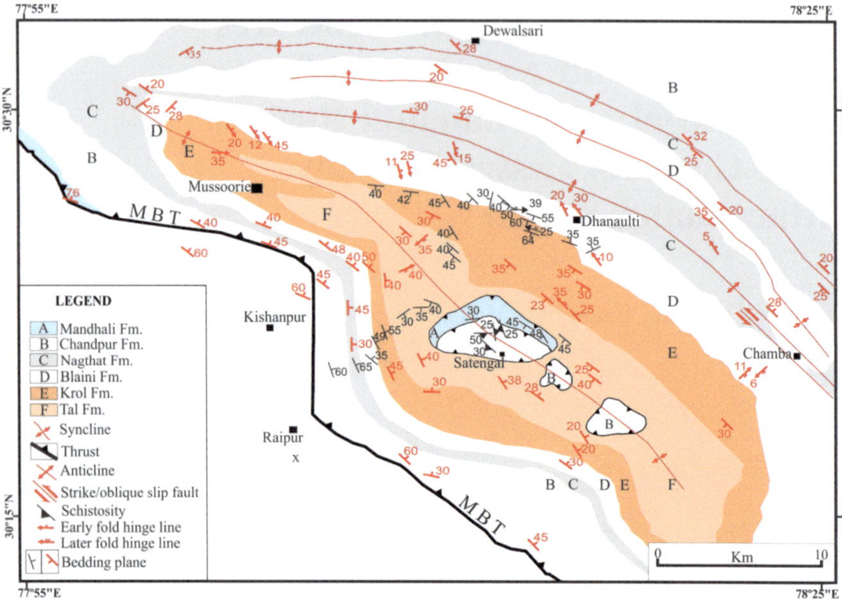

Fig. 1 Structural map of the central portion of Mussoorie syncline is showing different structural data. Structural data in red colour are taken from previous works, and those in black colour indicates those collected in the present study. The map is reproduced from Dubey and Jayangondaperumal (2005)

and the Nagthat Formation belong to the Jaunsar Group and the remainder to the Mussoorie Group (Fig. 1).

Raman Spectroscopic study of carbonaceous material of samples collected across the axis of Mussoorie Syncline yields a deformation temperature <300 °C (Célérier et al. 2009a, b). This is overlain by less metamorphosed rocks of the Simur Group and constitutes partly the Garhwal nappe. The southern limb of the Mussoorie Syncline overrides the Siwalik zone along the MBT (Fuchs and Sinha 1978). Intense-sheared black Proterozoic Chandpur phyllites along the MBT near the Sahenshahi temple has been studied in detail by Bose and Mukherjee (submitted-a). The authors use Raman Spectroscopy on carbonaceous materials and obtain a deformation temperature <550 °C (Bose and Mukherjee, submitted-a). Further, ~6–49 MPa of flow stress and $\sim 10^{-15}$–10^{-16} s^{-1} of strain rate has been obtained from the MBT zone rocks from this region that reveals a top-to-S/SW fore-shear in field and under an optical microscope (Bose et al. submitted-a).

According to Srivastava et al. (2011), the Sategal area deformed four times. The first three deformation phases are attributed to ductile deformations and the last one a brittle deformation. The Mussoorie Syncline, a structural part of the Krol Nappe, consists of Tal-, Krol-, Blaini-, Nagthat-, Chandpur-, and Subathu Formation. In Dhanaulti and nearby areas, mainly Blaini-, Tal-, and Krol Formation crop out. The main meso-scale structures in this area are primary bedding planes, slaty cleavages, axial planes of different generations of folds (Srivastava et al. 2011), and thrusts/ faults (Dubey 2005). The pop-up klippen structure is indicated by the rocks of Chandpur, Mandhali and Subathu Formation at the Satengal area (Dubey and Jayangondaperumal 2005).

The Baliana Formation (/Chandpur Formation) and the Mandhani Formation (/ Saklana Formation) are separated by the Ringalgarh Thrust. The pop-up klippen developed since the MBT bifurcates at south into an oblique ramp and into a blind thrust that generated two low-angle backthrusts (Dubey and Jayangondaperumal 2005).

At Satengal and in the adjoining areas, folding happened thrice. The main schistosity trends NW and NE. The NW trending earlier tight to isoclinal F_1 folds are characterized by deformed quartz veins and quartzo-feldspathic layers. The usually NE trending F_2 folds developed on the limbs of the F_1 folds with open to isoclinal geometries. F_3 chevron folds occur in phyllites of Chandpur Formation near Satengal and trends E-W (Srivastava et al. 2011).

3 Present Work

This section addresses details of S_0 and S_1 planes. Brittle shear detail is also presented from the following three traverses: (i) Mussoorie up to Kathu Khal (Surkunda Devi temple) along the Mussourie-Chamba road; (ii) Kathu-Khal to Satengal; and (iii) ~NE-SW traverse was taken from Raipur to Silla.

3.1 S_0-Plane (Fig. 2)

Primary sedimentary bedding planes (S_0) are commonly observed in the study area. These S_0 planes can be identified based on lithology change and color variation. From L-1 to L-6, dip of S_0 ranges 40°–42° with ~SW dip direction. The dip direction changes from SW to NW in and around L-7. Dip ranges 25°–65° with a ~SW dip direction between L-8 and L-11. S_0 dips oppositely towards NE and N with 30° dip at L-12 and L-13, south of Dhanaulti.

The S_0-planes near the Satengal region between L-15 to 18 show 40°–45° dip, and dip direction towards SW. At L-24 grey coloured quartzites dip N, from L-28 to 31 the dip of the S_0 varies 35°–60° with dip direction NW to E, in between L-32 and 33 dip direction varies from SE to S. S_0 dips toward SW direction in locations 33–37.

Slaty Cleavage (S_1) (Fig. 2):

The exposed rocks in and around Dhanaulti are slaty. Phyllites are observed in the Satengal region. A pervasive, parallel foliation (S_1) of fine-grained platy minerals occurs in these rocks, which generally develop perpendicular to the direction of

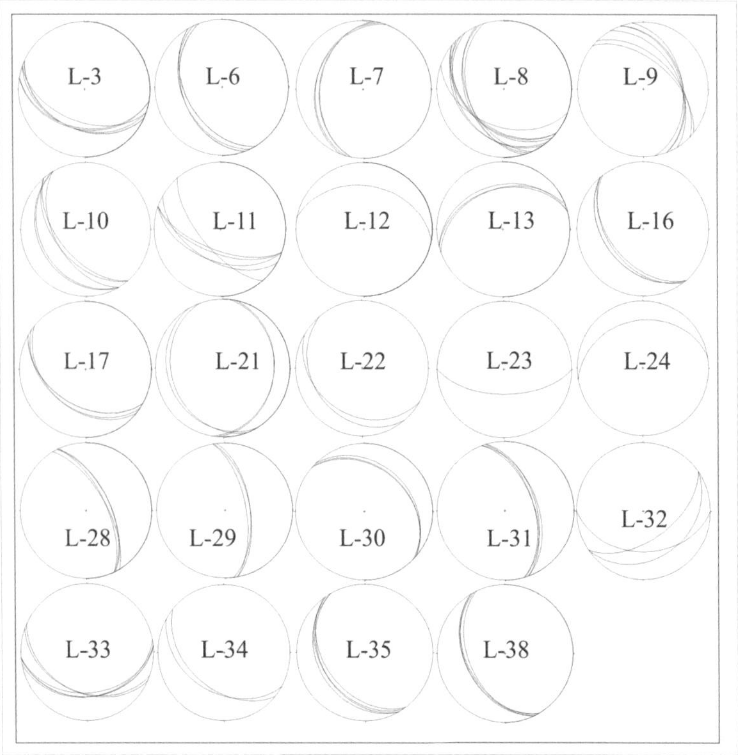

Fig. 2 Stereo-plots of planer features S_0 and S_1 on equal area net. L-21 to L-24 shows S_1-plots, and the rest represents S_0-plots

maximum stress. Usually this S_1 plane parallels the S_0 planes if not folded or sheared. In Dhanaulti region, the S_1 planes dip 30°–40° towards SW. In Satengal region, S_1 dips 25°–50° and the dip direction varies widely from SW to S to SE to N.

3.2 *Shear Fractures*

Shear fractures (Fig. 3) are very common structural features observed in the study area. The P-planes at L-28 and 36 generally dip in opposite direction and with the M/Y shear fracture this makes a low-angle. Kinematically, such shear planes can usually be correlated with low-angle thrusts. Within the same setting, the Riedel fractures can be represented as low-angle normal faults. The R'-planes/antithetic faults make a high-angle with the Y plane. The angles between the P, R and $R'$with the Y-plane can depend on the rock's physical properties and the stress regime.

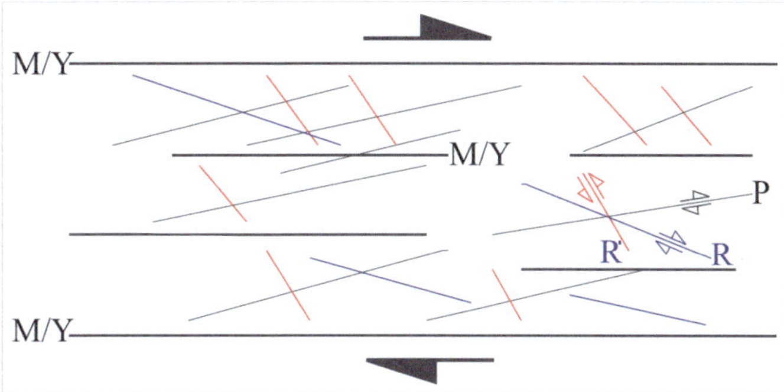

Fig. 3 Nomenclatures for shear fractures developed in a top-to-right shear zone. M/Y, Main shear plane; P: shear fractures; R: Riedel shears; R′: fracture. From Petit (1987)

3.3 Traverse-Wise Description

The first ~E-W traverse was taken along the Mussoorie-Chamba road from Mussoorie up to Kathu Khal (Surkunda Devi temple). A vertical shear zone (Fig. 4) exists at L-2 on a N140°S striking road cut section. Around 13 km east of L-3, top-to-SE brittle shear (Fig. 5a) was observed at L-7. Top-to-NW down shear (Fig. 6a) was observed ~650 m SE from L-6. At L-7, top-to-NE up brittle shear (Fig. 6c), i.e., back deformation, occurs in slates. About 600 m east from L-7, the same top-to-NE up brittle shear (Fig. 6d) occurs in slates at L-8. A top-to-NE sub-horizontal brittle shear (Fig. 6b) exists in slate close to L-9. Around 300 m south of L-9, a top-to-SW up shear, also can be called a fore shear that is common in Himalayan orogen in its various units (e.g., Mukherjee 2010a, b, 2012, 2013, 2014; Mukherjee and Koyi 2010a, b) exists in slate. Top-to-NW shear (Fig. 5b) also exists.

Fig. 4 Brittle vertical fault zone. Y- and P-planes well developed

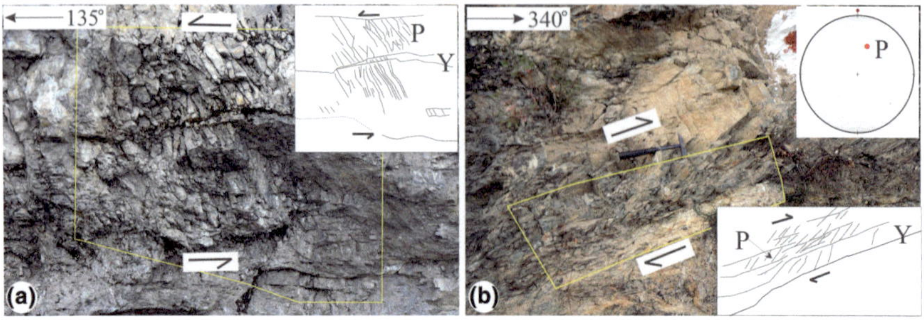

Fig. 5 **a** Field photograph of vertical exposure of quartzite showing top-to-SE shear. Y- and P-planes are stereo-plotted in inset. **b** Top-to-SW up brittle shear. Pole of P-plane (25° dip towards 10°) is plotted in inset

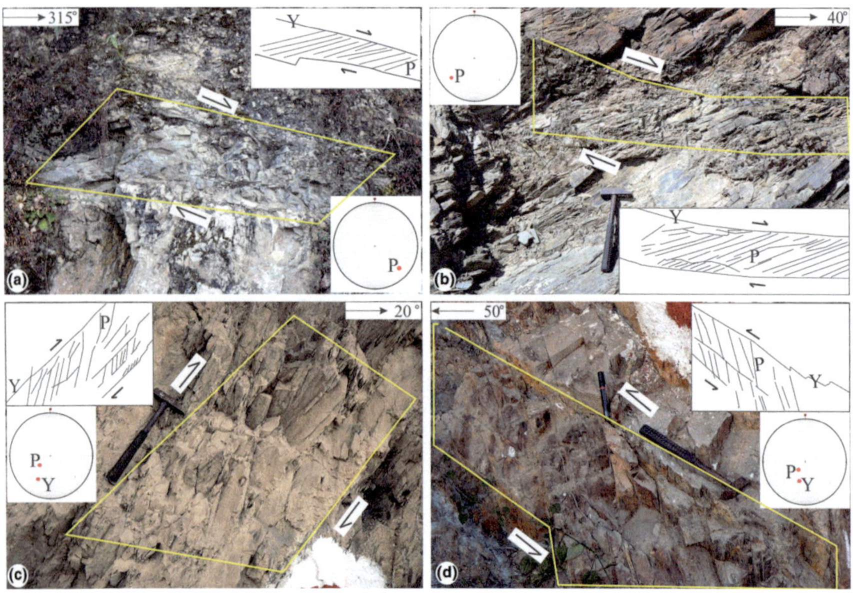

Fig. 6 Brittle shear Y- and P-planes observed in a vertical rock section. Poles of brittle planes are plotted inside insets. **a** Note a sharp contact between slate at top and quartzite at bottom. Top-to-NW brittle shear localized in slate due to high accumulation of shear strain. P-plane attitude: 10° dip towards 140°. **b** Shear fractures developed on slate. Top-to-NE brittle shear sense. The P-plane dips 14° and towards 225°. **c** Y- (dip 34° towards 200°) and P-planes (dip 65° towards 240°) developed on slate. Top-to-NE up shear. **d** Top-to-NE up shear. Y-plane: dip 35° towards 190°, P-plane: dip 70° towards 230°

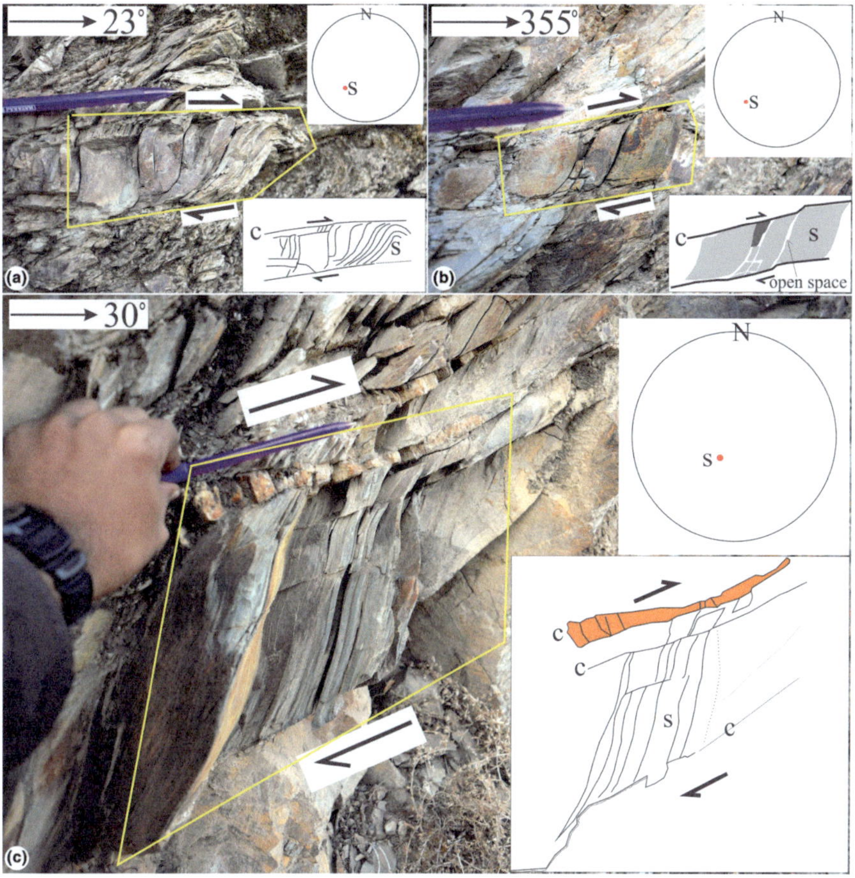

Fig. 7 a, b Sub-vertical sections at L-12 showing S- and C-fabrics developed in slates. top-to-NE up ductile shear. Stereo-plot of S planes in inset. Fractures developed along the S-planes in slates. In particular, in **c** the C-planes are occupied by secondary quartz mineralization. Here the pen length is 13 cm

Slate is the dominant rock type at L-12 which is ~2 km SW to Dhanaulti. Here the rocks deformed in a ductile-brittle regime. In response to ductile deformation, S- and C-fabric developed prominently in slates. Secondary mineralization of quartz is noted at few places along the shear planes that connote back-shear (Fig. 7a–c). The dip and dip-direction of S-fabric indicates a top-to-NE up shear (back-structure). Brittle shear fractures (Fig. 8c, d) also develop at L-13 showing both top-to-NE up and horizontal shear senses.

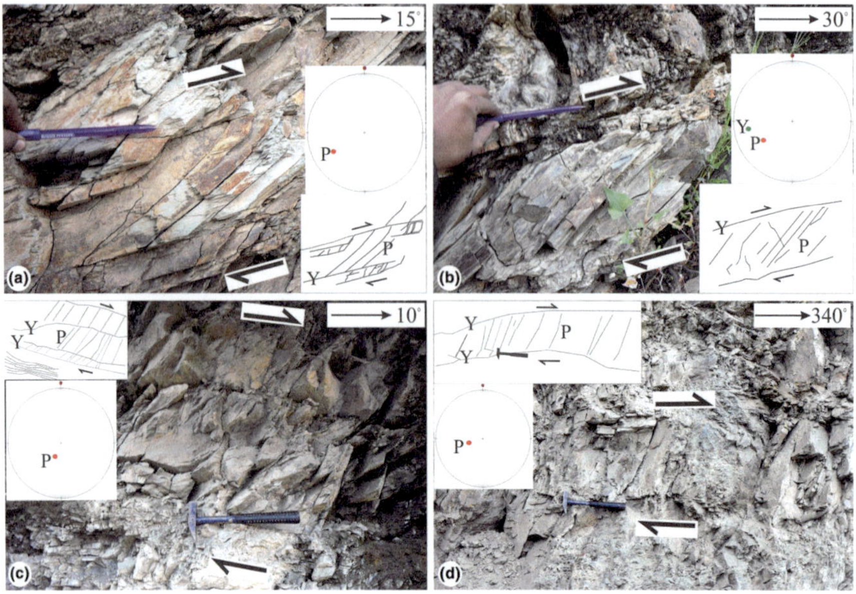

Fig. 8 Brittle shear planes in vertical sections close to L-12. **a** P-plane dips 30° towards 248°, top-to-NNE shear. **b** P-plane dips 40° towards 238°, top-to-NE shear. **c** P-plane dips 70° towards 190°, top-to-NNE shear. **d** P plane dips 65° towards 255°, at L-13. Top-to-NNW horizontal shear

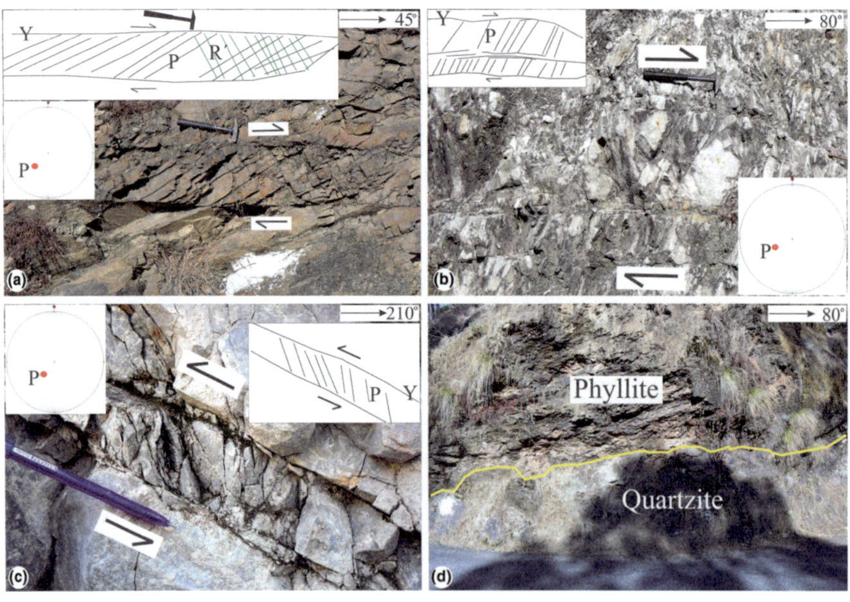

Fig. 9 **a** Clear-cut top-to-NE brittle shear observed in vertical section of slates. P-plane dips 46° towards 230°. **b** Top-to-NEE shear developed in white Ringalgarh quartzite. P-plane dips 58° towards 252°. **c** Top-to-NE up shear prominent in white quartzite. P-plane dips 65° towards 232°. **d** Sharp contact between quartzite and phyllite noted at L-19

Fig. 10 **a** Top-to-NE down shear in quartzite. Y-plane dips 70° towards 78°, P-planes dips 15° towards 80°. **b** Top-to-SE down shear in shale. Y-plane dips 22° towards 38°, P-plane dips 28° towards 320°. **c** Field photograph showing vertical sensed of shear Y-, P-, R-, and R$^{/}$-planes are marked. **d** Top-to-SE down shear in bedded quartzite. Y-plane dips 22° towards 180°; P-plane dips 36° towards 286°. **a**, **b**: sub-vertical sections; **c**, **d**: inclined sections

The second traverse was taken from Kathu-Khal to Satengal. At L-14, top-to-NE horizontal shear (Fig. 9a) was observed in slates. Near Satengal, at L-17 top-to-NE shear exists in quartzites. Top-to-NE up shear (Fig. 9b, c) occur at L-17 and 18 in white quartzites.

The third ~NE-SW traverse was taken from Raipur to Silla. At L-28 shear fractures are quite prominent in the quartzites showing top-to-NE down slip (Fig. 10a). About 1.5 km NE of L 28, top-to-SE down shear (Fig. 10b) occurs in black shale. Vertical slip (Fig. 10c) exists at L-31. Top-to-SE down shear sense (Fig. 10d) was observed at L-32. At L-33, top-to-SW down shear sense (Fig. 11a) was observed on quartzite. At L-34, top-to-NW up shear (Fig. 11b) was observed in quartzites, Near L-34, a top-to-NE up shear (Fig. 11c) was observed in white quartzite. Top-to-SW down shear (Fig. 11d) was observed at L-36 in black shale. All the observed shear senses are shown along the field traverses presented in Fig. 12.

Cross-sections (Figs. 13 and 14) were prepared along lines a–b close to Dhanaulti and along the line X–Y. The lines a–b and X–Y can be found in Fig. 12. The conventional method is used to draw the cross section. As we do not know the actual curvature at the limbs therefore straight lines are drawn to construct the fold. As sharp hinge bearing chevron folds were noted in the smaller scale (Figure with authors), a large fold of such geometry looks plausible (Fig. 14).

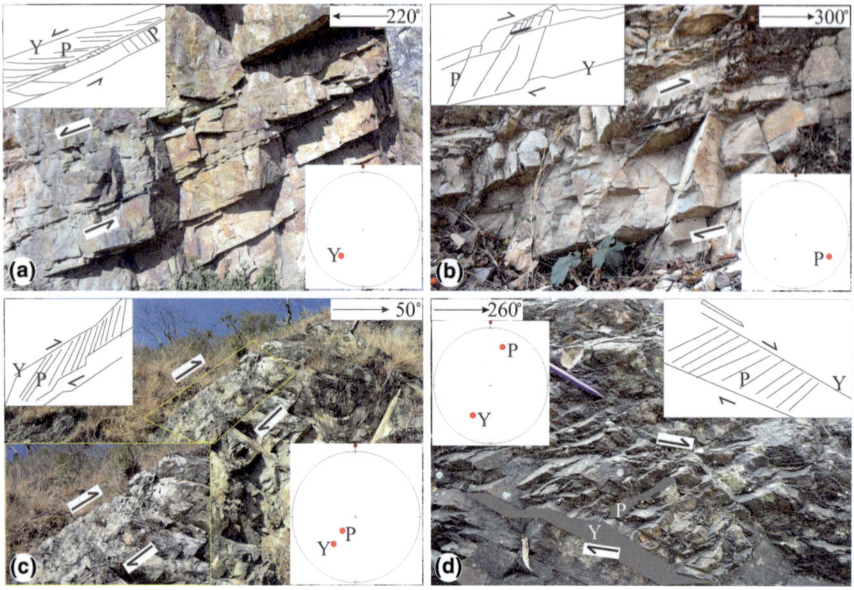

Fig. 11 **a** Shear fractures developed on quartzite showing top-to-SW down brittle shear sense. Y plane attitude dips 22° towards 223°; **b** top-to-NW up shear. P-plane dips 23° towards 218°. The rock type is quartzite. **c** Top-to-NE up shear. Y-plane dips 33° towards 226°, P-plane dips 62° towards 226°. Shear fractures developed quartzite. **d** Shear fractures developed on shale showing top- to-SW down shear sense. Y-plane dips 31° towards 222°, P-plane dips 15° towards 18°

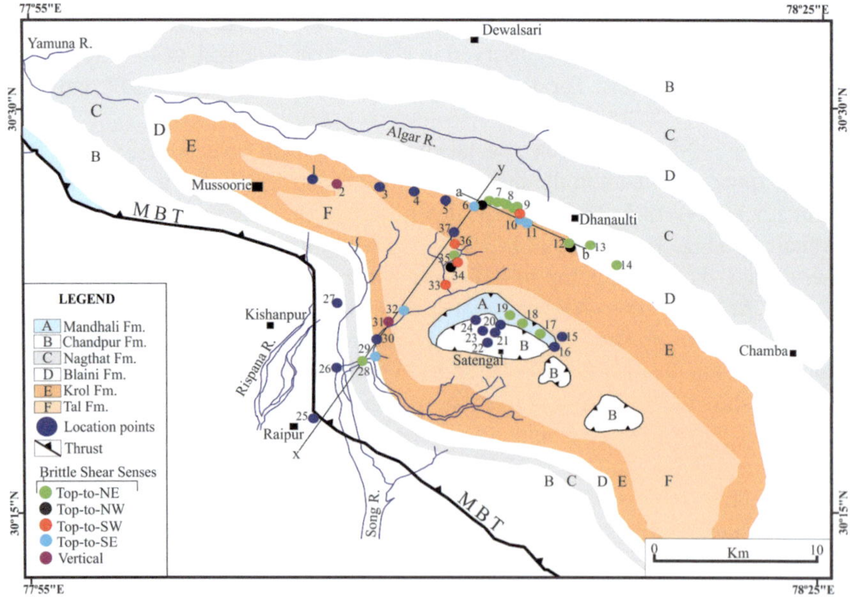

Fig. 12 Distributions of different brittle shear senses are shown in the map. The map is reproduced from Dubey and Jayangondaperumal (2005)

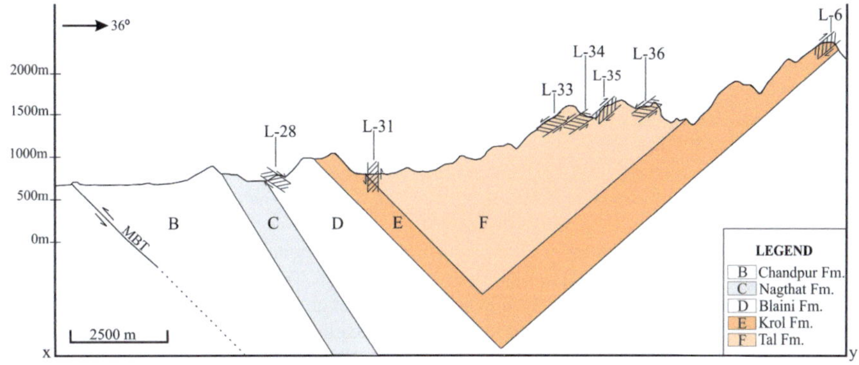

Fig. 13 Cross-section of the Mussoorie syncline is drawn along the X–Y line of Fig. 12. The X–Y line trends N36°E. B: Chandpur Fm, C: Nagthat Fm, D: Blaini Fm, E: Krol Fm, F: Tal Fm

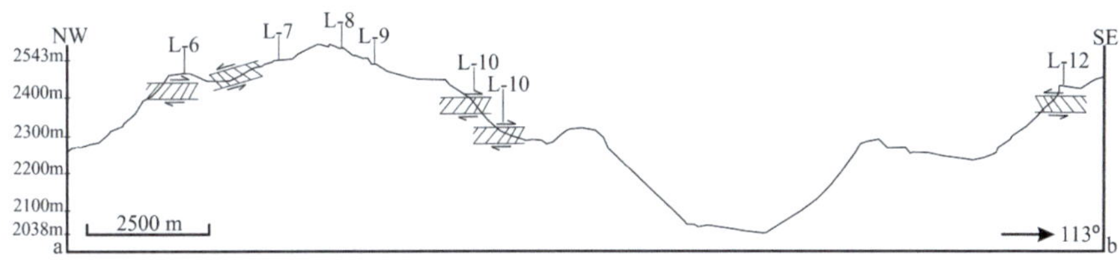

Fig. 14 A NW-SE profile along the line a–b of Fig. 13

This preliminary study reports Himalayan back-shear (top-to-NE) and arc-parallel shear (top-to-NW) for the first time in terms of meso-scale structures. We have been presently working on these issues in detail.

Acknowledgements This work was a part of the M.Sc. thesis of S. Mahato. S. Mukherjee thanks IIT Bombay for providing a sabbatical in 2017. N. Bose acknowledges Junior Research Fellowship of UGC, grant number: F.2-2/98(SA-1). The Springer proofreading team is thanked for assistance. Consult Banerjee et al. (2019) for tectonic updates on arc-parallel deformation from Sikkim Lesser Himalaya, and Dutta et al. (submitted) from that in Mohand region, Siwalik. This will establish link between chapters in the book. Mukherjee (2019) summarizes this work.

References

Auden JB (1937) The structure of the Himalaya in Garhwal. Records of the Geological Survey of India 71, 407–433

Banerjee S, Bose N, Mukherjee S (2019) Field structural geological studies around Kurseong, Darjeeling-Sikkim Himalaya, India. In: Mukherjee S (ed) Tectonics and structural geology: Indian context. Springer International Publishing AG, Cham, pp 425–440. ISBN 978-3-319-99340-9

Bell TH, Rubenach MJ, Fleming PD (1986) Porphyroblast nucleation, growth and dissolution in regional metamorphic rocks as a function of deformation partitioning during foliation development. Journal of Metamorphic Geology 4, 37–67

Bose N, Dutta D, Mukherjee S (2018) Role of grain-size in phyllonitisation: insights from mineralogy, microstructures, strain analyses and numerical modeling. Journal of Structural Geology 112, 39–52

Bose N, Mukherjee S (Submitted-1) Estimation of deformation temperature, flow stress and strain rate from the Main Boundary Thrust rocks, Garhwal Lesser Himalaya, India. Journal of Earth System Science

Bose N, Mukherjee S (Submitted-2) Field documentation and genesis of the back-structures from a part of the Garhwal Lesser Himalaya. Uttarakhand, India. Tectonic implications. In: Sharma, Villa IM, Kumar S (eds) Crustal Architecture and Evolution of the Himalaya-Karakoram-Tibet Orogen. Geological Society of London Special Publications

Célérier J, Harrison TM, Beyssac O, Herman F, Dunlap WJ, Webb AAG (2009a) The Kumaun and Garwhal Lesser Himalaya, India: Part 2. Thermal and deformation histories. Geological Society of America Bulletin 121, 1281–1297

Célérier J, Harrison TM, Webb AAG, Yin A (2009b) The Kumaun and Garhwal Lesser Himalaya, India: Part 1. Structure and stratigraphy. Geological Society of America Bulletin 121, 1262–1280

Dubey AK (2014) Understanding an orogenic belt. Springer, Cham. ISBN 978-3-319-05587-9

Dubey AK, Jayangondaperumal R (2005) Pop-up klippen in the Mussoorie Syncline Lesser Himalaya: evidence from field and model deformation studies. In: Saklani RS (ed) Himalaya (Geological Aspects), Satish Serial Publishing House, Delhi 3, pp 203–222

Dutta D, Biswas T, Mukherjee S (Submitted) Orogen-parallel compression in NW Himalaya: Evidence from structural and paleostress studies of brittle deformation from the pebbles of Upper Siwalik conglomerates, Uttarakhand, India. Journal of Earth System Science

Fuchs G, Sinha AK (1978) The tectonics of the Garhwal-Kumaun Lesser Himalaya. Jahrbuch der Geologischen Bundesanstalt 121, 219–241

Jain AK (1972) Overthrusting and emplacement of basic rocks in Lesser Himalaya, Garhwal, UP. Journal of Geological Society of India 13, 226–237

Jayangondaperumal R, Dubey AK, Sangode SJ, Sathyanarayana KV (2001) Superimposed folding, finite strain and magnetic lineation in the Mussoorie Syncline, Lesser Himalaya: implications for regional thrusting and the Indian Plate motion. Himalayan Geology 22, 207–216

Long S, McQuarrie N (2010) Placing limits on channel flow: insights from the Bhutan Himalaya. Earth and Planetary Science Letters 290, 375–390

Mandl G (1999) Faulting in brittle rocks: an introduction to the mechanics of tectonic faults. Springer Science and Business Media

Marshak S (1988) Kinematics of orocline and arc formation in thin-skinned orogens. Tectonics 7, 73–86

Marshak S (2009) Essentials of geology, 4th edn. WW Norton and Company, pp 18. ISBN: 9780471405214

Middlemiss CS (1887) Crystalline and metamorphic rocks of the Lower Himalaya, Garhwal and Kumaun. Records of the Geological Survey of India 20, 134–143

Mukherjee S (2010a) Structures in meso- and micro-scales in the Sutlej section of the Higher Himalayan Shear Zone, Indian Himalaya. e-Terra 7, 1–27

Mukherjee S (2010b) Microstructures of the Zanskar Shear Zone. e-Journal: Earth Science India 3, 9–27

Mukherjee S (2012) Tectonic implications and morphology of trapezoidal mica grains from the Sutlej section of the Higher Himalayan Shear Zone, Indian Himalaya. The Journal of Geology 120, 575–590

Mukherjee S (2013) Higher Himalaya in the Bhagirathi section (NW Himalaya, India): its structures, backthrusts and extrusion mechanism by both channel flow and critical taper mechanisms. International Journal of Earth Sciences 102, 1851–1870

Mukherjee S (2014) Mica inclusions inside Host Mica Grains from the Sutlej section of the Higher Himalayan Crystallines, India-Morphology and Constrains in Genesis. Acta Geologica Sinica 88, 1729–1741

Mukherjee S (2019) Introduction to "Tectonics and Structural Geology: Indian Context". In: Mukherjee S (ed) Tectonics and structural geology: Indian context. Springer International Publishing AG, Cham, pp 1–5. ISBN: 978-3-319-99340-9

Mukherjee S, Koyi HA (2010a) Higher Himalayan Shear Zone, Zanskar section-microstructural studies and extrusion mechanism by a combination of simple shear and channel flow. International Journal of Earth Sciences 99, 1083–1110

Mukherjee S, Koyi HA (2010b) Higher Himalayan Shear Zone, Sutlej section-structural geology & extrusion mechanism by various combinations of simple shear, pure shear & channel flow in shifting modes. International Journal of Earth Sciences 99, 1267–1303

Passchier CW, Trouw RAJ (2005) Microtectonics. Springer, Berlin, pp 1–366. ISBN10 3-540-64003-7

Petit JP (1987) Criteria for the sense of movement on fault surfaces in brittle rocks. Journal of Structural Geology 9, 597–608

Powell CM, Conaghan PJ (1973) Plate tectonics and the Himalayas. Earth and Planetary Science Letters 20, 1–12

Saklani PS (1978) Deformation and tectonism of Mukhem area, Lesser Himalaya. Tectonic Geology of the Himalaya 15, 42

Shanker R, Ganesan TM (1973) A note on the Garhwal Nappe. Himalayan Geology 3, 72–82

Srivastava HB, Sinha LK, Katiyar V (2011) Mesoscopic structures from the area around Satengal, Lesser Garhwal Himalaya. Journal of Scientific Research 55, 25–34

Valdiya KS (1975) Lithology and age of the Tal Formation in Garhwal, and implication on stratigraphic scheme of Krol Belt in Kumaun Himalaya. Geological Society of India 16, 119–134

Valdiya KS (1980) Geology of Kumaun Lesser Himalaya. Wadia Institute of Himalayan Geology

Field Structural Geological Studies Around Kurseong, Darjeeling-Sikkim Himalaya, India

Saikat Banerjee, Narayan Bose and Soumyajit Mukherjee

1 Introduction

Mallet (1874) divided the metamorphic rocks of the Darjeeling-Kurseong region (see Table 1) into two Formations: Darjeeling Formation of high grade garnetiferous schist, mica gneiss and the phyllitic Daling Group of rock. The pioneering geological inputs provide a detailed account of the lithology and deformation of this region (e.g., Sinha Roy 1973; Acharyya and Ray 1977; Acharyya 1989). Ray (1947) has described the Senchal Series of rocks, which includes both the Darjeeling and the Daling rocks. In the study area, the Gondwana rocks are overlaid by metamorphic rocks of the Daling- and the Darjeeling Group/Formation (Fig. 1; Lahiri 1973). Mukhopadhyay and Gangopadhyay (1971) worked in the both side of Teesta valley and found more or less the same lithology. According to Jangpangi (1972), each of the rock units are separated by thrust faults. On the other hand, Mallet (1874) suggested that the contact between the Daling and the Darjeeng is gradational. As per Ray (1947), there is no distinct separation between Daling and Gonwana at Tindharia. But Singh (1972) suggests near Tindharia the Daling rocks overlie the Gondwana rocks and they are separated by the NE trending Tindharia Thrust. The Daling Group consists of mainly slates and phyllites. The metamorphic-grade of phyllite elevates near the Tindharia Thrust (Singh 1972). This could be the effect of shear heating of the fault (Mukherjee and Mulchrone 2013; Mulchrone and Mukherjee 2015, 2016; Mukherjee 2017a, b; Mukherjee and Khonsari 2017, 2018; Mukherjee and Agarwal, submitted). The Daling Group can be recognized as quartz-chlorite schist. The Darjeeling Group can be identified as garnetiferous mica schist and mica gneiss. Especially, in the Darjeeling–Bijonbari area, such rock types are exposed (Sen 1971). The main composition of the gneiss in the Darjeeling Group

S. Banerjee · N. Bose · S. Mukherjee (✉)
Department of Earth Sciences, IIT Bombay, Powai, Mumbai 400076, Maharashtra, India
e-mail: soumyajitm@gmail.com

S. Mukherjee (ed.), *Tectonics and Structural Geology: Indian Context*,
Springer Geology, https://doi.org/10.1007/978-3-319-99341-6_16

Table 1 The stratigraphic succession of Darjeeling-Kurseong

Mallet (1874) Auden (1935)	Ray (1947)			Wager (1939)		Gansser (1964)
Darjeeling gneiss	Senchal series	Pelitic group (Darjeeling gneiss)	Typical coarse mica-gneiss	Daling series	Injected gneiss (Darjeeling) Dalings, belongs to Mt. Everset Pelitic series	Darjeeling gneiss
			Flaggy garnetiferous mica- schist and quartzite			
			Carbonaceous mica-schist garnet rich mica-schist andcalc-schist			Garnetiferous mica-schist
Transition zone		Slate-greywacke Group (daling)	Golden and silvery mica-schist			Daling schist
Daling series			Greywacke-schist			
			Slate phyllite with occasional quartzite, and greywacke-schist			

Reproduced from Lahiri (1973)

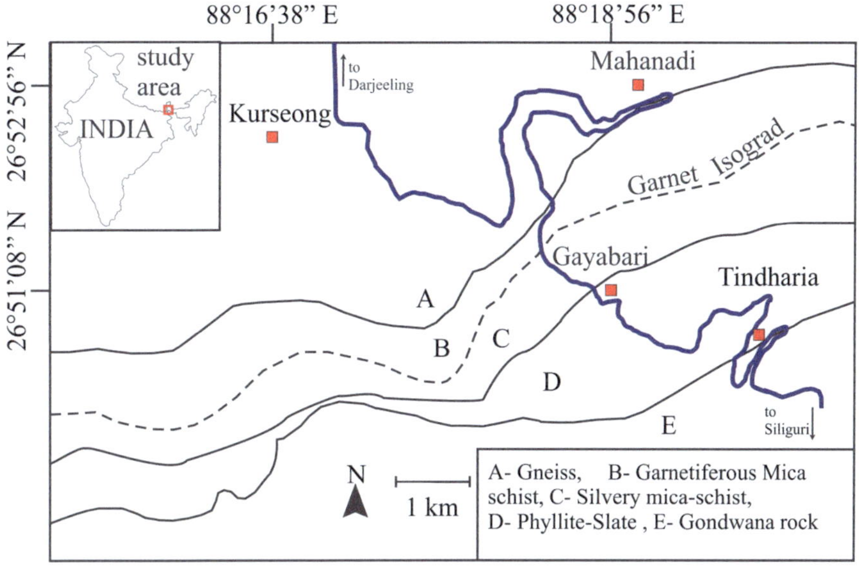

Fig. 1 Lithological map reproduced from Lahiri and Gangopadhway (1974). The garnet isograd is shown

is feldspar, biotite and occasional muscovite (Sen 1971). The Lesser Himalayan phyllites connote the sediments of the northern boundary of the Indian plate (Acharyya et al. 2017). However, different authors provided different names for the similar litho-/structural-units. For example, Bhattacharyya and Mitra (2009), and their subsequent works report that the following lithounits are present in the current study area (from S to N): Siwalik Formation, Gondwana/Buxa Formation, Daling Group with Lingtse Gneiss intrusions, Paro Gneiss, Darjeeling/Kanchenjunga Gneiss. This has been followed in Figs. 2, 3 and 4.

2 Structures

Most of the previous workers document two deformation phases. Mukhopadhyay and Gangopadhyay (1971) describe three phases of deformations. The F_1 tight folds are characterized by long limbs, axial plane schistosity and mineral lineation (Sen 1973). The inter-limb angles are <35° and these folds are mostly rootless and with smooth round hinge (Mukhopadhyay and Gangopadhyay (1971)). The D_1 is commonly confined within the banded semi-pelitic rocks (Lahiri 1973). The L_1 mineral lineations at the eastern part of the Teesta river plunge NE, but that at the

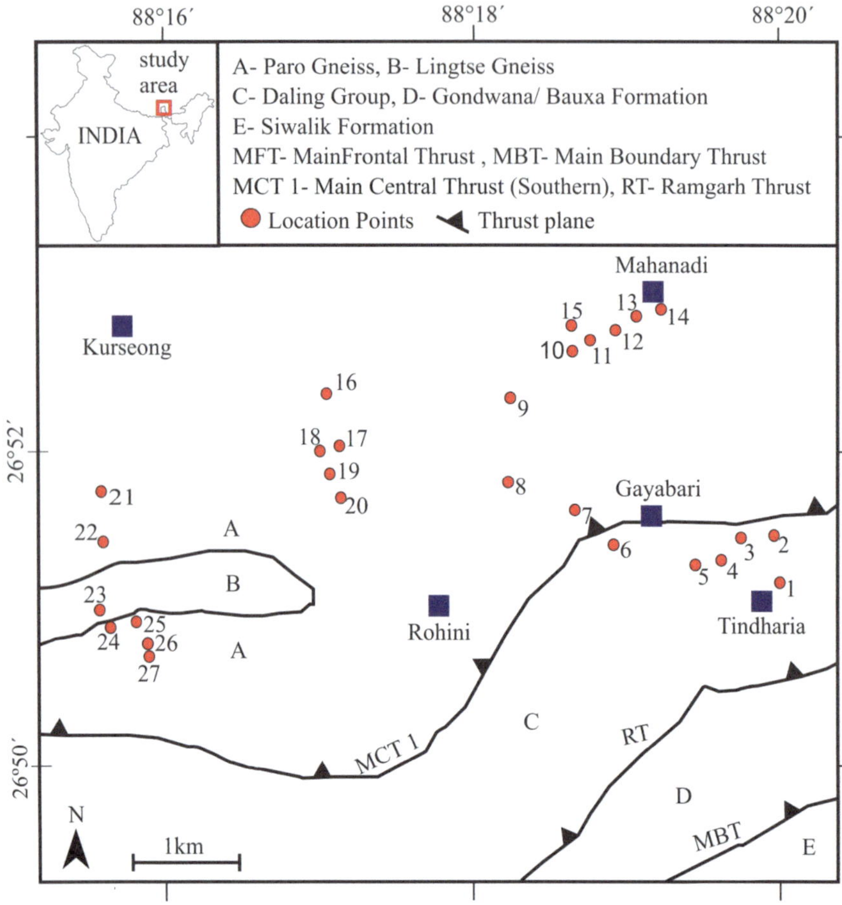

Fig. 2 Location map of the study area

western part plunges NW (Mukhopadhyay and Gangopadhyay (1971)). Minor open folds on the axial plane schistosity of the F_1 folds define the F_2 folding (Sen 1971). In Daling rocks, D_2 displays polyclinal open minor folds and pucker cleavages. The S_2 schistosity is at high-angle to the S_1 (Lahiri 1973). According to Mukhopadhyay and Gangopadhyay (1973), the second generation structures are SE plunging fold axes. Mukhopadhyay and Gangopadhyay (1973) has described the third generation structures in terms of disharmonic- and often symmetric folds. Rarely chlorite occurs along the schistosity. These fold axes plunge towards N or S. Bose et al. (2014) report a fourth phase of folding (F4) from this region.

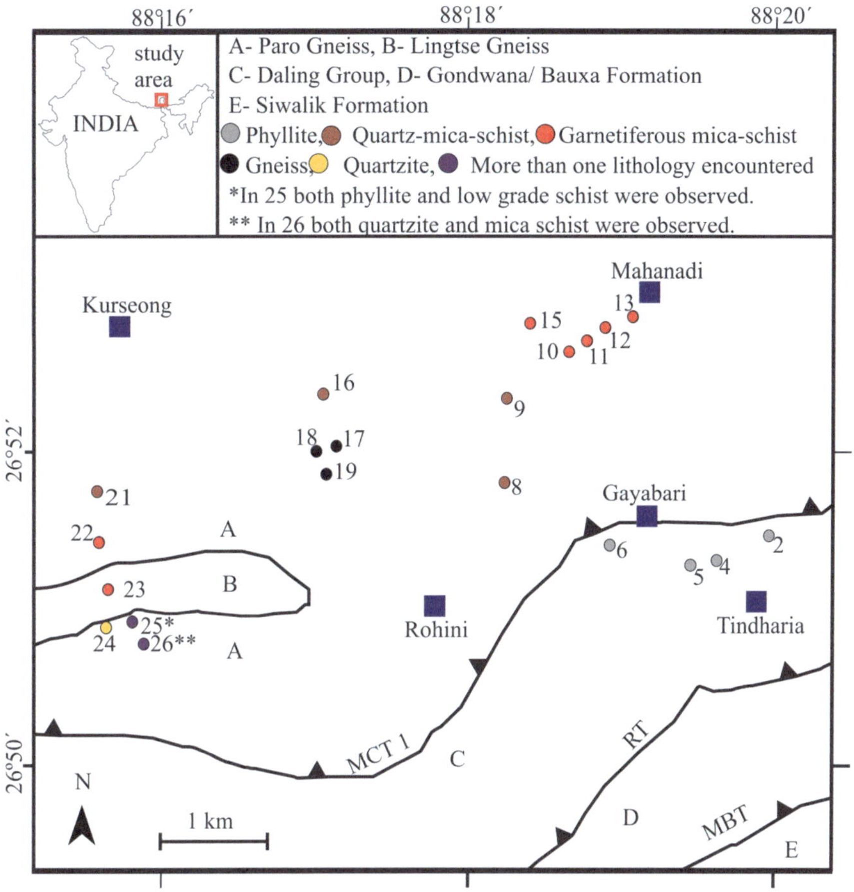

Fig. 3 The map shows lithology encountered at different locations

3 Metamorphism

Ray (1947) observes chlorite grade systematically altering to sillimanite grade towards north. While in the western part of the Teesta valley, one can observe chlorite to sillimanite transition, in eastern part staurolite and sillimanite do not appear as the index minerals (Mukhopadhyay and Gangopadhyay 1973). The term reverse metamorphism was used by Jangpangi (1972) to describe the inverted metamorphism in this region. Lahiri (1973) documents recrystallized quartz and biotite. Garnet took the longest time span to grow. Thin-section study of garnet has reveal its syn- to post-kinematic nature with respect to the F_1 and the F_2 folding. Recrystallized chlorite along the axes of the F_3-folds connotes local retrogression (Mukhopadhyay and Gangopadhyay 1973). Ray (1947) observes the metamorphic grade change towards north following a Barrovian sequence. Singh (1972) confirms from chemical

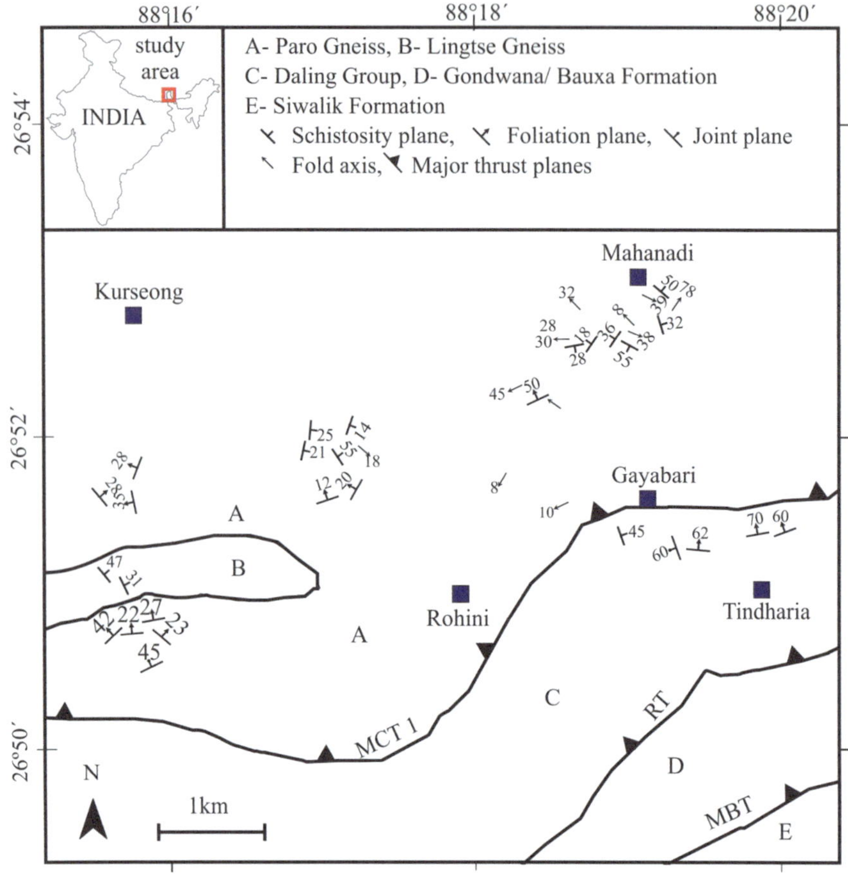

Fig. 4 Structural data plotted on the study area map

analyses that both the Daling and Darjeeling rocks are of same origin but due to heterogeneity in bulk composition the metamorphic products are different.

4 Lithologies

In the field area (Fig. 2), the lithology varies along the N-S transect. A change in metamorphic grade was also observed. Starting with phyllites near Tindharia in the southern part of the field area, up to gneissic grade of rock was encountered near Kurseong (Fig. 3). According to the classification of Mallet (1874), phyllites belong to the Daling Group of the (Inner-) Lesser Himalaya. The other rock-types, quartz-mica-schist, garnet bearing mica-schist, mica-gneiss and quartzite, belong to the Darjeeling Formation.

Phyllite (Daling Group):

Phylliteis the type lithology around the Tindharia Railway Station, in between L-1 and 5 (Fig. 4). Overall dip direction of the phylliteis towards north. The same lithology was also encountered between L-25 and 31 in the Kurseong Pankhabari road-cut section. Grain size of phyllite increases for 3 km along the road-cut section towards north. A gradual change in grade from phyllite to quartz-biotite schist happens after L-5 proceeding towards north.

Quartz-mica-schist (Daling Group):

Mica schists occur at places between Gayabari and Mahanadi railway stations (L-7, L-8 in Fig. 4). The grey coloured quartz-mica schists were defined by the spaced cleavage domain of biotite and coarse microlithon of quartz grains. Concordant quartz vein within the quartz mica schist occur near a landslide zone (L-9) at ~2.5 km from the Gayabari Railway station towards Mahanadi. The same quartz-mica schist also occur in L-16 (~2.3 km from the Kurseong railway station in Kurseong–Rohini road), and in L-21 (~2.1 km from the Kurseong railway station in the Kurseong–Pankhabari road). L-24, 25 and is characterized by low-grade mica schists.

Garnet bearing mica schist (Darjeeling Formation):

Garnet appears in the cleavage domain of mica schists after L-9 towards Mahanadi. In the Tindharia-Kurseong section, garnet-bearing mica schists are black and are crenulated. Garnet also occurs within the mica schists along the Kurseong-Pankhabari road section. In L-25 and L-26 greyish brown coloured garnet bearing mica-schists are found.

Mica gneiss (Darjeeling Formation):

In the Tindharia-Kurseong section, 2 km before the Mahanadi railway station, light-coloured gneissic rocks crop out along with garnetiferous mica schist. The light-coloured units also consist of garnets. In the Kurseong–Rohini section, gneiss and mica schist occur in bands.

Quartzite (Daling Group):

The quartzite unit is overlaid by shinning greenish phyllite. Prominent foliation planes dip towards N. In L-25 and 26, more than one lithology was encountered.

5 Structures

Foliation planes in the Tindharia–Kurseong section dip towards N, NE, and SW. In Kurseong–Rohini section, they dip towards E and NW, and in the Kurseong–Pankhabari section towards NW, N and NNE (Figs. 4 and 5).

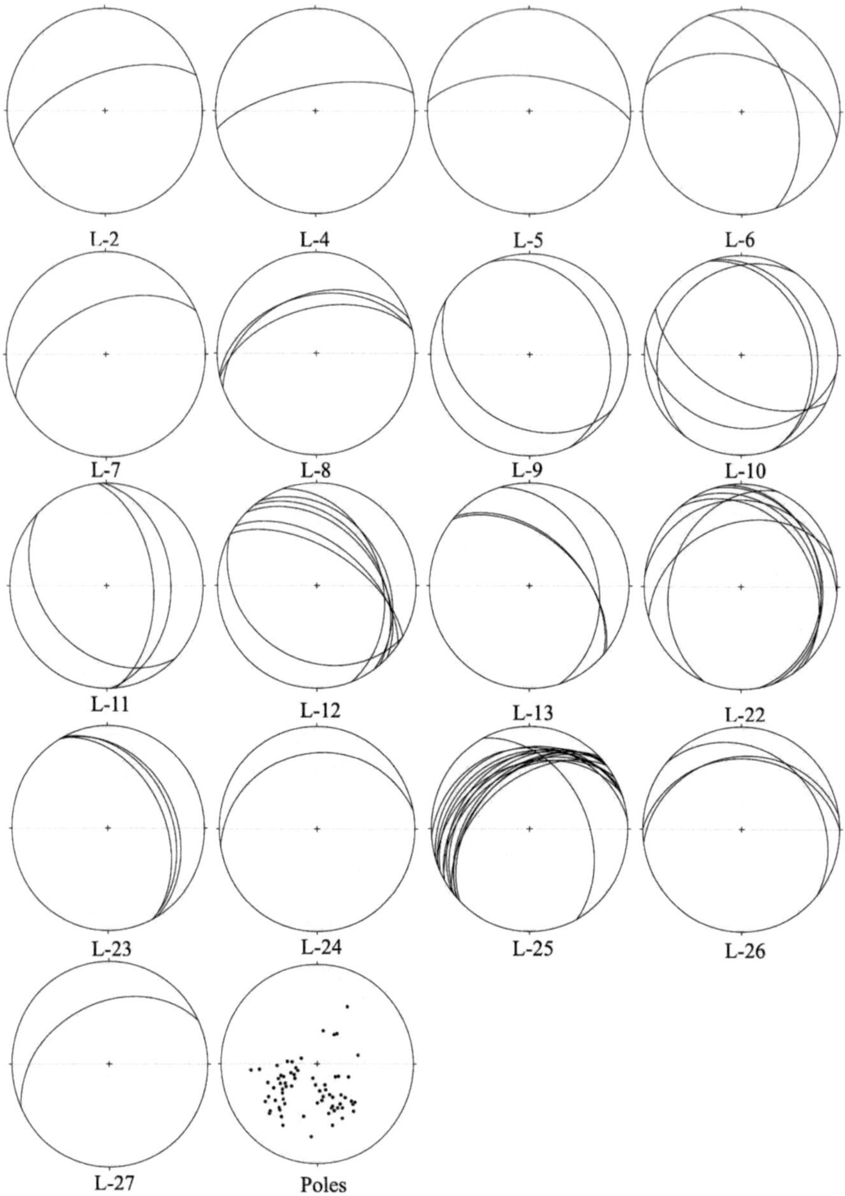

Fig. 5 Location-wise stereo-plots of planar features (joints, schistosity and other foliation planes). Poles of planar data shown

5.1 Folds

Tight, isoclinal and open folds are noted in the field area. Tight isoclinal asymmetric folds are found in light-coloured quartz rich gneissic bands in L-11 to -14 in Darjeeling Formation (Figs. 6 and 7). Open folds are observed in quartz mica schist (L-9) before reaching Mahanadi. Fold axes plunge in variable amount: 8°–38° broadly in three directions: NE, SW and SE.

5.2 Shear Senses

Three shear senses are noted in mica schist and garnet bearing quartz mica schist of Darjeeling Formation: top-to-S, top-to-E and top-to-W. The inclination of the P-planes with respect to the Y-planes are used to deduce the slip sense conclusively, even in absence of any slipped marker layer (e.g., Passchier and Trouw 2005; Mukherjee 2014a, b, 2015). Out of these, only the former shear sense has been reported by the previous authors from this and other Himalayan segments, which is rather common in the collisional mountains as the "fore-shear" (Mukherjee 2007,

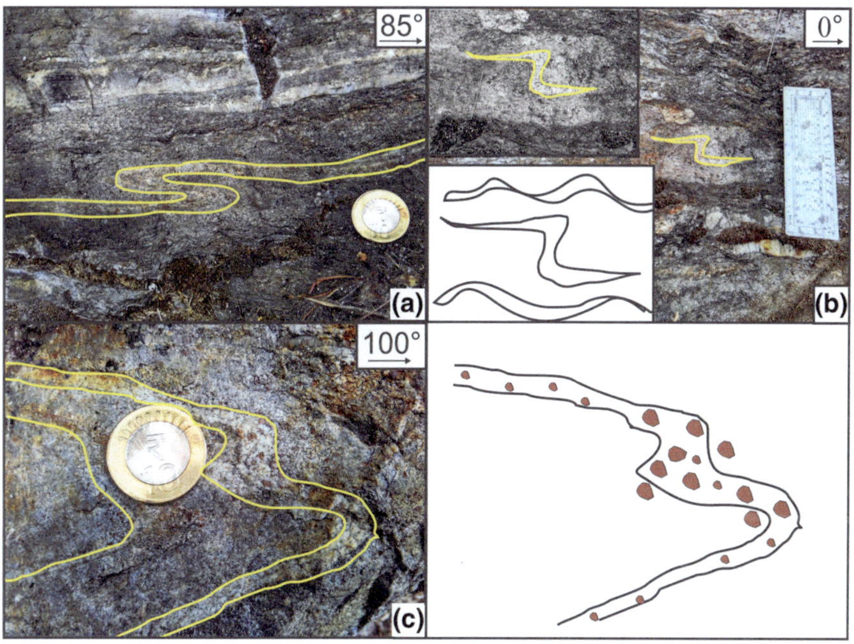

Fig. 6 Folds in gneiss in sub-vertical sections. **a** Folded light quartz rich garnetiferous mica-gneissic and dark micaceous layer in L-13. **b** Tight fold in lighter coloured gneissic layer. A rootless fold is noted at L-15

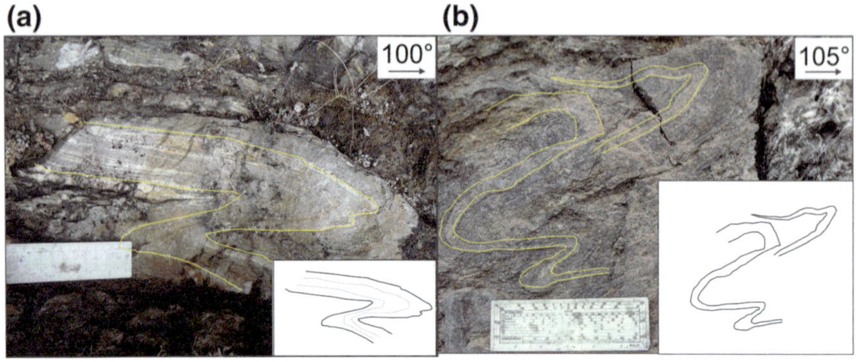

Fig. 7 Tight folds in sub vertical sections of light colouredgarnetiferous quartz mica gneiss. **a** Isoclinal fold in L-10, **b** secondary Z-fold in L-10

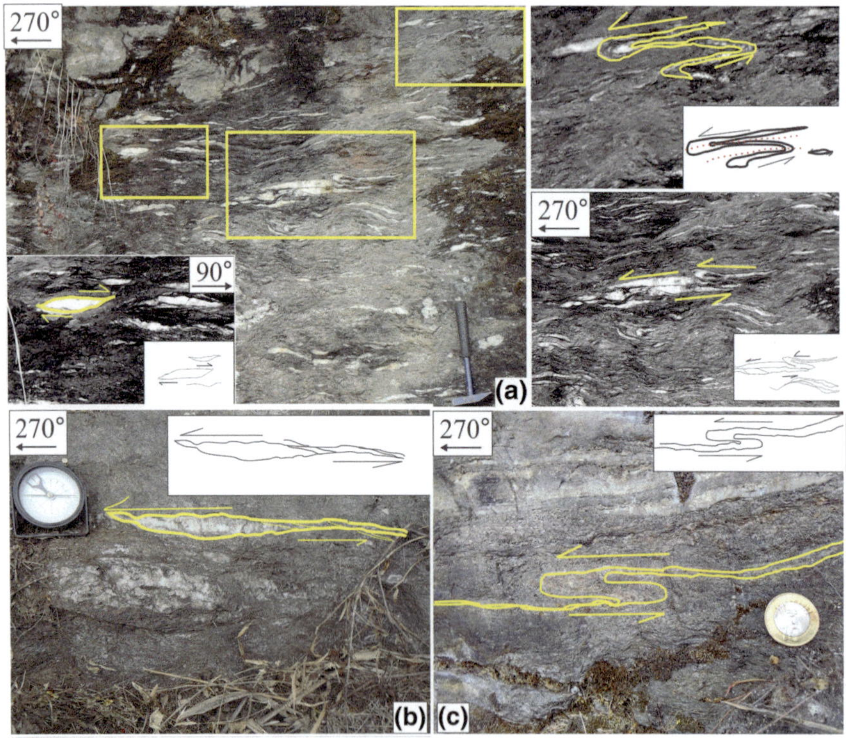

Fig. 8 Top-to-west ductile sheared quartz body within non-foliated micaceous rock in sub-vertical sections. **a** Two shear senses observed in quartz vein within the non-foliated rock at L-10. **b** Sheared quartz vein within the gneissic rock at L-18. Intrafolial folding in garnet-rich light colored band within a non-foliated micaceous rock

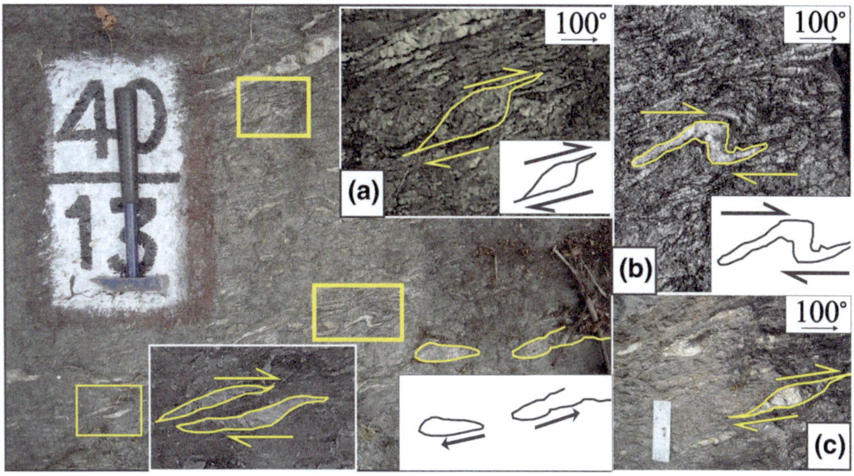

Fig. 9 Top-to-E sheared quartz vein in garnet-bearing mica schist in a sub-vertical section. **a** Lens-shaped quartz vein (L-14). **b** Rootless fold (L-14). **c** Quartz lens (L-10)

Fig. 10 Top-to-S shear sense observed in sub-vertical sections. **a** Quartz fish at L-6. **b** Brittle Y- and P-planes in low grade schist in L-6

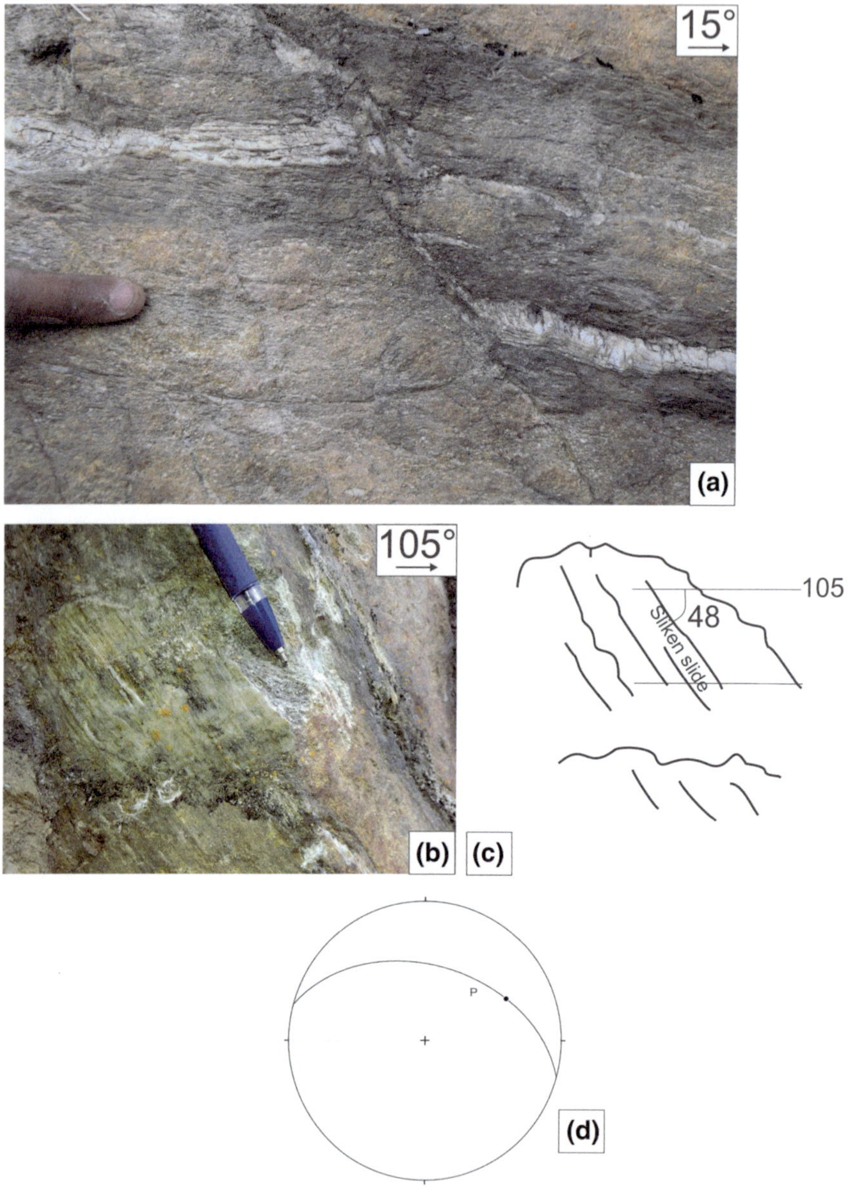

Fig. 11 **a** Small scale normal fault observed in L-6 marked by displacement of quartz vein seen in a sub vertical section, **b** fault plane containing slicken slide within it, **c** sketch of the slicken slide lineation on the fault plane. plunge 34° towards 62°E, **d** stereo-plot of the fault plane and the lineation on it

2012, 2013a, b, c; Mukherjee and Koyi 2010a, b). The rest of the shear senses are the new findings through this work. Intrafolially folded quartz veins show a top-to-W shear around L-10 (Mukherjee et al. 2015; Fig. 8). Top-to-E shear in L-14 is marked by rootless intrafolial folds, lens-shape mineral fish of quartz (e.g., Mukherjee 2011), fracture planes in quartz vein (L-19), around Rohiniroad (Fig. 9). Asymmetric eye/augen-shaped clasts (Fig. 3 and 11a) and brittle Y- and P-planes in L-6 (Fig. 10) characterize the top-to-S shear. Orogen-parallel shear (top-to-E and -W) are also new findings in this work. Mahato et al. (2019) also report the same from the Lesser Himalaya from Garhwal region, Uttarakhand. Orogen parallel deformation is also recently further reported from Mohand area, Siwalik (Dutta et al. submitted).

5.3 Small Scale Normal Faults

In L-6 ($\sim$950 m before the Gayabari railway station), a single small-scale NNE dipping normal fault (Fig. 11) was observed within the high-grade phyllitic rock unit marked by the slip of a narrow quartz vein. The vein also shows a small-scale pinch-and-swell geometry.

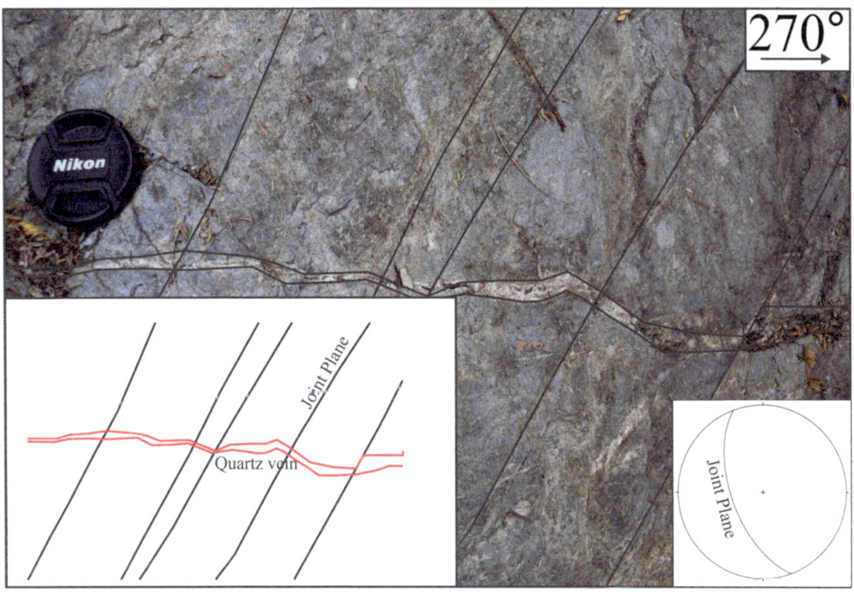

Fig. 12 Joint planes in phyllitic rock cross-cutting quartz vein in L-5, observed in a sub-vertical section

5.4 Foliations and Joints (Fig. 12)

Foliation planes in the Tindharia–Kurseong section dip towards N, NE, SW. In Kurseong–Rohini section, they dip towards E and NW, and in the Kurseong–Pankhabari section towards NW, N and NNE. Two sets of joint planes were seen in phyllite at L-27 (~2.8 km before the location Pankhabari in the Kurseong–Pankhabari road). One set of joint planes dip towards NW.

Acknowledgements This work was a part of SB's M.Sc. thesis. SM acknowledges the research sabbatical for the year 2017 and the CPDA support from IIT Bombay. Annett Buettener, Helen Ranchner and the Springer proof-reading team are thanked for assistance. Mukherjee (2019) summarizes this work.

References

Acharyya SK (1989) The Daling Group, its nomenclature, tectonostratigraphy and structural grain: with notes on their possible equivalents. Geological Survey of India 5–13 (Special Publication 22)

Acharyya SK, Ghosh S, Mandal N, Bose S, Pande K (2017) Pre-Himalayan tectono-magmatic imprints in the Darjeeling-Sikkim Himalaya (DSH) constrained by 40Ar/39Ar dating of muscovite. Journal of Asian Earth Sciences 146, 211–220

Acharyya SK, Ray KK (1977) Geology of the Darjeeling–Sikkim Himalaya. Guide to Excursion No. 3. In: Fourth international Gondwana symposium (Calcutta). Geological Survey of India, pp 1–25

Auden JB (1935) Traverses in the Himalaya. Records of Geological survey of India 69, 123–167

Bhattacharyya K, Mitra G (2009) A new kinematic evolutionary model for the growth of a duplex —an example from the Rangit duplex, Sikkim Himalaya, India. Gondwana Research 16, 697–715

Bose S, Mandal N, Acharyya SK, Ghosh S, Saha P (2014) Orogen-transverse tectonic window in the Eastern Himalayan fold belt: a superposed buckling model. Journal of Structural Geology 66, 24–41

Dutta D, Biswas T, Mukherjee S (Submitted) Orogen-parallel compression in NW Himalaya: evidence from structural and paleostress studies of brittle deformation from the pebbles of Upper Siwalik conglomerates, Uttarakhand, India. Journal of Earth System Science

Jangpangi BS (1972) Some observations on the stratigraphy and reverse metamorphism in Darjeeling Hills. Himalayan Geology 2, 356–370

Lahiri S (1973) Some observations on structure and metamorphism of the rocks of Kurseong–Tindharia region, Darjeeling district, West Bengal. Himalayan Geology 3, 365–371

Lahiri S, Gangopadhway PK (1974) Structure pattern in rocks in Pankhabari–Tindharia region Darjeeling district WB with special reference its bearing on stratigraphy. Himalayan Geology 4, 151–170

Mahato S, Mukherjee S, Bose N (2019) Documentation of brittle structures (back shear and arc-parallel shear) from Sategal and Dhanaulti regions of the Garhwal Lesser Himalaya (Uttarakhand, India). In: Mukherjee S (ed) Tectonics and structural geology: Indian context. Springer International Publishing AG, Cham, pp 411–423. ISBN 978-3-319-99340-9

Mallet FR (1874) On the geology and mineral resources of the Darjiling District and the western Duars. Memoirs of the Geological Survey of India 11, 1–50

Mukherjee S (2007) Geodynamics, deformation and mathematical analysis of metamorphic belts, NW Himalaya. Ph.D. thesis, Indian Institute of Technology Roorkee, pp 1–267

Mukherjee S (2011) Mineral fish: their morphological classification, usefulness as shear sense indicators and genesis. International Journal of Earth Sciences 100, 1303–1314
Mukherjee S (2012) Tectonic implications and morphology of trapezoidal mica grains from the Sutlej section of the Higher Himalayan Shear Zone, Indian Himalaya. The Journal of Geology 120, 575–590
Mukherjee S (2013a) Higher Himalaya in the Bhagirathi section (NW Himalaya, India): its structures, backthrusts and extrusion mechanism by both channel flow and critical taper mechanisms. International Journal of Earth Sciences 102, 1851–1870
Mukherjee S (2013b) Deformation microstructures in rocks. Springer Geochemistry/Mineralogy, Berlin. pp 1–111. ISBN 978-3-642-25608-0
Mukherjee S (2013c) Channel flow extrusion model to constrain dynamic viscosity and Prandtl number of the Higher Himalayan Shear Zone. International Journal of Earth Sciences 102, 1811–1835
Mukherjee S (2014a) Review of flanking structures in meso- and micro-scales. Geological Magazine 151, 957–974
Mukherjee S (2014b) Atlas of shear zone structures in meso-scale. Springer Geology, Cham, pp 1–124. ISBN 978-3-319-0088-6
Mukherjee S (2015) Atlas of structural geology. Elsevier, Amsterdam. ISBN: 978-0-12-420152-1
Mukherjee S (2017a) Shear heating by translational brittle reverse faulting along a single, sharp and straight fault plane. Journal of Earth System Science 126(1)
Mukherjee S (2017b) Review on symmetric structures in ductile shear zones. International Journal of Earth Sciences 106, 1453–1468
Mukherjee S (2019) Introduction to "Tectonics and Structural Geology: Indian Context". In: Mukherjee S (ed) Tectonics and structural geology: Indian context. Springer International Publishing AG, Cham, pp 1–5. ISBN: 978-3-319-99340-9
Mukherjee S, Khonsari MM (2017) Brittle rotational faults and the associated shear heating. Marine and Petroleum Geology 88, 551–554
Mukherjee S, Khonsari MM (2018) Inter-book normal fault-related shear heating in brittle bookshelf faults. Marine and Petroleum Geology 97,45–48
Mukherjee S, Koyi HA (2010a) Higher Himalayan Shear Zone, Sutlej Section—structural geology & extrusion mechanism by various combinations of simple shear, pure shear & channel flow in shifting modes. International Journal of Earth Sciences 99, 1267–1303
Mukherjee S, Koyi HA (2010) Higher Himalayan Shear Zone, Zanskar Section—microstructural studies & extrusion mechanism by a combination of simple shear & channel flow. International Journal of Earth Sciences 99, 1083–1110
Mukherjee S, Mulchrone KF (2013) Viscous dissipation pattern in incompressible Newtonian simple shear zones: an analytical model. International Journal of Earth Sciences 102, 1165–1170
Mukherjee S, Punekar JN, Mahadani T, Mukherjee R (2015) A review on intrafolial folds and their morphologies from the detachments of the western Indian Higher Himalaya. In: Mukherjee S, Mulchrone KF (eds) Ductile shear zones: from micro- to macro-scales. Wiley Blackwell, pp 182–205
Mukhopadhyay MK, Gangopadhyay PK (1971) Structural characteristics of rocks around Kalimpong, West Bengal. Himalayan Geology 1, 213–230
Mukhopadhyay MK, Gangopadhyay PK (1973) Rotational Garnets And Quartz Micro-Fabric in Si and Se: An Example from Metamorphic Rocks around Takdah, Darjeeling District, West Bengal. Himalayan Geology 3, 162–175
Mulchrone KF, Mukherjee S (2015) Shear senses and viscous dissipation of layered ductile simple shear zones. Pure and Applied Geophysics 172, 2635–2642
Mulchrone KF, Mukherjee S (2016) Kinematics and shear heat pattern of ductile simple shear zones with 'slip boundary condition'. International Journal of Earth Sciences 105, 1015–1020
Passchier CW, Trouw, RAJ (2005) Microtectonics, 2nd edn. Springer, Berlin. pp 1–366. ISBN-10 3-540-64003-7

Ray S (1947) Zonal metamorphism in the eastern Himalaya and some aspects of local geology. Quarterly Journal of the Geological, Mining and Metallurgical Society of India 19, 117–140
Sen A (1971) Preliminary observations on the Banded Darjeeling gneiss occurring around Sukhiapokhri, Darjeeling district, West Bengal. Himalayan Geology 1, 276–278
Sen A (1973) Structural features of the rocks in Sukhiapokhri-Bijanbari region, Darjeeling District, West Bengal. Himalayan Geology 3, 357
Singh NK (1972) Petrochemistry of the rocks around Kurseong. Darjeeling District, WBHimalayan Geology 2, 502–514
Sinha Roy S (1973) Gondwana pebble-slate in the Rangit Valley tectonic window, Darjeeling Himalayas and its significance. Bulletin-Geological Society of India 14, 31–39
Wager LR (1939) The Lachi Series of Northern Sikkim. Records of the Geological Survey of India 74, 171–188

Pb—Isotopic Characterization of Major Indian Gondwana Coalfields: Implications for Environmental Fingerprinting and Gondwana Reconstruction

Rajeev Kumar, Joy Gopal Ghosh, S. S. Patel, Avijit Das, S. Sengupta, K. V. S. S. Krishna and D. Guha

1 Introduction

Lead (Pb) isotopic ratios (IR) can be exploited as a mighty tracer for the global atmospheric Pb dispersal derived from natural as well as anthropogenic sources. Geological origin controls the characteristic isotopic composition of Pb, which is a unique property. This property is used to trace Pb sources in the environment (Kylander et al. 2010). Since the phasing out of leaded gasoline from India in the year 2000, coal—combustion remains the principal source of Pb and mercury pollution in the environment, which in turn is responsible for severe health problems in the Indian scenario.

Pb as a trace element can be present in coal in the form of its own minerals such as galena (PbS) or clausthalite (PbSe) or as an admixture of Ba minerals, in the pyrite crystalline structure substituting for Fe, or can be bonded to organic matter in low rank coals (Swaine 1990; Finkelman 1994). The geological changes and the atmospheric inputs during coalification also control the origin of trace elements in the coal. The interactions in the milieu of geological, biochemical and physio—chemical factors are controlling factors for concentration and distribution of trace elements in coals. Hence, it is evident that the trace elements in coal may have diverse origins (Swaine 1990). Knowledge of the mode of occurrence of a hazardous element, such as lead, is of great interest because it may help us to understand its behaviour during coal conversion processes and its potentially hazardous activity. Moreover, interest in lead is enhanced by the fact that by

R. Kumar (✉) · S. S. Patel · A. Das · K. V. S. S. Krishna · D. Guha
G & IG Division, CHQ, GSI, 15 Kyd Street, Kolkata 700016, India
e-mail: rajkrdes@gmail.com

J. G. Ghosh · S. Sengupta
Geological survey of India, N.I.T, N.H. 5P, Faridabad 121001, India

S. Mukherjee (ed.), *Tectonics and Structural Geology: Indian Context*, Springer Geology, https://doi.org/10.1007/978-3-319-99341-6_17

comparing LIRs in ores, contaminant sources and contaminated sites, it can be used as a tracer for environmental management (Díaz-Somoano et al. 2007, 2009).

Apart from environmental application, Pb isotopes have the potential for stratigraphic correlation also. The high precision radiometric dates constrain the absolute age of Formation. Likewise, Sr—isotopes, $\delta^{13}C$ and $\delta^{18}O$ measurements provide meticulous correlation constrains (Veizer et al. 1999; Melchin and Holmden 2006; Jones et al. 2010). It is well established that geological origin controls the characteristic isotopic composition of Pb, which is a unique property. In the ambit of this, the LIRs can be used for stratigraphic correlation and reconstruction also. The occurrence of coal globally dates during the Gondwana period. The coal deposits from India and the deposits from Antarctica can be similarly correlated in global context of Gondwana reconstruction on the basis of LIRs.

The present work aims at the measurement of high precision LIRs of Indian Gondwana coals by Multi-Collector ICPMS housed at the Geological survey of India, Kolkata. This dataset will help in establishing a background for identifying relative contribution of coal, especially those used in coal-based thermal plants for the lead pollution in India. Not only this, these data set can be extended to have wider geological implications such as Gondwana reconstruction.

2 Regional Geology

The sedimentary strata that accumulated in different areas of peninsular India between Permo-Carboniferous and Lower Cretaceous are referred as Gondwanas (Veevers and Tewari 1995). The Gondwana deposits mark resumption of sedimentation in the peninsular India in Late Carboniferous after a long hiatus since the Proterozoic.

Formation of Gondwana basins in India as well as in other continents records a major tectonic event of the Earth, which has been interpreted as driven by the episodic release of Pangean induced heat. The Gondwana successions are preserved in a number of discrete basins in the peninsular India. The structural basins are relics of the original depositional master basin that was disrupted during and after deposition (Veevers and Tewari 1995). The basins define three linear belts along the present-day river valleys of (1) Narmada–Son–Damodar (NSD), (2) Pranhita–Godavari and (3) Mahanadi (Fig. 1). The latter two river valleys are subparallel to each other and meet the ENE–WSW trending NSD valley and the present-day eastern margin of the subcontinent at high angle (Biswas 1999; Mishra et al. 1999; Acharyya 2000).

Gondwana Basins of India account for nearly 99% of coal resource of the country. The basins occur along major river valleys either as discrete bodies or are unified by post-Permian strata and are named after the Rivers Damodar, Son, Mahanadi, Godavari etc. or the linear hill ranges like Satpura and Rajmahal (Fig. 1). Throughout the Gondwana sequence in the Peninsular India the Barakar Formation is the predominant coal bearing horizon. Indian Gondwana coal are

Fig. 1 Location of sampling points from the Indian Gondwana Coalfields covered in present study

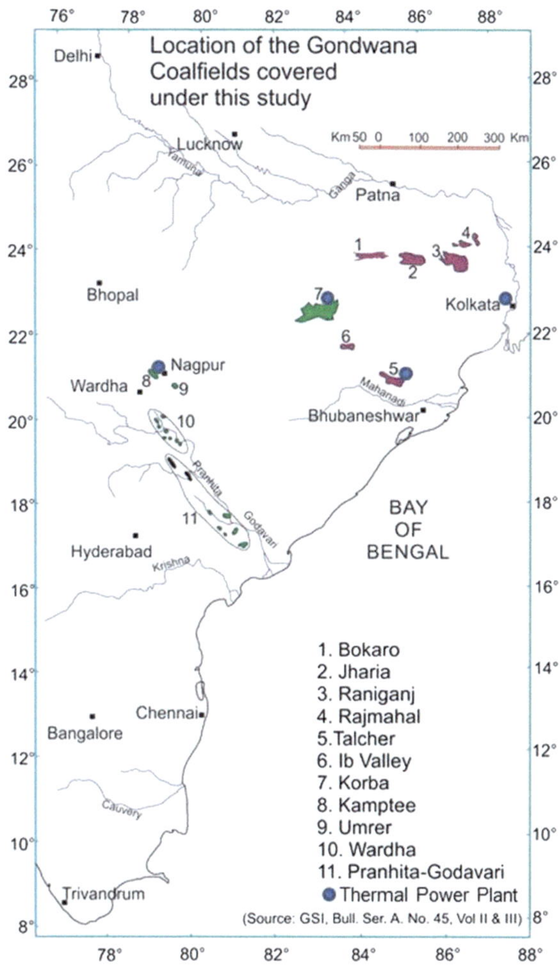

mostly bituminous to sub—bituminous and occurs as seams inter -banded with sediments. They typically have high ash content in range of 35–50% (Raja Rao 1982, 1983, 1987; Mukherjee 2019).

3 Methodology

3.1 Sample Collection

Coal samples have been collected from actively mined Gondwana coal deposits spread all over India. The coalfields covered in the purview of this study are Talcher and Ib Valley Coalfield (40 Samples), Rajmahal Coalfield (20 samples), Wardha

Valley Coalfield (18 samples), Godavari Valley Coalfield (18 samples) and Korba Coalfield (14 samples). Over all 110 coal samples have been analysed in this study covering the E, S and central Indian Coalfields with the objective to delineate inter as well as intra—seam variations, if at all present which can be traced after analyses. All the samples are bituminous to sub—bituminous coal and feed the major thermal power plants (TPPs) for power generation. In addition to the coal samples, fly ash samples of major TTPs in the corresponding area have also been analysed for Pb-isotopic composition. Figure 1 shows the disposition of Indian Gondwana Coal deposits and location of the samples collected.

3.2 Analytical Procedure

The physical and chemical processing methodology adopted for the determination of Pb—isotopic composition for coal samples was dry ashing and mineralization process as per Díaz-Somoano et al. (2007). The coal samples were pulverised in agate motor and dried to constant weight at room temperature. After this step the samples were finally dried in a muffle furnace (~90 °C; 2 h) to remove the moisture. One gm of coal sample was taken and then mineralized under controlled conditions using a programmable furnace by the dry ashing with a temperature increase of 1 °C per minute to a maximum temperature of 500 °C. The total mineralization time was 10 h after attaining the required temperature. The ash—content was determined gravimetrically. The ash was then dampened with one drop of Milli-Q deionized water before treating with the 5 mL of concentrated HF and 0.5 mL of concentrated $HClO_4$ in savillex vials and left overnight. The mixture of acids was evaporated to near dryness and this step was repeated. After attaining the incipient moisture—content, the residue was dissolved in 2 mL of concentrated HNO_3, transferred to a 100 mL volumetric flask, filled to the mark with Milli-Q deionised water and transferred to a 100 mL PP Nalgene bottle. This stock solution was later diluted to 2% (v/v) HNO_3 for determination of the LIRs. All the experiments were performed with considerable number of duplicate and procedural blanks to determine possible contamination from reagents and general handling.

The fly ash samples were dried to remove moisture content and then 0.2 g of sample has been digested using an acid solution of $HCl:HNO_3:HF::3:3:2$ using Anton Paar Multiwave 3000 Microwave digestion system. The mixture of acids and fly ash was evaporated to near dryness and this step was repeated again (Zhou et al. 2014). The residue was dissolved in 2 ml of concentrated HNO_3, transferred to a 100 ml volumetric flask, filled to the mark with MilliQ water and transferred to a 100 ml HDPE bottle.

These chemically processed samples of coal and fly ash have been analysed for LIRs using Nu Plasma II MC ICPMS housed at LAMCI facility of G & IG Division of the Geological Survey of India, Kolkata. The instrument is equipped with 22 collectors (17 F Collectors and 5 Ion Counters). Routine measurements made in static mode using the Faraday detectors with 10^{11} Ω resistors and Ion counting

Table 1 MC ICP MS operating conditions during data acquisition

Instrument/model	MC ICPMS/Nu plasma II
Collection mode	Static
Data acquisition	Isotope ratio mode
Sample cone	Ni (1 mm orifice)
Skimmer cone	Ni (0.7 mm orifice)
Collector types used	Six Faradays
Coolant gas flow	13 L/min
Auxiliary (Ar)	0.87–0.9 L/min
Nebuliser	33–34 psi
RF power	1300 W
Standards used	SRM 981 and SRM 983
Rinse time between two samples	4–5 min with 5% HNO_3

channels with ion yield ~1.0 mv per ppm. The typical operating parameters of measurement are provided in Table 1.

Samples and standards were adjusted to a Pb/Tl ratio of 9:1 (after Weiss et al. 2004).The concentration of Tl solution was maintained at 8 ppb. The analytical protocol followed during the course of analyses was standard-sample-standard bracketing method using two Standards (SRM 981 & 983) after every five to seven unknown solutions. NIST SRM-981 are $^{206}Pb/^{204}Pb = 16.934 \pm 0.007$, $^{207}Pb/^{204}Pb = 15.486 \pm 0.012$, and $^{208}Pb/^{204}Pb = 36.673 \pm 0.033$ (n = 15), respectively, which matches well with the recalibrated magnitudes: 16.9322, 15.4855, and 36.6856, respectively. Similarly, for NIST SRM 983 ratios are $^{204}Pb/^{206}Pb = 0.000298$, $^{207}Pb/^{206}Pb = 0.070906$, and $^{208}Pb/^{206}Pb = 0.013095$ (n = 20) respectively, which matches well with the recalibrated magnitudes: 0.000371, 0.071201, and 0.013619, respectively. (Todt et al. 1996; Belinda et al. 2000; Tanimizu and Ishikawa 2006 references rearranged with oldest at first and youngest at last). The contribution of ^{204}Hg in the measurement of ^{204}Pb was periodically monitored and mathematical correction was performed in the data acquisition program (called as the *NICE: Nu Instruments Calculation Editor: File* in Nu Plasma II) to account for any small contribution from Hg on the signal at ^{204}Pb (typically <300 counts s^{-1}, or below 0.4% of the total ^{204}Pb signal). The precision of the measurements by MC-ICP-MS was below 0.07% for all measured isotope ratios. The corresponding blank Pb-content value obtained during the analyses span has been in the range of 10–11 ng ml^{-1} (White et al. 2000; Das et al. 2016 this reference attached as protocols followed are similar to this for our Laboratory).

4 Results

It is a general practise to use $^{207}Pb/^{206}Pb$ (or $^{206}Pb/^{207}Pb$) and $^{208}Pb/^{206}Pb$ ratios in variation plots for environmental studies and $^{208}Pb/^{204}Pb$, $^{207}Pb/^{204}Pb$ and $^{206}Pb/^{204}Pb$ ratios for geochronological and geochemical studies. Since the

Table 2 Lead isotope composition of Indian Gondwana Coal—deposits and fly ash determined in this study and other coal deposits from published literature

Samples	$^{207}Pb/^{206}Pb$				$^{208}Pb/^{206}Pb$				Pb—content (mg kg^{-1})			
	Mean	Max	Min	RSD	Mean	Max	Min	RSD	Mean	Max	Min	RSD
Talcher & Ib valley coalfields (n = 40)	0.8220	0.8531	0.7215	0.0266	2.0863	2.1655	1.9484	0.0192	40.94	72.73	15.40	0.34
Rajmahal coalfield (n = 20)	0.8022	0.8467	0.7748	0.0233	2.0104	2.0560	1.9746	0.0140	74.60	160.89	23.48	0.44
Godavari valley coalfield (n = 18)	0.7761	0.8219	0.7150	0.0449	2.0382	2.0767	1.9858	0.0130	237.74	565.91	54.67	0.79
Wardha valley coalfield (n = 18)	0.8303	0.8731	0.8050	0.0170	2.0735	2.0998	2.0373	0.0069	180.98	437.92	39.13	0.59
Korba coalfield (n = 14)	0.8514	0.8845	0.8050	0.0270	2.1442	2.2231	2.0725	0.0209	18.43	42.10	3.20	0.68
East Indian coals (Kumar et al. 2016b, 35th IGC)	0.8263	0.8007	0.8518		2.0837	2.1606	2.0067		238	440	37	
Kolaghat fly ash	0.8726 ± 0.0007	–	–	–	2.1542 ± 0.0016	–	–	–	–	–	–	–
Bandel fly ash	0.8608 ± 0.0005	–	–	–	2.1382 ± 0.0008	–	–	–	–	–	–	–
Korba fly ash	0.8478 ± 0.0010	–	–	–	2.1070 ± 0.0002	–	–	–	–	–	–	–
Talcher fly ash	0.8668 ± 0.0001	–	–	–	2.1534 ± 0.0007	–	–	–	–	–	–	–
Khaiperkheda fly ash	0.8757 ± 0.0001	–	–	–	2.1706 ± 0.0003	–	–	–	–	–	–	–
Raniganj coalfield (Das et al. 2016)	0.8162	0.8275	0.8007	0.0097 (Std Dev)	2.0602	2.0875	2.0175	0.230 (Std Dev)	–	–	–	–
Jharia coalfield (Das et al. 2016)	0.8396	0.8573	0.8191	0.0148 (Std Dev)	2.0946	2.1308	2.0543	0.0280 (Std Dev)	–	–	–	–

(continued)

Table 2 (continued)

Samples	$^{207}Pb/^{206}Pb$				$^{208}Pb/^{206}Pb$				Pb—content (mg kg^{-1})			
	Mean	Max	Min	RSD	Mean	Max	Min	RSD	Mean	Max	Min	RSD
Chinese Coal (Zhang et al. 2007; Cheng and Hu 2010)	0.8479	0.8661	0.8326	–	2.0892	2.141	1.99	–	40	400	–	–
World Coal (Díaz-Somoano et al. 2009)	0.8322	0.8546	0.8036	–	2.0567	2.1013	1.9822	–	–	–	–	–
Indonesian Coal (Díaz-Somoano et al. 2009)	0.8449	0.8474	0.8417	–	2.0929	2.1013	2.0878	–	–	–	–	–
Chinese fly ash (Chen et al. 2005)	0.868	0.867	0.855	–	2.13	2.157	2.103	–	–	–	–	–

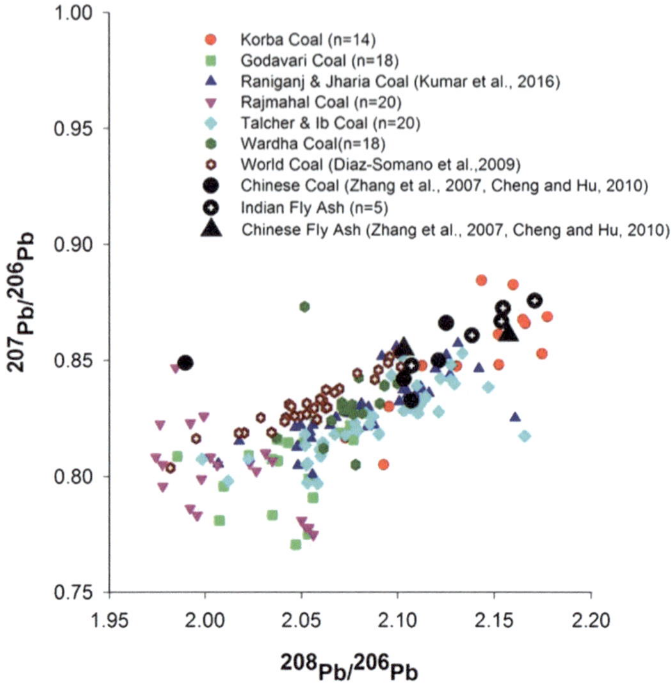

Fig. 2 Compiled three ratio plot of Indian Coal and fly ash in comparison to the published values of coal and fly ash elsewhere

materials analysed in the present work are coal and fly ash, hence, it is apposite to represent the results in terms of $^{207}Pb/^{206}Pb$ (or $^{206}Pb/^{207}Pb$) and $^{208}Pb/^{206}Pb$ ratios. The results obtained in the present wok as well as dataset available from published literature are summarised in Table 2. The compiled three ratio plot of Indian Coal and Fly ash in comparison to the published values of coal and fly ash elsewhere are shown in Fig. 2.

4.1 *LIRs of Coal Samples*

The coal deposits of Talcher and Ib valley are situated in the eastern part of India and follow the Mahanadi graben along which the coal has developed. Total 40 samples have been analysed from these coal—deposits collected from working seams and act as feed to the National Thermal Power Corporation (NTPC) Super Thermal Power plant at Talcher. These coal also act as feed in the Bandel and Kolaghat Thermal Power Plant. The $^{207}Pb/^{206}Pb$ ratio range between 0.7215 and 0.8531 with a mean value of 0.8220 whereas the $^{208}Pb/^{206}Pb$ ratio ranges between 1.9484 and 2.1655 with a mean value of 2.0863. The Pb-concentration measured of these coal—deposits lie between 15.40 and 72.73 mg kg^{-1} with a mean value of 40.94 mg kg^{-1}.

The coal-deposits of Rajmahal are situated in the eastern part of India where the coal development has taken place along the main Rajmahal—Purnea Basin. The analysed coal samples (n = 20) from Rajmahal Coalfield have isotopic signatures in range of 0.7748–0.8467 with mean value of 0.8022 and 1.9746–2.0560 with mean value of 2.0104 for the $^{207}Pb/^{206}Pb$ and $^{208}Pb/^{206}Pb$ ratios respectively. This coalfield is also the supplier of coal to Bandel and Kolagahat Thermal Power Plant for power generation.

The coal deposits of Godavari Valley have been developed along the course of river Godavari. On the Gondwana Basin's map of India (Fig. 1); in fact two coalfields have a linear extension from NW to SE. The NE extension of the coal-field is along Wardha River and hence, known as Wardha Valley Coalfield and the SE extension is known as the East Godavari Valley Coalfield. The 18 coal samples analysed from Godavari Valley Coalfield have $^{207}Pb/^{206}Pb$ ratios in range of 0.7150–0.8219 with mean value of 0.7761 and $^{208}Pb/^{206}Pb$ ratios in range of 1.9858–2.0767 with mean value of 2.0382. The concentration of Pb in this coalfield obtained in range 238–566 mg kg^{-1} with a mean value of 238 mg kg^{-1}.

The 18 coal samples analysed from Wardha Valley Coalfield have $^{207}Pb/^{206}Pb$ ratios in range of 0.8050–0.8731 with mean value of 0.8303 and $^{208}Pb/^{206}Pb$ ratios in range of 2.0373–2.0998 with mean value of 2.0735. The concentration of Pb in this coalfield ranges 39–438 mg kg^{-1} with a mean value of 181 mg kg^{-1}.

The Korba Coalfield forms the central Indian Gondwana Coal deposits. Quarriable potentiality of the seam of Barakar Formation and its occurrence at shallow depth (8.80 m) further enhances its importance. It is 16.82–18.74 m thick and occurs in the western part of the area only in between two NNE-SSW faults. The fourteen samples analysed from this coalfield has isotopic composition in range 0.8050–0.8845 with mean value of 0.8514 and 2.0725–2.2231 with mean value of 2.1442 for the $^{207}Pb/^{206}Pb$ and $^{208}Pb/^{206}Pb$ ratios respectively. The Pb—concentration of these coal samples are in range of 3.2–42 mg kg^{-1} with mean value of 18.5 mg kg^{-1}.

4.2 *LIRs of Fly Ash Samples*

During the course of this work fly ash samples have also been analysed from the power plants using the coal deposits as feed. The Kolaghat, Bandel and NTPC Talcher Termal Power Plants have their source in coalfields of Raniganj, Jharia, Talcher, Ib and Rajmhal. The NTPC Korba has its source for power generation in Korba Coalfield. The coals of Wardha valley coalfields act as feed to the Khaiperkheda Thermal Power Plant. The results obtained are presented in Table 2.

5 Discussion

As is evident from the previous section that the LIRs of Indian Gondwana coal—deposits ranges 0.7150–0.8845 for 207/206 Pb and 1.9484–2.2231 for 208/206 Pb with Pb concentration ranges 3.2–566 mg kg^{-1}. The variability of the Pb—content and its isotopic composition can be attributed to high and variable ash—content of the coal seams including its pyrite—content and/or more radiogenic behaviour. It might be worth noting that the data obtained in the present work compares well with the records from the Chinese coals (Shanghai, Nanjing and Huinan Coalfield) that have Pb—content varying 40–400 ppm with a $^{206}Pb/^{207}Pb$ IR variation of 0.8326–0.8661 and $^{208}Pb/^{206}Pb$ IR values of 1.99–2.141 (Zhang et al. 2007; Cheng and Hu 2010).

Similarly the LIRs obtained in the fly ash analyses ranges 0.8478–0.8757 for 207/206 Pb and 2.1070–2.1706 for 208/206 Pb. These values have an increasing trend with respect to the feed coal Pb isotopic compositions when compared to the respective coal—deposits. Also the dataset obtained complies well with the published literature which reports the Pb isotopic composition for Chinese fly ash, which ranges 0.855–0.867 for 207/206 and 2.103–2.157 for 208/206, respectively (Chen et al. 2005).

5.1 LIRs in Fingerprinting

Uranium can be bonded to the organic material of the coal and also to the inorganic components (Swaine 1990). In this case, the isotopic composition i.e. LIRs of the coal is a result of mixing of the isotope ratios of common Pb in organic material at the time of its sedimentation and the radioactive products of the subsequently transported U and Th. Because of the predominance of ^{238}U in natural isotopic mixtures, the fingerprints of ^{206}Pb are substantially affected by systems highly enriched in U. This anomalous isotopic composition of Pb and the influence of U on the Pb IR are the properties used to characterize and ultimately "fingerprinting/tracing" the source.

The present work reports and reviews the isotopic composition i.e., the LIRs of the Indian Gondwana basin Coal deposits. These dataset aim at establishing a background for identifying relative contribution of coal, especially those used in coal—fired thermal plants for the Pb pollution in India. To test this hypothesis, the LIRs data generated by Das et al. (2016) for the different anthropogenic contributors of Pb in Kolkata City and LIRs reported by Kumar et al. (2016c) in the dust storm aerosols of Indian subcontinent has been used for tracing/fingerprinting the pollutant in the township of Kolkata. The dataset obtained by Das et al. (2016) are on a gamut of food items (such as Rice, Dal, Spices, Vegetables, Fish, Meat, Chicken and Herbs), street dust, diesel and rainwater of Kolkata city. By applying binary mixing model on these dataset along with data reported by Kumar et al.

(2016a, b) has shown that coal—combustion is not the major contributor in the atmosphere of Kolkata City instead the diesel is the principal contributor in the form of vehicular exhaust. Thus using the LIR in coal samples from East Indian Gondwana coal deposits, a logical clue for "tracing/fingerprinting" the Pb pollution source in the Indian scenario particularly for the Kolkata City has been achieved (Das et al. 2016; Kumar et al. 2017).

It is further recommended that LIRs of environmental milieu/matrices should be taken up in different power—hub Indian cities and the inventory on LIRs of Indian coal should be compared for getting a novel qualitative as well as quantitative "fingerprinting" of environmental pollution and contribution from anthropogenic sources.

5.2 LIRs in Gondwana Reconstruction

Apart from these environmental aspect, the LIR data set generated in the present work particularly of the Mahanadi and Godavari valley coal—deposits can have implications on the Gondwana reconstruction.

The Gondwana sequence in the East Antarctica is restricted to the area around Beaver Lake in the Lambert graben. The sequence is known as the Amery Group (Mond 1972) and represents >4000 m thick sequence of coal—bearing conglomerate—sandstone—minor shale association. Broadly, the sequence is >400 m thick Upper Permian Radok Conglomerate in the basal part, followed by $\sim$3000 m thick coals bearing Bainmedart Coal Measures which is overlain by around 750 m thick Triassic Flagstone Bench Formation. The lithofacies and palynology of the rocks of the Amrey Group has been studied (Playford 1990; Webb and Fielding 1993; McLoughlin and Drinnan 1997), but the information on lead- isotopic composition of this Group of rocks is not yet reported.

In the Gondwana reconstruction, there is a consensus that the $\sim$1.0 Ga mobile belts of the Eastern Ghats in India and the Eastern Reyner Complex in Antarctica were parts of a suture zone associated with Rhodinia Formation. However, these belts occur parallel to the boundaries of the continents and hence using them in Gondwana reconstruction could be superfluous. The Gondwana basins of the Amery Group of the Lambert Graben and those of the Mahanadi and Godavari basins in India that are aligned transverse to the continental margins are favoured for a tighter Gondwana reconstruction. Available literature postulates correlation of Lambert Graben with Godavari basin on the basis of regional geology and coal petrology (Holdgate et al. 2005) while the correlation model with Mahanadi basin is favoured based on geophysics and bathymetry (Stagg 1985; Mishra et al. 1999) (Fig. 3). Since geophysics and bathymetry gets much influenced due to post-breakup events in the seafloor, a stronger correlation of the Lambert Graben with either the Mahanadi or the Godavari basin would require further constraints. The distinctive Pb—isotopic compositional variation between the coals from the Mahanadi basin and the coals from the Godavari—Wardha basin as obtained in the

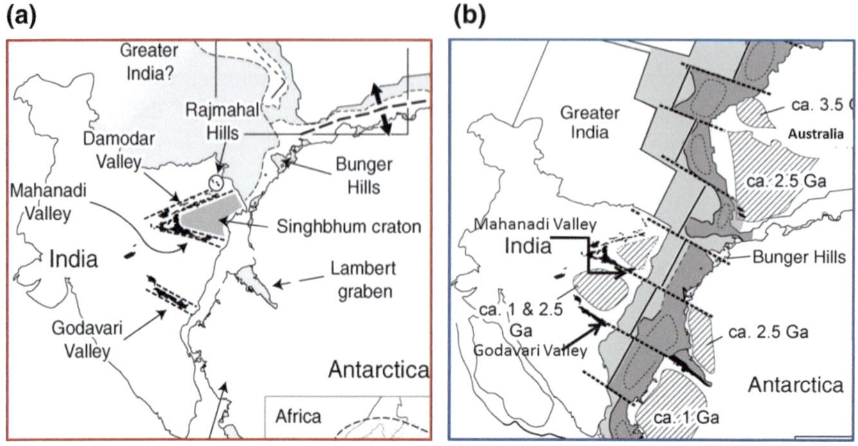

Fig. 3 Proposed reconstruction models of Permian-Triassic Gondwana illustrating separation of Indian and Antarctic, disposition of rifted coalfields of Mahanadi Valley, Godavari Valley and Lambert graben (Antarctica) shown. **a** After Stagg (1985) and Boger and Wilson (2003); **b** After Harrowfield et al. (2005)

present work and similar data set from Amrey Group and its coal deposit could offer a good constrain to achieve a better fit between India and Antarctica in a Gondwana framework. Moreover, the possible reconstruction—hypotheses proposed between Indian and East Antarctica can also be tested with the aid of LIRs in coal of these two presently widely separated terrains (Stagg 1985; Boger and Wilson 2003; Harrowfield et al. 2005).

6 Conclusions

Pb is present in different modes of occurrences in different coals. The evaluation of LIRs obtained from coals is helpful in discriminating between the different lead species. The present study is an attempt to create a comprehensive and systematic database for the Indian Gondwana coal deposits. Hence, the data in terms of LIRs of coal are unknown unique isotopic signature of Indian Gondwana coal. The LIR of Indian Gondwana coals varies 0.7150–0.8845 for 207/206 Pb and 1.9484–2.2231 for 208/206 Pb with 3.2–566 mg kg^{-1} of Pb concentration. The variability of the Pb—content and its isotopic composition can be attributed to high and variable ash—content of the coal seams including its pyrite content as well as more radiogenic behaviour. Similarly the LIRs obtained in the fly ash analyses ranges 0.8478–0.8757 for 207/206 Pb and 2.1070–2.1706 for 208/206 Pb. These values have an increasing trend with respect to the feed coal Pb isotopic compositions when compared to the respective coal—deposits. Explicitly, the inventory generated in terms of LIRs and Pb—content in Indian Gondwana coals would contribute to establish a background for identifying relative contribution of coal, especially those used in coal—fired thermal plants for the lead pollution in India. Not only this, these unique isotopic signatures of Pb can be used for regional implications as chemical tracers (Sr—isotopes, $\delta^{13}C$ and $\delta^{18}O$ measurements are generally used) for stratigraphic correlation and Gondwana reconstruction also.

Acknowledgements This work is the outcome of the research project (RP/CHQMIV/2014/114) approved by Ministry of Mines and to be taken up by the Geological Survey of India, Kolkata. The author would like to thank the Director General, Geological Survey of India, Kolkata, for all sort of administrative supports and permission to publish this research outcome. The West Bengal Power Development Corporation, Kolkata and different regional ancillaries of Coal India Ltd. are acknowledged for their fruitful discussions and granting permission to collect samples from Thermal Power Plants and collieries respectively. Soumyajit Mukherjee edited and reviewed this manuscript. This work is summarized in Mukherjee (2019).

References

Acharyya SK (2000) Tectonic setting and nature of the Gondwanic Indian crust. Proc. Vol., Int. Seminar, Precambrian Crust in Eastern and Central India. Geological Survey of India, Special Publication 57, 1–8

Biswas SK (1999) A review on the evolution of rift basins in India during Gondwana with special reference to western Indian basins and their hydrocarbon prospects. In: Sahni A, Loyal RS (eds) Gondwana assembly: new issues and perspectives. proceedings of indian national science academy special issue 65, pp 261–283

Boger SD, Wilson CJL (2003) Brittle faulting in the Prince Charles Mountains, East Antarctica: cretaceous transtensional tectonics related to the break-up of Gondwana. Tectonophysics 367, 173–186

Chen JM, Tan MG, Li YL, Zhang YM, Lu WW, Tong YP, Zhang GL, Li Y (2005) A lead isotope record of Shanghai atmospheric lead emissions in total suspended particles during the period of phasing out of leaded gasoline. Atmospheric Environment 39, 1245–1253

Cheng H, Hu Y (2010) Lead (Pb) isotopic fingerprinting and its applications in lead pollution studies in China: a review. Environmental Pollution 158, 1134–1146

Das A, Krishna KVSS, Kumar R, Das A, Sengupta S, Ghosh JG (2016) Tracing lead contamination in foods in the city of Kolkata, India. Environmental Science and Pollution Research 23, 22454–22466

Díaz-Somoano M, Suárez-Ruiz I, Alonso JIG, Ruiz Encinar J, López-Anton MA, Martínez-Tarazona MR (2007) Lead isotope ratios in Spanish coals of different characteristics and origin. International Journal of Coal Geology 71, 28–36

Díaz-Somoano M, Kylander ME, Lopez-Anton MA, Suarez-Ruiz I, Martínez Tarazona MR, Ferrat M, Kober B, Weiss DJ (2009) Stable lead isotope compositions in selected coals from around the world and implications for present day aerosol source tracing. Environmental Science and Technology 43, 1078–1085

Finkelman RB (1994) Modes of occurrence of potentially hazardous elements in coal: levels of confidence. Fuel Processing Technology 39, 21–34

Flem B, Grimstvedt A, Cook N (2000) Lead isotope determinations by inductively-coupled plasma mass spectrometry (ICP-MS): potential of sector field instruments. NGU-BULL 436, 203–207

Harrowfield M, Holdgate RG, Wilson CJL (2005) Tectonic significance of the Lambert graben, East Antarctica: Reconstructing the Gondwanan rift. Geology 33, 197–200

Holdgate GR, McLoughlin S, Drinnan AN, Finkelman RB, Willett JC (2005) Inorganic chemistry, petrography and palaeobotany of Permiancoals in the Prince Charles Mountains, East Antarctica. International Journal of Coal Geology 63, 156–177

Jones DS, Maloof AC, Hurtgen MT, Rainbird RH, Schrag DP (2010) Regional and global chemostratigraphic correlation of the early Neoproterozoic Shaler Supergroup, Victoria Island, Northwestern Canada. Precambrian Research 181, 43–63

Kumar R, Patel SS, Manjhi JK (2016a) Pb-isotopic characterization of major Indian Coalfields and their potential as heavy metal pollutants. GSI Unpublished Report

Kumar R, Patel SS, Ghosh JG (2016b) Variability of Pb-isotopes in the East Indian Gondwana Coal deposits: its influence on Kolkata street dust. 35TH IGC ABSTRACT

Kumar S, Aggarwal SG, Malherbe J, Barre JPG, Berial S, Gupta PK, Donard (2016c) Tracing dust transport from Middle-East over Delhi in March 2012 using metal and lead isotope composition. Atmospheric Environment 132, 179–187

Kumar R, Patel SS, Das A, Ghosh JG, Sengupta S, Krishna KVSS, Guha D (2017) Variability of Pb-isotopes in the East Indian Gondwana Coal deposits: It's influence on Kolkata street dust. Indian Journal of Geoscience 71(4), 575–588

Kylander ME, Klaminder J, Bindler R (2010) Natural Lead Isotope Variations in the Atmosphere. Earth and Planetary Science Letters 290, 44–53

McLoughlin S, Drinnan AN (1997) Revised stratigraphy of the Permian Bainmedart coal measures, northern Prince Charles Mountains, East Antarctica. Geological Magazine 134, 335–353

Melchin MJ, Holmden CE (2006) Carbon isotope chemostratigraphy in Arctic Canada: sea-level forcing of carbonate platform weathering and implications for Hirnantian global correlation. Palaeogeography, Palaeoclimatology, Palaeoecology 234, 186–200

Mishra DC, Chandra Sekhar DV, Venkata Raju DCh, Vijay Kumar V (1999) Crustal structure based on gravity-magnetic modelling constrained from seismic studies under Lambert Rift, Antarctica and Godavari and Mahanadi rifts, India and their interrelationship. Earth and Planetary Science Letters 172, 287–300

Mond A (1972) Permian sediments of the Beaver Lake area, Prince Charles Mountains. In: Adie RJ (ed) Antarctic geology and geophysics. Universitetsforlaget, Oslo, pp 585–589

Mukherjee S (2019) Introduction to "Tectonics and Structural Geology: Indian Context". In: Mukherjee S (ed) Tectonics and structural geology: Indian context. Springer International Publishing AG, Cham, pp 1–5. ISBN: 978-3-319-99340-9

Playford G (1990) Proterozoic and Palaeozoic palynology of Antarctica: a review. In: Taylor TN, Taylor EL (eds) Antarctic paleobiology—its role in the reconstruction of Gondwana. Springer Verlag, New York. Plumstead, pp 50–70

Raja Rao CS (1982) Coalfields of India. Vol-II, coal resources of Tamil Nadu, Andhra Pradesh, Orissa and Maharashtra. Geological Survey of India Bulletin Series A, 45

Raja Rao CS (1983) Coalfields of India. Vol-III, Coal resources of Madhya Pradesh and Jammu and Kashmir. Geological Survey of India Bulletin Series A, 45

Raja Rao CS (1987) Coalfields of India. Vol-IV, Part I. Coal resources of Bihar. Geological Survey of India Bulletin Series A, 45

Stagg HMJ (1985) The structure and origin of Prydz Bay and MacRobertsonShelf. East Antarctica: Tectonophysics 114, 315–340

Swaine DJ (1990) Trace elements in coal. Butterworths and Co. Publishers

Tanimizu M, Ishikawa T (2006) Development of rapid and precise Pb isotope analytical techniques using MC-ICP-MS and new results for GSJ rock reference samples. Geochemical Journal 40, 121–133

Todt W, Cliff RA, Hanser A, Hofmann AW (1996) Evaluation of a $^{202}Pb/^{205}Pb$ double spike for high precision lead isotope analysis in earth processes: reading the isotopic code. American Geophysical Union, pp 429–437

Veevers JJ, Tewari RC (1995) Gondwana master basin of peninsular India between Tethys and the interior of the Gondwanaland province of Pangea. Memoir of Geological Society of America 187, 1–73

Veizer J, Ala D, Azmy K, Bruckschen P, Buhl D, Bruhn F, Carden GAF, Diener A, Ebneth S, Godderis Y, Jasper T, Korte C, Pawellek F, Podlaha O, Strauss H (1999) 87Sr/86Sr, 13C and 18 O evolution of Phanerozoic seawater. Chemical Geology 161, 59–88

Webb A, Fielding R (1993) Permo-Triassic sedimentation within the Lambert Graben, northern Prince Charles Mountains, East Antarctica. In: Findlay RH, Uruug R, Banks MR, Veevers JJ (eds) Gondwana eight: assembly, evolution and dispersal. A. A. Balkema, Rotterdam, pp 357–369

Weiss D, Kober B, Dolgopolova A, Gallagher K, Spiro B, Le Roux G, Mason TFD, Kylander M, Coles BJ (2004) Accurate and precise Pb isotope ratio measurements in environmental samples by MCICP- MS. International Journal of Mass Spectrometry 232, 205–215

White WM, Albaredeb F, Teloukb P (2000) High-precision analysis of Pb-isotope by multicollector ICP-MS. Chemical Geology 167, 257–270

Zhang GL, Yang FG, Zhao WJ, Zhao YG, Yang JL, Gong ZT (2007) Historical change of soil Pb content and Pb isotope signatures of the cultural layers in urban Nanjing. CATENA 69, 51–56

Zhou C, Liu G, Cheng S, Fang T, Lam PKS (2014) The environmental geochemistry of trace elements and naturally radionuclides in a coal gangue brick-making plant. Scientific Reports 4, 6221

MIX
Papier aus verantwortungsvollen Quellen
Paper from responsible sources
FSC® C105338

FSC
www.fsc.org

If you have any concerns about our products,
you can contact us on
ProductSafety@springernature.com

In case Publisher is established outside the EU,
the EU authorized representative is:
Springer Nature Customer Service Center GmbH
Europaplatz 3, 69115 Heidelberg, Germany

Printed by Libri Plureos GmbH
in Hamburg, Germany